BURN-IN TESTING

ITS QUANTIFICATION AND OPTIMIZATION

Books by Dimitri Kececioglu, Ph.D., P.E.

Reliability Engineering Handbook, Volume 1
Reliability Engineering Handbook, Volume 2
Reliability & Life Testing Handbook, Volume 1
Reliability & Life Testing Handbook, Volume 2
Maintainability, Availability and Operational
  Readiness Engineering Handbook, Volume 1
Environmental Stress Screening
  — Its Quantification, Optimization and Management
Burn-in Testing — Its Quantification and Optimization

BURN-IN TESTING
ITS QUANTIFICATION AND OPTIMIZATION

Dimitri Kececioglu, Ph.D., P.E.
Department of Aerospace and Mechanical Engineering
The University of Arizona

Feng-Bin Sun, Ph.D.
Corporate Reliability Engineering Department
Quantum Corporation

> To join a Prentice Hall PTR Internet
> mailing list, point to
> http://www.prenhall.com/mail_lists/

Prentice Hall PTR
Upper Saddle River, New Jersey 07458
http://www.prenhall.com

Library of Congress Cataloging-in-Publication Data

Kececioglu, Dimitri
  Burn-in testing: its quantification and optimization/Dimitri
Kececioglu, Feng-Bin Sun.
    p. cm. — (Reliability series)
  Includes bibliographical (p.   ) references and index.
  ISBN 0-13-324211-0
  1. Electronic apparatus and appliances—Testing. 2. Environmental
testing. I. Sun, Feng-Bin. II. Title. III. Sreies: Kececioglu,
Dimitri. Reliability series.
  TK7870.K363 1997
  621.3815'48—dc21   97-13930
                                                                     CIP

Production Editor: *Kerry Reardon*
Acquisitions Editor: *Bernard M. Goodwin*
Cover Designer: *Lundgren Graphics*
Cover Design Director: *Jerry Votta*
Marketing Manager: *Dan Rush*
Manufacturing Manager: *Alexis R. Heydt*

 ©1997 Prentice Hall PTR
Prentice-Hall, Inc.
A Simon & Schuster Company
Upper Saddle River, New Jersey 07458

Prentice Hall books are widely used by corporations and government agencies for
training, marketing, and resale. The publisher offers discounts on this book when ordered
in bulk quantities.
For more information contact:
             Corporate Sales Department
             Phone: 800-382-3419 Fax: 201-236-7141
             E-mail: corpsales@prenhall.com
or write: Prentice Hall PTR
             One Lake Street
             Upper Saddle River, N.J. 07458

All rights reserved. No part of this book may be reproduced, in any form or by any means,
without permission in writing from the publisher.

Printed in the United States of America

10   9   8   7   6   5   4   3   2   1

ISBN 0-13-324211-0

Prentice-Hall International (UK) Limited, *London*
Prentice-Hall of Australia Pty. Limited, *Sydney*
Prentice-Hall Canada Inc., *Toronto*
Prentice-Hall Hispanoamericana, S.A., *Mexico*
Prentice-Hall of India Private Limited, *New Delhi*
Prentice-Hall of Japan, *Tokyo*
Simon & Schuster Asia Pte. Ltd., *Singapore*
Editora Prentice-Hall do Brasil, Ltda., *Rio de Janeiro*

To my wonderful wife Lorene,
daughter Zoe,
and son John.

— Dimitri B. Kececioglu

To my wife Le-Ling and daughter Spring.
To my dear parents.
To all my teachers.

— Feng-Bin Sun

TABLE OF CONTENTS

PREFACE.. xxiv
CHAPTER 1– INTRODUCTION................................. 1
  1.1– WHY BURN-IN?... 1
  1.2– HOW BURN-IN WORKS 5
  1.3– A BRIEF REVIEW OF BURN-IN HISTORY.............. 6
  1.4– WHAT THIS BOOK OFFERS 10
  REFERENCES 11
  ADDITIONAL REFERENCES................................. 12
CHAPTER 2– BURN-IN DEFINITIONS, CLASSIFICATIONS, DOCUMENTS AND TEST CONDITIONS 15
  2.1– BURN-IN DEFINITIONS 15
  2.2– THE DIFFERENCE BETWEEN BURN-IN AND ENVIRONMENTAL STRESS SCREENING (ESS) 16
  2.3– BURN-IN METHODS AND THEIR EFFECTIVENESS ... 16
  2.4– BURN-IN DOCUMENTS 19
  2.5– BURN-IN TEST CONDITIONS SPECIFIED BY MIL-STD-883C... 19
  2.6– TEST TEMPERATURE 25
  REFERENCES 27
CHAPTER 3– FREQUENTLY ENCOUNTERED TERMINOLOGIES AND ACRONYMS IN BURN-IN TESTING... 29
  3.1– BURN-IN TERMINOLOGIES............................ 29
  3.2– BURN-IN ACRONYMS 56
  REFERENCES 66
CHAPTER 4– PHENOMENOLOGICAL OBSERVATIONS AND THE PHYSICAL INSIGHT OF THE FAILURE

PROCESS DURING BURN-IN................... 69

4.1– INTRODUCTION.. 69

4.2– THE CONVENTIONAL BATHTUB CURVE CONCEPT.. 69

4.3– THE S-SHAPED *CDF* PATTERN....................... 72

4.4– THE UNIFIED-FIELD FAILURE THEORY AND THE ROLLER-COASTER FAILURE RATE CURVE.. 79

4.5– PHYSICAL EXPLANATION OF THE FAILURE PATTERN DURING BURN-IN 84

    4.5.1– THE STRESS-STRENGTH INTERFERENCE CONCEPT 84

    4.5.2– THE STRESS-STRENGTH INTERFERENCE AND COMPONENT FAILURE PATTERNS............. 88

        4.5.2.1– BIMODAL STRENGTH DISTRIBUTION.. 88

        4.5.2.2– BEHIND THE BATHTUB CURVE: STRESS-STRENGTH INTERFERENCE.... 90

        4.5.2.3– BIMODAL STRENGTH DISTRIBUTION AND THE CORRESPONDING S-SHAPED CDF PATTERN.......................... 94

        4.5.2.4– FAILURE RATE CURVE AS DERIVED FROM STRESS-STRENGTH INTERFERENCE AND A PHYSICS-OF-FAILURE MODEL.................... 106

REFERENCES... 111

CHAPTER 5– MATH MODELS DESCRIBING THE FAILURE PROCESS DURING BURN-IN AND THEIR PARAMETERS' ESTIMATION................. 115

5.1– RELIABILITY MODELS FOR A MIXED WEIBULL POPULATION 115

5.2– BATHTUB CURVE MODELS 117

    5.2.1– MODEL 1 – TWO-PARAMETER WEIBULL MODEL... 117

    5.2.2– MODEL 2 – THREE-PARAMETER MODEL...... 117

CONTENTS

- 5.2.3– MODEL 3 – SIX-PARAMETER WEIBULL MODEL 118
- 5.2.4– MODEL 4 – FIVE-PARAMETER MODEL 118
- 5.3– SELECTION OF THE MODELS FOR BURN-IN TESTS 119
- 5.4– PARAMETER ESTIMATION FOR THE BIMODAL MIXED POPULATION 119
  - 5.4.1– JENSEN'S GRAPHICAL METHOD FOR THE WELL-SEPARATED CASE 120
  - EXAMPLE 5-1 120
  - SOLUTION TO EXAMPLE 5-1 123
  - 5.4.2– BAYESIAN METHOD FOR THE WELL-SEPARATED CASE 125
    - 5.4.2.1– BAYESIAN METHOD FOR THE UNGROUPED DATA CASE 126
    - EXAMPLE 5-2 127
    - SOLUTION TO EXAMPLE 5-2 127
    - 5.4.2.2– BAYESIAN METHOD FOR THE GROUPED DATA CASE 135
    - EXAMPLE 5-3 140
    - SOLUTION TO EXAMPLE 5-3 140
  - 5.4.3– EML ALGORITHM FOR BOTH THE WELL-SEPARATED AND THE WELL-MIXED CASES .. 155
    - 5.4.3.1– THE MLE FOR UNGROUPED NON-POSTMORTEM DATA 155
    - 5.4.3.2– THE MLE FOR GROUPED NON-POSTMORTEM DATA 156
    - EXAMPLE 5-4 157
    - SOLUTION TO EXAMPLE 5-4 157
    - EXAMPLE 5-5 158
    - SOLUTION TO EXAMPLE 5-5 158

EXAMPLE 5-6 160
SOLUTION TO EXAMPLE 5-6 160
5.4.4– THE LEAST-SQUARES METHOD FOR THE BATHTUB MODEL 162
EXAMPLE 5–7 164
SOLUTION TO EXAMPLE 5–7 164
EXAMPLE 5–8 164
SOLUTION TO EXAMPLE 5–8 164
EXAMPLE 5–9 165
SOLUTION TO EXAMPLE 5–9 165
REFERENCES 165
ADDITIONAL REFERENCES 166
APPENDIX 5A 167
APPENDIX 5B 173
APPENDIX 5C 187
APPENDIX 5D 208
CHAPTER 6– BURN-IN TIME DETERMINATION USING A QUICK CALCULATION APPROACH 211
CHAPTER 7– BURN-IN TIME DETERMINATION BASED ON THE BIMODAL TIMES-TO-FAILURE DISTRIBUTION 221
7.1– THE SUBPOPULATION TRUNCATION APPROACH .. 221
EXAMPLE 7–1 225
SOLUTION TO EXAMPLE 7–1 225
7.2– THE BURN-IN TIME CORRESPONDING TO A ZERO SLOPE POINT OF THE FAILURE RATE CURVE 229
7.3– THE BURN-IN TIME FOR A SPECIFIED ERROR ON THE ZERO FAILURE RATE CURVE SLOPE 229
EXAMPLE 7–2 230

CONTENTS

SOLUTION TO EXAMPLE 7-2 230
EXAMPLE 7-3 231
SOLUTION TO EXAMPLE 7-3 231
EXAMPLE 7-4 232
SOLUTION TO EXAMPLE 7-4 232
7.4- THE BURN-IN TIME CORRESPONDING TO A SPECIFIED FAILURE RATE GOAL 233
EXAMPLE 7-5 234
SOLUTION TO EXAMPLE 7-5 234
EXAMPLE 7-6 234
SOLUTION TO EXAMPLE 7-6 234
7.5- THE BURN-IN TIME CORRESPONDING TO A SPECIFIED RELIABILITY GOAL 235

    7.5.1- APPLICATION OF EQ. (7.12) USING THE BATHTUB MODEL 1 236

    EXAMPLE 7-7 237

    SOLUTION TO EXAMPLE 7-7 237

    EXAMPLE 7-8 237

    SOLUTION TO EXAMPLE 7-8 237

    7.5.2- APPLICATION OF EQ. (7.12) USING THE BIMODAL MIXED-WEIBULL MODEL 237

    EXAMPLE 7-9 238

    SOLUTION TO EXAMPLE 7-9 238

7.6- GRAPHICAL DETERMINATION OF THE BURN-IN PERIOD FROM THE FAILURE RATE AND RELIABILITY FUNCTIONS 238

EXAMPLE 7-10 239
SOLUTION TO EXAMPLE 7-10 240
REFERENCES 245

APPENDIX 7A 247

APPENDIX 7B 252

CHAPTER 8– MEAN RESIDUAL LIFE (MRL) CONCEPT AND ITS APPLICATIONS TO BURN-IN TIME DETERMINATION 255

8.1– INTRODUCTION 255

8.2– MATHEMATICAL DEFINITION OF MRL AND ITS RELATIONSHIP TO THE FAILURE RATE FUNCTION 256

8.3– MRL COMPLETELY DETERMINES A LIFE DISTRIBUTION 258

8.4– EMPIRICAL MRL FUNCTIONS 260

    8.4.1– FOR A COMPLETE, OR NONCENSORED, RANDOM SAMPLE 261

    8.4.2– FOR A CENSORED RANDOM SAMPLE 263

8.5– MEAN RESIDUAL LIFETIMES FOR SOME FREQUENTLY USED LIFE DISTRIBUTIONS 266

    8.5.1– THE WEIBULL DISTRIBUTION 266

    8.5.2– THE GAMMA DISTRIBUTION 267

    8.5.3– THE EXPONENTIAL DISTRIBUTION 268

    8.5.4– THE TRUNCATED NORMAL DISTRIBUTION .. 269

    8.5.5– THE LOGNORMAL DISTRIBUTION 270

    8.5.6– THE RAYLEIGH DISTRIBUTION 271

    8.5.7– THE MIXED-LIFE DISTRIBUTION 272

8.6– THE EFFECT OF BURN-IN ON THE MRL ASSUMING A BATHTUB-SHAPED FAILURE RATE CURVE 274

8.7– APPLICATION OF THE MRL CONCEPT TO THE OPTIMUM BURN-IN TIME DETERMINATION 278

    8.7.1– OPTIMUM BURN-IN TIME FOR A PRESPECIFIED MRL 278

        8.7.1.1– APPLICATION OF EQ. (8.57) USING

THE BATHTUB MODEL 1	280
EXAMPLE 8–1	280
SOLUTION TO EXAMPLE 8–1	281
EXAMPLE 8–2	281
SOLUTION TO EXAMPLE 8–2	281
8.7.1.2– APPLICATION OF EQ. (8.57) USING THE BIMODAL, MIXED-WEIBULL MODEL	281
EXAMPLE 8–3	282
SOLUTION TO EXAMPLE 8–3	282
8.7.2– OPTIMUM BURN-IN TIME FOR THE MAXIMUM MRL	282
8.7.2.1– APPLICATION OF EQ. (8.61) USING THE BATHTUB MODEL 1	283
EXAMPLE 8–4	283
SOLUTION TO EXAMPLE 8–4	283
EXAMPLE 8–5	284
SOLUTION TO EXAMPLE 8–5	284
8.7.2.2– APPLICATION OF EQ. (8.61) USING THE BIMODAL, MIXED-WEIBULL MODEL	284
EXAMPLE 8–6	285
SOLUTION TO EXAMPLE 8–6	285
8.8– DETERMINING THE OPTIMUM BURN-IN TIME DIRECTLY FROM THE EMPIRICAL MRL FUNCTION OR PLOT	286
8.8.1– OPTIMUM BURN-IN TIME FOR A SPECIFIED MRL GOAL	286
8.8.2– OPTIMUM BURN-IN TIME FOR THE MAXIMUM MRL	287
EXAMPLE 8–7	289

SOLUTIONS TO EXAMPLE 8–7	291
EXAMPLE 8–8	295
SOLUTIONS TO EXAMPLE 8–8	297
EXAMPLE 8–9	301
SOLUTIONS TO EXAMPLE 8–9	301
EXAMPLE 8–10	305
SOLUTIONS TO EXAMPLE 8–10	308
8.9– FURTHER COMMENTS	312
REFERENCES	313
ADDITIONAL REFERENCES	314
APPENDIX 8A	319
APPENDIX 8B	322
APPENDIX 8C	323
APPENDIX 8D	326
APPENDIX 8E	328
APPENDIX 8F	330
APPENDIX 8G	333
CHAPTER 9– BURN-IN TIME DETERMINATION FOR THE MINIMUM COST	337
9.1– COST MODEL 1	338
EXAMPLE 9–1	341
SOLUTION TO EXAMPLE 9–1	342
9.2– COST MODEL 2	344
EXAMPLE 9–2	347
SOLUTION TO EXAMPLE 9–2	348
9.3– COST MODEL 3	352
EXAMPLE 9–3	359

CONTENTS

SOLUTION TO EXAMPLE 9–3	359
9.4– COST MODEL 4	363
EXAMPLE 9–4	367
SOLUTIONS TO EXAMPLE 9–4	368
REFERENCES	372
ADDITIONAL REFERENCES	372
CHAPTER 10– BURN-IN QUANTIFICATION AND OPTIMIZATION USING THE BIMODAL MIXED-EXPONENTIAL DISTRIBUTION	375
10.1– INTRODUCTION	375
10.2– THE MIXED-EXPONENTIAL LIFE DISTRIBUTION	375
10.3– THE BIMODAL MIXED-EXPONENTIAL LIFE DISTRIBUTION	377
10.4– THE OPTIMUM BURN-IN TIME FOR A SPECIFIED POST-BURN-IN MISSION RELIABILITY	382
EXAMPLE 10–1	383
SOLUTION TO EXAMPLE 10–1	383
10.5– THE OPTIMUM BURN-IN TIME FOR A SPECIFIED MEAN RESIDUAL LIFE (MRL) GOAL	384
EXAMPLE 10–2	387
SOLUTION TO EXAMPLE 10–2	387
10.6– THE OPTIMUM BURN-IN TIME FOR A SPECIFIED POST-BURN-IN FAILURE RATE GOAL	388
EXAMPLE 10–3	391
SOLUTION TO EXAMPLE 10–3	391
10.7– THE OPTIMUM BURN-IN TIME FOR A SPECIFIED BURN-IN EFFICIENCY	393
EXAMPLE 10–4	394
SOLUTION TO EXAMPLE 10–4	395

10.8– THE OPTIMUM BURN-IN TIME FOR A SPECIFIED POWER FUNCTION	396
EXAMPLE 10–5	397
SOLUTION TO EXAMPLE 10–5	397
10.9– THE OPTIMUM BURN-IN TIME FOR THE SPECIFIED BURN-IN RISKS	398
10.9.1– DEFINITION OF BURN-IN RISKS AND THE CORRESPONDING BURN-IN INTERVAL	398
10.9.2– RESTRICTIONS ON THE FAILURE RATES FOR A VALID BURN-IN INTERVAL	399
EXAMPLE 10–6	400
SOLUTION TO EXAMPLE 10–6	400
EXAMPLE 10–7	401
SOLUTION TO EXAMPLE 10–7	401
10.10– THE NUMBER AND COST OF FAILURES	402
10.11– THE OPTIMUM BURN-IN TIME FOR THE MINIMUM COST	404
10.11.1– COST MODEL FORMULATION	409
10.11.2– COST MODEL MINIMIZATION	410
EXAMPLE 10–8	414
SOLUTIONS TO EXAMPLE 10–8	414
10.12– BAYESIAN APPROACH TO BURN-IN	416
10.12.1– WHY THE BAYESIAN APPROACH?	416
10.12.2– THE JOINT PRIOR DISTRIBUTION FOR PARAMETERS OF THE BIMODAL MIXED-EXPONENTIAL LIFE DISTRIBUTION	417
10.12.3– THE BAYESIAN COST MODEL AND THE OPTIMAL DURATION OF BURN-IN	421
10.12.4– OTHER BURN-IN MEASURES BASED ON BAYESIAN RESULTS	425

CONTENTS

    EXAMPLE 10-9 428
    SOLUTION TO EXAMPLE 10-9 428
  REFERENCES 432
  ADDITIONAL REFERENCES 432
  APPENDIX 10A 433

CHAPTER 11– THE TOTAL-TIME-ON-TEST (TTT) TRANSFORM AND ITS APPLICATION TO BURN-IN TIME-DETERMINATION 437

  11.1– INTRODUCTION 437
  11.2– THE TTT TRANSFORMS, THE SCALED TTT TRANSFORMS AND THE SCALED TTT PLOTS 438
    11.2.1– THE TTT STATISTICS AND THE SCALED TTT STATISTICS 438
    11.2.2– THE TTT TRANSFORM AND THE SCALED TTT TRANSFORM 440
    11.2.3– THE TTT PLOT FOR A GIVEN SAMPLE 442
  11.3– GEOMETRICAL PROPERTIES OF THE SCALED TTT TRANSFORMS 443
  11.4– THE SCALED TTT TRANSFORMS OF SOME FREQUENTLY USED LIFE DISTRIBUTIONS 445
    11.4.1– THE EXPONENTIAL LIFE DISTRIBUTION ... 445
    11.4.2– THE WEIBULL LIFE DISTRIBUTION 446
    11.4.3– THE GAMMA LIFE DISTRIBUTION 449
    11.4.4– THE TRUNCATED NORMAL LIFE DISTRIBUTION 451
    11.4.5– THE LOGNORMAL LIFE DISTRIBUTION 453
    11.4.6– THE MIXED-LIFE DISTRIBUTION 456
  11.5– SCALED TTT PLOTS FOR COMPLETE AND INCOMPLETE LIFE DATA 458
    11.5.1– COMPLETE, OR NONCENSORED, LIFE DATA 458

11.5.2– TIME-TERMINATED LIFE DATA............. 461

11.5.3– FAILURE-TERMINATED LIFE DATA.......... 465

11.6– DESIRED STRUCTURE OF THE OBJECTIVE FUNCTIONS TO BE OPTIMIZED USING THE *TTT* TRANSFORM AND THE GRAPHICAL OPTIMIZATION PROCEDURE....................... 468

11.7– OPTIMUM BURN-IN TIME DETERMINATION USING THE *TTT* TRANSFORM FOR THE MINIMUM COST OR FOR THE MAXIMUM PROFIT 472

11.7.1– THE BERGMAN-KLEFSJO'S COST MODEL .. 472

EXAMPLE 11-1 477

SOLUTION TO EXAMPLE 11-1 478

11.7.2– THE WEISS-DISHON'S MODEL 1 – THE COST MODEL 481

EXAMPLE 11-2 484

SOLUTION TO EXAMPLE 11-2 484

11.7.3– THE WEISS-DISHON'S MODEL 2 – THE PROFIT MODEL 488

EXAMPLE 11-3 490

SOLUTION TO EXAMPLE 11-3 490

11.8– OPTIMUM BURN-IN TIME DETERMINATION USING THE *TTT* TRANSFORM FOR THE MAXIMUM POST-BURN-IN MEAN RESIDUAL LIFE (MRL).............................. 492

11.9– BURN-IN TIME DETERMINATION DIRECTLY FROM A SAMPLE OF OBSERVED TIMES TO FAILURE 495

11.9.1– GENERAL PROCEDURE 495

11.9.2– REFINED PROCEDURE 497

EXAMPLE 11-4 501

SOLUTION TO EXAMPLE 11-4 503

CONTENTS

EXAMPLE 11-5 512
SOLUTION TO EXAMPLE 11-5 512
REFERENCES 522
ADDITIONAL REFERENCES 522
APPENDIX 11A 524

CHAPTER 12- ACCELERATED BURN-IN TESTING AND BURN-IN TIME REDUCTION ALGORITHMS 527

12.1- INTRODUCTION 527
12.2- BURN-IN TIME REDUCTION USING AN ELEVATED TEMPERATURE - THE ARRHENIUS MODEL 528

    12.2.1- THE ACCELERATION FACTOR WHEN THE LIFE DISTRIBUTION IS EXPONENTIAL 528

    12.2.2- THE ACCELERATION FACTOR WHEN THE LIFE DISTRIBUTION IS NOT EXPONENTIAL 530

    12.2.3- BURN-IN TIME REDUCTION USING AN ELEVATED TEMPERATURE 531

    EXAMPLE 12-1 531
    SOLUTION TO EXAMPLE 12-1 531

    12.2.4- A WORD OF CAUTION ON THE USE OF THE ARRHENIUS ACCELERATION MODEL 532

    12.2.5- THE PHYSICAL MEANING OF ACTIVATION ENERGY 533

    12.2.6- EXPERIMENTAL DETERMINATION OF THE ACTIVATION ENERGY 534

    EXAMPLE 12-2 536
    SOLUTION TO EXAMPLE 12-2 539

    12.2.7- THE PITFALL OF THE ARRHENIUS MODEL AND THE TEMPERATURE DEPENDENCY OF THE ACTIVATION ENERGIES 541

12.2.8– MODIFICATION AND PARAMETER ESTIMATIONS OF THE ACCELERATION EQUATION 544

    12.2.8.1– FIRST MODIFIED ARRHENIUS MODEL AND ITS PARAMETERS' ESTIMATION 544

    EXAMPLE 12–3.................................. 548

    SOLUTION TO EXAMPLE 12–3................ 549

    12.2.8.2– SECOND MODIFIED ARRHENIUS MODEL AND ITS PARAMETERS' ESTIMATION 553

    EXAMPLE 12–4.................................. 556

    SOLUTION TO EXAMPLE 12–4................ 556

12.2.9– BURN-IN TIME REDUCTION USING AN ELEVATED TEMPERATURE AND THE MODIFIED ARRHENIUS MODEL.............. 559

EXAMPLE 12–5.. 560

SOLUTION TO EXAMPLE 12–5 561

12.3– BURN-IN TIME REDUCTION USING AN ELEVATED VOLTAGE BIAS STRESS – THE INVERSE POWER LAW MODEL 563

EXAMPLE 12–6.. 564

SOLUTION TO EXAMPLE 12–6 564

12.4– BURN-IN TIME REDUCTION USING AN ELEVATED TEMPERATURE PLUS AN ELEVATED VOLTAGE BIAS – THE COMBINATION MODEL 565

    12.4.1– WHEN THE TEMPERATURE DEPENDENCY OF THE ACTIVATION ENERGY IS NOT CONSIDERED – THE CONVENTIONAL COMBINATION MODEL 565

    EXAMPLE 12–7.. 566

    SOLUTION TO EXAMPLE 12–7 566

    12.4.2– WHEN THE TEMPERATURE DEPENDENCY

CONTENTS

OF THE ACTIVATION ENERGY IS CONSIDERED – THE MODIFIED COMBINATION MODEL 566

EXAMPLE 12-8 568

SOLUTION TO EXAMPLE 12-8 568

12.5– OPTIMUM BURN-IN TIME DETERMINATION BASED ON THE TEST RESULTS AT A HIGHER STRESS LEVEL 569

REFERENCES 570

ADDITIONAL REFERENCES 570

CHAPTER 13– ACCELERATED BURN-IN USING TEMPERATURE CYCLING 573

13.1– WHY TEMPERATURE CYCLING? 573

13.2– TEMPERATURE PROFILE AND THE MODEL FOR THE ACCELERATION FACTOR 577

13.3– EQUIVALENT ACCELERATION FACTOR EVALUATION USING EQ. (13.6) 580

13.4– EQUIVALENT ACCELERATION FACTOR EVALUATION USING EQ. (13.7) 584

13.5– APPLICATION OF THE AGING ACCELERATION MODELS 590

EXAMPLE 13-1 590

SOLUTION TO EXAMPLE 13-1 591

13.6– OPTIMUM NUMBER OF THERMAL CYCLES FOR A SPECIFIED FIELD $MTBF$ GOAL 593

13.6.1– USING THE MIXED-EXPONENTIAL LIFE DISTRIBUTION 594

EXAMPLE 13-2 596

SOLUTION TO EXAMPLE 13-2 597

13.6.2– USING THE WEIBULL LIFE DISTRIBUTION . 597

EXAMPLE 13-3 600

SOLUTION TO EXAMPLE 13-3 601
13.7– A SUMMARY OF SOME USEFUL THERMAL FATIGUE LIFE PREDICTION MODELS FOR ELECTRONIC EQUIPMENT........................ 601
   13.7.1– WIRE FATIGUE LIFE PREDICTION 602
   13.7.2– WIRE BOND FATIGUE LIFE PREDICTION .. 603
   13.7.3– DIE FRACTURE AND FATIGUE LIFE PREDICTION 606
   13.7.4– DIE AND SUBSTRATE ADHESION FATIGUE LIFE PREDICTION 607
   13.7.5– SOLDER JOINT FATIGUE LIFE PREDICTION 609
   13.7.6– COMMENTS.................................... 611
13.8– CONCLUSIONS 611
REFERENCES 612
APPENDIX 13A... 614
APPENDIX 13B... 619
CHAPTER 14– GUIDELINES FOR BURN-IN QUANTIFICATION AND OPTIMUM BURN-IN TIME DETERMINATION 621
14.1– INTRODUCTION 621
14.2– GENERAL PROCEDURE............................. 621
14.3– TIMES-TO-FAILURE DATA COLLECTION 622
14.4– INITIAL DATA ANALYSIS........................... 624
14.5– PARAMETRIC BURN-IN DATA ANALYSIS AND OPTIMUM BURN-IN TIME DETERMINATION 626
   14.5.1– PARAMETRIC BURN-IN DATA ANALYSIS.... 626
   14.5.2– OPTIMUM BURN-IN TIME DETERMINATION 629
14.6– NON-PARAMETRIC BURN-IN DATA ANALYSIS AND OPTIMUM BURN-IN TIME DETERMINATION 632

14.6.1– NON-PARAMETRIC BURN-IN DATA ANALYSIS 632

14.6.2– BURN-IN TIME OPTIMIZATION OBJECTIVE FUNCTION DEVELOPMENT 635

14.6.3– OPTIMUM BURN-IN TIME DETERMINATION 635

14.7– BURN-IN TIME JUSTIFICATION AND ADJUSTMENT 635

14.8– ACCELERATED BURN-IN AND BURN-IN TIME CONVERSION FROM ONE STRESS TO ANOTHER STRESS 638

REFERENCES 640

INDEX 641

ABOUT THE AUTHORS 661

PREFACE

This book results from a long-standing and growing demand from industry to quantify and optimize burn-in testing. Burn-in and environmental stress screening (ESS) have become the primary approaches in modern electronic industry to precipitate and eliminate latent defects in products which are introduced mainly during the manufacturing, assembling and packaging processes. Many successful applications of burn-in and ESS have shown that scientifically planned and conducted burn-in and ESS will yield effective screening and thereby provide failure-free products with specified and desirable post-burn-in, or post-screening, reliability, failure rate, and mean life goals.

A distinction is made here between burn-in and ESS. Burn-in is a process of powering a component at a specified operating or accelerated, *constant* temperature stress though temperature cycling and/or power cycling are sometimes applied to accelerate the burn-in process. ESS is an accelerated *process* of stressing a product, prior to its being shipped, in continuous cycles between predetermined environmental extremes, primarily temperature cycling, plus random vibration. Generally speaking, burn-in is conducted primarily at the component level, while ESS is conducted at the assembly, module and system levels.

As the second of our two sister texts, "Environmental Stress Screening (ESS) – Its Quantification, Optimization and Management" and "Burn-in Testing – Its Quantification and Optimization", this book attempts to summarize the knowledge derived from the studies of numerous workers in the field. It presents a comprehensive coverage of the subject from the basic definition of burn-in to state-of-the-art concepts and applications such as the mean residual life (MRL) concept, total-time-on-test (TTT) transforms and accelerated burn-in. Emphasis is given to the quantification, or mathematical and statistical description and formulation, of the failure process during burn-in, and the determination of the optimum burn-in time for prespecified post-burn-in (conditional) mission reliability, failure rate and mean life goals, for minimum total cost or for maximum total profit goal, and others.

This book is intended to serve two kinds of readers: the practicing engineers, including reliability engineers, reliability and life testing engineers, and product assurance engineers, and the advanced undergraduate and graduate students. As an advanced topic in Reliability Engineering and Life Testing, this book is an ideal sequel to the "Reliability Engineering Handbook," by Dr. Dimitri Kececioglu, published by Prentice Hall, Inc., Upper Saddle River, New Jersey 07458, Vol. 1, 720 pp. and Vol. 2, 568 pp., in 1991, and now in its Fifth Printing, and to the "Reliability and Life Testing Handbook," by Dr. Dimitri Kececioglu, also published by Prentice Hall, Vol. 1, 960 pp., 1993, and Vol. 2, 900 pp., 1994. It can be used as an one-semester college textbook for those students who already took at least one course in Reliability Engineering.

Chapter 1 is an introduction to burn-in testing and its objective, working mechanisms, economic benefits, types, and historical background, and includes a brief summary of the major features of this book.

Chapter 2 presents the basic definition of burn-in, its comparison with ESS, various burn-in methods and their effectiveness, burn-in documents, and a summary of various test conditions and temperatures of burn-in.

Chapter 3 presents a comprehensive list of the terminologies and acronyms that are frequently encountered in burn-in literature and documents.

Chapter 4 offers a physical insight of the failure pattern and a physical quantification of the failure process during burn-in. After reviewing various failure patterns during burn-in, such as the bathtub-shaped failure rate curve, the S-shaped CDF, and the roller-coaster failure rate curve, stress-strength interference theory – the physical law behind these patterns is presented. It is shown how the stress-strength interference between a constant external stress and a bimodally distributed internal strength can result in an S-shaped CDF pattern, and how a time-dependent failure rate function can be derived from the stress-strength interference principle by incorporating a time-dependent and distributed internal strength and a distributed external stress.

Chapter 5 presents five (5) general mathematical models including a mixed-Weibull distribution model and four (4) appropriate bathtub curve models to describe the bimodal failure process during burn-in. Applicability and selection rules of these models for any specific burn-in situation are discussed. Great efforts are spent and details are given to the methods of parameter estimation for the bimodal mixed population case, such as Jensen's graphical, the Bayesian, the MLE, and the least-squares methods. Numerical examples are given for each presented method.

Chapter 6 presents a simple and quick calculation approach for burn-in time determination based on the observed average failure rate's plot. This chapter serves as a warm-up prelude for the readers who favor intuitive ideas about burn-in time determination using the early portion of the conventional bathtub curve.

Chapter 7 provides methods to determine the optimum burn-in time for a specified post-burn-in failure rate goal, for zero failure rate slope with and without error, and for a post-burn-in mission reliability goal. These methods are based on the bimodal times-to-failure distribution assumption. Mathematical expressions are developed when different failure models, such as those presented in Chapter 5, are used. Numerical examples are given to illustrate their applications.

Chapter 8 presents a comprehensive coverage of the mean residual life (MRL) concept, and its applications in burn-in time determination. The mathematical definition of MRL, its relationship with the failure rate function, the uniqueness of the life distribution corresponding to a given MRL, empirical MRL functions for various data formats, and MRL expressions for frequently used life distributions are presented sequentially. The effect of burn-in on MRL is discussed in detail. Methodologies of determining the optimum burn-in time for a specified MRL, or for the maximum MRL with or without a distribution assumption (therefore parametric or non-parametric in nature), are presented and illustrated with numerical examples.

Chapter 9 covers several representative cost models in terms of burn-in time, such as Jensen's general model, Plesser's model, Washburn's model, etc. The procedure to determine the optimum burn-in time for the minimum cost is discussed and illustrated with numerical examples.

Chapter 10 presents an extensive coverage of the quantification and optimization of burn-in using the bimodal mixed-exponential life distribution. This distribution has several unique features both mathematically and physically. The closed-form solution of the optimum burn-in time for a specified post-burn-in mission reliability, for a specified MRL goal, for a specified failure rate goal, for a specified burn-in efficiency, for a specified burn-in power function, and for the specified burn-in risks are derived and illustrated with examples. A life-cycle-cost model is developed based on Renewal Theory and the bimodal mixed-exponential life distribution assumption. A procedure to determine the optimum burn-in time for the minimum cost is given and illustrated by an example. To overcome the difficulty in determining the optimum burn-in time starting with the unknown parameters of the mixed distribution, a Bayesian approach is presented, which uses the *beta* and *exponential* priors for the unknown parameters of the bimodal mixed-exponential life distribution, respectively. The optimum burn-in time is then determined to minimize the expected total cost with respect to the joint prior distribution of the mixed life distribution parameters.

Chapter 11 presents a very useful concept, the TTT transform and the TTT plot, and the principle of using the TTT transform to graphically solve certain types of optimization problems with a particular structure of the objective function. The mathematical definition of the TTT transform, the TTT plot for a given sample of various data formats, the geometrical properties of the TTT transforms, the TTT transforms of frequently used life distributions, and the desired struc-

PREFACE

ture of the objective functions to be optimized using the TTT transform are presented sequentially. The application of the TTT transform in optimum burn-in time determination, with known life distributions using various cost, profit and multistate models, are discussed and illustrated with numerical examples. With a sample of observed times to failure, whose life distribution is not known, the TTT plot can also be applied to determine the optimum burn-in time graphically for the minimum cost and for the maximum MRL. The general procedure and a refined procedure are given and illustrated with numerical examples.

Chapter 12 covers the accelerated burn-in tests using the Arrhenius model for high constant temperature stress, the Inverse Power Law model for voltage bias stress, and the Combination model for combined constant temperature and voltage or any other nonthermal stress. Considering the temperature dependency of the activation energy, the modified versions of the conventional Arrhenius model and Combination model are derived. Procedures for determining the activation energy experimentally, and for estimating the parameters of the two modified Arrhenius models are presented. Examples are given to illustrate how a shorter burn-in time can be achieved using accelerated stresses.

Chapter 13 quantifies the accelerated burn-in process under temperature cycling. The justification of applying temperature cycling in burn-in is discussed first. A general aging acceleration model for a typical temperature profile used in temperature cycling burn-in is derived based on each one of the two modified Arrhenius models. An example is given which illustrates the use of the developed aging acceleration models. Next, a math model is derived, which is based on the acceleration factor and the life distribution at the burn-in stress level, to determine the optimum thermal cycling time or the optimum number of thermal cycles for a given temperature profile and a specified field $MTBF$ goal. Finally, several useful physical models for thermal fatigue life prediction, for various failure mechanisms which may occur during thermal cycling, are summarized.

Chapter 14 gives a systematic summarizing guideline on *how to apply* the burn-in quantification and optimization theories and techniques presented in this book. A general step-by-step procedure is presented and illustrated by a vivid flow chart. Then, each step is further explained in detail with figures and tables showing the input and output of each step, procedures and methodologies needed, and their corresponding locations in the book. Therefore, this chapter serves as a map, or a compass, of using this book, we hope!

In developing this book, a number of computer programs in FORTRAN language were written. These are listed as appendices at the end of each chapter. A conscientious effort has been made throughout to give credit to original sources. For the occasional oversights that may have developed during the "baking process" the authors offer their apologies.

ACKNOWLEDGMENTS

The authors are deeply indebted to their families, particularly their wives and children, for their understanding, encouragement and support during the development of this book. Thanks also go to many of our colleagues in industry and to graduate students at The University of Arizona for their inspiration and help in writing this book.

Dimitri B. Kececioglu
Feng-Bin Sun
Tucson, Arizona
March 28, 1997

Chapter 1

INTRODUCTION

1.1 WHY BURN-IN?

Reliability engineers have long recognized an inherent characteristic in many types of equipment to exhibit a decreasing failure rate during their early operating life. Intuitively, a relatively high early failure rate, that decreases with time until it eventually levels off, can be explained by the inherent variability of any production process.

Table 1.1 [1] lists the possible failure mechanisms which are unique to each process step involved in manufacturing a microelectronic device. A failure mechanism will be eventually exhibited as a failure because the device will fail to meet its specified performance characteristics. Common microcircuit failures can be classified with respect to failure as follows:

1. Degradation of performance characteristics.

2. Shorts.

3. Opens.

4. Intermittents.

Not all failure mechanisms appear in equal measure in failed devices, as shown in Table 1.2 [1] for two microcircuit technologies used in manufacturing memories.

The 'substandard' portion of the production of identical parts can be expected to fail early, and they do so quickly. The failures of these substandard parts are labeled "early life failures." Experience shows that semiconductors, prone to fail early, will usually fail within the first 1,000 operating hours under use conditions [2; 3]. After that the

Table 1.1 – Failure mechanisms versus process steps of manufacturing a microelectronic device [1].

Process step	Failure mechanism introduced as a reliability influencing variable
Slice preparation	Dislocations and stacking faults. Nonuniform resistivity. Irregular surfaces. Cracks, chips, scratches (general handling damage). Contamination.
Passivation	Cracks and pin holes. Nonuniform thickness.
Masking	Scratches, nicks, blemishes in the photomask. Misalignment. Irregularities in photoresist pattern (line widths, spaces, pinholes).
Etching	Improper removal of oxide. Undercutting. Spotting (etch splash). Contamination (photoresist, chemical residue).
Diffusions	Improper control of doping profiles.
Final seal	Poor hermetic seal. Incorrect atmosphere sealed in package. Broken or bent external leads. Cracks, voids in kovar-to-glass seals. Electrolytic growth of metals or metallic compounds across glass seals between leads and metal case. Loose conducting particles in packages. Improper marking.

Table 1.1 – Continued.

Process step	Failure mechanism introduced as a reliability influencing variable
Metalization	Scratched or smeared metalization (handling damage). Thin metalization due to insufficient deposition or due to oxide steps. Oxide contamination, material incompatibility. Corrosion (chemical residue). Misalignment and contaminated contact areas. Improper alloying temperature or time.
Die separation	Improper die separation resulting in cracked or chipped die.
Die bonding	Voids between header and die. Overspreading and/or loose particles of eutectic solder. Poor die-to-header bond. Material mismatch.
Wire bonding	Overbonding or underbonding. Material incompatibility or contaminated bonding pad. Plaque formation. Insufficient bonding pad area or spacings. Improper bonding procedure or control. Improper bond alignment. Cracked or chipped die. Excessive loops, sags, or lead length. Nicks, cuts and abrasions on leads. Unremoved pigtails.

Table 1.2– Non-package malfunction summary for bipolar and MOS memory device technologies [1].

Failure classification	Percentage	
	Bipolar	MOS
Surface	29.00	45.09
Oxide	14.00	24.86
Diffusion	1.00	9.83
Metalization	21.00	1.74
Bond	5.00	4.05
Interconnection	29.00	4.05
Die (mechanical)	—	1.74
Degraded input circuitry	1.00	8.64
Total	100.00	100.00

failure rate stabilizes, perhaps for as long as 25 years, before beginning to increase again as the components go into wear out. These failures, termed "infant mortalities," can be as high as 10% in a new, unproven technology and as low as 0.01% in a proven technology.

In the interest of eliminating early life failures, and thereby increasing the reliability of the equipment in the field, the engineering team will first look at the manufacturing process, making all necessary improvements. However, once this has been accomplished to the extent possible, the only way to improve the product's reliability is to eliminate the substandard parts prior to their being assembled into the equipment or system. Burn-in is a technique to accomplish this goal.

Burn-in test assures that substandard components, which do not meet their failure rate, mean life or reliability goal, are identified by subjecting them to high temperature, and at times in conjunction with other high stresses such as voltage, wattage, vibration, etc. This temperature and the additional stresses are higher than use condition stresses, and usually near their rated capacity or higher, but preferably not in excess of 20% above their rated capacity. Die-related faults, which are a function of time, bias, current and temperature accelerating factors in electronic components, can be detected in a relatively short period of time through burn-in, since this process activates the time-temperature dependent failure mechanisms. However, it has to be assured that if temperatures and other stresses as high as 20% above rated capacity are used, the integrity of these components is not compromised and degraded. These substandard components in their produced, or purchased, lot need to be removed to ensure that the surviving components yield a relatively constant and acceptably low failure rate, before they go into the next assembly stage to become part of equipment or systems. The sooner a faulty component is detected, the cheaper it is to replace or repair it. For example, a correction at the IC fabrication line costs about 50 cents; at the board level it costs about $5.00; at the system level, about $50.00; and as a field failure, about $500.00 [2; 3]. IC suppliers also use burn-in to constantly monitor, review and modify the fabrication processes, refine the circuit's design and correct design deficiencies.

1.2 HOW BURN-IN WORKS

Most burn-in is based on MIL-STD-883, Methods 1005 and 1015, because of the standardization provided by these defined conditions, irrespective of whether the application (usage conditions) and product classification is commercial, industrial or military/ aerospace. For example, the integrated circuits (IC's) are operated under the maximum electrical conditions with an elevated temperature of typically 125°C for 48, 96, 160 or 240 hours, depending on the failure mechanisms and the post-burn-in reliability objectives.

However, one can not apply these burn-in conditions indiscriminately. Instead, a comprehensive understanding of the components to be burned-in is needed, particularly in the following aspects:

1. The internal construction and fabrication of the components.

2. The circuit functions and how the schematic relates to the circuit layout.

3. How many circuit nodes can be actually activated and stressed.

4. The fault coverage.

5. The possible failure modes and the corresponding mechanisms.

6. The acceleration factors that will precipitate those failure modes or faults.

Several types of burn-in methods have spun off from these considerations, mostly based on device functionality and technology. These methods are the following:

1. Test during burn-in (TDBI).

2. High-voltage (cell) stress burn-in.

3. Static burn-in.

4. Dynamic burn-in.

5. The combinations of the above.

The test procedures for these methods will be presented in the next chapter.

Generally, TDBI is effective for uncovering intermittent defects (which are typically random), detecting soft errors during prototype evaluation (used during product and process development), and for effectively burning-in dynamic RAMs. It is not cost-effective for the majority of VLSI circuits. It is too expensive for widespread application. High-voltage stress burn-in is effective, in conjunction with thermal acceleration, for uncovering MOS devices with defective oxides. Static burn-in is recommended for the elimination of low-activation energy infant mortalities. Dynamic burn-in, in conjunction with high-voltage stress, is recommended for high-activation energy infant mortalities. However, no matter what a kind of burn-in method is applied, the best one is that which accelerates the known failure mechanisms and stresses the devices most completely, with the simplest circuitry that costs the lowest.

1.3 A BRIEF REVIEW OF BURN-IN HISTORY

The earliest components that were burned-in were vacuum tubes [4]. Since they were operated above 150°C, they were not burned-in for very long, nor in a hostile environment. With the advent of transistors in the **early 1950's**, the question arose as to how long these transistors will survive. Many engineers, in the transition between thinking in a voltage mode and a current mode, regarded the transistor as a vacuum tube that behaved erratically, leaked like a sieve, and lacked a filament. This uncertainty led to the use of extended operational life tests of 1,000 hours, the standard on small samples, and the electrical tests made at 100, 250, 500 and 1,000 hours. In this era, reliability as a discipline barely existed for electronic components, with the exception of those intended for military airframe ('aerospace' as a term did not exist then) and battlefield users of electronic equipment, and statisticians.

In the point-contact and grown junction era, yields amounted to 2% or 3%, and no manufacturer would ship warranted products at the price of about $50 each without some assurance of a reasonable Acceptance Quality Limit (AQL). The disconcerting potential for return-generated debit memos led to the establishment of QA sampling. Among other stressing operations, such as high temperature storage, temperature cycling, acceleration, and mechanical shock, the most practical stress from a manufacturing point of view is burn-in. Burn-in results showed that the failure performance of semiconductors also followed the bathtub-curve pattern, except that the front end of the curve was deeper, and the 'part you sit in,' or the so-called useful-life period, was magnitudes longer than previously expected.

In the **late 1950's**, the Joint Electronic Devices Engineering Council's (JEDEC's) registration and standardization of specifications gained wide acceptance, and a degree of order entered the burn-in field, with a temperature standardized at 125°C, and a duration of the conditioning at 168 hours for no better reason than that this is an average week's time, and fits well into production schedules. Conditions were stabilized on either of the following:

1. *Operating or junction temperature bias*, where the device was biased into its operating region and the junctions raised to some temperature, usually 150° to 170°C, while the case was held at room temperature.

2. *High Temperature Reverse Bias (HTRB)*, where the junctions were reverse biased and the ambient temperature raised to a range of 125° to 150°C.

Junction temperature burn-in was designed to expose "infant mortality" by stressing to maximum temperature and electrical conditions, while the HTRB aimed at exposing surface defects ("inversions") in PNP devices. HTRB required burn-in, after which the bias remained applied and temperature was reduced to the room temperature of 25°C, followed by an electrical test which was needed since the inversion effect in PNP devices disappeared within a few hours.

Junction temperature bias equipment was a large open rack with slant trays populated by hundreds of "deep throat" sockets designed for continued insertion and removal of devices. The slant allowed free-air circulation, sometimes forced, to maintain the required 25°C case temperature. HTRB used an industrial oven, with edge-card connectors mounted inside, wired through the oven wall to another edge-card connector outside. This connector allowed a "performance" or "personality" board to be plugged in and electrical bias applied. A "socket board," with sockets for approximately 20 devices, plugged into the inside connector. In an 8-cubic-foot oven, this arrangement accommodated approximately 600 transistors for burn-in. Since HTRB involved little dissipation, maintaining close temperature gradients throughout the oven posed no significant problem. Even a few linear feet per minute of air movement maintained the ±3% generally accepted industry standard. Power for HTRB simply consisted of commercial grade power supplies with reasonable regulation, moderate amperage, and overvoltage protection.

Junction temperature burn-in proved more complex. With maximum power requirements for each device, each rack required kilowatts of power. One major manufacturer used banks of welding generators to provide the 500 amperes required at a reasonable cost. These generators met the ±20% regulation standard considered acceptable for this type of conditioning. Oscillation presented a big problem, and efforts to control it, where desirable or possible, consisted largely of strategically placed ceramic capacitors, plus judicious and often innovative positioning of ferrite beads as close to the body of the transistor as possible.

With the advent of integrated circuits in the **1960's**, engineers reasoned that what worked for discrete semiconductors would also work for discretes bundled together on a single chip of silicon. Questions then arose, such as, "How do you get maximum dissipation from a junction shielded from the outside world by other junctions?" and "Does HTRB designed for PNP transistors really applies to an NMOS world?" Moreover, contrasted with discrete semiconductor devices, industry standardization for integrated circuits never materialized. Despite serious efforts by the Electronic Industries Association (EIA), standards remained far from finalized by the mid-1960's. Since industry cannot tolerate a vacuum, Joseph Brauer at Rome Air Development Center filled industry's need by creating MIL-STD-883 [5] which forms the foundation for standards used throughout the industry today.

Early 1972 saw the introduction of "on-board drivers" which are located on the burn-in board and can drive approximately 75 devices. Simple circuitry and good reliability made this a practical approach. With dynamic burn-in of TTL (transistor-transistor logic), engineers recognized the need to monitor on-board drivers. A simple pilot light bulb attached to the board served to indicate the presence of DC and clocking signals. Short circuits were not identified as a problem, except in higher power linear devices, probably because the bond wires would not carry a sustained short.

By the **mid-1970's**, the advent of memory devices caused burn-in technology to begin shifting from a board containing devices for conditioning to a chamber with associated drive electronics. As driver requirements grew more complex and drivers gravitated outside the burn-in chambers, more highly automated bonding technologies began producing devices that could sustain short circuits of large magnitudes. Consequently, shorted devices sometimes burned up high power drivers. The need to avoid this catastrophe led to the development of monitors to detect a shorted condition, shut off the driver, and sound an alarm. This approach protects drivers, but in a production situation, it requires a full-time attendance waiting for shorts and taking prompt actions. This problem was taken care of by two alternative approaches: socket isolation and row isolation.

Late **1970's** saw the introduction of a new evolving technology, Burn-In Test Systems (BITS), which combines the time consuming functional testing with burn-in, and offers the following advantages:

1. Dramatic reduction in test equipment requirements.

2. Ability to test with multiple patterns thousands of times, rather than just once, and at varying temperatures.

3. Ability to detect and report soft failures as well as temperature-dependent failures.

4. Ability to optimize burn-in duration on a lot-by-lot basis and tailor burn-in time to the requirements of the lot.

5. Confidence that the burned-in components will operate under specific conditions with no defective units escaping from burn-in.

From **1980's on**, the burn-in techniques and their applications have become more and more mature and cost effective. Component manufacturers benefited a lot from burn-in in this more and more competitive world market. New burn-in techniques come out from different perspectives such as the individual-temperature/voltage-control burn-in methodology [6]. The use of conventional environmental chambers for burn-in of high-power or high-frequency devices has several drawbacks. Putting high-power devices into a large chamber having a temperature-controlled air stream can lead to uncertain junction temperatures. This uncertainty arises from bulk airflow variations, local deviations in airflow around devices and variations in device power dissipation. This results in an ill-defined case-to-ambient thermal resistance. An alternative technology was developed in the **early 1990's** which permits large scale burn-in of semiconductor devices without the use of a chamber. The system directly controls the case, or package, temperature. *By direct contact with the case of an IC device, temperature is controlled through conductive heat transfer rather than by air or liquid convection.* Note that when the chamber is used, the device junction temperature is a product of power dissipation and the sum of junction-to-case and case-to-ambient thermal resistance. In general, junction-to-case thermal resistance is a relatively well-characterized constant whereas case-to-ambient thermal resistance is a variable. Since the magnitude of case-to-ambient thermal resistance is high compared to the junction-to-case thermal resistance, it is the largest contributor to variations in junction temperature at a given power level. Direct control of device case temperature, as opposed to device ambient temperature control, yields a much more predictable and stable junction temperature.

In the mean time, the manufacturing techniques advanced significantly. The overall defect rates of the integrated circuits decreased rapidly. According to [7], the failure rates of IC's for **1983** were well below 1% per thousand hours. In **1992**, the defect rates fell below 100 ppm. Numbers such as these prompted some engineers to intend to turn down the flame of burn-in testing. As Charles Packard of Unisys Defense Systems, Higgins, Minnesota, remarked [7] "Burn-in is probably doing more harm than good." Some people also pointed out to data based on an IES study [8] which suggested that failures caused by testing and handling, such as physical damage and ESD failures, were nearly two orders of magnitude higher than actual burn-in failures. People began to wonder "Is burn-in burned out?" Despite advances in reliability and test methods, many reliability experts and test engineers are skeptical about eliminating burn-in and have urged caution. As is being remarked by some people, the competitive pressure could cause manufacturers to rush shipments and deliver bad parts. More importantly, the industry doesn't have enough data yet on many new processes; consequently, removing burn-in for these processes would be very risky.

1.4 WHAT THIS BOOK OFFERS

The objective of this book is to systematically summarize the knowledge derived from the studies of numerous investigators in this field. It presents comprehensive coverage of the subject from the basic definition of burn-in to the state-of-the-art concepts and applications, such as

1. stress-strength interference theory applied to various failure patterns during burn-in,

2. bathtub curve models and mixed-life-distribution models for burn-in data analysis,

3. mean residual life (MRL) concept and its applications in optimum burn-in time determination,

REFERENCES

4. Renewal Theory and the Bayesian method in burn-in cost model development and minimization,

5. TTT transforms and TTT plots and their applications to optimum burn-in time determination,

6. accelerated burn-in using elevated temperature, elevated voltage bias and temperature cycling,

7. temperature dependency of the activation energy and the modified Arrhenius model, modified Combination model, and the average equivalence integration models for high temperature burn-in, high temperature-voltage burn-in and temperature-cycling burn-in, respectively.

Emphasis is given to the quantification, or mathematical and statistical description and formulation, of the failure process during burn-in, and the determination of the optimum burn-in time to meet various engineering specifications, such as the specified post-burn-in (conditional) mission reliability, failure rate and mean residual life goals, and for minimum total cost or for maximum total profit goal, and others.

Throughout this book, mathematical derivations are provided wherever the authors believe necessary. To enhance the understanding of various concepts and methodologies presented in this book, their logical relationships, and their practical applications, a guideline chapter has been designed and presented in Chapter 14 to give a systematic review and summary of the subjects presented in the whole book. It is our sincere hope that this last chapter does provide a useful "compass" for our dear readers including teachers and their students.

REFERENCES

1. J. E. Arsenault and J. A. Roberts, *Reliability and Maintainability of Electronic Systems*, Computer Science Press, 584 pp., 1980.

2. Buck, C. N., "Improving Reliability," *Quality*, pp. 58-60, February 1990.

3. Posedel, R. J., "Burn-in – the Way to Reliability," *Quality*, pp. 22-23, August 1982.

4. Hanlon, E., "Burn-in," *Circuits Manufacturing*, pp. 56-68, July 1980.

5. MIL-STD-883C, Test Methods and Procedures for Microelectronics, August 1983.

6. Jones, E. and Sheppe, R., "Alternate Approach to Traditional Burn-in," *Evaluation Engineering*, pp. 16-24, October 1991.

7. Romanchik, D., "Burn-in: Still a Hot Topic," *Test & Measurement World*, pp. 51-54, January 1992.

8. *IES Sixth National Symposium and Workshop – ESSEH*, Institute of Environmental Sciences, Mount Prospect, Illinois, 1990.

ADDITIONAL REFERENCES

1. Kececioglu, Dimitri B. and Sun, Feng-Bin, *Environmental Stress Screening – Its Quantification, Optimization and Management*, Prentice Hall, Inc., Englewood Cliffs, N.J., 530 pp., 1995.

2. Kececioglu, Dimitri B., *Reliability Engineering Handbook*, Prentice Hall, Inc., Englewood Cliffs, N.J., Vol.1, 720 pp. and Vol.2, 568 pp., 1991.

3. Jensen, Finn and Peterson, N. E., *Burn-in*, John Wiley & Sons, Inc., 167 pp., 1982.

4. Kamath, A. R. R., Keller, A. Z. and Moss, T. R., "An Analysis of Transistor Failure Data," *5th Symposium on Reliability Technology*, Bradford, September 1978.

5. Weiss, G. H. and Dishon, M., "Some Economic Problems Related to Burn-in Programs," *IEEE Transactions on Reliability*, Vol. R-20, No. 3, pp. 190-195, August 1971.

6. Plesser, K. T. and Field, T. O., "Cost-optimized Burn-in Duration for Repairable Electronic Systems," *IEEE Transactions on Reliability*, Vol. R-26, No. 3, pp. 195-197, August 1977.

7. Washburn, L. A., "Determination of Optimum Burn-in Time : a Composite Criterion," *IEEE Transactions on Reliability*, Vol. 38, No. 2, pp. 193-198, June 1989.

8. Marcus, Richard and Blumenthal, Saul, "A Sequential Screening Procedure," *Technometrics*, Vol. 16, No. 2, pp. 229-234, May 1974.

9. Ninomiya, T. and Harada, K., "Multilayer Debugging Process (a New Method of Screening)," *IEEE Transactions on Reliability*, Vol. R-21, No. 4, pp. 224-229, November 1972.

10. Marko, D. M. and Schoonmaker, T. D., "Optimizing Spare Module Burn-in," *Proceedings of the Annual Reliability and Maintainability Symposium*, pp. 83-86, 1982.

11. Nguyen, D. G. and Murthy, D. N. P., "Optimal Burn-in Time to Minimize Cost for Products Sold Under Warranty," *IIE Transactions*, Vol. 14, No. 3, pp.167-172, 1982.

ADDITIONAL REFERENCES

12. Sultan, T. I., "Optimum Burn-in Time: Model and Application," *Microelectronics and Reliability*, Vol. 26, No. 5, pp. 909-916, 1986.
13. Chandrasekaran, R., "Optimal Policies for Burn-in Procedures," *Operations Research*, Vol. 14, No. 3, pp. 149-160, 1977.
14. Kuo, W. and Kuo, Y., "Facing the Headaches of Early Failures: A State-of-the-Art Review of Burn-in Decisions," *Proceedings of the IEEE*, Vol. 71, No. 11, pp. 1257-1266, 1983.
15. Kuo, W., "Reliability Enhancement Through Optimal Burn-in," *IEEE Transactions on Reliability*, Vol. R-33, No. 2, pp. 145-156, June 1984.
16. Chi, D. H. and Kuo, W., "Burn-in Optimization under Reliability and Capacity Restrictions," *IEEE Transactions on Reliability*, Vol. 38, No. 2, pp. 193-198, June 1989.
17. Buck, C. N., Cartalano, L. J. and Shalvoy, C. E., "Justifying the Cost of Test During Burn-in," *Electronics Test*, pp. 31-35, April 1988.
18. Shaw, M., "Recognizing the Optimum Burn-in Period," *Quality and Reliability Engineering International*, Vol. 3, No. 4, pp. 259-263, 1987.
19. Wager, A. J., Thompson, D. L. and Forcier, A. C., "Implications of a Model for Optimum Burn-in," *Proceedings of the Annual Reliability Physics Symposium*, pp. 286-291, 1983.
20. Vassilev, V., "Cost Effectiveness of Burn-in Procedures of Semiconductor Devices and Integrated Circuits," *Proceedings of the 7th Symposium on Reliability in Electronics*, Vol. 11, pp. 543-553, 1988.
21. Whitbeck, C. W. and Leemis L. M., "Components vs System Burn-in Techniques for Electronic Equipment," *IEEE Transactions on Reliability*, Vol. 38, No. 2, pp. 206-209, June 1989.
22. Coleman, G. and Williams, R. W., "Justifying Burn-in – an Analysis," *Electronic Production*, pp. 57-61, February 1980.
23. Foster, R. C., "How to Avoid Getting Burned with Burn-in," *Circuits Manufacturing*, pp. 56-61, August 1976.
24. Holcomb, D. P. and North, J. C., "An Infant Mortality and Long-Term Failure Rate Model for Electronic Equipment," *AT & T Technical Journal*, Vol. 64, No. 1, pp. 15-31, January 1985.
25. Wurnik, F. and Pelloth, W., "Functional Burn-in for Integrated Circuits," *Microelectronics and Reliability*, Vol. 30, No. 2, pp. 265-274, 1990.
26. Slunsky, L., "Monitored Burn-in as the Method of Quality Control in Electronic Components and Devices Production," *Proceedings of the 7th Symposium on Reliability in Electronics*, Vol. 11, pp. 627-633, 1988.

27. Loranger, J. A., "The Case for Component Burn-in: the Gain is Well Worth the Price," *Electronics*, pp. 73-78, January 1975.

28. Hamilton, H. E., "An Overview – VLSI Burn-in Considerations," *Evaluation Engineering*, pp. 16-20, February 1992.

29. Hnatek, E. R., "A Realistic View of VLSI Burn-in," *Evaluation Engineering*, pp. 80-89, February 1989.

30. Smith, B., "Burn-in," *Quality*, pp. 30-32, August 1988.

31. Parsons, R., "Semiconductor Device Burn-in, Is There a Future?," *Quality and Reliability Engineering International*, Vol. 2, pp. 255-257, 1986.

Chapter 2

BURN-IN DEFINITIONS, CLASSIFICATIONS, DOCUMENTS AND TEST CONDITIONS

2.1 BURN-IN DEFINITIONS

In MIL-STD-883C [1], Method 1015.3, "Burn-in Test," burn-in is defined as follows:

Burn-in is a test performed for the purpose of screening or eliminating marginal devices, those with inherent defects or defects resulting from manufacturing aberrations which cause time and stress dependent failures.

Kuo and Kuo [2] define the burn-in as follows:

Burn-in is a stress operation that combines the appropriate electrical conditions with the appropriate thermal conditions to accelerate the aging of a component or device. In other words, burn-in is a process which operates the electronic components or systems under electrical and thermal conditions to demonstrate the real life of the components or systems in a compressed time.

2.2 THE DIFFERENCES BETWEEN BURN-IN AND ENVIRONMENTAL STRESS SCREENING (ESS)

There have been a lot of confusions both in the industry and in the literature about the terms "Burn-in" and "ESS." "Burn-in" and "ESS" have been used interchangeably by many people. In fact, they are two very relevant but different concepts.

Burn-in is a generally lengthy process of *powering* a product at a specified *constant* temperature.

ESS evolved from burn-in techniques but is a considerably advanced process. Generally, ESS is an accelerated process of stressing a product in continuous cycles between predetermined environmental extremes, primarily temperature cycling plus random vibration.

The misconception is due to the wrong assumption that historical "burn-in" procedures conducted on electronic equipment, currently in the inventory, are as cost effective as the ESS temperature cycling and random vibration screens. This assumption was refuted by the findings of the Institute of Environmental Sciences (IES) [3]. Table 2.1 illustrates the differences between burn-in and ESS procedures. It may be seen that burn-in can be regarded as a special case of ESS where the temperature change rate for thermal cycling is zero and vibration is sinusoidal if ever used.

2.3 BURN-IN METHODS AND THEIR EFFECTIVENESS

According to Flaherty [4] and Hnatek [5], the methods used to stress all the nodes in a device depend on the IC's complexity and its failure mechanisms. Therefore, they fall into the following categories:

1. **Static Burn-In**

    The simplest type of burn-in is static, or steady-state, burn-in. Static burn-in maintains a steady-state bias on each device under high temperature for a number of hours to accelerate the migration of impurities to the surface so that a potential failure will occur. A static system is cheaper and simpler, and is useful with contamination-related failure mechanisms. It is, however, less effective than dynamic burn-in for large scale integration (LSI) and very large scale integration (VLSI) devices.

TABLE 2.1– Comparison of burn-in test and *ESS* procedure.

Criteria	Burn-in	ESS
Temperature	Operating or accelerated	Cycled from high to low operating
Vibration	Sinusoidal (if used)	Random, normally 20–2,000 Hz
Temperature rate of change	Usually constant, but sometimes cycled	5°C per minute minimum
Length of time	Normally 168 hours or less	• 10 or 5 minutes perpendicular to each axis of orientation for vibration, and • 10 to 20 cycles for temperature cycling.

2. **Dynamic Burn-In**

    Dynamic burn-in uses power source voltage, clock signals, and address signals to ensure all internal nodes are reached during temperature stressing. It is more effective at detecting early failures in complex devices. It is also more expensive and requires more dedicated burn-in boards. It should be noted that a dynamic burn-in system can also be used for static burn-in, but not the other way around.

3. **Test During Burn-In**

    A subset of dynamic burn-in is Test During Burn-In (TDBI). TDBI adds functional testing and, possibly, monitoring of component outputs to show how they are responding to specific input stimuli. TDBI is the most comprehensive burn-in technique, especially when coupled with scan-based technology. It has been used primarily for dynamic random access memories (DRAMs) but is applicable for all large memories due to their long electrical test times. Normally, the electrical testing is performed after burn-in to detect failures. TDBI is not appropriate for EPROM's (erasable programmable ROM's), microprocessors and other VLSI circuits.

    A typical TDBI is performed in the following manner:

Step 1. The devices are operated at an elevated temperature (125°C) and voltage (7 to 7.5 V) for an extended period of time, while all devices-under-test (DUT's) are subjected to functional testing using a complex test pattern.

Step 2. The DUT's are operated for a short duration at a lower temperature (70°C) and voltage (5.5 V) during which parametric testing is performed.

Step 3. Repeat Steps 1 and 2 for 4 to 8 hours or longer.

4. **High-Voltage Stress Tests**

   High-voltage stress tests are categorized as burn-in screens because of their device-aging-acceleration features due to the application of voltage, time and temperature. For these tests, the distribution of the voltage stress throughout the IC is accomplished by carefully designed dynamic and functional operation. IC memory suppliers have used high-voltage stress tests in lieu of dynamic burn-in as a means to uncover oxide defects in MOS IC's. Some suppliers use high-voltage stress tests in conjunction with either dynamic burn-in or TDBI.

   A typical high-voltage stress test involves cycling through all addresses (for a memory) using selected memory data patterns for 2 seconds, both high and low (logically speaking), with 7.5 V forcing function being applied (for a 5-V rated part, for example).

Note that for VLSI devices which have a unique set of characteristics significantly different from small-scale integrated (SSI), medium-scale integrated (MSI) and large-scale integrated (LSI) devices, burn-in needs to be refined since both stress coverage and test coverage are required to develop an effective burn-in method for these devices.

Foster [6;7] classified the generally used burn-in methods into the following categories:

1. Elevated temperature plus power – the cheapest but the least effective method.

2. Elevated temperature plus power with all inputs reverse biased, or the so-called High Temperature Reverse Bias (HTRB) – a method with moderate cost and reasonable effectiveness for most devices.

3. Elevated temperature, power, dynamic excitation of inputs, and full loading of all outputs – an effective and expensive method.

4. Optimum biasing combined with temperature in the range of 200 to 300°C, or the so-called High-Temperature Operating Test

(HTOT) – an expensive and difficult-to-carry-out method which is not applicable to plastic devices due to the high temperature involved.

No matter how the burn-in methods are classified, one thing is certain. Each failure mechanism has a specific activation energy that dominates the effectiveness of each burn-in method. Burning-in components, by applying high voltage to their pins, accelerates the time-to-failure of oxide defects (weak oxide, pin holes, uneven layer growth, etc.) typically found in MOS devices. High temperature also accelerates these and other defects, such as ionic contamination and silicon defects. Table 2.2 [5] is a summary of failure mechanisms accelerated by various popular burn-in methods based on the activation energies for these failure mechanisms. Table 2.3 [5] is a summary of the effectiveness for these burn-in methods versus the major technology categories.

2.4 BURN-IN DOCUMENTS

Several military documents define burn-in standards which have been used throughout the industry. Among these, MIL-STD-781, issued in 1967, has been used to demonstrate the reliability of production electronic equipment. This standard has needed substantial improvements in the area of burn-in application as pointed out by many researchers. Consequently, a revised version of MIL-STD-785 lists burn-in as one of the eight (8) major tasks comprising a reliability program, and has been identified with major cost and reliability improvements. In addition, MIL-STD-883 and MIL-HDBK-217 have become a focal point in system concepts for ensuring a successful reliability demonstration test. Industry generally accepts MIL-STD-883 as the basis for most burn-in conditioning done by manufacturers of industrial electronic equipment. However, there are some criticisms concerning its ineffectiveness, expensiveness and possible damages to the equipment [8]. Also, there is little adequate theory to permit the calculation of the optimum burn-in time.

2.5 BURN-IN TEST CONDITIONS SPECIFIED BY MIL-STD-883C

In MIL-STD-883C [1], the following six (6) basic test conditions are specified for the burn-in test:

1. **Test Condition A – Steady State, Reverse Bias**

TABLE 2.2– Failure mechanisms accelerated by various burn-in methods [5].

Failure mechanisms	Static burn-in forward bias	Static burn-in reverse bias	Dynamic burn-in pulsed	High voltage stress burn-in	TDBI	TDBI plus high voltage
Leakage current	•	•		•		
Surface instability/ ionic contamination		•		•		
Step coverage	•		•			
Oxide defects		•	•	•		•
Stress coverage					•	•
Epitaxial & crystal defects			•			
Metalization defects	•		•			
Junction anomalies	•		•			
Inversion		•				
Channeling		•				
Wire bond problems	•		•			
Tunneling				•		
Metal migration	•			•		
Refresh problems for DRAM (charge storage)				•[†]		
GaAs ohmic contact resistance (such as Au/Ag /Ni/AuGe/Ni contacts)		•				

[†] Effective at low temperature.

TABLE 2.3– Effectiveness of various burn-in methods versus technology categories [5].

Technology categories	Static burn-in forward bias	Static burn-in reverse bias	Dynamic burn-in pulsed	High voltage stress burn-in	TDBI	TDBI plus high voltage
Linear ICs						
Contamination		•				
Step coverage /leakage	•					
CMOS logic						
Contamination		•				
Step coverage /leakage	•					
TTL logic (Std, S, H, LS, ALS, AS, F)	•					
Memories						
All			•		•	
MOS			•	•	•	•†
Microprocessors			•	•		
GaAs digital ICs:						
(1) SSI	•					
(2) MSI, LSI, & RAM			•			

† Effective for DRAM refresh failure mechanism.

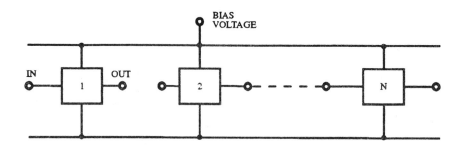

Fig. 2.1– Steady state burn-in test for Test Conditions A, B and C [1].

This test condition is illustrated in Fig. 2.1 and is suitable for use on all types of circuits, both linear and digital. In this test, as many junctions as possible will be reverse biased to the specified voltage.

2. **Test Condition B – Steady State, Forward Bias**

   This test condition is also illustrated in Fig. 2.1 and can be used on all digital type circuits and some linear types. In this test, as many junctions as possible will be forward biased to the specified voltage.

3. **Test Condition C – Steady State, Power and Reverse Bias**

   This test condition is also illustrated in Fig. 2.1 and can be used on all digital type circuits and some linear types where the inputs can be reverse biased and the output can be biased for maximum power dissipation or vice versa.

4. **Test Condition D – Parallel or Series Excitation**

   This test condition is typically illustrated in Fig. 2.2 and is suitable for use on all circuit types. Parallel or series excitation, or any combination thereof, is permissible. However, all circuits must be driven with an appropriate signal to simulate, as closely as possible, circuit application and all circuits should have the maximum load applied. The excitation frequency should not be less than 60 Hz.

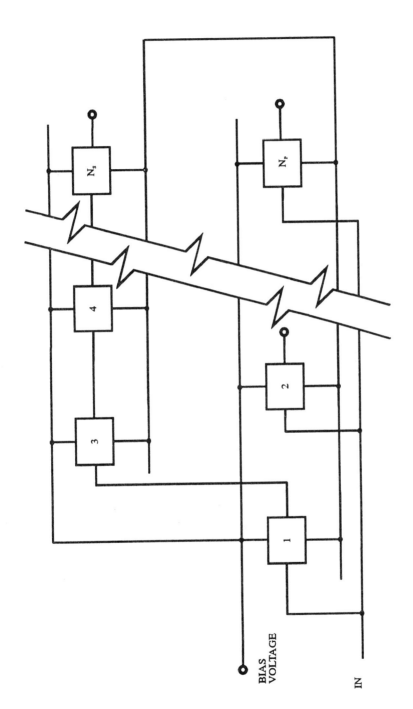

Fig. 2.2– Typical parallel, series excitation for Test Condition D [1].

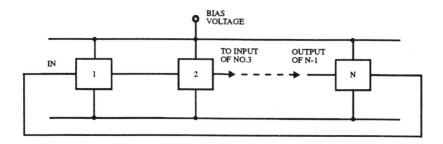

Fig. 2.3– Ring oscillator for Test Condition E [1].

5. **Test Condition E – Ring Oscillator**

    This test condition is illustrated in Fig. 2.3, with the output of the last circuit normally connected to the input of the first circuit. The series will be free running at a frequency established by the propagation delay of each circuit and associated wiring, and the frequency shall not be less than 60 Hz. In the case of circuits which cause phase inversion, an odd number of circuits shall be used. Each circuit in the ring shall be loaded to its rated maximum capacity.

6. **Test Condition F – Temperature-Accelerated Test**

    Under this test condition, microcircuits are subjected to bias(es) at an ambient test temperature, typically from 151°C to 300°C, which considerably exceeds their maximum rated junction temperature. It is generally found that microcircuits will not operate properly at these elevated temperatures in their applicable procurement documents. Therefore, special attention should be given to the choice of bias circuits and conditions to assure that important circuit areas are adequately biased without incurring any damaging overstresses to other areas of the circuit. To properly select the accelerated test conditions, it is recommended that an adequate sample of devices be exposed to the intended high temperature while measuring voltage(s) and current(s) at each device terminal to assure that the applied electrical stresses do not induce damaging overstresses. Note that Test Condition F should not be applied to Class S devices.

Table 2.4 can be used to establish the alternate time and temperature values. This table is based on the following two regression equations for Class B and Class S [9;10], respectively:

For Class B:

$$T_b = 4.303 \times 10^{-4} \, e^{\frac{5,106.8}{273.15+T^*}}, \tag{2.1}$$

where

T_b = burn-in time in hours,

and

T^* = burn-in temperature, °C.

For Class S:

$$T_b = 6.454 \times 10^{-4} \, e^{\frac{5,106.8}{273.15+T^*}}, \tag{2.2}$$

where

T_b = burn-in time in hours,

and

T^* = burn-in temperature, °C.

Any time-temperature combination which is contained in Table 2.4 for the appropriate class may be used for the applicable test condition. The test conditions, duration and temperature, selected prior to test should be recorded and shall govern for the entire test.

2.6 TEST TEMPERATURE

Unless otherwise specified, the ambient burn-in test temperature shall be 125°C minimum for conditions A through E (except for hybrids, see Table 2.4). At the supplier's option, the test temperature for Conditions A through E may be increased and the test time reduced according to Table 2.4. Since case and junction temperature will, under normal circumstances, be significantly higher than ambient temperature, the circuit employed should be so structured that the maximum rated junction temperature for test or operation shall not exceed 200°C for Class B or 175°C for Class S.

The specified test temperature is the minimum actual ambient temperature to which all devices in the working area of the chamber shall be exposed. This should be assured by making whatever adjustments

TABLE 2.4– Recommended burn-in times and temperatures for various test conditions [1].

Minimum temperature T^*, °C	Minimum burn-in time, hr		Test condition	Minimum reburn-in time, hr
	Class S	Class B		
100	–	352	Hybrids only	24
105	–	300	Hybrids only	24
110	–	260	Hybrids only	24
115	–	220	Hybrids only	24
120	–	190	Hybrids only	24
125	240	160	A–E	24
130	208	138	A–E	21
135	180	120	A–E	18
140	160	105	A–E	16
145	140	92	A–E	14
150	120	80	A–E	12
175	–	48	F	12
200	–	28	F	12
225	–	16	F	12
250	–	12	F	12

are necessary in the chamber profile, loading, location of control or monitoring instruments, and the flow of air or other suitable gas or liquid chamber medium.

REFERENCES

1. MIL-STD-883C, *Test Methods and Procedures for Microelectronics*, August 1983.
2. Kuo, W. and Kuo, Y., "Facing the Headaches of Early Failures: A State-of-the-Art Review of Burn-in Decisions," *Proceedings of the IEEE*, Vol. 71, No. 11, pp. 1257-1266, November 1983.
3. Institute of Environmental Sciences, *Environmental Stress Screening Guidelines for Assemblies*, 940 East Northwest Highway, Mount Prospect, IL 60056, 1984.
4. Flaherty, J. M., "A Burn-in' Issue – IC Complexity," *Test & Measurement World*, pp. 61-64, October 1993.
5. Hnatek, E. R., "A Realistic View of VLSI Burn-in," *Evaluation Engineering*, pp. 80-89, February 1989.
6. Foster, R. C., "Why Consider Screening, Burn-in, and 100-Percent Testing for Commercial Devices," *IEEE Transactions on Manufacturing Technology*, Vol. MFT-5, No. 3, pp. 52-58, September 1976.
7. Foster, R. C., "How to Avoid Getting Burned with Burn-in," *Circuits Manufacturing*, Vol. 16, No. 8, pp. 56-61, 1976.
8. Romanchik, D, "Burn-in: Still a Hot Topic," *Test & Measurement World*, pp. 51-54, January 1992.
9. Peck, D. S. and Zierdt, C. H., Jr., "The Reliability of Semiconductor Devices in the Bell System," *Proceedings of the IEEE*, Vol. 62, No. 2, pp. 185-211, February 1974.
10. MIL-STD-883B, *Test Methods and Procedures for Microelectronics*, November 1980.

Chapter 3

FREQUENTLY ENCOUNTERED TERMINOLOGIES AND ACRONYMS IN BURN-IN TESTING

3.1 BURN-IN TERMINOLOGIES

The following is a collection of definitions, in alphabetical order, based on the references in this chapter [1 through 11], for some frequently used terms in burn-in literature and documents:

Accelerated aging – Increasing certain stress parameters, such as temperature or voltage, above their normal operating values to observe deterioration in a relatively short period, and infer to expected service life under normal conditions in a longer calendar age.

Accelerated test – A test which uses stress levels higher than normal to reduce test time.

Acceleration factor – The multiplier by which the calendar time, or other usage, of an accelerated test will be factored to calculate the equivalent test time, or usage, under less stress or normal stress.

Acceleration spectral density – Also called *power spectral density*, a measure of the variance or mean-squared acceleration per unit of spectrum bandwidth representing the spectral distribution of the average energy.

Acceleration test – A test where the test article is subjected to an acceleration force in one plane.

Acceptance test – A test used to demonstrate compliance of a product to specified criteria as a condition of acceptability for "next" usage (next assembly, customer acceptance, etc.).

Acoustic vibration – Excitation of an article by aerodynamic pressure fluctuations or acoustically propagated noise.

Activation energy – An energy input, in electron volt units, which is required to cause a molecule of a constituent to participate in the reaction and which determines the relationship between a component's temperature and its lifetime, for certain temperature dependent kinds of failure modes of chemical nature. This energy may be determined experimentally by observing the times to failure of different batches of components at different temperatures.

Aging/Conditioning/Life-aging/Product maturing – Accumulation of stressful usage, experienced by a device whose failure distribution is a function of that usage.

Airborne – A classification for the most general aircraft equipment environment.

Airborne inhabited (AI) – A classification for the aircraft equipment environment found in inhabited areas that does not experience environmental extremes.

Airborne uninhabited (AU) – A classification for the aircraft equipment environment found in cargo storage areas, and wing and tail installations in which the following typical conditions are often present:

1. Contamination from engine exhaust, hydraulic fluid, oil.
2. Extreme pressure.
3. Temperature cycling.
4. Vibration cycling.

Ambient environment – Room conditions with no special controls imposed except for worker comfort.

Ambient test – A test, in contrast with "Environmental test," which is conducted under existing static conditions of the laboratory, factory floor, or office environment in which no environmental condition is applied.

Arrival quality – The percent of products found to be defect-free upon installation, or first use, at the customer's usage site.

Assembly/Module – A number of parts joined together to perform a specific function and capable of disassembly, such as a printed circuit board. An assembly of parts designed to function in conjunction with similar or different modules when assembled into a unit; e.g., printed circuit assembly, power supply module, core memory module.

Assembly level – The level at which a screen is applied, such as module, unit, system, etc.

Attribute – Qualitative property that a product has or does not have.

Attributes testing – A test procedure to classify items under test according to qualitative rather than quantitative characteristics, such as **Go/No-Go testing**.

Autopsy – Failure analysis to determine the root cause of a removed item's last wearout failure when this is the primary reason for the item being decommissioned or scrapped. Secondary failure modes and degradation may be noted.

Average outgoing quality level (AOQL) – Average percent defective of product that could leave an inspection, including both accepted and rejected but screened lots.

Axis – The direction along which a mechanical stress is applied.

Batch – A definite quantity of some product or material produced under homogeneous conditions. Compare with **Lot**.

Bathtub curve – A classical and graphical description for the life-cycle failure pattern of a product, which plots cumulative operating time or cycles versus the life-cycle instantaneous failure rate or hazard rate. Three phases are often claimed from the plot; i.e., the early life period during which the failure rate starts from a relatively high level and decreases and levels off at a stable and low level, the useful life period during which the failure rate stays at a relatively constant and low level, and finally the wearout period during which the failure rate increases significantly.

Benign environment – Conditions external to a unit which induce either minimum or no stress on it.

Bimodal distribution – A mixed distribution composed of two different subpopulations whose proportions sum up to one (1).

Black-box testing – Testing, contrast to "white-box testing" and "Gray-box testing," which is conducted for acceptability of the external output given specified external inputs and without concern for intermediate internal functions.

Board level – A level at which circuit boards are screened prior to sub- or final assembly. Components are in place and the product can be operated with proper interconnections. The only added complexity is usually in packaging.

Bogey test – Also called "pre-stress testing," a single-step, go/no-go, mechanical part test to a predetermined number of cycles run at a prespecified stress level, that is calculated to equal some given percentile of the worst case customer usage for one lifetime. This kind of test is intended to detect or screen manufacturing defects and used as the first stress step of some types of **Step-stress tests**.

Broad-band vibration – Vibration which contains significant levels within a broad frequency range, as opposed to narrow-band vibration, where the significant vibration levels are concentrated around a single frequency or a very narrow band of a frequency range.

Bulk defect – A time or temperature dependent, intrinsic or processing imperfection, impurity, or crystalline defect in the semiconductor material.

Burn-in – The process of continuously powering a product, usually at a constant temperature, in order to accelerate the aging process.

Burn-in chamber – An environmental chamber used to perform burn-in. The chamber volumes usually range from two (2) cubic feet, for temperature and batch versatility, to thirty two (32) cubic feet, for large manufacturing batch lots.

Burn-in effectiveness – The percentage increase of temperature-induced failures that will be experienced in a given time for a specified product at a given higher temperature than at a lower reference temperature.

Chamber/Temperature chamber – A cabinet in which products are placed to subject them to a stress or stresses. This usually consists of an insulated unit equipped with air circulation and temperature conditioning equipment with which ambient temperature can be varied rapidly and controlled in a prescribed manner.

Circuit board – An assembly containing a group of interconnected parts which are mounted on a single board.

Class B device – Classification for a microelectronic item which has been screened in a manner corresponding to its intention for use, where maintenance and replacement are difficult and expensive, and where reliability is slightly less imperative than in a Class S device. Such a device has passed a level of environmental stress screening, which is imposed by a customer, that is less rigorous than that of a Class S device. Contrast with a **Class S device**.

Class S device – Classification for a microelectronic item which has been screened corresponding to its intention for use, where maintenance and replacement are either extremely difficult or impossible, as in a satellite, where reliability is imperative. Such a device has passed the most rigorous level of environmental stress screening imposed by a customer. "S" in the title stands for outer space. Contrast with a **Class B device**.

Cold cracking – Crack defects caused by exposure to low temperature.

Cold soak – The part of the cycle of **Environmental testing** at the lowest temperature. Contrast with **Temperature soak**.

Cold solder connection/Cold solder joint/ High resistance joint – A failure mode in which a soldered electrical connection has solidified unevenly, forming poorly conductive microscopic cracks, which may lead to open or intermittent open failures in the field under slight mechanical stress.

Cold starts – Application of power following low-temperature stabilization.

Combined environmental test – A test performed during which more than one environmental stress is imposed.

Commercial – Military specifications are not imposed.

Common cause failures – Several items malfunctioning due to a single cause.

Component – A nonrepairable throw-away item (e.g., integrated circuit, resistor, capacitor, diode, transistor, transformer, hybrid, etc.).

Component defect – A defect caused by the failure of a component to meet its design specifications.

Component level – A level at which components are screened before being placed on circuit boards. Components may be screened by the manufacturer or by the user upon receipt.

Conditioning – Screening for a specified period of time.

Control chart – A graphical method for evaluating whether a process is in a state of statistical control; i.e., the plotted statistical measures are within the prespecified control limits.

Control limit – Limit on a control chart for judging whether a statistical measure, obtained from the sample, falls within acceptable bounds.

Corrective action – A process of correcting the root cause of a defect or failure.

Corrective action effectiveness – A percentage estimate of the fraction of the failures of a particular type, that a corrective action was to eliminate, and that will no longer be present in a new design.

Cost effectiveness – Monetary difference between the cost and the benefits derived from it.

Cost of performance – Cost of conformance plus cost of failure.

Crazing – Minute crack, or cracks, on or near the surface of materials, such as ceramics and plastics, caused by different rates of expansion or contraction of different layers.

Creep – A gradual change in the dimensions of a material under a mechanical load, or high stress, usually at high temperature.

Damage – Reduction in strength due to stress degradation, often expressed as the ratio of the "number of usage cycles applied" to the "number of usage cycles to failure."

Damaging overstress – An induced or natural overstress which exceeds the design capability of the article being screened, or under test, and which causes partial or catastrophic failure of the item.

Defect/Latent defect – A flaw in an item that would eventually prevent it from meeting its functional requirements when operating within its specified environment and within its expected lifetime. Stated equivalently, any non-conformance of a unit of product with specified requirements.

Defective/Defective unit – A unit of a product which contains one or more defects.

Defect density – Average number of latent defects per item. Symbols, such as D_{IN}, D_{OUT}, D_R and D_O, stand for incoming, outgoing, remaining and observed defect densities, respectively.

Defect-free – That portion of a test or screening sequence which must be completed without the appearance of a defect.

Degradation – Gradual deterioration in performance as a function of time and/or stress.

Degradation factor – The factor by which the *inherent reliability* is reduced due to environmental stresses the equipment experiences when being manufactured, transported, maintained, and used throughout its life cycle.

Degradation failure – A deterioration of one or more parameters beyond specification limits in two or more partial steps, such as electronic part drift, lubricant aging, inelastic or plastic strain, metal corrosion, and wearing away by abrasion.

Dependent failure – Failure caused by the failure of an associated item or items.

Derating – Upgrading component reliability by intentionally reducing the stress level in the application of an item.

Derating factor – Factor by which an equipment is derated to achieve a reliability **Safety margin**.

Design defect – A defect caused by a faulty product or process design.

Design deficiency – One or more of the following unreliability causes:

1. Actual mistakes in the design.
2. Incapable state-of-the-art components.
3. Oversights.
4. Unavoidable complexities and undesirable conditions.
5. Unforeseen material incompatibilities and unforeseen conditions.
6. Unknown environments.

Design fault – Faults generated during design and not corrected.

Design margin – The self-imposed restriction on a design more severe than either specified or operational use requirements.

Design of experiments (DOE) – A branch of applied statistics dealing with planning, conducting, analyzing, and interpreting controlled tests to evaluate the effects and interactions of various factors which control the value of one or more parameters.

Designed test – A controlled test to evaluate a parameter or group of parameters.

Design weakness – The inability of a product to survive such rigors as handling, transportation and service.

Detectable failure – A failure that can be detected with 100% test detection efficiency.

Detection efficiency/Test detection efficiency – A characteristic of a test measured by the ratio of the number of failure modes detected to the total number of potential failure modes.

Dynamic burn-in – The process of applying source voltage, clock signals, and address signals to ensure that all internal nodes are reached during temperature stressing.

Electromigration – A failure mode for a very large-scale integrated circuit due to high current densities and operating temperatures which cause or enlarge vacancies or *puddles* in the aluminum crystal lattice grain boundaries of conductor leads. This failure mode has symptoms of open contacts and bond lifts, and shorts or leaks between and within layers and junctions.

Endurance strength/Fatigue strength – Maximum repetitive or cyclic damage that a material can withstand before fracturing. For mechanical components it corresponds to the strength distribution at and beyond the knee of the $S - N$ diagram.

Environmental cycle – A single complete submittal of the equipment to all of the specified environments.

Environmental sensitivity – The change in a specified parameter of a part, assembly, unit, or system that results from exposure to the environment.

Environmental stress screening/Stress screening – A process in which 100% of products are subjected to one or more stresses (jointly or separately), such as thermal cycling, random vibration, high temperature burn-in, electrical stress, thermal shock, sine-fixed frequency-vibration, low temperature, sine-wave-swept-frequency vibration, altitude or humidity, etc., with the intent of forcing latent defects to surface as early failures.

Environmental testing – A process in which a sample of a product is subjected to environmental stimulation. Testing can be used in a variety of ways:

1. Testing on a prototype at extreme temperatures to confirm the range in which it was designed to function.
2. Testing on a randomly selected product at extreme temperature ranges to confirm continuing design and production process compliance.
3. Life-testing a product to determine its mean time between failures ($MTBF$).
4. Testing the product under the simulated environments which the product will encounter in transportation and operation.

Equipment power on-off cycle – The state during which an electronic item goes from zero electrical activation level to its normal design system activation level and back again to the zero activation level.

Escapes – A proportion of incoming defect density, referred to as D_{OUT}, which is not detected by a screen test and which is passed on to the next level.

ESSEH – Environmental stress screening of electronic hardware.

ESS **profile** – The sequence and duration of stress environments to which items are to be subjected.

Experiment – Same as **Designed test**.

Factorial experiment – An experiment in which all possible treatment combinations formed from two or more factors, each being studied at two or more levels, are examined to estimate the main effects and the interactions.

Factory failure cost savings – Costs that are avoided by the use of screening techniques to reduce the number of failures occurring at the manufacturer's facility.

Failure – Lack of proper operation. Failures can be classified into the following most common types:

> Critical failure – A case where the product is unable to operate under the conditions it is expected to.
>
> Non-critical failure – Failure that occurs only outside the normal operating range of the product.
>
> Hard failure – A critical failure where the product stops and does not resume function.
>
> Soft failure – A critical or non-critical failure where the product stops functioning under certain conditions but then resumes under others.
>
> Infancy failure – Failure that occurs in the early stages of a product's life.
>
> Electrical failure – An electrical malfunction caused by electrical components, connections, switches or related devices.
>
> Mechanical failure – A mechanical malfunction caused by cracking, fracture, displacement or misalignment of assemblies.
>
> Process failure – Failure caused by the manufacturing process.

Failure-free period – A continuous period of time during which an item is to operate without failure while under environmental stresses.

Failure-free test – A test to determine if an equipment can operate without failure for a predetermined time period under specific stress conditions.

Fallout – Failures observed during, or immediately after, and attributed to, stress screens.

Fatigue ductility coefficient – The minimum true strain required to cause failure in one reversal cycle of a test which applies cyclic tension.

Fatigue ductility exponent – A material fatigue property, nearly a constant ranging from about −0.50 to −0.70 for metals, which is equal to the log of the fraction of a given plastic strain range which will cause a tenfold increase in the median fatigue life.

Fatigue failure – The breakage failure of a component subjected to cyclic stresses, due to the spreading of a fracture which starts from a weak point.

Fatigue resistance – Resistance to metal crystallization and embrittlement from flexing, which would lead to metal fracture.

Fatigue strength – See **Endurance strength**.

Fatigue strength coefficient – A material fatigue property, ranging between about 100 ksi and 500 ksi for heat-treated steels, which is equal to the true stress required to cause a fracture in one reversal of a destructive cyclic tension test.

Fatigue strength deviation – A measure of inherent scatter in fatigue strength, which is equal to the standard deviation of the fatigue strength from the mean fatigue strength over a range of fatigue lives.

Fatigue strength exponent/Basquin's exponent – A material fatigue property, ranging from about −0.05 to −0.12, which is equal to the log of the fraction of a given true stress range which will cause a tenfold increase in the median fatigue life.

Fatigue strength limit – The maximum alternating stress, ranging between 20 ksi and 200 ksi, below which there will be no fatigue failure regardless of the number of operating cycles.

Fault detection – One or more tests conducted to determine whether or not any malfunctions or faults are present in a unit.

Fault detection coverage/Fault detection efficiency /Diagnostic coverage – Percentage of failures which can be detected and diagnosed correctly.

Fault isolation – Tests conducted to isolate faults within the unit under test.

Field – The place where a product is ultimately used.

Field failure cost savings – Costs that are avoided by the use of screening techniques to reduce the number of failures occurring during the operational life of the equipment.

Field replaceable unit (FRU) – See **Line replaceable unit (LRU)**.

Field warranty rate – The rate of failure during the time a warranty is in effect.

Final assembly level – A screening level at which *ESS* is applied to finished products.

Fixed sinusoidal vibration – Vibration excitation with a constant level and frequency, and with a waveform of a sinusoid.

Fractional defective – Proportion of defective units expressed as a decimal rather as a percentage.

Fractional factorial design/Fractional replicates – A factorial experiment in which only an adequately chosen fraction of the treatment combinations required for the complete factorial experiment is selected to be run.

Freak failures – Anomalous component, subsystem, or system failures usually distributed in time, centered near a specified percentage of useful life such as 10%, on less than or equal to another specified percentage of a population, such as 20%.

Functional test – A test which measures a limited number of critical parameters to assure that the test article is operating properly.

Fixture/Fixturing – An intermediate structure to attach and secure items on a shaker or within a chamber in preparation for running screens.

Ghost – An **Intermittent** class of failure, the cause of which can neither be diagnosed nor effectively repaired.

Go/No-Go – The binomial attribute condition or state of operability of a unit which can only have the following two parameters:

1. Go or functioning properly.
2. No-go or not functioning properly.

Go/No-Go testing – Test method using a fixed measure for *inspection by attributes* to determine whether or not a measurement conforms with specifications.

Ground benign environment (GRB) – A classification for the least stressful component or system environment.

Ground fixed environment (GRF) – A classification for a maintained installation condition somewhat less ideal than **Ground benign** but more ideal than **Ground mobile**, such as the permanent installation in air traffic control, in communication facilities, in ground support equipment, in racks with adequate cooling air, in radar, in unheated buildings, etc.

Ground mobile environment (GRM) – A classification for an equipment environment which is more severe than **Ground fixed** environment mainly due to the following reasons:

1. Less uniform maintenance.
2. More limited cooling supply.
3. Natural outdoor environmental stresses.
4. Shock.
5. Vibration.

Hard failure – A failure, in contrast with **Intermittent** and **Ghost failure**, which is always repeatable or continuously observable.

Heat endurance – The total time at a specific heat condition which a material can withstand before failing a specific physical test.

Heat shock – A form of **Thermal shock** due to sudden exposure to a high temperature for a short period.

High reliability program – A series of tasks performed to assist in developing an equipment which has an extremely low probability of failure.

High temperature reverse bias (HTRB) – Fault avoidance methodology in which burn-in of diodes and transistors is conducted with the junctions reverse biased, in order to force any failures to occur which are likely to be caused by ion migration in bonds of dissimilar metals.

High-voltage stress tests – Categorized as burn-in screens due to their voltage, time and temperature device accelerated aging features, the high-voltage stress tests apply carefully-designed dynamic and functional operations to accomplish the distribution of the voltage stress throughout the integrated circuits.

Hot start – Application of power following high temperature stabilization.

Imperfection – A departure of a quality characteristic from its intended level or state without any association with conformance to specification requirements or to the usability of a product or service.

Incipient failure – A failure which has not occurred at a time when the damage, which must lead to failure, is first detected, but which will undoubtedly occur during the useful life of the specific item in an operating environment.

Infant failures – Failures which occur early in the operating life of a component, module, or unit.

Inherent defect – A failure or defect that is a function of the intended design application of the item, when operated in its intended operational and logistic support environment.

Intermittent failure – A non-permanent item failure which lasts for a limited time period, followed by recovery of its ability to perform within specified limits.

Latency – A state in which a fault remains undetected; i.e., being **Incipient**.

Latency time – Time period between the occurrence of a failure to its detection.

Latent defect/Random defect – An inherent or induced weakness, not detectable by ordinary means, which will either be precipitated as an early failure under environmental stress screening conditions or will eventually fail in the intended use environment.

Level of assembly – The hierarchy of items within a larger item which are functionally or physically grouped and considered together for various purposes, such as system, subsystem, module, board, and component.

Life-cycle testing – A process through which a small percentage of products are subjected to stresses similar to those they will experience during their lives. This process is usually used to determine a product's anticipated life expectancy and its *MTBF*.

Life test – A test performed on a group of items (parts, assemblies, units, or systems), continued until a specified percentage of this group fails or some predetermined minimum operating time has elapsed.

BURN-IN TERMINOLOGIES

Limiting quality level (LQL) – See **Lot tolerance percent defective.**

Line replaceable unit (LRU) – A unit normally removed and replaced as a single item which consists of assemblies (SRU's), accessories, and components that collectively perform a specific functional operation.

Lot – A group of products manufactured or processed under substantially the same conditions.

Lot tolerance percent defective (LTPD) – Value of lot percent defective on an *operating characteristic curve* corresponding to the value of the customer's risk.

Main effect – A term describing the contribution of a factor at each level to process responses averaged over all levels of other factors in the experiment.

Malfunction – Inability to perform a required operation, in particular, an **Intermittent failure.**

Manufacturing defect – A flaw caused by in-process errors or uncontrolled conditions during assembly, test, inspection, or handling during manufacturing.

Margin of safety – See **Safety margin.**

Marginal failure – A failure which may cause minor injury in a military system, minor property damage, or minor system damage which will result only in delay, loss of availability, or mission degradation.

Mean residual life – The expected value of the remaining life of a component at a specified age.

Module – An assembly of parts usually packaged in a plug-in form for ease of maintenance of the next higher level of assembly, such as printed circuit assembly, power supply module, core memory module, etc. It is designed to function in conjunction with similar or different modules when assembled into a unit.

Monitoring/Condition monitoring/ Performance monitoring – Observing or keeping track of output measurements, or obtaining data by any appropriate means, to signal changes of operating conditions or of outputs, and the necessity of corrective actions to eliminate any potential failures.

Multiple failures – The simultaneous occurrence of two or more *independent* failures.

Multiple fault condition – More than one fault occurring during the same fault detection process.

Next higher level effect – The consequences a failure mode has on the operation, functions, or status of the items in the next higher level of assembly.

Nonchargeable failure – A failure which is either irrelevant, or relevant, but caused by a condition prespecified as not the responsibility of any given organization, and therefore not charged.

Nonconformance/Nonconformity – A departure of quality characteristic from its intended level, or state, that occurs with a severity which is sufficient enough to cause an associated product, or service, not to meet a specification requirement.

Nonconforming unit – A unit of product containing at least one nonconformity.

Non-environmental stress screening failures – Failures which occur during **Environmental stress screening** but are not considered as screened **Incipient failures**, such as misinterpreted failures, repair-induced failures, operator-induced failures, dependent failures, and failures due to improper facility installation, etc.

Nonrelevant failures – Equipment failures which are not expected to occur during field service.

Notching – A technique of reducing the input power spectral density (PSD) level over a small frequency bandwidth, in particular around an equipment resonant frequency, so that the equipment is not damaged due to overstress at resonant frequency.

On-off cycling – Switching of equipment primary power on and off at specified intervals.

Orthogonal axis vibration – Excitation applied at right angles to an axis.

Parameter drift – A form of wearout in which electrical and mechanical components, which are designed to certain tolerance limits, drift out of or exceed their tolerance limits, as a function of operating time and stress level; however, the product recovers its normal function after a rest period.

Part – An element of a subassembly, or an assembly, so constructed as to be impractical to further disassemble for maintenance.

Part fraction defectives – The number of defective parts contained in a part population divided by the total number of parts in the population expressed in parts defective per million parts (PPM).

Patent defect – An inherent, or induced, weakness which can be detected by inspection, functional test, or other defined means without the need for stress screens.

Percent defective – The number of defective units, divided by the number of units of the product times 100.

Periodic conformance test – A test performed at regular intervals to verify continued compliance with specified requirements.

Power cycling/Operational cycling – The process of continuously turning the product on and off at predetermined intervals. Power cycling adds internal heat and ages the product much faster than continuous power on and off. Power cycling also allows monitoring of products. This provides diagnostic information used to determine reasons for soft failures.

Precipitation of defects – The process of transforming a latent defect into a patent defect through the application of stress screens.

Predefect-free – That portion of a screening sequence prior to the disclosure of a defect. Contrast with **Defect-free**.

Prestress screening – Eliminating *incipient failures* which would probably occur during the early life period by applying an environmental and/or use stress at a peak abuse level.

Prestress test – See **Bogey test**.

Printed circuit board (PCB) – Same as circuit board.

Printed wiring board (PWB) – Same as circuit board.

Producibility – Ease of fabrication and assembly.

Product assurance – A discipline that assures that products are designed and manufactured to be highly reliable, easy to maintain and safe to operate.

Production flaw – A latent defect in a product, such as a poorly soldered connection, a loose screw, or a contaminant like scrap wire, foreign particles, dust, moisture, etc.

Production lot – A group of items manufactured under essentially the same conditions and processes.

Production reliability test – A test performed to measure the reliability of production items.

Production sampling – Sample testing during production to determine compliance to specifications.

Productivity – Value achieved for the resources expended.

Proof-of-design test – A test performed during the design cycle to demonstrate compliance with specified design criteria.

Qualification test – A test to verify that a product's design meets contractual requirements.

Quality assurance – The total effort of *quality control* and *quality engineering*.

Quality defect – Any failures, errors, flaws, interruptions, and faults occurring in a product, its subassemblies, and *field replaceable units* during manufacturing and early field operation but before the end of the specified early life or promised warranty period. Contrast with **Reliability defect**.

Quality engineering (QE) – Analysis of the manufacturing system at all stages of development to maximize ultimate process and product quality.

Quality function deployment (QFD) – A structured method to identify customer needs, translate them into a realizable product or service parameters, and guide the implementation process in such a way as to bring about a competitive advantage.

Quality management – The totality of functions involved in the determination and achievement of quality.

Quasi-random vibration(QRV) – QRV may be described as an equally-spaced line spectrum whose fundamental frequency varies randomly with time within a restricted frequency range and, in turn, causes a random fluctuation in the amplitudes of the spectral lines. The fundamental frequency fluctuates sufficiently to produce an essentially continuous spectrum when averaged over a suitable interval.

Random vibration – Vibration excitation where magnitude and frequency are specified by a probability distribution function.

Relevant failure – Equipment failure which can be expected to occur in field service. Contrast with **Irrelevant failure**.

Reliability defects – Any failures, errors, flaws, interruptions, and faults occurring in a product, its subassemblies, and **Field replaceable units** after the end of the specified early life or promised warranty period. Contrast with **Quality defect**.

Reliability demonstration – A test to determine a product's reliability, failure rate, $MTBF$, or percentile life.

Repair time – The period required to locate the cause of failure, perform the repairs required, and verify that the repair action was effective.

Resonant dwell – Vibration excitation of a test article at one of its resonant frequencies for a sustained period of time.

Run-in – A process, often confused with **Burn-in** and **ESS**, of simulating or duplicating but not exceeding the stresses normally expected in the customer environment to *surface* all non-working functions.

Safety allowance – Same as **Design margin**.

Safety factor – Same as **Design margin**.

Safety margin – Difference between the designed-in strength (stress at failure) and the operating stress level expressed in units of the rating, or the safety factor minus one (1) in mechanical component design.

Sample test – A test performed on a limited number of items during the production process.

Screen effectiveness – Generally, a measure of the capability of a screen to precipitate latent defects to failure. Sometimes used specifically to mean screening strength.

Screen level – The assembly level at which an ESS program is implemented; such as component, board, subassembly or final assembly.

Screen parameters – Parameters in screening strength equations which relate to screening strength, such as vibration g-levels, temperature rate of change, duration, etc.

Screenable latent defect – A latent defect which has an inherent failure rate of greater than 10^{-3} failures per hour under field stress conditions.

Screening – A process, or combination of processes, applied to 100% of a lot, or group of like items, to identify and eliminate defects.

Screening attrition costs – Costs associated with replacing and re-screening parts that fail during part level environmental screening, or repairing items at a higher level of assembly that fail during higher level environmental screening.

Screening complexity – The degree to which a product is screened, such as the following:

1. Static – The application of a single stress to an unpowered product.
2. Dynamic – Power, and on occasion an additional screen, applied to a product during the initial screen.
3. Exercised – Power and monitoring, or loading inputs and outputs, are applied during the screen. The application of signals or loads is sometimes necessary to avoid undue product damage.
4. Full functional – The product is operated during screening as though in actual use, with all circuitry exercised.
5. Monitored – Providing monitoring devices that either display or record screen performance which is also helpful in providing field failure analysis data.

The screen complexity is dependent on the need to identify or force specific types of failures, the product's anticipated operating environment, the level of assembly at which the product will be stressed and, in general, consideration of the degree of stress that is both necessary and cost-justifiable to gain a measurable result.

Screening cost effectiveness – The dollar difference between the cost of screening and the benefits derived.

Screening-induced degradation – See **Stress screen degradation**.

Screening regimen – A combination of stress screens applied to any level of complexity, such as module, unit or system, identified in the order of application.

BURN-IN TERMINOLOGIES

Screening sequence – Chronological order of activities to conduct *ESS*.

Screening strength – The probability that a specific screen will precipitate a latent defect to failure, given that a latent defect susceptible to the screen is present. Symboled *SS*, screening strength is usually measured by the percentage of total defects identified by a specific screen.

Seasoning – A process, often confused with **Burn-in**, of energizing a consumable item at its rated usage stress for a period of time equal to a specified percentage of its average rated laboratory life, often chosen to be 1%, in order to reduce fluctuation.

Secondary damage – Damage to an item due to failure of another item within the same configuration.

Secondary damage effects – Consequences indirectly caused by the interaction of a damage mode with a system, subsystem, or one of their components.

Secondary failure – A dependent item failure either indirectly or directly due to the failure of another item.

Selection and placement – The process of systematically selecting the most effective stress screens and placing them at the appropriate levels of assembly.

Sensor – A device which observes a physical condition of interest and outputs an appropriate electrical signal related to the condition.

Shakedown – A form of **Run-in** using actual operation under rigorous but typical usage merely to show that a unit is capable of operation.

Shaker – A machine usually powered by electricity, hydraulic pressure, or compressed air, providing a controlled source of vibration to a product.

Shock test – Subjection of a test article to one or more singly applied acceleration pulses at a frequency of less than 1 Hz.

Shop replaceable unit (SRU)/Shop replaceable assembly – An assembly or any combination of parts, subassemblies, and assemblies mounted together, normally capable of independent operation in a variety of situations and repairable at maintenance levels above the field, such as on the bench, in a depot repair facility, or in a logistic center.

Sine vibration/Sine fixed frequency vibration – The vibratory excitation is sinusoidal with a prespecified fixed frequency.

Sine wave, swept frequency vibration – The vibratory excitation is applied with a gradually increasing or decreasing strength and frequency.

Single-frequency vibration – Sine vibration at only one frequency.

S-N curve – A graph of applied stress, S, versus the corresponding cyclic fatigue life, N.

Sneak circuit analysis – A procedure of identifying *latent* circuit paths which may cause occurrence of *unwanted* functions or *inhibit* desired functions assuming all components are functioning properly.

Sneak condition – Latent paths, timing indications, or labels in electrical hardware or computer logic.

Sneak indicators – False or ambiguous system status that can result from improper connection or control of display devices and their sensors.

Sneak labels – Potential causes of erroneous operator actions which can arise from a lack of precise nomenclature on controls or operating consoles.

Sneak paths – Latent current paths that cause an unwanted function to occur or inhibit a desired function independent of component failures.

Sneak software analysis – Analyzing software sneak conditions using electrical/software analytical analogies.

Sneak timing – Inappropriate system response which may be due to incompatible hardware or logic sequences.

Soft failure – See **Intermittent failure**. Contrast with **Hard failure**.

Solid failure – A form of **Hard failure** reported by a maintenance specialist who observes a user perceived continuous, permanent, or easy-to-duplicate failure condition, during corrective maintenance, in correcting that failure.

Sonic fatigue – Mechanical wearout due to vibration transmitted through a gaseous medium.

Sonic vibration – Potentially stressful mechanical vibration waveforms induced sonically in the frequency band between 1 Hz and 20 kHz.

Spectrum analyzer bandwidth – The interval separating the upper and lower frequencies observed at any given time during verification of a random vibration spectrum.

Standard environmental profile – A *cookbook*, or standard testing environment described in a general reference, which is not tailored to a specific application. Contrast with **Tailored environmental profile**.

Static/Steady-state burn-in – A process of maintaining a steady-state bias on each device under a high temperature for a number of hours to accelerate the migration of impurities to the surface so that a potential failure will occur.

Steady-state burn-in – See **Static burn-in**.

Step stress – An experimental process of increasing stress levels incrementally so that the effects of each step may be evaluated.

Stimulus – Any energy applied to a device to produce a measurable response.

Strain – Solid body deformation measured by changed distances between points on the body.

Strain gage – Any device which can sense the deformation of a solid body and transmit the signal into a measurable output.

Strength – The inherent ability of resisting failures. The stress magnitude at failure.

Stress – Electrical, thermal or mechanical forces caused by a stimulus being applied to a product.

Stress screen degradation – Reduction of component strength in a good component due to the application of a stress screen.

Stress screening – The process of applying mechanical, electrical and/or thermal stresses to an equipment item for the purpose of precipitating latent part and workmanship defects to early failure.

Stress-strength classifications – Three categories are usually classified for stress and strength according to the randomness and dependence:

1. Deterministic.
2. Random, fixed.
3. Random, independent.

Stress-strength interference – A probabilistic approach for quantifying the failure process, in particular, the probability of a component with a known distributed strength surviving a given distributed stress application.

Subassembly level – The stage of production prior to final assembly, but more advanced than the circuit board level. The goal of screening at this level is to find latent defects in packaging and interconnections.

Substandard item – An item which may perform all its functions, but not all to the minimum specified requirements.

Swept sinusoidal vibration – Vibration excitation with amplitude progressively varying over a specified frequency range.

System/Equipment – A group of units interconnected or assembled to perform some overall electronic function (e.g., electronic flight control system, communications system).

Tailoring – The process of selecting an initial stress screening program and the subsequent changes to the methods and severities of the screens to achieve maximum effectiveness.

Tailored environmental profile – Duplicating a known or anticipated field environmental profile as closely as possible. Contrast with **Standard environmental profile**.

Target flaw – The latent defect or incipient failure type that the corresponding specific stress screen is chosen to precipitate.

Temperature cycling – A process through which a product is subjected to a predetermined temperature change rate between established temperature extremes for a specified period of time. This process can verify that the product will operate at various temperatures, as well as its ability to withstand the rates of change between these extremes. In ESS thermal cycling, the ability to develop fast rates of changes will determine how many cycles are required to force the greatest number of latent defects into failure.

Temperature shock – Repeatedly subjecting an item (part, module, unit, system) to high and low temperature limits with a high rate of change between them, such as equal to or greater than 25°C per minute.

Temperature soak – Portion of the thermal cycle in an environmental chamber at which the highest stable temperature is induced. Contrast with **Cold soak**.

Temperature stabilization period – This is the amount of time required to reach thermal equilibrium, at which item and chamber temperatures are identical, starting from the end of the prior stabilization period.

Test – A process of subjecting an item to conditions designed to determine whether or not the item meets specified performance requirements, such as qualification test, reliability demonstration test, acceptance test, etc. It should be pointed out that the conditions of a test may duplicate those of a screen. However, the intent of a screen is to precipitate failures while the intent of a test is to demonstrate how well the item meets its specified requirements. The goal of a test is usually zero failures. The goal of a screen is to force defects to surface during the screen, rather than during subsequent testing or operation.

Test detection efficiency – A measure of test thoroughness, or coverage, which is expressed as the fraction of patent defects detectable, by a defined test procedure, to the total possible number of patent defects which can be present. Symbol **DE** is used synonymously as the probability of detection.

Test during burn-in (TDBI) – A subset of dynamic burn-in which adds functional testing and, possibly, monitoring of component outputs to show how they are responding to specific input stimuli.

Test specification – A document defining tests to be performed and the specified parameter limits.

Test strength (TS) – The product of screening strength and test detection efficiency. The probability that a defect will be precipitated by a screen and detected in a test.

Test-to-failure – Testing continued until the item ceases to function within specified parameter limits.

Thermal aging – A form of **Accelerated aging** by exposing items to given thermal conditions, or to a programmed series of conditions, for a given period.

Thermal conductivity – The rate of heat transfer through a material.

Thermal cycling – Same as **Temperature cycling**.

Thermal endurance – Minimum time required at a selected temperature for a material, or system of materials, to deteriorate to some predetermined minimum acceptable level of electrical, mechanical or chemical performance under prescribed test conditions.

Thermal equilibrium – Item temperature condition in which the rate of change is less than 2°C/hr.

Thermal fatigue – Mechanical failure mode of materials subjected to the stress of alternating cycles of heating and cooling, including loss of temper, off-tolerance states and work hardening.

Thermal gradient stresses – Cyclically cumulative, or one-shot, mechanical stresses within an integral assembly caused by an unmatched distribution of either heat or thermal expansion between adjacent parts of an assembly, such as printed circuit board warping and integrated circuit walk-out from sockets.

Thermal rating – The temperature range within which a material will function without unacceptable degradation.

Thermal resistance – The opposition to heat flow through a material.

Thermal shock – Subjecting a material to rapid and wide range of changes in temperature to ascertain its ability to withstand those rapid changes.

Thermal survey – The measurement of thermal response characteristics, such as time to equilibrium, etc, at points of interest within an equipment when temperature extremes are applied to the equipment.

Tolerance – The allowable variation in a quality characteristic within which an item is judged to be acceptable, and equal to the difference between the upper limit and lower limit.

Tolerance limits – Minimum and/or maximum permissible parameter values for an item, as determined by nominal and basic values and allowances.

BURN-IN TERMINOLOGIES

Transducer – A device which changes the format of a sensor's output in order to facilitate using it, but does not perform analog-digital conversions.

Transparent failure – A fault tolerant failure causing no obvious performance degradation.

Treatment – A combination of levels of different factors involved in an experiment.

Treatment combination – One set of levels of all factors involved in running an experiment.

Troubleshooting time – The period required to isolate the fault at the system, unit, module or part level.

Ultrasonic vibration – Potentially stressful mechanical vibration waveforms in the frequency band above 20 kHz. Contrast with **Sonic vibration.**

Unilateral tolerance – Tolerance which allows variations in one positive or negative direction from the specified dimension.

Unit – A self-contained collection of parts and/or assemblies within one package performing a specific function or group of functions, and removable as a single package from an operating system, such as an autopilot computer, VHF communications transmitter, voltmeter, etc.

Variable data – Data which contains a measured value as opposed to go-no-go indication.

Vibration – Periodic movement of an item.

Vibration survey – The measurement of vibration response characteristics, at points of interest within an equipment, when vibration excitation is applied to the equipment.

Warranty – Explicit or implied guarantee from the seller to the purchaser that an item will perform as specified, or implied, for a minimum specified period of time or usage.

Warranty claim – Action started by the equipment user for authorized warranty repair, replacement, or reimbursement made from the local dealer or manufacturer.

Warranty maintenance – Any corrective maintenance during the warranty period of the product or during a promised free maintenance period after repair.

Warranty period – Calendar time during which the warranty is in effect.

Wearout – The process of attrition which results in an increasing failure rate.

Wearout failure – Malfunction due to equipment deterioration caused by environmental stresses such as abrasion, radiation, fatigue and creep; or corrosion and other chemical reactions.

Wearout life/Wearout period/Wearout failure period – The total most appropriate life of units after useful life during which the equipment's failure rate increases above the failure rate during useful life.

Workmanship defects – Defects caused by human error during fabrication and assembly.

Yield – The probability that an equipment is free of screenable latent defects when offered for acceptance.

3.2 BURN-IN ACRONYMS

The following is a collection of some acronyms, arranged in alphabetical order, and their corresponding descriptions, based on [1 through 11], which may be encountered in burn-in literature and documents:

AC – Alternating current.

AFC – Amortized facility cost.

Ag – Silver.

AGREE – Advisory Group on the Reliability of Electronic Equipment.

AI – Airborne inhabited.

AIF – Airborne-inhabited fighter.

AIM – Avalanche-induced migration.

Al – Aluminum.

ALC – Air Force Logistics Center.

ALSTTL – Advanced low-power Schottky transistor-transistor logic.

BURN-IN ACRONYMS

Am – Americium.

AOQ – Average outgoing quality.

AOQL – Average outgoing quality limit (or level).

AQL – Acceptance quality level or average quality level.

AQR – Arrival quality reporting.

ARL – Acceptance reliability levels.

As – Arsenic.

ASICs – Application-specific integrated circuits.

ASTTL – Advanced Schottky transistor transistor logic.

ATE – Automatic test equipment.

Au – Gold.

AU – Airborne uninhabited.

AUF – Airborne-uninhabited fighter.

bit – Binary digit.

BIT – Built-in test.

BITE – Built-in test equipment.

BOC – Best operational capability.

B.S. – British Standards.

CCCs – Ceramic chip carriers.

CCD – Charge coupled device.

CDE – Chance defective exponential.

CERDIPs – Ceramic dual-in-line packages.

CIM – Computer integrated manufacturing.

Cl – Chlorine.

CML – Current mode logic.

CMOS – Complementary metal oxide semiconductor.

CMOS/SOS – CMOS fabricated on silicon or sapphire.

CND – Cannot duplicate.

COB – Chip on board.

COE – Coefficient of thermal expansion.

COQ – Cost of quality.

COS – Chip on substrate.

Cr – Chromium.

Cu – Copper.

CVD – Chemical vapor deposition.

DC – Direct current.

DE – Detection efficiency.

DFR – Decreasing failure rate.

DFRA – Decreasing failure rate average.

DID – Diffusion induced dislocations.

DIPs – Dual-in-line packages.

DMOS – Double-diffused MOS.

DMRL – Decreasing mean residual life.

DOA – Dead on arrival.

DOE – Design of experiments.

DR – Discrimination ratio.

DRAM – Dynamic random access memory.

DTC – Direct test cost.

DTL – Diode transistor logic.

DUT – Device under test.

EAPROM – Electrically alterable, programmable read-only memory.

EBIC – Electron-beam-induced current.

ECL – Emitter coupled logic.

EDAX – Energy dispersive analysis of X-rays.

EED – Emitter edge dislocation.

EEPROMs – Electrically erasable PROM's (E^2PROM's).

EMI – Electromagnetic interference.

EOS – Electrical overstress.

EPROMs – Erasable PROM's.

ESD – Electrostatic discharge.

ESS – Environmental stress screening.

ESSEH – Environmental stress screening of electronic hardware.

ETM – Electrothermomigration.

eV – Electron volts.

FAMECA/FMECA – Failure modes, effects, and criticality analysis.

FAMOS – Floating-gate avalanche MOS.

FBT – Functional board tester.

FET – Field effect transistor.

FFAT – Failure-free acceptance tests.

FFT – Failure-free tests.

FIT – Failures in unit time (1 failure per 10^9 device-hours).

FL – Fault location.

FLOTOX – Floating oxide.

FMA – Failure modes analysis.

FMEA – Failure modes and effects analysis.

FOM – Figure of merit.

FPHP – Flat plate heat pipe.

FPPM – Failed parts per million.

FRACAS – Failure reporting and corrective action system.

FRB – Failure review board.

FRU – Field replaceable unit.

FTA – Fault tree analysis.

FTCs – Flip tab carriers.

FTTL – Fast transistor-transistor logic.

Ga – Gallium.

GaAs – Gallium arsenide.

GB – Ground benign.

GF – Ground fixed.

GIDEP – Government Industry Data Exchange Program.

GM – Ground mobile.

GRB – Ground benign environment.

GRF – Ground fixed environment.

GRM – Ground mobile environment.

HAST – Highly accelerated stress tests.

HC – Heat concentration.

HCMOS – High-speed CMOS.

HMOS – High-speed MOS.

HD – Heat dissipation.

HDSCs – High-density signal carriers.

HEMT – High electron mobility transistors (also known as MODFET's).

HMOS – Highly scaled MOS.

HTCMOS – High-speed TTL-compatible CMOS.

HTOT – High-temperature operating tests.

HTRB – High-temperature reverse bias.

HTTL – High-speed transistor-transistor logic.

IC – Integrated circuit.

ICA – In-circuit analyzer.

ICs – Integrated circuits.

ICT – In-circuit tester.

IDMRL – Increasing then decreasing mean residual life.

IEEE – Institute of Electrical and Electronic Engineers.

IES – Institute of Environmental Sciences. A professional organization of engineers and scientists involved in the study and application of climatic testing. The IES is a valuable source for ESS studies and cost analysis data.

IFR – Increasing failure rate.

IFRA – Increasing failure rate average.

IIL – Integrated injection logic.

IMPATT – Impact avalanche and transit time.

IMRL – Increasing mean residual life.

I/O – Input/output.

ISL – Integrated Schottky logic.

I^{3}L – Isoplanar integrated injection logic.

I-V – Current-voltage (characteristic).

LBS – Loaded board shorts.

LCC – Life-cycle cost.

LCCCs – Leadless ceramic chip carriers.

LCMs – Liquid-cooled modules.

LCRU – Least-cost replaceable unit.

LED – Light-emitting diode.

LQL – Limiting quality level.

LRU – Line replaceable unit.

LSC – Logistic support cost.

LSTTL – Low-power Schottky transistor-transistor logic.

LTTL – Low-power transistor-transistor logic.

LSCS – Logistic support cost savings.

LSI – Large-scale integration.

LTPD – Lot tolerance percent defective.

MCMs – Multi-chip modules.

MIPS – Million instructions per second.

MLE – Maximum likelihood estimate.

MNOS – Metal nitride oxide semiconductor.

MOS – Metal oxide semiconductor.

MRL – Mean residual life.

MSI – Medium-scale integration.

MTBF – Mean time between failures.

MTBR – Mean time between removals.

MTBUR – Mean time between unscheduled removals.

NBUE – New better than used in expectation.

NBU – New better than used.

NFF – No fault found.

NMOS – N-channel metal oxide semiconductor.

NPL – Natural process limit.

NTF – No trouble found.

NWU – New worse than used.

NWUE – New worse than used in expectation.

OC – Operating characteristic.

OEM – Original equipment manufacturer.

PA – Procuring activity.

PCB – Printed circuit board.

PCC – Plastic chip carriers.

PED – Plastic encapsulated device.

PEP – Production engineering phase.

PFMEA – Process failure modes and effects analysis.

PGA – Pin grid array.

PICs – Power integrated circuits.

PIN – P intrinsic N.

PIND – Particle impact noise detector.

PLCCs – Plastic leaded chip carriers.

PM – Performance monitoring.

PM – Preventive maintenance.

PMOS/pMOS – MOS which will conduct through a p-channel.

PPM – Parts per million.

PQFPs – Plastic quad flat packs.

PROMs – Programmable read-only memory.

ps – Picosecond (10^{-12} seconds).

PSD – Power spectral density.

PSG – Phosphosilicate glass.

PSR – Power stress ratio.

Pt – Platinum.

PTHs – Plated through holes.

PWA – Printed wiring assembly.

PWB – Printed wiring board.

QA – Quality assurance.

QC – Quality control.

QE – Quality engineering.

QFD – Quality function deployment.

QFP – Quad flat pack.

QPL – Qualified parts list.

RAM – Random access memory.

RAM – Reliability, availability and maintainability.

RAS – Reliability, availability and serviceability.

RCM – Reliability-centered maintenance.

REDR – Recombination-enhanced defect reactions.

RFI – Radio frequency interference.

RGM – Reliability growth management.

RH – Relative humidity.

RISE – Reliability improvement of selected equipment.

ROI – Return on investment.

ROM – Read-only memory.

RQL – Rejectable quality level.

R&R – Remove and replace.

RSER – Residual SER.

RTD – Resistance temperature detector.

RTOK – Retest OK.

RTV – Room temperature vulcanized.

SAW – Surface acoustic wave.

SCPs – Single-chip packages.

SEM – Scanning electron microscope.

SEMs – Standard electronic modules.

SER – Soft error rates.

Si – Silicon.

SIMs – Super integration modules.

Si$_3$N$_4$ – Silicon nitride.

SiO$_2$ – Silicon dioxide.

SIP – Single-in-line package.

SOP – Small outline (integrated circuit) package.

SOR – System operational reliability.

Sn – Tin.

SQL – Specified quality level.

SRAM – Static RAM.

SRD – Standard reporting designation.

SRU – Shop replaceable unit.

SS – Screening strength.

SSI – Small-scale integration or stress-strength interference.

STTL – Schottky transistor-transistor logic.

TAAF – Test, analyze and fix.

TAB – Tape automated bonding.

TAC – Thermal accommodation coefficient.

TCM – Thermal conduction module.

TDBI – Test during burn-in.

TDDB – Time-dependent dielectric breakdown.

TEC – Thermal expansion coefficient.

TEM – Transmission electron microscope.

THB – Temperature humidity bias.

Ti – Titanium.

TS – Test strength.

TTL – Transistor-transistor logic.

TTL-LS – Transistor-transistor logic (low-power Schottky).

TTT – Total time on test.

TWT – Traveling wave tube.

ULSI – Ultra-large-scale integration.

UUT – Unit under test.

UV – Ultra violet.

VLSI – Very-large-scale integration.

VMOS – V-groove MOS (or vertical MOS if stated).

VSR – Voltage stress ratio.

WUC – Work unit code.

REFERENCES

1. Omdahl, T. P., *Reliability, Availability, and Maintainability (RAM) Dictionary*, ASQC Quality Press, 361 pp., 1988.

2. Posedel, R. J., "Burn-in: the Way to Reliability," *Quality*, pp. 22-23, August, 1982.

3. Hnatek, E. R., "A Realistic View of VLSI Burn-in," *Evaluation Engineering*, pp. 80-89, February 1980.

4. Flaherty, J. M., "A Burn-in' Issue – IC Complexity," *Test & Measurement World*, pp. 61-64, October 1993.

5. Amerasekera, E. A. and Campbell, D. S., *Failure Mechanisms in Semiconductor Devices*, John Wiley & Sons, New York, 205 pp., 1987.

6. DOD-HDBK-344 (USAF), *Environmental Stress Screening of Electronic Equipment*, Department of Defense, Washington, DC, October 20, 1986.

REFERENCES

7. Thermotron Industries, *The Environmental Stress Screening Handbook*, 291 Kollen Park Drive, Holland, MI 49423, 36 pp., 1988.

8. Saari, A. E. et al, *Environmental Stress Screening*, RADC-TR-86-149, September 1986.

9. Tustin, W., "Recipe for Reliability: Shake and Bake," *IEEE Spectrum*, pp. 37-42, December 1986.

10. Pellicone, V. H. and Popolo, J., *Improved Operational Readiness Through Environmental Stress Screening*, RADC-TR-87-225, Final Technical Report, November 1987.

11. Institute of Environmental Sciences, *Environmental Stress Screening Guidelines for Assemblies*, 940 East Northwest Highway, Mount Prospect, IL 60056, 1984.

Chapter 4

PHENOMENOLOGICAL OBSERVATIONS AND THE PHYSICAL INSIGHT OF THE FAILURE PROCESS DURING BURN-IN

4.1 INTRODUCTION

Burn-in, as a stressing operation, predisplays the early life period of a product in a compressed time by using an appropriate elevated level for failure-stimulating stress(es) such as temperature, voltage, humidity, etc. The inherent failure characteristics of the unburned-in products during the early stage of their field operation are expected to be revealed during burn-in. This chapter will first examine the observed failure patterns of a product during burn-in and then explore the physical processes behind these failure patterns.

4.2 THE CONVENTIONAL BATHTUB CURVE CONCEPT

Reliability engineers have observed, for a long time, a decreasing failure rate phenomenon as a whole during the early operating life of many types of products, before it levels off to a stabilized low level. This can date back to the early 1950's when the first bathtub curve was

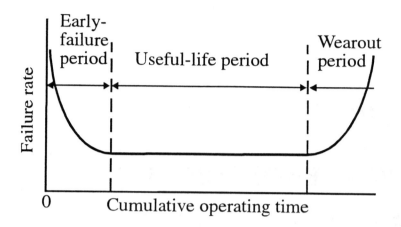

Fig. 4.1– A typical bathtub curve.

developed by Mr. Clifford M. Ryerson during 1951-1952 based on actual factory and field data [1]. Shown in Fig. 4.1 is a typical bathtub curve which is composed of three failure periods.

The first part of the bathtub curve, known as *the early/initial failure , infant mortality, burn-in, debugging, or running-in period* , typically exhibits a decreasing failure rate. During this period, the weak components that were marginally functional are weeded out. The early failures may be caused by the following:

1. Poor manufacturing techniques, poor workmanship, poor quality control, such as poor welds or seals, poor solder joints, poor connections, dirt or contamination on surfaces or in materials, chemical impurities in metals or insulation, voids, cracks, thin spots in insulation or protective coatings, and incorrect positioning of parts, etc.

2. Insufficient burning-in, debugging and breaking in.

Systematically conducting effective burn-in can eliminate most of the early failures and prevent them from occurring during the warranty period.

The long and fairly flat portion of the failure rate curve is called *useful life, chance failure, random failure, stable failure, or intrinsic failure period.* During this period, failures seem to occur in a random fashion at a uniform or constant rate which does not seem to depend on how long the product has been operating. Chance failures may be caused by the interference, or overlap, of the designed-in strength

THE CONVENTIONAL BATHTUB CURVE CONCEPT

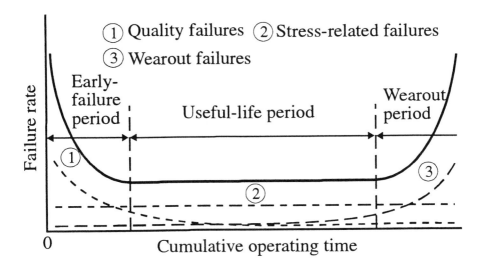

Fig. 4.2– The decomposition of the bathtub curve.

and the experienced stress during operation, due to insufficient design safety factors or margins, occurrence of higher than expected random loads, occurrence of lower than expected random strength, etc. Rationally providing an adequate design margin and effectively controlling the random stress fluctuation can significantly improve the useful-life-period reliability.

The final part of the curve, where the failure rate increases with operating time, is known as the wearout failure period. During this period, degradation failures occur at an ever increasing pace. Wearout failures are caused by inherent strength deterioration due to a series of physical and chemical processes, such as aging, corrosion or oxidation, insulation breakdown or leakage, ionic migration of metals in vacuum or on surfaces, creep, frictional wear, fatigue, shrinkage and cracking in plastics, etc. Timely preventive maintenance and replacements can improve the product's operational reliability and availability.

Figure 4.2, on the other hand, shows the decomposition of the bathtub curve which turns out to be the sum of three separate overlapping failure distributions. The infant mortality period is characterized by a high, but rapidly decreasing, failure rate of components that are (1) high, quality-relevant, failure components, (2) constant, stress-related, failure components, and (3) low wearout failure components.

The useful life period is characterized by a constant failure rate of components that are (1) low and decreasing quality-relevant failure components, (2) constant, stress-related failure, components, and (3) low but increasing wearout failure components.

The combination of all these components results in a fairly constant failure rate, because the decreasing quality-relevant failures and increasing wearout failures tend to offset each other, while the stress-related failures exhibit a relatively large amplitude.

The wearout period is characterized by an increasing failure rate of components that are (1) negligible, quality-relevant, failure components, (2) constant, stress-related, failure components, and (3) initially low but rapidly increasing wearout failure components.

The bathtub curve got its name in 1959 at a Reliability Training Course in Atlantic City, New Jersey, which was jointly sponsored by IRE and ASQC and organized by Mr. Ryerson [1]. The key of this training course for reliability engineers was that the bathtub curve characteristic had to be earned, it could never be assumed without life test data to prove its validity. As pointed out by [1], the bathtub characteristic does not appear automatically and indeed may not be achieved at all without intense reliability effort in all areas of design and manufacturing. When understood in its original concept as a primary objective of Reliability Engineering and Product Assurance, the bathtub curve retains its position as a valuable technical tool.

Though there were some criticisms in 1980's about the validity of the traditional bathtub curve model as a general rule of describing the component's failure rate pattern, a lot of effort, in reliability literature, has been devoted to the parametric modeling and evaluation of the bathtub curve. They will be discussed extensively in the next chapter.

4.3 THE S-SHAPED *CDF* PATTERN

For a long time, people have observed an S-shaped cumulative distribution function (*cdf*) for many electronic components and equipment [2 through 15]. Figure 4.3 is a *cdf* plot based on the experimental data of CMOS transistors from RCA [7; 14; 15]. Figure 4.4 is another *cdf* plot on Weibull probability plotting paper based on the results from an extensive test on alloy specimens taken from turbine blade forgings [14; 16].

It will be shown quantitatively in the next chapter that the typical S-shaped *cdf* is a direct, visual indication of the bimodality of the times-to-failure distribution. That is, for a well-separated case, the first knee of the curve separates the components' failures into two subpopulations (bimodal); i.e., the weak, or freak, subpopulation which fails earlier and the strong, or main, subpopulation which fails later. In other words, failure bimodality graphically corresponds to an S-shaped

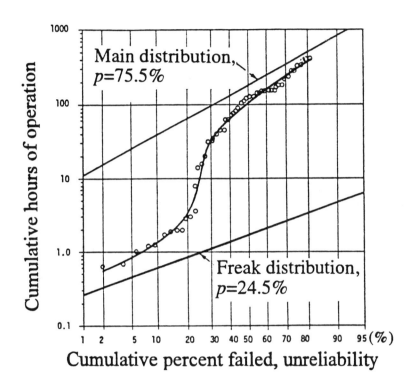

Fig. 4.3– A *cdf* plot based on the experimental data of CMOS transistors from RCA [7; 14; 15].

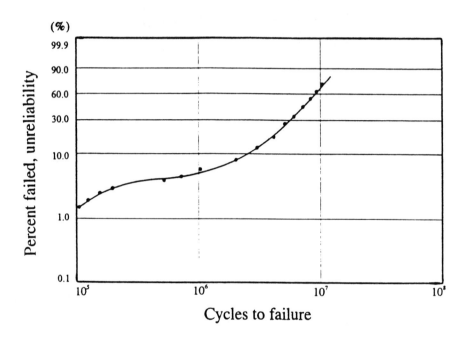

Fig. 4.4– A *cdf* plot on the Weibull probability plotting paper based on the results from an extensive test on alloy specimens taken from turbine blade forgings [14; 16].

cdf on the probability plotting paper. The plateau, or the knee, provides a hint of the necessary burn-in duration.

One might predict a double-S-shaped *cdf* for the trimodal distribution of the times to failure, and an $(n-1)$-S-shaped *cdf* for the n-modal failure time distribution, which is indeed the case for the well-separated population. Figure 4.5 is a *cdf* plot generated from a trimodal life distribution which corresponds to a *pdf* and a failure rate function shown in Figs. 4.6 and 4.7, respectively.

The trimodal failure pattern is typically observed from the non-screened equipment with non-burned-in components. As components with bimodal failure distributions go through various production and handling processes and are finally assembled into an equipment, the failure susceptibility of some of the components is altered drastically. As a consequence of this, a new weak subpopulation of components, known as infant mortality subpopulation, is born. This new subpopulation is composed of components from both the freak and the main subpopulations which fail much earlier. Thus, we are dealing with two types of early failures; i.e., infant mortality and freak failures. The objective of component burn-in is to eliminate the freak subpopulation. However, at the equipment and system level, one cannot assure that burn-in will eliminate the early failures because the infant mortality subpopulation may be created during the manufacturing processes. To eliminate all early failures in equipment or systems, a component-level burn-in plus a system-level environmental stress screening is recommended.

To summarize, an equipment or a system with a trimodal failure distribution, or a double-S-shaped *cdf*, is composed of the following three groups of components:

1. Infant mortalities which are components that fail due to the weakness introduced by the system assembly process.

2. Freak components which fail due to the inherent defects, or flaws, in the components.

3. Strong components which fail due to wearout failure mechanisms.

Typically, the "infant mortality" components comprise 0.1% to 5% of the production [14] and fail before 200 hr of operation, and often quite early, say, after 10 to 20 hr of operation [15]. The "freaks" comprise 2% to 20% of the production [14] and fail within the time span of 1,000 to 20,000 hr [15]. The "strong" components comprise more than 85% of the production [14] and fail very late, typically beyond 10 years.

Mathematically, an S-shaped or a multiple-S-shaped *cdf* can be represented by a bimodal or multimodal mixed life distribution. The

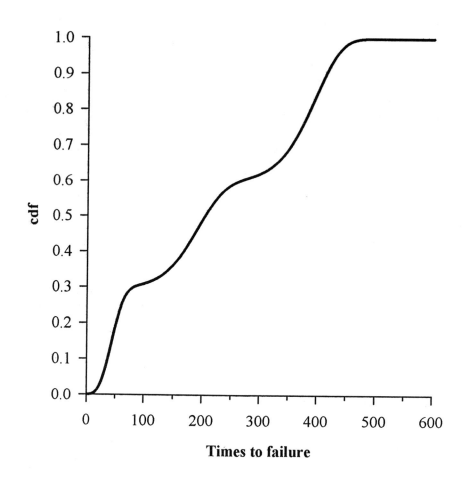

Fig. 4.5– A *cdf* plot generated from a trimodal life distribution.

THE S-SHAPED CDF PATTERN

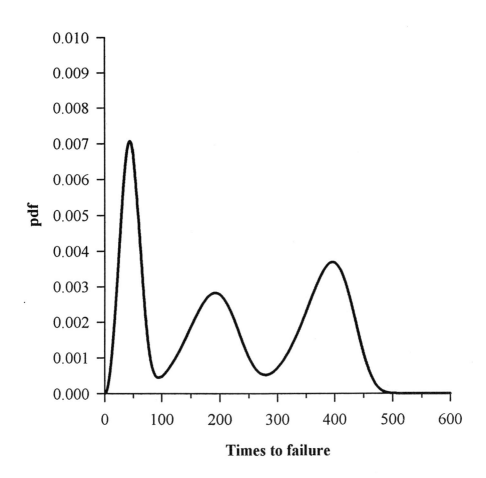

Fig. 4.6– The *pdf* corresponding to the *cdf* plot given by Fig. 4.5.

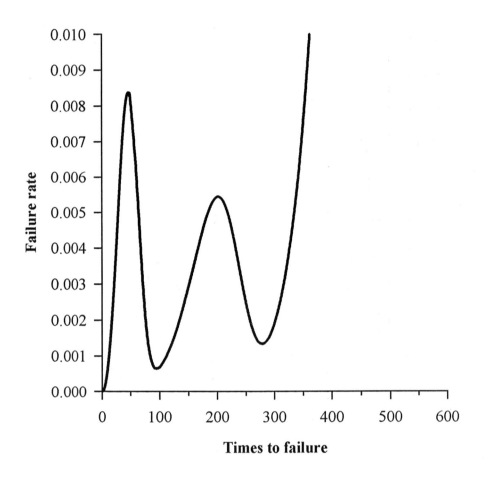

Fig. 4.7– The failure rate function corresponding to the *cdf* plot given by Fig. 4.5.

mathematical modeling and parameter estimation techniques will be detailed in the next chapter.

4.4 THE UNIFIED-FIELD FAILURE THEORY AND THE ROLLER-COASTER FAILURE RATE CURVE

In 1981, Mr. K. L. Wong [16] revealed, based on an extensive historical data study, that the instantaneous failure rate of electronic equipment decreases with respect to the operational age during its useful life and that the conventional bathtub curve for electronic equipment is an exception rather than the rule. It was believed that the failure mechanisms causing the operational failures are the same as those during the infant mortality period; i.e., aside from the long term gross wearout failures, all other failures develop from flaws. Therefore, the same failure theory, the so-called Unified Field (Failure) Theory as proposed by Wong [16], is applicable throughout all phases of equipment operation. This theory states the following:

"Flaws are built into a piece of equipment. Stresses, either external or internal, bring the flaws to a state that causes an item to malfunction. This process occurs during manufacturing, test and field operation."

As the flaws become failures and are identified and eliminated, the decreasing flaw population gives the failure rate of equipment an exponentially decreasing characteristic. Figure. 4.8, which is taken from [17], was used by Wong [16] as one of the key support evidences to his Unified Field (Failure) Theory. A decreasing failure rate is shown even after 10,000 hr of operation of Honeywell digital air data computer used on commercial airlines. Based on the extensive field and test data gathered over 8 years from avionics equipment on board military and commercial aircraft by Honeywell, it was concluded [17] that the conventional "bathtub" curve is no longer considered applicable to solid-state electronics; furthermore, the wearout portion of the curve (if there is one for solid-state electronics) appears to extend well beyond the useful operational life of aircraft electronic products.

Later, Wong [18] further explored the shape of electronic components' failure rate curves in general and observed that the decreasing failure rate curve is often decorated by one or more failure humps. Meanwhile investigators studying the semiconductor device burn-in subscribe to the concept of freak failures which manifest themselves as a *hump* on the failure rate curve [18], such as those shown in Fig. 4.7. Wong [18] believes that the failure humps along the way on the

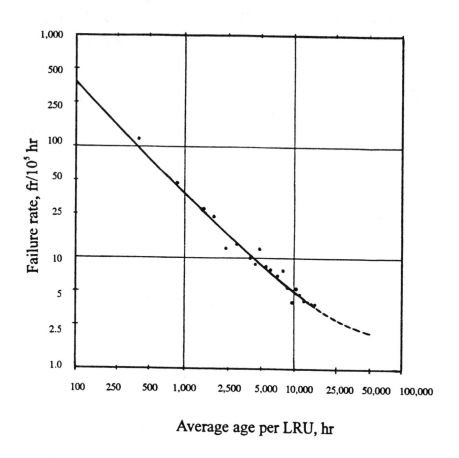

Fig. 4.8– Electronic component failure rate versus average age in hr [17].

THE ROLLER-COASTER FAILURE RATE CURVE

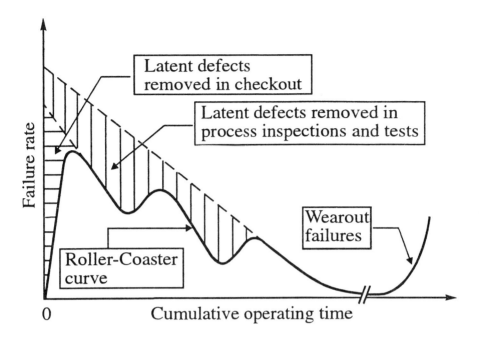

Fig. 4.9– A typical Roller-Coaster curve [18; 19].

failure rate curve came from failures of small flaws left over from major flaw groups due to limitations in quality inspection and test. The inspection and test processes can lower the failure rate curve as a result of identifying and removing some of the defects or flaws. However, when inspection and test reach their limits, small flaws will remain in the system and emerge as humps on the failure rate curve. Failures from dominating flaw groups will cause large humps. Screening data as well as operational data also revealed that, indeed, one can expect electronic systems to have generally decreasing failure rate curves with one or more failure humps on each of them. Wong [18; 19; 20] calls the failure rate curves with failure humps the *Roller-Coaster Curves*. According to Wong [18; 19], a typical Roller-Coaster curve appears like that shown in Fig. 4.9.

Kececioglu [21, pp. 24-29, Fig. 16; 22, p. 187, Fig. 4.26] presents an actual Roller-Coaster-shaped failure rate curve for 5xSK and KSK centrifugal pumps under both preventive and corrective maintenance as shown in Fig. 4.10.

Figure 4.11 is another illustration of actual Roller-Coaster-shaped

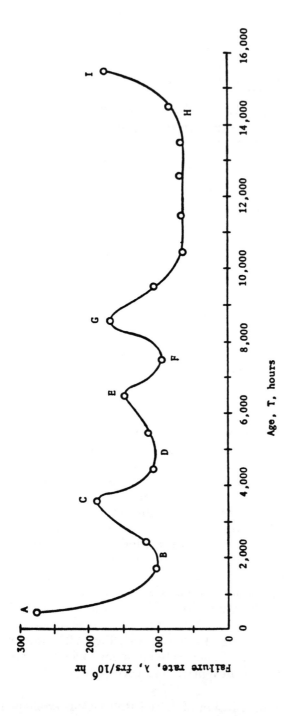

Fig. 4.10– An actual Roller-Coaster-shaped failure rate curve for 5xSK and KSK centrifugal pumps under both preventive and corrective maintenance [21; 22].

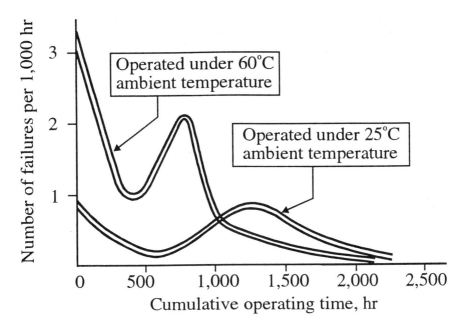

Fig. 4.11– Failure rate curve for AVR-200.

failure rate curves, at both normal ambient temperature of 25°C and at an elevated ambient temperature of 60°C, for the AVR-200 which is a 178-part transistorized receiver set [19; 20].

After exploring much data, Wong [19] concluded that the failure rate of electronics does not stay constant with respect to age, instead, the failure rate decreases as it ages and follows a Roller-Coaster track. If this conclusion is generally true, the reliability demonstration and other relevant reliability methodologies may need modification, and will become much more complicated.

Mathematically, however, the so-called Roller-Coaster failure rate curve concept is equivalent to the S-shaped *cdf* idea. This is so because the corresponding *cdf* curve for the Roller-Coaster-shaped failure rate curve will be S-shaped or multiple-S-shaped. On the other hand, an S-shaped or multiple-S-shaped *cdf* curve corresponds to a Roller-Coaster-shaped failure rate curve with one or more humps. This has been verified graphically by the correspondence between Fig. 3.5 and Fig. 3.7.

4.5 PHYSICAL EXPLANATION OF THE FAILURE PATTERN DURING BURN-IN

In the previous three sections, the phenomenological observations of the failure process during burn-in were reviewed. No matter which curve one believes in, or is committed to, the conventional Bathtub curve, the S-shaped *cdf* curve, or the Roller-Coaster failure rate curve, the fact that the failure rate of the components or equipment during burn-in is decreasing *as a whole* (humps may be present) is common to each of these three curves, and has long been observed and well accepted by most reliability practitioners and researchers. Now let's look into the failure process and explore its initiation and propagation from the stress/strength point of view.

4.5.1 THE STRESS-STRENGTH INTERFERENCE CONCEPT

If one agrees to characterize a component by using its failure rate function, $\lambda(T)$, and estimates its parameters from burn-in data, or from life testing data, the result is just a macroscopic model. Such a model is merely a statistical representation of the component's operational performance but not directly related with what happens inside the component, the black box. To delve into the black box, one needs to postulate various hypotheses about the microscopic behavior of the component and conduct a series of tests to prove their validity. One generally-accepted hypothesis of this nature is the *stress-strength interference* (SSI) theory [23]. Let s denote the applied stress and S denote the component's inherent strength. Here stress and strength are two terms used in a quite general context. For mechanical components, the terms *stress* and *strength* are self-explanatory, for example, tensile stress and yield strength, shear stress and shear strength, etc. For nonmechanical components, the stress may be the applied power, voltage, temperature, humidity, etc. Correspondingly, strength is the stress level that causes the component to fail, such as dielectric breakdown voltage, failure causing temperature, etc. Note that both stress and strength have to be expressed in the same measurement unit. It can be safely stated that whatever the failure's nature is and whenever the failure occurs, it is always due to the occurrence of an event that causes the applied stress to be greater than the inherent strength; i.e., when $s > S$.

Because of the imperfections, nonuniformity and defects introduced during the manufacturing process, the strength of a batch of components will be statistically distributed rather than being identically single valued. Similarly, the magnitude of the applied stress varies over a considerable range in practice due to the stochastic nature of the stress applied in an actual application. For example, the mechanical stresses applied to a structural part in a spacecraft during its entire mission are stochastic in nature, as are the electrical input signals into a capacitor in a filter circuit.

Let $f_s(s)$, $F_s(s)$, $R_s(s)$, $f_S(S)$, $F_S(S)$ and $R_S(S)$ represent the *pdf's*, *cdf's* and survival functions for stress, s, and strength, S, respectively. Then, the component reliability is given by

$$R = P(s < S) = P(S > s), \quad (4.1)$$

or

$$R = \int_0^\infty \left[f_S(S) \int_0^S f_s(s) ds \right] dS = \int_0^\infty \left[f_s(s) \int_s^\infty f_S(S) dS \right] ds, \quad (4.2)$$

which can be rewritten equivalently as

$$R = \int_0^\infty f_S(S)\, F_s(S)\, dS = \int_0^\infty f_s(s)\, R_S(s)\, ds. \quad (4.3)$$

Equation (4.2), or Eq. (4.3), is generally referred to as the *stress-strength interference model*. Figure 4.12 illustrates a distributed stress, a corresponding distributed strength and their interference.

Note that Eqs. (4.2) and (4.3) do not reflect the time dependence of the distributions for stress and strength and the corresponding component reliability. Generally, the distributions of the applied stress and of the component's strength do not stay fixed or stationary during operation, but vary with operating time. For example, the strength deteriorates due to component's usage and its own inherent wearout mechanisms. The stochastic loading often yields a non-stationary stress spectrum whose distribution parameters vary with time. Figure 4.13 is an example where the strength of a component under stationary random loading degrades with time.

Let $S(t)$ and $s(t)$ denote the strength and stress at any time t. Let $f_{S(t)}[S(t)]$, $F_{S(t)}[S(t)]$, $R_{S(t)}[S(t)]$, $f_{s(t)}[s(t)]$, $F_{s(t)}[s(t)]$ and $R_{s(t)}[s(t)]$ denote the *pdf's*, *cdf's* and the survival functions of the strength and stress distributions at any time t. Then, Eqs. (4.1), (4.2) and (4.3) can be rewritten as

$$R(t) = P[s(t) < S(t)] = P[S(t) > s(t)], \quad (4.4)$$

$$R(t) = \int_0^\infty \left[f_{S(t)}[S(t)] \int_0^{S(t)} f_{s(t)}[s(t)]\, ds(t) \right] dS(t)$$

$$= \int_0^\infty \left[f_{s(t)}[s(t)] \int_{s(t)}^\infty f_{S(t)}[S(t)]\, dS(t) \right] ds(t), \quad (4.5)$$

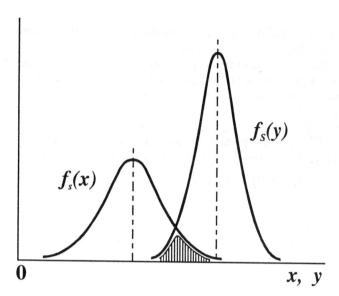

Fig. 4.12– Distributed stress, distributed strength and their interference.

and

$$R(t) = \int_0^\infty f_{S(t)}[S(t)]\, F_{s(t)}[S(t)]\, dS(t) = \int_0^\infty f_{s(t)}[s(t)]\, R_{S(t)}[s(t)]\, ds(t).$$

(4.6)

Thus, given the stress and strength distributions, $f_{s(t)}[s(t)]$ and $f_{S(t)}[S(t)]$, the component's reliability or survival probability at any time t can be evaluated using Eqs. (4.5) and (4.6). For some special stress-strength distribution pairs, such as normal-normal and lognormal-lognormal, etc., closed form reliability solutions are readily available. However, for other stress-strength distribution pairs, it is usually difficult or impossible to analytically solve R or $R(t)$. Then, a computer-aided numerical evaluation is needed to accomplish this task.

For more information about the stress-strength interference theory and its applications, the readers are referred to [24 through 36].

Next we will study the life-cycle failure behavior of the components using the stress-strength interference approach.

PHYSICAL EXPLANATION OF THE FAILURE PATTERN

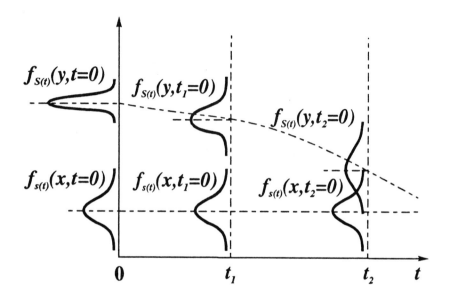

Fig. 4.13– Interference between a stationary random stress and the component's random strength that deteriorates with time.

4.5.2 STRESS-STRENGTH INTERFERENCE AND COMPONENT FAILURE PATTERNS

4.5.2.1 BIMODAL STRENGTH DISTRIBUTION

Jensen and Petersen [14] proposed a bimodal mixed distribution with the following *pdf* as a basic model to represent the strength distributions of non-burned-in components:

$$f_S(S) = p_1\, f_{S_1}(S) + p_2\, f_{S_2}(S), \tag{4.7}$$

where

$p_1 =$ proportion of the 1st subpopulation,
$f_{S_1}(S) =$ strength *pdf* of the 1st subpopulation,
$p_2 =$ proportion of the 2nd subpopulation,
$f_{S_2}(S) =$ strength *pdf* of the 2nd subpopulation,

and

$p_1 + p_2 = 1$ or 100%.

This distribution is shown in Fig. 4.14 where the entire strength distribution is composed the following two parts:

1. A *freak distribution* with a proportion of p_1 and a *pdf* of $f_{S_1}(S)$, on the left side, representing a subpopulation of the weak or substandard components with strength values much smaller than the specified.

2. A *main distribution* with a proportion of p_2 and a *pdf* of $f_{S_2}(S)$, on the right side, representing a subpopulation of the strong or standard components with strengths around the designed or specified value.

Though other distribution patterns may be observed in reality, such as a unimodal pattern with no freak distribution or a multimodal pattern with two or more freak distributions, the bimodal nature of strength distributions is typical enough for non-burned-in electronic, electro-mechanical and mechanical components [14]. It is expected that an effective burn-in will eliminate those substandard components governed by the freak strength distribution.

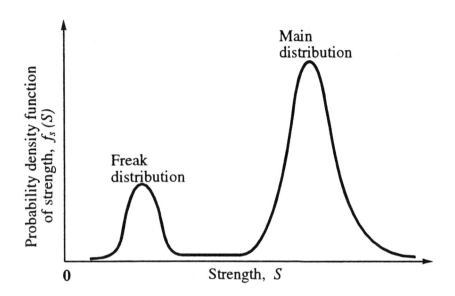

Fig. 4.14– The bimodal mixed strength distribution as a basic model to represent the strength distributions of non-burned-in components.

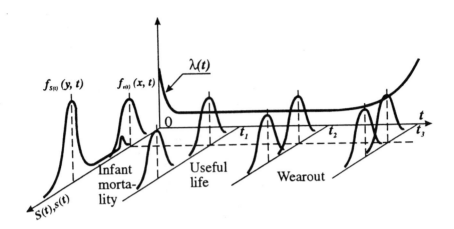

Fig. 4.15– Stress-strength distribution interference and the bathtub curve.

4.5.2.2 BEHIND THE BATHTUB CURVE: STRESS-STRENGTH INTERFERENCE

Figure 4.15, adapted from [27], gives a graphical insight about the failure process behind the bathtub curve; i.e., the stress-strength interference throughout the whole life cycle. During the early life period, the interference is mainly between the freak strength distribution and the applied stress (including burn-in stress) distribution. The weak or substandard components fail due to their low freak strengths. After this infant mortality period, hopefully, the freak subpopulation gets eliminated and the main distribution governs the component's residual strength. During the second phase, the useful life period, the interference is mainly between the applied stress distribution and the main strength distribution. The failures occur mainly due to the external random overstresses. Finally, as the wearout period arrives, the component's strength deteriorates significantly. As a consequence, the interference between the applied stress and the main degrading strength grows and the failure probability increases causing more failures to occur.

Lewis and Chen [37] used the stress-strength interference theory to derive a heuristic failure rate for a component subjected to repetitive loading whose application frequency is Poisson distributed in time, or equivalently, whose times between applications are exponentially distributed. The objective of their work is to bring out the origins of the bathtub curve in the coupled effects of the following three (3) factors:

1. Capacity variability (distributed freak strength during the infant mortality period).

2. Load variability (random overstress during the useful life period).

3. Capacity deterioration (the strength degradation during the wearout period).

The following four assumptions have been made for the formulation [37]:

1. Ordinarily, there is no stress imposed on the component. The stress applications occur instantaneously at random time intervals governed by a homogeneous Poisson process with intensity (mean) ξ. The load duration is negligible.

2. The stress magnitudes are independently and identically distributed (i.i.d.) with a *pdf* of $f_s(s)$, *cdf* of $F_s(s)$ and survival function of $R_s(s)$.

3. The strength decreases with time due to aging according to the following relationship:

$$S(t) = S_i - g(t), \tag{4.8}$$

where

$S(t)$ = strength at time t, a random variable,
S_i = initial strength, or strength at $t = 0$, $S(0)$,

and

$g(t)$ = monotonically increasing function of time t.

4. S_i is a random variable with the *pdf* of $f_{S_i}(S_i)$.

For a fixed value of S_i, or equivalently, for a fixed value of $S(t)$ calculated from Eq. (4.8), the component's instantaneous conditional failure rate at time t is given by

$\lambda(t|S_i) \Leftrightarrow \lambda(t|S(t))$,

$\Leftrightarrow$ probability that a component, having survived a cumulative operating time t, will fail in the following unit time interval,

$\Leftrightarrow$ probability that a component whose strength has deteriorated without disturbance (no failure or replacement; ie., surviving) for a cumulative operating time t will fail in the following unit time interval,

$\Leftrightarrow \lim_{\Delta t \to 0} P\{[\text{the stress application appears in time interval } (t, t+\Delta t)] \cap [\text{the stress amplitude is greater than the component strength at time } t]\}/\Delta t$,

$\Leftrightarrow \lim_{\Delta t \to 0} P\{[\text{number of stress applications in } (t, t+\Delta t) \geq 1] \cap [s > S(t)]\}/\Delta t$.

Assume that the stress and the strength are statistically independent, though it may not be the case in reality. Then,

$P\{[\text{number of stress applications in } (t, t+\Delta t) \geq 1] \cap [s > S(t)]\}$
$= P[\text{number of stress applications in } (t, t+\Delta t) \geq 1] \cdot P[s > S(t)]$,

and

$\lambda(t|S_i) = \lim_{\Delta t \to 0} \dfrac{P[\text{\# of stress applications in } (t, t+\Delta t) \geq 1] \cdot P[s > S(t)]}{\Delta t}$.

Now,

$P[\text{number of stress applications in } (t, t+\Delta t) \geq 1]$
$= \sum_{i=1}^{\infty} e^{-\xi \Delta t} \dfrac{(\xi \Delta t)^i}{i!}$,
$= 1 - \sum_{i=0}^{0} e^{-\xi \Delta t} \dfrac{(\xi \Delta t)^i}{i!}$,
$= 1 - e^{-\xi \Delta t} \dfrac{(\xi \Delta t)^0}{0!}$,
$= 1 - e^{-\xi \Delta t}$,

and

$$P[s > S(t)] = \int_{S(t)}^{\infty} f_s(s) \, ds = R_s[S(t)].$$

Therefore,

$$\lambda(t|S_i) = \lim_{\Delta t \to 0} \frac{(1 - e^{-\xi \Delta t}) R_s[S(t)]}{\Delta t},$$

$$= R_s[S(t)] \lim_{\Delta t \to 0} \frac{1 - e^{-\xi \Delta t}}{\Delta t}.$$

Applying L'Hospital's rule; i.e., taking the first derivative of both the numerator and the denominator with respect to Δt, yields

$$\lambda(t|S_i) = R_s[S(t)] \lim_{\Delta t \to 0} \frac{0 + \xi \, e^{-\xi \Delta t}}{1},$$

or

$$\lambda(t|S_i) = \xi \, R_s[S(t)]. \tag{4.9}$$

Correspondingly, the conditional reliability at time t for a given fixed value of S_i is

$$R(t|S_i) = e^{-\int_0^t \lambda(\tau|S_i) \, d\tau} = e^{-\int_0^t \xi \, R_s[S(\tau)] \, d\tau},$$

or

$$R(t|S_i) = e^{-\xi \, \mu_s(S_i,t)}, \tag{4.10}$$

where

$$\mu_s(S_i, t) = \int_0^t R_s[S(\tau)] \, d\tau, \tag{4.11}$$

which is a function of both initial strength S_i and time t.

Therefore, the marginal reliability, *pdf* and failure rate are given, respectively, by

$$R(t) = \int_0^{\infty} R(t|S_i) \, f_{S_i}(S_i) \, dS_i,$$

or, from Eq. (4.8),

$$R(t) = \int_0^{\infty} e^{-\xi \, \mu_s(S_i,t)} \, f_{S_i}(S_i) \, dS_i, \tag{4.12}$$

$$f(t) = -\frac{dR(t)}{dt} = \int_0^\infty \xi \left[\frac{\partial \mu_s(S_i,t)}{\partial t}\right] e^{-\xi\, \mu_s(S_i,t)}\, f_{S_i}(S_i)\, dS_i,$$

$$= \xi \int_0^\infty R_s[S(t)]\, e^{-\xi\, \mu_s(S_i,t)}\, f_{S_i}(S_i)\, dS_i,$$

or

$$f(t) = \xi \int_0^\infty R_s[S_i - g(t)]\, e^{-\xi\, \mu_s(S_i,t)}\, f_{S_i}(S_i)\, dS_i, \qquad (4.13)$$

and

$$\lambda(t) = \frac{f(t)}{R(t)} = \xi\, \frac{\int_0^\infty R_s[S_i - g(t)]\, e^{-\xi\, \mu_s(S_i,t)}\, f_{S_i}(S_i)\, dS_i}{\int_0^\infty e^{-\xi\, \mu_s(S_i,t)}\, f_{S_i}(S_i)\, dS_i}. \qquad (4.14)$$

Equation (4.14) represents a failure rate model in which the infant mortality, useful life and aging process are expressed explicitly in terms of strength variability, stress variability and strength deterioration. Figure 4.16, generated from Eq. (4.14), shows a typical failure rate pattern for a component whose inherent strength and external stress are governed by the following parameters:

$$\frac{\overline{S}}{\overline{s}} = 1.5,$$

$$\rho_{S_i} = \frac{\sigma_{S_i}}{\overline{S_i}} = 0.1,$$

$$g(t) = 0.1\, S_i \left(\frac{t}{t_0}\right)^m,$$

$$\xi\, t_0 = 100,$$

$$m = 4,$$

and

$$\rho_s = \frac{\sigma_s}{\overline{s}} = 0.15.$$

4.5.2.3 BIMODAL STRENGTH DISTRIBUTION AND THE CORRESPONDING S-SHAPED CDF PATTERN

Jensen and Petersen [14] studied the failure rate pattern assuming a constant external stress, s, a bimodal initial strength distribution, $f_{S_i}(S_i)$, and a linearly decreasing function for strength deterioration, $S(t)$. Note that in reality, $S(t)$ may be non-linear depending on the actual failure physics. The proposed bimodal initial strength pdf, $f_{S_i}(S_i)$, is a mixture of two density functions; i.e., the freak/weak strength subpopulation $f_{S_{i1}}(S_i)$ with proportion p and the main/strong strength

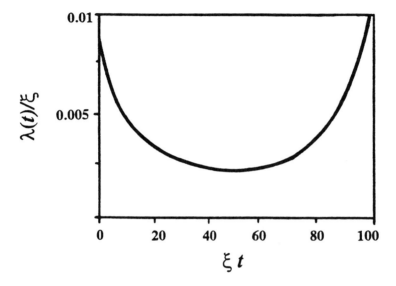

Fig. 4.16– The bathtub-shaped failure rate curve generated from Eq. (4.14) which is derived from the stress-strength interference theory.

subpopulation $f_{S_{i2}}(S_i)$ with proportion $(1-p)$. This mixture distribution, shown in Fig. 4.17, is expressed as follows:

$$f_{S_i}(S_i) = p\, f_{S_{i1}}(S_i) + (1-p)\, f_{S_{i2}}(S_i), \quad S_i \geq \gamma, \tag{4.15}$$

where

$$p + (1-p) \equiv 1 = 100\%,$$

and

$$\gamma = \text{minimum or threshold strength}.$$

For more information about mixed distributions, the readers are referred to the next chapter.

The component strength is degrading linearly with an initial-strength-dependent slope, $b(S_i)$, given by the following assumed form [14]:

$$b(S_i) = -\frac{1}{(S_i - s)^m}, \quad m > 0, \quad S_i > s, \tag{4.16}$$

where m is a component strength-deteriorating constant. Note that the absolute value of this slope is inversely proportional to the initial strength value; i.e., the weaker the component is initially, the faster it deteriorates, and the stronger the component is initially, the slower it deteriorates. Then, the instantaneous strength at any time t, $S(t)$, is

$$S(t) = S_i + b(S_i)\, t = S_i - \frac{t}{(S_i - s)^m}. \tag{4.17}$$

Failure occurs when $S(t)$ decreases to the magnitude of the applied stress s; i.e., when

$$S(t_f) = S_i - \frac{t_f}{(S_i - s)^m} = s, \tag{4.18}$$

where

$$t_f = \text{time to failure}.$$

Solving Eq. (4.18) for t_f yields the time to failure for a fixed value of initial strength S_i, or

$$t_f = (S_i - s)^{m+1}. \tag{4.19}$$

However, since S_i is a random variable with pdf, $f_{S_i}(S_i)$, given by Eq. (4.15), t_f is also a random variable with a pdf given by the following:

$$f_{t_f}(t_f) = f_{S_i}(S_i)\Big|_{S_i = S_i^{-1}(t_f)} |J|, \tag{4.20}$$

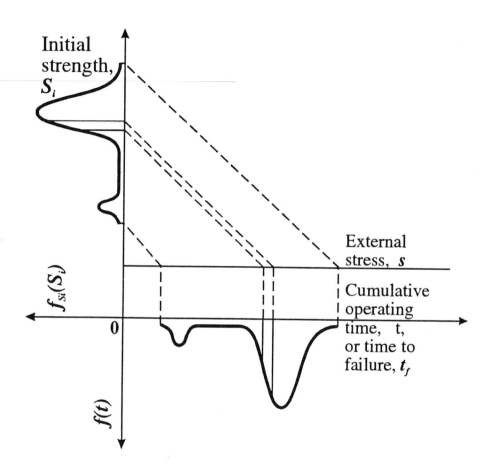

Fig. 4.17 – Bimodal initial strength distribution, its linear deterioration and the corresponding bimodal times-to-failure distribution.

where

$S_i^{-1}(t_f)$ = inverse function of S_i in terms of t_f,

or

$$S_i^{-1}(t_f) = s + (t_f)^{\frac{1}{m+1}}, \text{ from Eq. (4.19)}, \qquad (4.21)$$

and

J = Jacobian of variable transform given by Eq. (4.21),

$$= \frac{dS_i}{dt_f},$$

$$= \frac{dS_i^{-1}(t_f)}{dt_f},$$

or

$$J = \left(\frac{1}{m+1}\right)(t_f)^{-\frac{m}{m+1}}. \qquad (4.22)$$

Substituting Eqs. (4.15), (4.21) and (4.22) into Eq. (4.20) yields the pdf of the times to failure; i.e.,

$$f_{t_f}(t_f) = f_{S_i}(S_i)\Big|_{S_i=s+(t_f)^{\frac{1}{m+1}}} \left|\left(\frac{1}{m+1}\right)(t_f)^{-\frac{m}{m+1}}\right|,$$

$$= [p\, f_{S_{i1}}(S_i) + (1-p)\, f_{S_{i2}}(S_i)]\Big|_{S_i=s+(t_f)^{\frac{1}{m+1}}}$$

$$\cdot \left(\frac{1}{m+1}\right)(t_f)^{-\frac{m}{m+1}},$$

or

$$f_{t_f}(t_f) = p\left\{\left(\frac{1}{m+1}\right)(t_f)^{-\frac{m}{m+1}} f_{S_{i1}}[s+(t_f)^{\frac{1}{m+1}}]\right\} + (1-p)$$

$$\cdot \left\{\left(\frac{1}{m+1}\right)(t_f)^{-\frac{m}{m+1}} f_{S_{i2}}[s+(t_f)^{\frac{1}{m+1}}]\right\}, \qquad (4.23)$$

or

$$f_{t_f}(t_f) = p\, f_{t_{f1}}(t_f) + (1-p)\, f_{t_{f2}}(t_f), \qquad (4.24)$$

where

$f_{t_{f1}}(t_f)$ = pdf of times to failure of the freak/weak subpopulation,

$$f_{t_{f1}}(t_f) = \left(\frac{1}{m+1}\right)(t_f)^{-\frac{m}{m+1}} f_{S_{i1}}[s+(t_f)^{\frac{1}{m+1}}], \qquad (4.25)$$

$f_{t_{f2}}(t_f)$ = pdf of times to failure of the main/strong subpopulation,

$$f_{t_{f2}}(t_f) = \left(\frac{1}{m+1}\right)(t_f)^{-\frac{m}{m+1}} f_{S_{i2}}[s+(t_f)^{\frac{1}{m+1}}]. \qquad (4.26)$$

PHYSICAL EXPLANATION OF THE FAILURE PATTERN

The reliability function of the component is given by

$$R_{t_f}(t_f) = \int_{t_f}^{\infty} f_{t_f}(\tau)\, d\tau,$$

$$= \int_{t_f}^{\infty} \left[p\, f_{t_{f1}}(\tau) + (1-p)\, f_{t_{f2}}(\tau) \right] d\tau,$$

$$= p \int_{t_f}^{\infty} f_{t_{f1}}(\tau)\, d\tau + (1-p) \int_{t_f}^{\infty} f_{t_{f2}}(\tau)\, d\tau,$$

or

$$R_{t_f}(t_f) = p\, R_{S_{i1}}[s + (t_f)^{\frac{1}{m+1}}] + (1-p)\, R_{S_{i2}}[s + (t_f)^{\frac{1}{m+1}}], \quad (4.27)$$

or

$$R_{t_f}(t_f) = p\, R_{t_{f1}}(t_f) + (1-p)\, R_{t_{f2}}(t_f), \quad (4.28)$$

where

$$R_{S_{i1}}[s + (t_f)^{\frac{1}{m+1}}] = \text{survival function of } S_{i1} \text{ at } [s + (t_f)^{\frac{1}{m+1}}],$$

$$= \int_{s+(t_f)^{\frac{1}{m+1}}}^{\infty} f_{S_{i1}}(S_i)\, dS_i,$$

$$R_{S_{i2}}[s + (t_f)^{\frac{1}{m+1}}] = \text{survival function of } S_{i2} \text{ at } [s + (t_f)^{\frac{1}{m+1}}],$$

$$= \int_{s+(t_f)^{\frac{1}{m+1}}}^{\infty} f_{S_{i2}}(S_i)\, dS_i,$$

$$R_{t_{f1}}(t_f) = R_{S_{i1}}[s + (t_f)^{\frac{1}{m+1}}], \quad (4.29)$$

and

$$R_{t_{f2}}(t_f) = R_{S_{i2}}[s + (t_f)^{\frac{1}{m+1}}]. \quad (4.30)$$

To understand the derivation of Eq. (4.27), one should be aware of the following relationship:

$$\frac{dZ[Y(x)]}{dx} = \left. \frac{dZ}{dY} \right|_{Y=Y(x)} \frac{dY(x)}{dx}.$$

Specifically in this case,

$$-\frac{dR_{S_{i1}}[s+(t_f)^{\frac{1}{m+1}}]}{dt_f}$$

$$= -\left. \frac{dR_{S_{i1}}(S_i)}{dS_i} \right|_{S_i = s + (t_f)^{\frac{1}{m+1}}} \frac{d[s+(t_f)^{\frac{1}{m+1}}]}{dt_f},$$

$$= f_{S_{i1}}[s + (t_f)^{\frac{1}{m+1}}] \left(\frac{1}{m+1}\right) (t_f)^{-\frac{m}{m+1}},$$

which turns out, when compared with Eq. (4.25), that

$$-\frac{dR_{S_{i1}}[s+(t_f)^{\frac{1}{m+1}}]}{dt_f} = f_{t_{f1}}(t_f),$$

and

$$-\frac{dR_{S_{i2}}[s+(t_f)^{\frac{1}{m+1}}]}{dt_f}$$
$$= -\frac{dR_{S_{i2}}(S_i)}{dS_i}\bigg|_{S_i=s+(t_f)^{\frac{1}{m+1}}} \frac{d[s+(t_f)^{\frac{1}{m+1}}]}{dt_f},$$
$$= f_{S_{i2}}[s+(t_f)^{\frac{1}{m+1}}] \left(\frac{1}{m+1}\right)(t_f)^{-\frac{m}{m+1}},$$

which turns out, when compared with Eq. (4.26), that

$$-\frac{dR_{S_{i2}}[s+(t_f)^{\frac{1}{m+1}}]}{dt_f} = f_{t_{2f}}(t_f).$$

Similarly, the *cdf* of the times to failure is

$$F_{t_f}(t_f) = 1 - R_{t_f}(t_f),$$

or

$$F_{t_f}(t_f) = p\, F_{S_{i1}}[s+(t_f)^{\frac{1}{m+1}}] + (1-p)\, F_{S_{i2}}[s+(t_f)^{\frac{1}{m+1}}], \quad (4.31)$$

or

$$F_{t_f}(t_f) = p\, F_{t_{f1}}(t_f) + (1-p)\, F_{t_{f2}}(t_f), \quad (4.32)$$

where

$$F_{S_{i1}}[s+(t_f)^{\frac{1}{m+1}}] = \text{cdf of } S_{i1} \text{ at } [s+(t_f)^{\frac{1}{m+1}}],$$
$$= \int_0^{s+(t_f)^{\frac{1}{m+1}}} f_{S_{i1}}(S_i)\, dS_i,$$
$$F_{S_{i2}}[s+(t_f)^{\frac{1}{m+1}}] = \text{cdf of } S_{i2} \text{ at } [s+(t_f)^{\frac{1}{m+1}}],$$
$$= \int_0^{s+(t_f)^{\frac{1}{m+1}}} f_{S_{i2}}(S_i)\, dS_i,$$
$$F_{t_{f1}}(t_f) = F_{S_{i1}}[s+(t_f)^{\frac{1}{m+1}}], \quad (4.33)$$

and

$$F_{t_{f2}}(t_f) = F_{S_{i2}}[s+(t_f)^{\frac{1}{m+1}}]. \quad (4.34)$$

PHYSICAL EXPLANATION OF THE FAILURE PATTERN 101

Finally, the failure rate function is given by

$$\lambda_{t_f}(t_f) = \frac{f_{t_f}(t_f)}{R_{t_f}(t_f)},$$
$$= \frac{p\, f_{t_{f_1}}(t_f) + (1-p)\, f_{t_{f_2}}(t_f)}{p\, R_{t_{f_1}}(t_f) + (1-p)\, R_{t_{f_2}}(t_f)}.$$

To express the failure rate in terms of the given information; i.e., the applied stress, s, the bimodal initial strength distribution, $f_{S_i}(S_i)$, and the strength deteriorating constant, m, substituting Eqs. (4.23) and (4.27) into the above equation yields

$$\lambda_{t_f}(t_f) = \frac{p\left\{\left(\frac{1}{m+1}\right)(t_f)^{-\frac{m}{m+1}} f_{S_{i1}}[s+(t_f)^{\frac{1}{m+1}}]\right\}}{p\, R_{S_{i1}}[s+(t_f)^{\frac{1}{m+1}}] + (1-p)\, R_{S_{i2}}[s+(t_f)^{\frac{1}{m+1}}]}$$
$$+ \frac{(1-p)\left\{\left(\frac{1}{m+1}\right)(t_f)^{-\frac{m}{m+1}} f_{S_{i2}}[s+(t_f)^{\frac{1}{m+1}}]\right\}}{p\, R_{S_{i1}}[s+(t_f)^{\frac{1}{m+1}}] + (1-p)\, R_{S_{i2}}[s+(t_f)^{\frac{1}{m+1}}]}.$$

(4.35)

For example, assume that the component's initial strength follows a bimodal mixed Weibull distribution with the following *pdf*:

$$f_{S_i}(S_i) = p\, \frac{\beta_1}{\eta_1}\left(\frac{S_i - \gamma_1}{\eta_1}\right)^{\beta_1 - 1} e^{-\left(\frac{S_i - \gamma_1}{\eta_1}\right)^{\beta_1}}$$
$$+ (1-p)\, \frac{\beta_2}{\eta_2}\left(\frac{S_i - \gamma_2}{\eta_2}\right)^{\beta_2 - 1} e^{-\left(\frac{S_i - \gamma_2}{\eta_2}\right)^{\beta_2}}, \quad (4.36)$$

where

$p = 0.05,$
$\beta_1 = 3.4, \quad \eta_1 = 6, \quad \gamma_1 = 2,$
$\beta_2 = 10, \quad \eta_2 = 30, \quad \gamma_2 = 2.$

This *pdf* is shown in Fig. 4.18. It is further assumed that the applied stress is a constant and equal to one, or $s = 1$. The strength deteriorating constant is $m = 2$. Then, the component's *pdf* and *cdf* of the times to failure, $f_{t_f}(t_f)$, $F_{t_f}(t_f)$, and the failure rate, $\lambda_{t_f}(t_f)$, can be obtained by substituting the given parameter values into Eqs. (4.23), (4.31) and (4.35), respectively, and they are plotted in Figs. 4.19, 4.20 and 4.21, respectively.

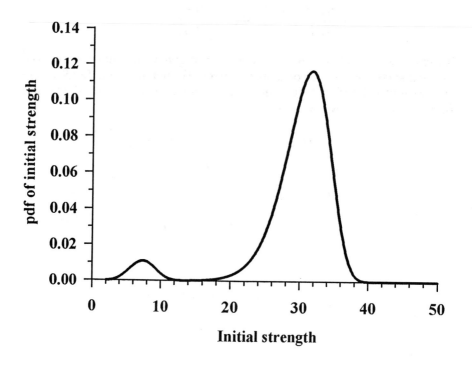

Fig. 4.18– The bimodal initial strength distribution with parameters: $p = 0.05$, $\beta_1 = 3.4$, $\eta_1 = 6$, $\gamma_1 = 2$, $\beta_2 = 10$, $\eta_2 = 30$ and $\gamma_2 = 2$.

PHYSICAL EXPLANATION OF THE FAILURE PATTERN

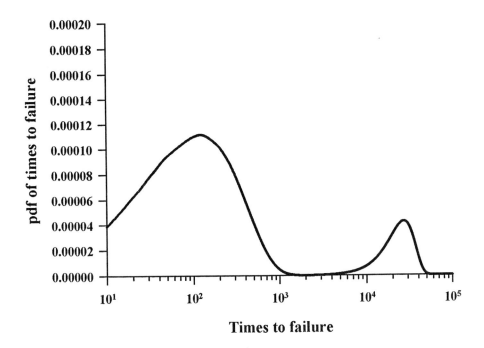

Fig. 4.19– Bimodal times-to-failure distribution corresponding to the bimodal strength distribution in Fig. 4.18.

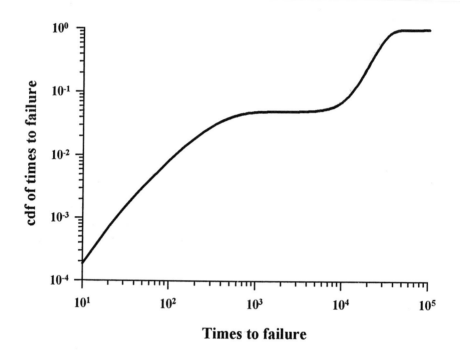

Fig. 4.20– S-shaped *cdf* of the times to failure corresponding to the bimodal strength distribution in Fig. 4.18.

PHYSICAL EXPLANATION OF THE FAILURE PATTERN

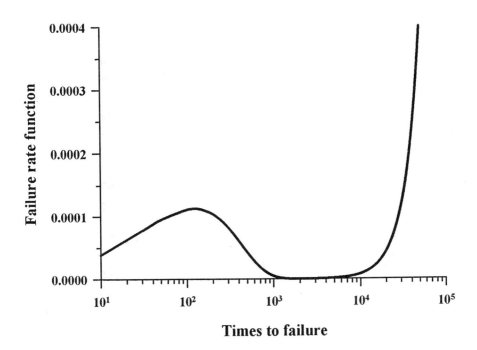

Fig. 4.21– The failure rate curve corresponding to the bimodal strength distribution in Fig. 4.18.

It may be seen, from Figs. 4.18 and 4.19, that a bimodal times-to-failure distribution is created corresponding to the bimodal strength distribution given by Eq. (4.36). A typical S-shaped *cdf* appears on the probability plotting paper indicating the bimodality of the times-to-failure distribution. Finally, the failure rate curve shows a pattern similar to the conventional bathtub curve, except that there is a distinctive hump appearing during the traditional early-life period, which was also observed by Wong [16; 18; 19; 20].

4.5.2.4 FAILURE RATE CURVE AS DERIVED FROM STRESS-STRENGTH INTERFERENCE AND A PHYSICS-OF-FAILURE MODEL

In additional to Jensen and Petersen [14], Loll [27] and Lewis and Chen [37] who studied the stress-strength interference and the corresponding failure rate pattern, Shen [38] systematically investigated the relationship between component failure rate and the stress-strength distributional characteristics. By considering some commonly encountered distributions, the quantitative relationships between a component's failure rate and the time dependence of its strength distributions are established.

Smith and Dietrich [39] presented a viable explanation of the reliability bathtub curve where the failure rate is governed by a physical process of the component's strength decay over time, the strength degrading due to crack propagation, and a distributed external stress.

A strength decay expression is derived, based on the Paris Law of crack growth and fracture mechanics, as follows [39]:

$$S(n) = V \left[\left(\frac{S_i}{V}\right)^{2W} - \zeta\, n \right]^{\frac{1}{2W}}, \qquad (4.37)$$

where

n = number of applied stress cycles,
$S(n)$ = residual strength of a component after n cycles of stress applications, a random variable,
S_i = initial strength of the component, a random variable,
$W = \dfrac{m}{2} - 1$,
m = fatigue strength exponent in the Paris Law which is expressed in the following differential form:
$\dfrac{da}{dn} = C\,(\Delta K)^m$,

a = crack length after n stress cycles,
C = fatigue strength decay rate,
ΔK = stress intensity factor,
$$V = \frac{K_c}{Y(a)\sqrt{\pi a}},$$
$Y(a)$ = geometric factor which is dependent on the crack length a,
K_c = fracture toughness,
π = 3.1415926,
$$\zeta = W\, C\, [Y(a)]^m\, \pi^{\frac{m}{2}}\, (s_r)^m,$$

and

s_r = stress range.

Then, the probability of failure, or the *cdf* of stress cycles to failure, is given by

$$F(n) = \int_0^\infty \left\{ f_{S(n)}[S(n)] \int_{S(n)}^\infty f_s(s)\, ds \right\} dS(n). \qquad (4.38)$$

Since the distribution of the component's strength at any number of stress cycles n, $f_{S(n)}[S(n)]$, is not known, but the distribution of the component's initial strength, $f_{S_i}(S_i)$, is usually known as a prior, then $f_{S(n)}[S(n)]$ needs to be derived from $f_{S_i}(S_i)$ and other relevant information using Eq. (4.37). For a given *pdf* of the initial component strength, $f_{S_i}(S_i)$, the *pdf* of the residual component strength after n stress cycles, $f_{S(n)}[S(n)]$, is given by

$$f_{S(n)}[S(n)] = f_{S_i}(S_i)\Big|_{S_i = S_i^{-1}[S(n)]}\, |J|, \qquad (4.39)$$

where

$S_i^{-1}[S(n)]$ = inverse function of S_i in terms of $S(n)$,

or

$$S_i^{-1}[S(n)] = V\left[\left(\frac{S(n)}{V}\right)^{2W} + \zeta\, n\right]^{\frac{1}{2W}}, \qquad (4.40)$$

from Eq. (4.37), and

J = Jacobian of the variable transform given by Eq. (4.40),
$$= \frac{dS_i}{dS(n)},$$

or

$$J = \left\{\left[\frac{S(n)}{V}\right]^{2W} + \zeta\, n\right\}^{\frac{1}{2W}-1} \left[\frac{S(n)}{V}\right]^{2W-1}. \quad (4.41)$$

Substituting Eqs. (4.40) and (4.41) into Eq. (4.39) yields

$$f_{S(n)}[S(n)] = f_{S_i}(S_i)\Big|_{S_i = V\left[\left(\frac{S(n)}{V}\right)^{2W} + \zeta\, n\right]^{\frac{1}{2W}}}$$

$$\cdot \left\{\left[\frac{S(n)}{V}\right]^{2W} + \zeta\, n\right\}^{\frac{1}{2W}-1} \left[\frac{S(n)}{V}\right]^{2W-1},$$

or

$$f_{S(n)}[S(n)] = \left[\frac{S(n)}{V}\right]^{2W-1} \left\{\left[\frac{S(n)}{V}\right]^{2W} + \zeta\, n\right\}^{\frac{1}{2W}-1}$$

$$\cdot f_{S_i}\left\{V\left[\left(\frac{S(n)}{V}\right)^{2W} + \zeta\, n\right]^{\frac{1}{2W}}\right\}.$$

$$(4.42)$$

Substituting Eq. (4.42) into Eq. (4.38) yields the *cdf* of stress cycles to failure, or

$$F(n) = \int_0^\infty \left[\frac{S(n)}{V}\right]^{2W-1} \left\{\left[\frac{S(n)}{V}\right]^{2W} + \zeta\, n\right\}^{\frac{1}{2W}-1}$$

$$\cdot f_{S_i}\left\{V\left[\left(\frac{S(n)}{V}\right)^{2W} + \zeta\, n\right]^{\frac{1}{2W}}\right\} \int_{S(n)}^\infty f_s(s)\, ds\, dS(n).$$

$$(4.43)$$

The *pdf* of stress cycles to failure, $f(n)$, can be obtained from

$$f(n) = \frac{dF(n)}{dn}, \quad (4.44)$$

and finally the failure rate function, $\lambda(n)$, can be obtained from

$$\lambda(n) = \frac{f(n)}{1 - F(n)}. \quad (4.45)$$

TABLE 4.1– Input parameters of stress distribution, strength distribution and the Paris Law for Figs. 4.22, 4.23 and 4.24.

Figure	Stress distribution	Initial strength distribution		% defectives	Paris Law parameters
		Freak	Main		
Fig. 4.22	Constant $s = 25$	Normal $\overline{S_{i1}} = 30$ $\sigma_{S_{i1}} = 3$	Normal $\overline{S_{i2}} = 60$ $\sigma_{S_{i2}} = 5$	$p = 10\%$	$m = 4.0$ $C = 4 \times 10^{-13}$ $K_c = 40$ $Y(a) = 1.2$
Fig. 4.23	Normal $\overline{s} = 40$ $\sigma_s = 4$	Lognormal $\overline{S_{i1}} = 40$ $\sigma_{S_{i1}} = 5$	Normal $\overline{S_{i2}} = 100$ $\sigma_{S_{i2}} = 10$	$p = 5\%$	$m = 3.0$ $C = 1 \times 10^{-12}$ $K_c = 25$ $Y(a) = 1.0$
Fig. 4.24	Normal $\overline{s} = 200$ $\sigma_s = 200$	Lognormal $\overline{S_{i1}} = 180$ $\sigma_{S_{i1}} = 120$	Lognormal $\overline{S_{i2}} = 330$ $\sigma_{S_{i2}} = 33$	$p = 7.5\%$	$W = 0.5$ $V = 150$ $\zeta = 1.0 \times 10^{-12}$

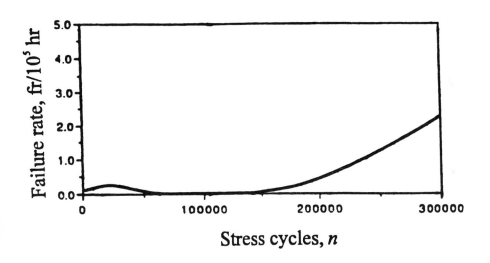

Fig. 4.22– The failure rate curve corresponding to the parameters of the stress distribution, strength distribution and the Paris Law given in the second row of Table 4.1.

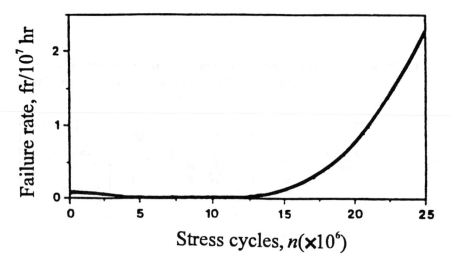

Fig. 4.23– The failure rate curve corresponding to the parameters of the stress distribution, strength distribution and the Paris Law given in the third row of Table 4.1.

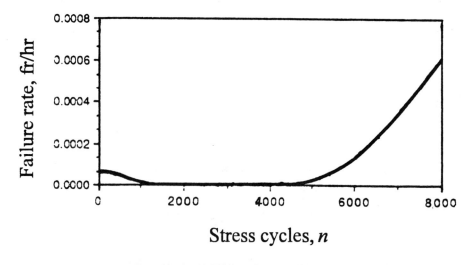

Fig. 4.24– The failure rate curve corresponding to the parameters of the stress distribution, strength distribution and the Paris Law given in the fourth row of Table 4.1.

Figures 4.22 through 4.24, taken from [39], represent the failure rate curves using Eq. (4.45), in conjunction with Eqs. (4.43) and (4.44), and are based on the input information listed in Table 4.1. Note that a bimodal mixed strength distribution with *pdf* in the form of Eq. (4.15) is assumed here.

It may be seen from Figs. 4.22, 4.23 and 4.24 that the failure rate curves, under the bimodal initial strength distribution assumption and a strength decay process governed by the Paris Law and a fracture mechanics model, exhibit the same pattern as that shown in Fig. 4.21 obtained using Jensen and Petersen's approach; i.e., a basic bathtub curve with a hump in the early-life period. This further ascertains that Wong's Roller-Coaster failure rate pattern [16; 18; 19; 20], that is characterized by a decreasing function with one or more humps in the early-life period, can be mathematically classified into the bimodal or multimodal, time decaying, mixed-strength-distribution-based failure rate category.

REFERENCES

1. Ryerson, C. M., "The Reliability Bathtub Curve is Vigorously Alive," *Proceedings Annual Reliability and Maintainability Symposium*, pp. 313-316, 1982.

2. Peck, D. S., "Semiconductor Device Life and System Removal Rates," *Proceedings of the Annual Symposium on Reliability*, pp. 593-601, 1968.

3. Peck, D. S., "The Analysis of Data from Accelerated Stress Tests," *Proceedings Annual Reliability Physics Symposium*, 1971.

4. Peck, D. S. and Zerdt, C. H., "The Reliability of Semiconductor Devices in the Bell System," *Proceedings of the IEEE*, Vol. 62, No. 2, pp. 185-211, February 1974.

5. Peck, D. S., "New Concepts About Integrated Circuits Reliability," *Proceedings Annual Reliability Physics Symposium*, pp. 1-6, 1978.

6. Peck, D. S., *Accelerated Testing Handbook*, Technology Associates, California, 1978.

7. Stitch, M., et al, "Microcircuit Accelerated Testing Using High Temperature Operating Tests," *IEEE Transactions on Reliability*, Vol. R-24, No. 4, pp. 238-250, 1975.

8. Edwards, J. R., et al, "VMOS Reliability," *Proceedings Annual Reliability Physics Symposium*, pp. 23-27, 1978.

9. Tarkaide, A. and Manabe, N., "RA System Using Process Failure Analysis for IC's," *Proceedings Annual Reliability and Maintainability Symposium*, pp. 1-6, 1977.

10. Hailberg, O., "Failure Rate as a Function of Time Due to Lognormal Life Distribution(s) of Weak Parts," *Microlectronics and Reliability*, Vol. 16, pp. 155-158, 1977.

11. Reynolds, F. H., "Accelerated Test Procedures for Semiconductor Components," *Proceedings Annual Reliability Physics Symposium*, pp. 166-178, 1977.

12. Reynolds, F. H. and Stevens, J. W., "Semiconductor Component Reliability in an Equipment Operating in Electromechanical Telephone Exchanges," *Proceedings Annual Reliability Physics Symposium*, pp. 7-13, 1978.

13. Keller, A. A. and Kamath, A. R. R., "Semiconductor Failure Analysis," *Symposium of the Society of Environmental Engineers*, Wembley, pp. 9-11, May 1979.

14. Jensen, F. and Petersen, N. E., *Burn-in*, John Wiley & Sons, New York, 167 pp., 1982.

15. Moltoft, J., "Behind the Bathtub Curve: A New Model and Its Consequences," *Microlectronics and Reliability*, Vol. 23, No. 3, pp. 489-500, 1983.

16. Wong, K. L., "Unified Field (Failure) Theory – Demise of the Bathtub Curve," *Proceedings Annual Reliability and Maintainability Symposium*, pp. 27-29, 1981.

17. Bezat, A. G. and Montague, L. L., "The Effect of Endless Burn-in on Reliability," *Proceedings Annual Reliability and Maintainability Symposium*, pp. 392-397, 1979.

18. Wong, K. L. and Lindstrom, D. L., "Off the Bathtub onto the Roller-Coaster Curve," *Proceedings Annual Reliability and Maintainability Symposium*, pp. 356-363, 1988.

19. Wong, K. L., "A New Environmental Stress Screening Theory for Electronics," *Proceedings Institute of Environmental Sciences*, pp. 218-224, 1989.

20. Wong, K. L., "Demonstrating Reliability and Reliability Growth with Environmental Stress Screening Data," *Proceedings Annual Reliability and Maintainability Symposium*, pp. 47-52, 1990.

21. Kececioglu, Dimitri B. and Hughes, R. C., "Design of Equipment to Optimize Reliability for Manufacturer's and Customer's Minimum Total Cost," *Allis-Chambers Research Report*, Submitted to the Office of Naval Research, Arlington, VA, under Contract Nonr-3931(00) (FBM), 103 pp., Feb. 1963.

REFERENCES

22. Kececioglu, Dimitri B., *Reliability Engineering Handbook*, Prentice Hall, Upper Saddle River, NJ 07458, Vol. 1, 688 pp., 4th printing, 1995.

23. Shooman, M. L., "Reliability Physics Models," *IEEE Transactions on Reliability*, Vol. R-17, No. 1, pp. 14-20, 1968.

24. Shooman, M. L., *Probabilistic Reliability: An Engineering Approach*, Second Edition, Robert E. Krieger Publishing Co., 1991.

25. Shaw, L., Shooman, M. L. and Schatz, R., "Time Dependent Stress-Strength Models for Non-electrical and Electrical Systems," *Proceedings Annual Reliability and Maintainability Symposium*, pp. 186-197, 1973.

26. Schatz, R., Shooman, M. L. and Shaw, L., "Application of Time Dependent Stress-Strength Models for Non-electrical and Electrical Systems," *Proceedings Annual Reliability and Maintainability Symposium*, pp. 540-547, 1974.

27. Loll, V., "Load-Strength Modeling of Mechanics and Electronics," *Quality and Reliability Engineering International*, Vol. 3, pp. 149-155, 1987.

28. Carter, A. D. S., *Mechanical Reliability*, John Wiley & Sons, New York, 146 pp., 1972.

29. Carter, A. D. S., *Mechanical Reliability*, Second Edition, Macmillan Education Ltd., 492 pp., 1986.

30. Kapur, K. L. and Lamberson, L. R., *Reliability in Engineering Design*, John Wiley & Sons, New York, 586 pp., 1977.

31. Haugen, E. B., *Probabilistic Approaches to Design*, John Wiley & Sons, New York, 323 pp., 1968.

32. Haugen, E. B., *Probabilistic Mechanical Design*, John Wiley & Sons, New York, 626 pp., 1980.

33. Ang, A. H. S. and Tang, W. H., *Probability Concepts in Engineering Planning and Design*, John Wiley & Sons, New York, 1975.

34. Kececioglu, Dimitri B., "Reliability Analysis of Mechanical Components and Systems," *Nuclear Engineering and Design*, Vol. 19, pp. 259-290, 1972.

35. Kececioglu, Dimitri B. and Haugen, E. B., "A Unified Look at Design Safety Factors, Safety Margins and Measures of Reliability," *Proceedings Seventh Annual Reliability and Maintainability Conference*, pp. 520-530, 1968.

36. Kececioglu, Dimitri B., *Engineering Design by Reliability*, Lecture Notes, The University of Arizona, Aerospace and Mechanical Engineering Department, Tucson, AZ, 600 pp., 1990.

37. Lewis, E. E. and Chen, Hsin-Chieh, "Load-Capacity Interference and the Bathtub Curve," *IEEE Transactions on Reliability*, Vol. 43, No. 3, pp. 470-475, September 1994.

38. Shen, K. C., "On the Relationship between Component Failure Rate and Stress-Strength Distributional Characteristics," *Microelectronics and Reliability*, Vol. 28, No. 5, pp. 801-812, 1988.

39. Smith, R. W. and Dietrich, D. L., "The Bathtub Curve: An Alternative Explanation," *Proceedings Annual Reliability and Maintainability Symposium*, pp. 241-245, 1994.

Chapter 5

MATH MODELS DESCRIBING THE FAILURE PROCESS DURING BURN-IN AND THEIR PARAMETERS' ESTIMATION

In the preceding chapter, the observed failure patterns during burn-in or in the early life period were examined and their inherent physical mechanisms were explored quantitatively. This chapter presents useful math models for the statistical description of the failure process during burn-in, and techniques for their parameters' estimation.

5.1 RELIABILITY MODELS FOR A MIXED-WEIBULL POPULATION

It has been shown [1; 2] that, if N identical components or equipment, from a mixed population which is composed of n different subpopulations, such as, $N_1, N_2, ... , N_n$, undertake a mission of T duration, starting the mission at age zero, the reliability function for this mixed population can be expressed by

$$R_{1,2,...,n}(T) = p_1 R_1(T) + p_2 R_2(T) + p_3 R_3(T) + ... + p_n R_n(T),$$

(5.1)

where

$$p_i = \frac{N_i}{N}, i = 1, 2, ..., n,$$

$$\sum_{i=1}^{n} N_i = N,$$

and

$$\sum_{i=1}^{n} p_i = 1.$$

The probability density function, or the distribution, of the times to failure for this mixed population is given by

$$f_{1,2,\ldots,n}(T) = p_1\, f_1(T) + p_2\, f_2(T) + p_3\, f_3(T) + \ldots + p_n\, f_n(T). \tag{5.2}$$

The corresponding failure rate function is given by

$$\lambda_{1,2,\ldots,n}(T) = \frac{p_1\, f_1(T) + p_2\, f_2(T) + \ldots + p_n\, f_n(T)}{p_1\, R_1(T) + p_2\, R_2(T) + \ldots + p_n\, R_n(T)}. \tag{5.3}$$

If only two subpopulations are involved, as is often the case encountered in burn-in tests, and the times to failure for each one of the subpopulations may be represented by an individual Weibull distribution, Eqs. (5.1) through (5.3) may be written as

$$R_{1,2}(T) = p_1\, e^{-(\frac{T-\gamma_1}{\eta_1})^{\beta_1}} + p_2\, e^{-(\frac{T-\gamma_2}{\eta_2})^{\beta_2}}, \tag{5.4}$$

$$f_{1,2}(T) = p_1\, \frac{\beta_1}{\eta_1}\, \left(\frac{T-\gamma_1}{\beta_1}\right)^{\beta_1-1} e^{-(\frac{T-\gamma_1}{\eta_1})^{\beta_1}}$$

$$+ p_2\, \frac{\beta_2}{\eta_2}\, \left(\frac{T-\gamma_2}{\beta_2}\right)^{\beta_2-1} e^{-(\frac{T-\gamma_2}{\eta_2})^{\beta_2}}, \tag{5.5}$$

and

$$\lambda_{1,2}(T) = \frac{p_1\, \frac{\beta_1}{\eta_1}(\frac{T-\gamma_1}{\beta_1})^{\beta_1-1} e^{-(\frac{T-\gamma_1}{\eta_1})^{\beta_1}} + p_2\, \frac{\beta_2}{\eta_2}(\frac{T-\gamma_2}{\beta_2})^{\beta_2-1} e^{-(\frac{T-\gamma_2}{\eta_2})^{\beta_2}}}{p_1\, e^{-(\frac{T-\gamma_1}{\eta_1})^{\beta_1}} + p_2\, e^{-(\frac{T-\gamma_2}{\eta_2})^{\beta_2}}}. \tag{5.6}$$

Since only two subpopulations are considered here, then

$$p_1 + p_2 = 1. \tag{5.7}$$

5.2 BATHTUB CURVE MODELS

The bathtub curve is a very effective way of describing the life characteristics of a population during the burn-in process. In Chapter 14, Volume 1 of [1], mathematical models covering the whole reliability bathtub curve, or parts thereof, are presented and analyzed extensively. These models are summarized here.

5.2.1 Model 1 — Two-parameter Weibull model

This model may be thought of as a truncated extreme-value distribution with a Weibull type parameterization. It can represent increasing, decreasing and bathtub-shaped failure rates for $\beta = 0.5$.

The failure rate function for this model is

$$\lambda(T) = (\frac{\beta}{\alpha^\beta}) T^{\beta-1} e^{(T/\alpha)^\beta}, \quad T \geq 0, \quad \alpha > 0, \quad \beta > 0, \tag{5.8}$$

the *pdf* is

$$f(T) = (\frac{\beta}{\alpha^\beta}) T^{\beta-1} e^{(T/\alpha)^\beta} e^{1-e^{(T/\alpha)^\beta}}, \tag{5.9}$$

and the reliability function is

$$R(T) = e^{1-e^{(T/\alpha)^\beta}}. \tag{5.10}$$

5.2.2 Model 2 — Three-parameter model

This model can also describe increasing, decreasing and bathtub-shaped failure rates.

The failure rate function for this model is

$$\lambda(T) = \delta T + \frac{\theta}{1+\beta T}, \quad T \geq 0, \quad \delta \geq 0, \quad \beta \geq 0, \quad \theta \geq 0, \tag{5.11}$$

the *pdf* is

$$f(T) = \frac{(1+\beta T)\delta T + \theta}{(1+\beta T)^{(\theta/\beta)+1}} e^{-\delta T^2/2}, \tag{5.12}$$

and the reliability function is

$$R(T) = \frac{e^{-\delta T^2/2}}{(1+\beta T)^{\theta/\beta}}. \tag{5.13}$$

5.2.3 Model 3 — Six-parameter model

This model's failure rate function is

$$\lambda(T) = \lambda_1 \, e^{-5T/T_e} + \lambda_2 + \lambda_3 \, e^{(T-T_w)/T_t}, \quad T \geq 0, \tag{5.14}$$

and its reliability function is

$$R(T) = exp\left\{-\left[\left(\frac{\lambda_1 T_e}{5}\right)\left(1 - e^{-5T/T_e}\right) + \lambda_2 T \right.\right.$$
$$\left.\left. + \lambda_3 T_t \left(e^{(T-T_w)/T_t} - e^{-T_w/T_t}\right)\right]\right\}, \tag{5.15}$$

where T_e, T_w and T_t are three time constants, and λ_1, λ_2 and λ_3 are three failure rates.

For $\quad 0 \leq T \leq T_e, \quad \lambda(T) \cong \lambda_1 \, e^{-5T/T_e} + \lambda_2$.
For $T_e < T < (T_w - 5T_t), \quad \lambda(T) \cong \lambda_2$.
For $\quad T \geq (T_w - 5T_t), \quad \lambda(T) \cong \lambda_2 + \lambda_3 \, e^{(T-T_w)/T_t}$.

5.2.4 Model 4 — Five-parameter model

This model can represent, in special cases, common distributions such as the Weibull ($k = 0$), exponential ($k = 1$, $c = 1$), extreme value ($k = 0$, $b = 1$), Makeham ($c = 1$, $b = 1$), and the bathtub curve ($c = 0.5$, $b = 1$).

The failure rate function is

$$\lambda(T) = k\,\lambda\,c\,T^{c-1} + (1-k)\,b\,T^{b-1}\beta\,e^{\beta T^b}, \tag{5.16}$$

where b and c are the scale parameters, and β and λ are the shape parameters with $b > 0$, $c > 0$, $\beta > 0$, $\lambda > 0$ and $0 \leq k \leq 1$.

The *pdf* is

$$f(T) = [k\,\lambda\,c\,T^{c-1} + (1-k)\,b\,T^{b-1}\beta\,e^{\beta T^b}]e^{-[k\lambda T^c + (1+k)(e^{\beta T^b}-1)]}. \tag{5.17}$$

The reliability function is

$$R(T) = exp[-k\,\lambda\,T^c - (1+k)(e^{\beta T^b}-1)]. \tag{5.18}$$

5.3 SELECTION OF THE MODELS FOR BURN-IN TESTS

Among the four models presented in Section 5.2, Model 1 is preferred not only for its simplicity, but also for its unique properties in describing particular cases. Also the parameters of this model can be obtained without much difficulty and without lack of precision. The other models either have too many parameters, thus causing difficulty in parameter estimation, or are too graphical in the estimation of their parameters, thus not so accurate. It is for this reason that only Model 1 is applied in this chapter. However readers are encouraged to try using the other three models as well, if the data warrant it.

5.4 PARAMETER ESTIMATION FOR THE BIMODAL MIXED POPULATION

From here on, only the case of the bimodal Weibull mixture model with a zero location parameter will be considered, then

$$F(T) = p F_1(T) + (1-p) F_2(T),$$

or, using the Weibull *pdf*,

$$F(T) = p\,[1 - e^{-(T/\eta_1)^{\beta_1}}] + (1-p)\,[1 - e^{-(T/\eta_2)^{\beta_2}}],$$

where $F_1(T)$ and $F_2(T)$ are the cumulative probability functions of the weak or substandard subpopulation and the main or standard subpopulation, respectively, and p is the proportion of the weak subpopulation.

Consider the case where N identical units are reliability tested with no replacement, and $r \leq N$ times to failure are registered, as $t_1, t_2, ..., t_r$. After they are plotted on Weibull probability paper, it is expected that a bimodal, mixed-Weibull distribution will be obtained. This distribution may be either well-separated or well-mixed in appearance. By well-separated, it is meant that the scale parameters for the two subpopulations are very different; in other words, the mixed *cdf* either has an obvious flat portion between the two subpopulations; e.g., the *cdf*'s plot is "S" shaped, or has an obvious intersection separating the two subpopulations. Needless to say, for the well-mixed case, the scale parameters of the two subpopulations are quite close to each other; in other words, the mixed *cdf* has no obvious flat portion or intersection. Readers can refer to [3] where the graphical behaviors of mixed-Weibull distributions are discussed quantitatively in great detail.

Different methods should be used to estimate the distribution parameters for these two cases. Four methods are introduced next.

5.4.1 JENSEN'S GRAPHICAL METHOD FOR THE WELL-SEPARATED CASE

If the two subpopulations are well separated, Jensen's graphical approach [4] is recommended. To determine the parameters of each subpopulation the following steps may be used:

1. Read off $\hat{p}$ from the Weibull plot at the ordinate corresponding to the flat portion, or the smallest slope portion, on the *cdf* curve.

2. Read off $\hat{\eta}_1$ at the abscissa corresponding to $MR = 0.632\,\hat{p}$, and $\hat{\eta}_2$ corresponding to $MR = \hat{p} + 0.632\,(1-\hat{p})$.

3. Draw an asymptotic line to the Weibull curve at the earliest time to failure. Then, $\hat{\beta}_1$ is the slope of this line.

4. Draw another asymptotic line to the Weibull curve at the right end of the time to failure. Then, $\hat{\beta}_2$ is the slope of this line. This procedure is shown in Fig. 5.1.

The theoretical basis for this method is as follows. For the well-separated case, η_1 is far away from η_2. So, it can be considered that, during the early failure period, chance failures have not occurred as yet, or $F_2(T) \cong 0$. Therefore,

$$F(T) \cong p\,F_1(T),$$

and

$$\hat{\eta}_1 \cong T|_{F(T)=0.632\,\hat{p}}.$$

During the chance-failure period, the early failures have almost disappeared, or $F_1(T) \cong 1$. Therefore,

$$F(T) \cong p + (1-p)\,F_2(T),$$

and

$$\hat{\eta}_2 \cong T|_{p+0.632\,(1-p)}.$$

For the reason that $\hat{p}$, $\hat{\beta}_1$ and $\hat{\beta}_2$ are so determined see Appendix 5A.

EXAMPLE 5–1

Failure times for 19 components ($r = 19$) and their median ranks are recorded in Table 5.1 from $N = 150$ CMOS components tested at 125°C and 5V. Determine the parameters of the two-Weibull mixture using Jensen's graphical method.

PARAMETER ESTIMATION FOR THE BIMODAL CASE

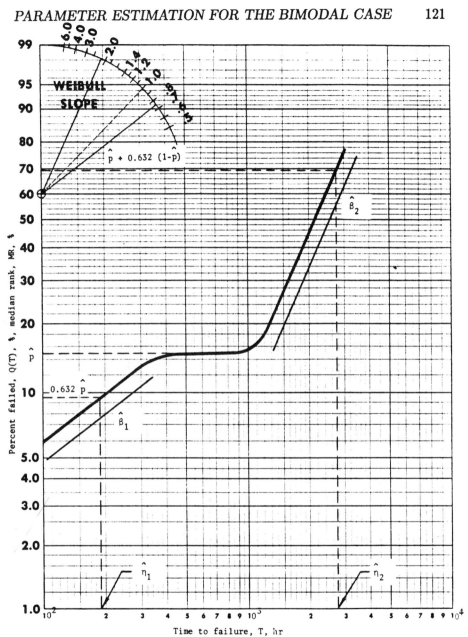

Fig. 5.1– Jensen's graphical method for a well-separated Weibull mixture.

Table 5.1– **Failure data and $MR's$ for $N = 150$ and $r = 19$, for Example 5–1.**

Rank, j	Time to failure, T_j, hr	Median rank, MR_j, %
1	100	0.47
2	200	1.13
3	250	1.79
4	420	2.46
5	420	3.13
6	588	3.79
7	588	4.45
8	588	5.12
9	708	5.78
10	1,044	6.45
11	2,892	7.11
12	2,892	7.78
13	3,396	8.44
14	3,396	9.11
15	3,997	9.77
16	3,997	10.44
17	3,997	11.10
18	4,165	11.77
19	4,500	12.43

Table 5.2– Failure data and $MR's$, after grouping, for Example 5–1.

Rank, j	Time to failure, T_j, hr	Median rank, MR_j, %
1	100	0.47
2	200	1.13
3	250	1.79
5	420	3.13
8	588	5.12
9	708	5.78
10	1,044	6.45
12	2,892	7.78
14	3,396	9.11
17	3,997	11.10
18	4,165	11.77
19	4,500	12.43

SOLUTION TO EXAMPLE 5–1

The failure data in Table 5.1 can be simplified by grouping them as given in Table 5.2.

By plotting the T_j versus the corresponding MR_j values in Table 5.2 on an appropriate Weibull probability paper (WPP), indexWPP an "S" shaped curve is obtained, as shown in Fig. 5.2. This indicates that the times-to-failure distribution of the CMOS components is a bimodal, well-separated Weibull mixture.

The weak subpopulation's percentage p can be read off the ordinate at the level corresponding to the flat portion on the *cdf* curve, which is

$$\hat{p} = 6.7\%.$$

The scale parameter of the weak subpopulation, η_1, can then be read off at the abscissa corresponding to

$$MR = 0.632\,\hat{p} = 0.632 \times 6.7\% = 4.23\%,$$

which is

$$\hat{\eta}_1 = 500 \text{ hr}.$$

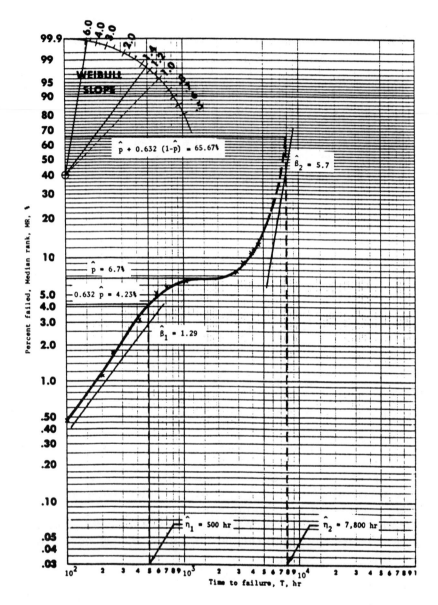

Fig. 5.2– Plot of data in Table 5.2 on WPP for parameter estimation by Jensen's graphical method, for Example 5–1.

The scale parameter of the strong subpopulation, η_2, can be read off at the abscissa corresponding to

$$MR = \hat{p} + 0.632(100 - \hat{p}) = 65.67\%,$$

which is

$$\hat{\eta}_2 = 7,800 \text{ hr}.$$

The shape parameter of the weak subpopulation, β_1, is the slope of the asymptotic line to the Weibull curve at the earliest time to failure, as shown in Fig. 5.2, which is

$$\hat{\beta}_1 = 1.29.$$

The shape parameter of the strong subpopulation, β_2, is the slope of the asymptotic line to the Weibull curve at its rightmost end, as shown in Fig. 5.2, which is

$$\hat{\beta}_2 = 5.7.$$

5.4.2 BAYESIAN METHOD FOR THE WELL-SEPARATED CASE

Sometimes one may not get a completely flat portion on the Weibull plot and the "knee" of the curve may be ill-defined. In that case a large error results in reading off the values of $\hat{p}$, $\hat{\eta}_1$, $\hat{\beta}_1$, $\hat{\eta}_2$ and $\hat{\beta}_2$ from the plot using Jensen's method. Kamath et al. [5] developed a consistent technique based on Bayesian inference to compute the probability of a certain failure belonging to the substandard or the standard subpopulation, and then determine the parameters of the whole mixed distribution. Kececioglu and Sun [6] present a combined-analytical-graphical Bayesian approach using the *fractional rank plotting method* to determine the parameters of a bimodally-mixed life distribution. This approach proves itself to be superior to the conventional separation plotting method since it makes full use of the information on the behavior of the distribution of the two subpopulations over the whole test duration. However, the conventional separation plotting method loses the strong distribution information in the early failure period and the weak distribution information in the chance failure period, by splitting the original sample into two independent subsamples. A detailed description of these methods and their comparisons are presented next.

5.4.2.1 BAYESIAN METHOD FOR THE UNGROUPED DATA CASE

Assume that the times to failure follow a bimodal, cumulative Weibull distribution, given by

$$F(T) = p\left[1 - e^{-(T/\eta_1)^{\beta_1}}\right] + (1-p)\left[1 - e^{-(T/\eta_2)^{\beta_2}}\right],$$

and that a test has been carried out on N items with r times to failure registered as $T_1, T_2, ..., T_r$. To apply a Bayes theorem it is necessary to determine a priori to which subpopulation a failure belongs. Lacking this information, it is reasonable to start off by letting the probability of each and every failure belonging to either the substandard or the standard subpopulation be 50%.

Having plotted the failure data, the first estimates of $\hat{p}, \hat{\eta}_1, \hat{\beta}_1, \hat{\eta}_2$ and $\hat{\beta}_2$ can be obtained by Jensen's method. If the test time is not long enough to show the standard distribution, then it is assumed that the standard subpopulation is exponential; i.e., $\hat{\beta}_2 = 1$. The value of $\hat{\eta}_2$ is computed by standard prediction techniques. Now the posterior probabilities can be calculated, using these initial estimates, from

$$P_1^i = \frac{f_1(T_i)}{f_1(T_i) + f_2(T_i)}, \qquad (5.19)$$

which gives the probability of failure i belonging to the first, or the substandard subpopulation.

Equation (5.19) results from the fact that the prior probability of the failure at T_i belonging to either the first or to the second subpopulation is 0.5, and the probability of existence of T_i for pdf $f_1(T_i)$ is given by $[f_1(T_i)]\Delta T$, and similarly for pdf $f_2(T_i)$ by $[f_2(T_i)]\Delta T$. Consequently, P_1^i is given by

$$P_1^i = \frac{0.5\,[f_1(T_i)]\Delta T}{0.5\,[f_1(T_i)]\Delta T + 0.5\,[f_2(T_i)]\Delta T},$$

where the $0.5's$ and the $\Delta T's$ cancel out to yield Eq. (5.19). In this equation, $f_1(T_i)$ and $f_2(T_i)$ are the $pdf's$ of the two subpopulations, respectively, given that a failure has occurred at time T_i. For the Weibull case they are

$$f_1(T_i) = \frac{\hat{\beta}_1}{\hat{\eta}_1}(T_i/\hat{\eta}_1)^{\hat{\beta}_1 - 1} exp[-(T_i/\hat{\eta}_1)^{\hat{\beta}_1}],$$

and

$$f_2(T_i) = \frac{\hat{\beta}_2}{\hat{\eta}_2}(T_i/\hat{\eta}_2)^{\hat{\beta}_2 - 1} exp[-(T_i/\hat{\eta}_2)^{\hat{\beta}_2}].$$

PARAMETER ESTIMATION FOR THE BIMODAL CASE   127

Then, the proportion of the substandard failures can be computed from

$$\hat{p} = \frac{\sum_{i=1}^{r} P_1{}^i}{N}. \tag{5.20}$$

Having done this, it can be decided how many and which failures belong to the first and to the second subpopulation. Then these two distributions can be plotted separately and their parameters can be found. However, it should be noticed that when plotting the second distribution, the time at truncation, or the time of the last weak failure, should be subtracted from the times-to-failure data if it is wished to establish that the standard population follows an exponential distribution. This is explained and proved in the next section.

EXAMPLE 5-2

The failure data are the same as those given in Example 5-1. Determine the parameters of the mixed-Weibull distribution using the Bayesian approach, taking the results obtained by Jensen's graphical method in Example 5-1 as the initial estimates.

SOLUTION TO EXAMPLE 5-2

The parameter estimates obtained in Example 5-1 by Jensen's graphical method were $\hat{p} = 6.7\%$, $\hat{\beta}_1 = 1.29$, $\hat{\eta}_1 = 500$ hr, $\hat{\beta}_2 = 5.7$, $\hat{\eta}_2 = 7,800$ hr. Therefore, the $pdf's$ of the weak and strong subpopulations, at any time T_i, are

$$f_1(T_i) = \frac{\hat{\beta}_1}{\hat{\eta}_1}(T_i/\hat{\eta}_1)^{\hat{\beta}_1-1} exp[-(T_i/\hat{\eta}_1)^{\hat{\beta}_1}],$$
$$= \frac{1.29}{500}(T_i/500)^{(1.29-1)} exp[-(T_i/500)^{1.29}],$$

or

$$f_1(T_i) = 0.00258\,(T_i/500)^{0.29} exp[-(T_i/500)^{1.29}],$$

and

$$f_2(T_i) = \frac{\hat{\beta}_2}{\hat{\eta}_2}(T_i/\hat{\eta}_2)^{\hat{\beta}_2-1} exp[-(T_i/\hat{\eta}_2)^{\hat{\beta}_2}],$$
$$= \frac{5.7}{7,800}(T_i/7,800)^{(5.7-1)} exp[-(T_i/7,800)^{5.7}],$$

or

$$f_2(T_i) = 0.00073\,(T_i/7,800)^{4.7}\,exp[-(T_i/7,800)^{5.7}],$$

respectively. Substitution of $f_1(T_i)$ and $f_2(T_i)$ into Eq. (5.19) yields the probabilities of all failure times belonging to the weak subpopulation, as given in Table 5.3. For example, for $i = 1$ with its corresponding life of $T_1 = 100$ hr,

$$f_1(T_1) = f_1(100\text{ hr}) = 0.00258\,(100/500)^{0.29}\,exp[-(100/500)^{1.29}],$$

or

$$f_1(100\text{ hr}) = 1.4270986 \times 10^{-3},$$

and

$$\begin{aligned} f_2(T_1) &= f_2(100\text{ hr}), \\ &= 0.00073\,(100/7,800)^{4.7}\,exp[-(100/7,800)^{5.7}], \end{aligned}$$

or

$$f_2(100\text{ hr}) = 9.3526781 \times 10^{-13}.$$

Then,

$$\begin{aligned} P_1^1 &= \frac{f_1(T_1)}{f_1(T_1) + f_2(T_1)}, \\ &= \frac{f_1(100\text{ hr})}{f_1(100\text{ hr}) + f_2(100\text{ hr})}, \\ &= \frac{1.4270986 \times 10^{-3}}{1.4270986 \times 10^{-3} + 9.3526781 \times 10^{-13}}, \end{aligned}$$

or

$$P_1^1 = 1.000000.$$

Similarly, for for $i = 10$ with its corresponding life of $T_{10} = 1,044$ hr,

$$\begin{aligned} f_1(T_{10}) &= f_1(1,044\text{ hr}), \\ &= 0.00258\,(1,044/500)^{0.29}\,exp[-(1,044/500)^{1.29}], \end{aligned}$$

or

$$f_1(1,044\text{ hr}) = 2.4083050 \times 10^{-4},$$

Table 5.3– Times to failure and their probabilities of belonging to the weak subpopulation, for Example 5–2.

Rank, i	Time to failure, T_i, hr	Belonging probability, P_1^i
1	100	1.000000
2	200	1.000000
3	250	1.000000
4	420	0.999999
5	420	0.999999
6	588	0.999995
7	588	0.999995
8	588	0.999995
9	708	0.999984
10	1,044	0.999762
11	2,892	0.039732
12	2,892	0.039732
13	3,396	0.002229
14	3,396	0.002229
15	3,997	0.000069
16	3,997	0.000069
17	3,997	0.000069
18	4,165	0.000026
19	4,500	0.000004

and
$$f_2(T_{10}) = f_2(1,044 \text{ hr}),$$
$$= 0.00073\,(1,044/7,800)^{4.7}\,exp[-(1,044/7,800)^{5.7}],$$
or
$$f_2(1,044 \text{ hr}) = 5.7388316 \times 10^{-8}.$$
Then,
$$P_1^{10} = \frac{f_1(T_{10})}{f_1(T_{10}) + f_2(T_{10})},$$
$$= \frac{f_1(1,044 \text{ hr})}{f_1(1,044 \text{ hr}) + f_2(1,044 \text{ hr})},$$
$$= \frac{2.4083050 \times 10^{-4}}{2.4083050 \times 10^{-4} + 5.7388316 \times 10^{-8}},$$
or
$$P_1^{10} = 0.999762.$$
Similarly, for $i = 11$ with its corresponding life of $T_{11} = 2,892$ hr,
$$f_1(T_{11}) = f_1(2,892 \text{ hr}),$$
$$= 0.00258\,(2,892/500)^{0.29}\,exp[-(2,892/500)^{1.29}],$$
or
$$f_1(2,892 \text{ hr}) = 2.8431296 \times 10^{-7},$$
and
$$f_2(T_{11}) = f_2(2,892 \text{ hr}),$$
$$= 0.00073\,(2,892/7,800)^{4.7}\,exp[-(2,892/7,800)^{5.7}],$$
or
$$f_2(2,892 \text{ hr}) = 6.8714476 \times 10^{-6}.$$
Then,
$$P_1^{11} = \frac{f_1(T_{11})}{f_1(T_{11}) + f_2(T_{11})},$$
$$= \frac{f_1(2,892 \text{ hr})}{f_1(2,892 \text{ hr}) + f_2(2,892 \text{ hr})},$$
$$= \frac{2.8431296 \times 10^{-7}}{2.8431296 \times 10^{-7} + 6.8714476 \times 10^{-6}},$$

or
$$P_1^{11} = 0.039732.$$

The estimate of the posterior proportion of the weak subpopulation, $\hat{p}$, can be calculated by substituting the P_1^i given in Table 5.3 into Eq. (5.20) with sample size $N = 150$. The result is

$$\hat{p} = 6.7226\%,$$

which turns out to be almost the same as that obtained by Jensen's graphical method.

From this it can be decided how many, and which failures, belong to the first and to the second subpopulation. Since

$$N \times \hat{p} = 150 \times 0.067226 = 10.0839,$$

the first 10 failures belong to the first (weak) subpopulation and the other 9 remaining failures plus $150-19 = 131$ survivals, or a total of 140 components, belong to the second (strong) subpopulation. This can also be demonstrated by checking the probability value of belonging to the first subpopulation of each failure in Table 5.3. It may be seen that the first 10 failures have very high *belonging probabilities* (almost equal to 1.0), while the remaining 9 failures have very low *belonging probabilities*. So, the conclusion can be drawn very easily, in support of the previous findings, that the first 10 failures do belong to the first subpopulation.

Now the times-to-failure data can be separated into two parts, with the first 10 failures of sample size 10 belonging to the first, and with the remaining 9 failures of sample size $150 - 10 = 140$ belonging to the second subpopulation. Then, the median ranks for each sample are recalculated, as given in Tables 5.4 and 5.5, respectively.

Plotting the T_j versus the corresponding MR_j values in Tables 5.4 and 5.5 on two appropriate WPP's yields two straight lines, as shown in Figs. 5.3 and 5.4, respectively. The shape and scale parameters for the two subpopulations are read off by the standard Weibull plotting technique, and found to be

$$\hat{\beta}_1 = 1.62,$$
$$\hat{\eta}_1 = 535 \text{ hr},$$
$$\hat{\beta}_2 = 4.3,$$

and

$$\hat{\eta}_2 = 8,250 \text{ hr}.$$

Table 5.4– Failure data and $MR's$ (after grouping) for the weak sample with $r = 10$ and $N_1 = 10$, for Example 5–2.

Rank, j	Time to failure, T_j, hr	Median rank, $MR_j, \%$
1	100	6.73
2	200	16.35
3	250	25.96
5	420	45.19
8	588	74.04
9	708	83.65
10	1,044	93.27

Table 5.5– Failure data and $MR's$ (after grouping) for the strong sample with $r = 9$ and $N_2 = 140$, for Example 5–2.

Rank, j	Time to failure, T_j, hr	Median rank, $MR_j, \%$
2	2,892	1.21
4	3,396	2.64
7	3,997	4.77
8	4,165	5.48
9	4,650	6.20

PARAMETER ESTIMATION FOR THE BIMODAL CASE

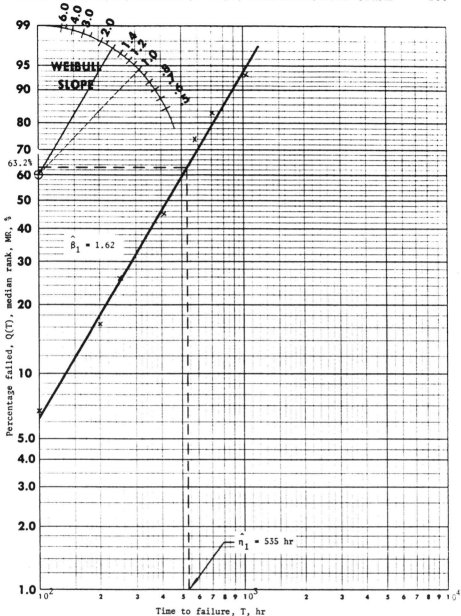

Fig. 5.3– Plot of the data in Table 5.4 on WPP for the weak subpopulation of Example 5-2.

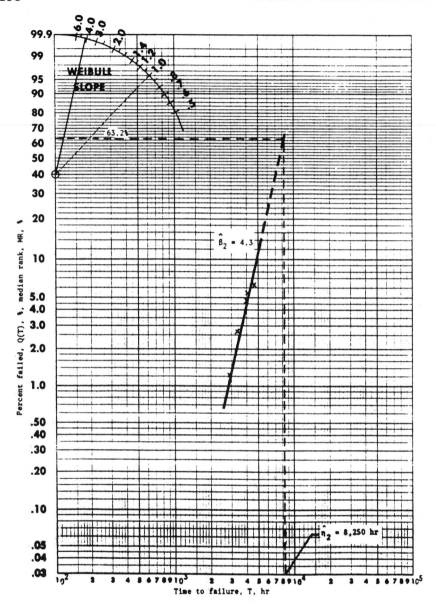

Fig. 5.4– Plot of the data in Table 5.5 on WPP for the strong subpopulation of Example 5–2.

Table 5.6– Parameter estimates obtained using Jensen's graphical method and the Bayesian approach, for Example 5–2.

Parameter	$\hat{p}$, %	$\hat{\beta}_1$	$\hat{\eta}_1$, hr	$\hat{\beta}_2$	$\hat{\eta}_2$, hr
Jensen's estimates	6.7000	1.29	500	5.7	7,800
Bayesian estimates	6.7226	1.62	535	4.3	8,250

A comparison is made in Table 5.6 between the parameter values obtained by Jensen's graphical method and those obtained by the Bayesian approach. It may be seen that there are significant differences in the η and β values. The Bayesian method is recommended because it is theoretically accurate.

5.4.2.2 BAYESIAN METHOD FOR THE GROUPED DATA CASE

Assume that the times to failure follow a bimodal, cumulative Weibull distribution, given by

$$F(T) = p\left[1 - e^{-(T/\eta_1)^{\beta_1}}\right] + (1-p)\left[1 - e^{-(T/\eta_2)^{\beta_2}}\right],$$

and that a test has been carried out on N items with $n_1, n_2, ..., n_i, ..., n_k$ failures in the time intervals (T_0, T_1), (T_1, T_2), ..., (T_{i-1}, T_i), ..., (T_{k-1}, T_k), respectively. Having plotted the failure data, the first estimates of p, β_1, η_1, β_2, and η_2 can be obtained by Jensen's graphical method as $\hat{p}^0$, $\hat{\beta}_1^0$, $\hat{\eta}_1^0$, $\hat{\beta}_2^0$, and $\hat{\eta}_2^0$. Now Bayes theorem can be applied to find the posterior estimates $\hat{p}$, $\hat{\beta}_1$, $\hat{\eta}_1$, $\hat{\beta}_2$, and $\hat{\eta}_2$.

Take the ith time interval (T_{i-1}, T_i) as an example. There are n_i failures occurring in this interval. It is desired to find out how many of them belong to the first subpopulation. There are $(n_i + 1)$ possibilities: 0 (no failures belong to the first subpopulation, and all n_i failures belong to the second subpopulation), 1 (only one failure belongs to the first subpopulation, and $(n_i - 1)$ failures belong to the second subpopulation), ..., n_i (all n_i failures belong to the first subpopulation, and no failures belong to the second subpopulation). The prior probability of each possibility can be taken as

$$P_i(l) = \frac{1}{n_i + 1}, \quad l = 0, 1, 2, ..., n_i, \quad i = 1, 2, ..., k, \tag{5.21}$$

assuming that each case is equally probable. The conditional probability that n_i failures occur in the interval (T_{i-1}, T_i), given that l failures belong to the first subpopulation, is

$$P_i(n_i|l) = C_{n_i}^l \; Q_1^l(\Delta T_i) \; Q_2^{n_i-l}(\Delta T_i),$$
$$l = 0, 1, 2, ..., n_i, \quad i = 1, 2, ..., k, \tag{5.22}$$

where

$$Q_1(\Delta T_i) = Q_1(T_i) - Q_1(T_{i-1}),$$
$$= e^{-(T_{i-1}/\hat{\eta}_1^0)^{\hat{\beta}_1^0}} - e^{-(T_i/\hat{\eta}_1^0)^{\hat{\beta}_1^0}},$$
$$Q_2(\Delta T_i) = Q_2(T_i) - Q_2(T_{i-1}),$$
$$= e^{-(T_{i-1}/\hat{\eta}_2^0)^{\hat{\beta}_2^0}} - e^{-(T_i/\hat{\eta}_2^0)^{\hat{\beta}_2^0}},$$

and

$$C_{n_i}^l = \frac{n_i!}{l! \, (n_i - l)!}, \quad l = 0, 1, 2, ..., n_i.$$

Therefore, the probability that there are j failures belonging to the first subpopulation, given that n_i failures have occurred in the time interval (T_{i-1}, T_i), is

$$P_i(j|n_i) = \frac{P_i(j) \; P_i(n_i|j)}{\sum_{l=0}^{n_i} P_i(l) \; P_i(n_i|l)},$$

$$= \frac{(\frac{1}{1+n_i}) \; C_{n_i}^j \; Q_1^j(\Delta T_i) \; Q_2^{n_i-j}(\Delta T_i)}{\sum_{l=0}^{n_i}(\frac{1}{1+n_i}) C_{n_i}^l \; Q_1^l(\Delta T_i) \; Q_2^{n_i-l}(\Delta T_i)},$$

$$= \frac{(\frac{1}{1+n_i}) \; C_{n_i}^j \; Q_1^j(\Delta T_i) \; Q_2^{n_i-j}(\Delta T_i)}{(\frac{1}{1+n_i}) \sum_{l=0}^{n_i} C_{n_i}^l \; Q_1^l(\Delta T_i) \; Q_2^{n_i-l}(\Delta T_i)},$$

PARAMETER ESTIMATION FOR THE BIMODAL CASE 137

or

$$P_i(j|n_i) = \frac{C_{n_i}^j \, Q_1^j(\Delta T_i) \, Q_2^{n_i-j}(\Delta T_i)}{\sum_{l=0}^{n_i} C_{n_i}^l \, Q_1^l(\Delta T_i) \, Q_2^{n_i-l}(\Delta T_i)},$$
$$j = 0, 1, 2, ..., n_i, \quad i = 1, 2, ..., k. \qquad (5.23)$$

Consequently, the expected number of failures belonging to the first subpopulation in the time interval (T_{i-1}, T_i) is

$$n_{i1} = \sum_{j=0}^{n_i} j \, P_i(j|n_i), \quad i = 1, 2, ..., k. \qquad (5.24)$$

The posterior estimate of p will then be

$$\hat{p} = \frac{\sum_{i=1}^{k} n_{i1}}{N}. \qquad (5.25)$$

Incidentally when $\Delta T_i \to 0$, or $T_{i-1} \to T_i$ and $n_i \to 1$, the ungrouped data case results. Then, the prior probability is

$$P_i(l) = \frac{1}{1+1} = 0.5, \quad l = 0, 1, \quad i = 1, 2, ..., k,$$

assuming that each case is equally probable. The conditional probability that this (one) failure occurs at time T_i, given that it does or does not belong to the first subpopulation, is

$$P_i(1|l) = \lim_{\Delta T_i \to 0} C_1^l \, Q_1^l(\Delta T_i) \, Q_2^{n_i-l}(\Delta T_i), \quad l = 0, 1, \quad i = 1, 2, ..., k,$$

or

$$P_i(1|l) = \lim_{\Delta T_i \to 0} C_1^l \, [f_1(T_i)\Delta T_i]^l [f_2(T_i)\Delta T_i]^{1-l},$$
$$= \begin{cases} \lim_{\Delta T_i \to 0} C_1^0 \, [f_1(T_i)\Delta T_i]^0 [f_2(T_i)\Delta T_i]^{1-0}, & \text{if } l = 0, \\ \lim_{\Delta T_i \to 0} C_1^1 \, [f_1(T_i)\Delta T_i]^1 [f_2(T_i)\Delta T_i]^{1-1}, & \text{if } l = 1, \end{cases}$$

or

$$P_i(1|l) = \begin{cases} \lim_{\Delta T_i \to 0} [f_2(T_i)\Delta T_i], & l = 0, \\ \lim_{\Delta T_i \to 0} [f_1(T_i)\Delta T_i], & l = 1. \end{cases}$$

Therefore, the posterior probability that the failure does or does not belong to the first subpopulation, given that one failure has occurred at T_i, is

$$P_i(j|1) = \frac{P_i(j)\,P_i(1|j)}{\sum_{l=0}^{1} P_i(l)\,P_i(1|l)},$$

$$= \begin{cases} \lim_{\Delta T_i \to 0} \dfrac{0.5[f_2(T_i)\Delta T_i]}{0.5[f_2(T_i)\Delta T_i]+0.5[f_1(T_i)\Delta T_i]}, & j = 0, \\ \lim_{\Delta T_i \to 0} \dfrac{0.5[f_1(T_i)\Delta T_i]}{0.5[f_2(T_i)\Delta T_i]+0.5[f_1(T_i)\Delta T_i]}, & j = 1, \end{cases}$$

or

$$P_i(j|1) = \begin{cases} \dfrac{f_2(T_i)}{f_2(T_i)+f_1(T_i)}, & j = 0, \\ \dfrac{f_1(T_i)}{f_2(T_i)+f_1(T_i)}, & j = 1. \end{cases}$$

The expected "number of failures" belonging to the first subpopulation at T_i is

$$\begin{aligned} n_{i1} &= \sum_{j=0}^{1} j\,P(j|1), \\ &= 0 \times P(0|1) + 1 \times P(1|1), \\ &= P(1|1), \end{aligned}$$

or

$$n_{i1} = \frac{f_1(T_i)}{f_2(T_i)+f_1(T_i)},$$

which turns out to be exactly the same as Eq. (5.19).

Similarly the posterior estimate of p is

$$\hat{p} = \frac{\sum_{i=1}^{k} \dfrac{f_1(T_i)}{f_2(T_i)+f_1(T_i)}}{N},$$

PARAMETER ESTIMATION FOR THE BIMODAL CASE

which is also the same as Eq. (5.20). Obviously ungrouped data is the special case of grouped data when $\Delta T_i \to 0$ and $n_i \to 1$ for $i = 1, 2, ..., k$.

Having found the posterior estimate of p, or $\hat{p}$, the sample sizes both for the weak and the strong subpopulations can be decided; i.e., $N_1 = \hat{p}\, N$ and $N_2 = N - N_1$. Instead of determining the parameters by splitting the original mixed sample into two separate subsamples at the time corresponding to the first N_1 failures, as done before, we will use the expected number of failures belonging to the first and to the second subpopulations in each time interval to get the corresponding failure ranks at the end of each time interval for both subsamples, respectively. These failure ranks are not integer now, but fractional, as will be seen in the next example. The median ranks can be calculated using these fractional ranks and be plotted versus the corresponding interval end-time on WPP to find the Weibull parameters for each subpopulation. We call this plotting approach the *fractional rank plotting method*. This plotting method is more reasonable and more logical than the conventional separation method because of the following:

1. The *separation plotting method* assumes that before the separation point there are no chance failures, and after the separation point there are no early failures. This is not true both in theory and in practice. Theoretically,

$$f_2(t) \neq 0, \quad \text{for any } 0 < t < T_{\text{separation}} > 0, \text{ if } \gamma_1 = 0,$$

and

$$f_1(t) \neq 0, \quad \text{for any } T_{\text{separation}} < t < \infty, \text{ if } \gamma_2 = 0.$$

Practically, the burn-in process can not eliminate all, or 100%, of the early failures before the separation time, and the chance failures can occur during the early failure period.

2. The *fractional rank plotting method* makes full use of the information on the behavior of the distribution of the two subpopulations over the whole test duration, which is exhibited by the raw test data of the mixed sample. But, the separation plotting method misses the strong distribution information of the early failure period and the weak distribution information of the chance failure period, by splitting the original sample into two independent subsamples.

Therefore, the *fractional rank plotting method* will yield more accurate parameter estimates, which are closer to the true values of the

Table 5.7– Failure data from life testing 90 units, for Example 5–3.

Time interval number, i	Time interval, $(T_{i-1}; T_i)$, hr	Failure number, n_i
1	(0; 600)	23
2	(600; 1,200)	9
3	(1,200; 1,800)	2
4	(1,800; 2,400)	3
5	(2,400; 3,000)	11
6	(3,000; 3,600)	6
7	(3,600; 4,200)	3
8	(4,200; 4,800)	6
9	(4,800; 5,400)	4
10	(5,400; 6,000)	3
11	(6,000; 6,600)	4

mixed population. An example is given next using the Bayesian approach for grouped data and the fractional rank plotting method presented above. The Kolmogorov-Smirnov (K-S) goodness-of-fit test is also applied to verify the accuracy of this method.

EXAMPLE 5–3

Given the life test data of Table 5.7, determine the mixed population's parameters using the Bayesian approach with both the *fractional rank plotting* and the *separation plotting methods*. Apply Jensen's graphical method to get the initial parameter estimates. Conduct the K-S goodness-of-fit test to compare the results of these three methods.

SOLUTION TO EXAMPLE 5–3

The median ranks at the end of each time interval, for the data in Table 5.7, are calculated and summarized in Table 5.8. For example, at the end of 5th time interval, or at $T_5 = 3,000$ hr, the cumulative number of failures is $r_5 = 48$. Then, the corresponding median rank is

$$MR_5 = \frac{r_5 - 0.3}{N + 0.4},$$
$$= \frac{48 - 0.3}{90 + 0.4},$$

PARAMETER ESTIMATION FOR THE BIMODAL CASE

Table 5.8– Failure data and $MR's$ at the end of each time interval in Table 5.7, for Example 5-3.

Time interval, i	Time at the end of each interval, T_i, hr	Cumulative failures, r_i	Median rank, MR_i, %
1	600	23	25.1
2	1,200	32	35.1
3	1,800	34	37.3
4	2,400	37	40.6
5	3,000	48	52.8
6	3,600	54	59.4
7	4,200	57	62.7
8	4,800	63	69.4
9	5,400	67	73.8
10	6,000	70	77.1
11	6,600	74	81.5

or

$$MR_5 = 0.528, \text{ or } 52.8\%.$$

After plotting the $T'_j s$ versus their $MR'_j s$ on WPP, an "S" shaped curve emerges, as shown in Fig. 5.5. Obviously this is a bimodal Weibull mixture. To start the Bayesian estimation, Jensen's graphical method is used to get the approximate, initial estimates of $\hat{p}^0$, $\hat{\beta}_1^0$, $\hat{\eta}_1^0$, $\hat{\beta}_2^0$ and $\hat{\eta}_2^0$.

The $\hat{p}^0$ value may be read off the MR at the "flat" portion level of the cdf curve, which is

$$\hat{p}^0 = 37\%.$$

The other parameters can be read off Fig. 5.5, in the way shown in Example 5-1, yielding

$$\hat{\beta}_1^0 = 0.8,$$
$$\hat{\eta}_1^0 = 530 \text{ hr},$$
$$\hat{\beta}_2^0 = 1.35,$$

and

$$\hat{\eta}_2^0 = 5,500 \text{ hr}.$$

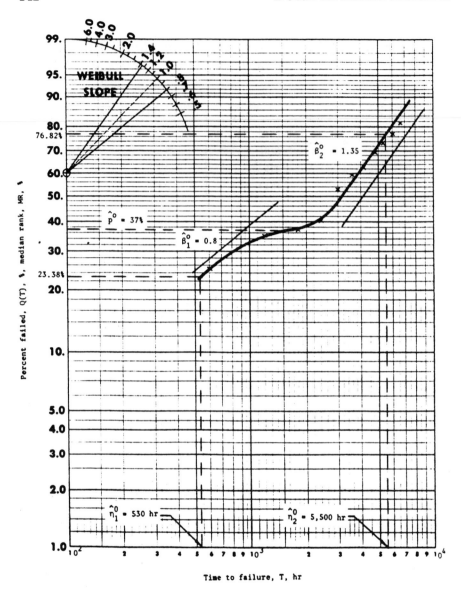

Fig. 5.5– Plot of the data in Table 5.8 on WPP, for initial parameter estimation, using Jensen's graphical method, for Example 5–3.

PARAMETER ESTIMATION FOR THE BIMODAL CASE

Therefore, the *pdf's* of the weak and strong subpopulations at any time T_i are

$$f_1(T_i) = \frac{\hat{\beta}_1^0}{\hat{\eta}_1^0}(T_i/\hat{\eta}_1^0)^{\hat{\beta}_1^0-1} exp[-(T_i/\hat{\eta}_1^0)^{\hat{\beta}_1^0}],$$

$$= \frac{0.8}{530}(T_i/530)^{(0.8-1)} exp[-(T_i/530)^{0.8}],$$

or

$$f_1(T_i) = 0.00151(T_i/530)^{-0.2} exp[-(T_i/530)^{0.8}],$$

and

$$f_2(T_i) = \frac{\hat{\beta}_2^0}{\hat{\eta}_2^0}(T_i/\hat{\eta}_2^0)^{\hat{\beta}_2^0-1} exp[-(T_i/\hat{\eta}_2^0)^{\hat{\beta}_2^0}],$$

$$= \frac{1.35}{5,500}(T_i/5,500)^{(1.35-1)} exp[-(T_i/5,500)^{1.35}],$$

or

$$f_2(T_i) = 0.000245(T_i/5,500)^{0.35} exp[-(T_i/5,500)^{1.35}].$$

Substitution of the initial parameter estimates obtained above into Eqs. (5.22), (5.23) and then into Eq. (5.24), yields the expected failure number belonging to the weak subpopulation in each time interval, as given in Table 5.9. For example, the total number of failures in the 3rd time interval $(T_2, T_3)=(1,200 \text{ hr}, 1,800 \text{ hr})$ is $n_3 = 2$ and $Q_1(\Delta T_3)$ and $Q_2(\Delta T_3)$ in Eqs. (5.22) and (5.23) are calculated as follows:

$$Q_1(\Delta T_3) = Q_1(T_3) - Q_1(T_2),$$
$$= Q_1(1,800) - Q_1(1,200),$$
$$= \left[1 - e^{-(\frac{1,800}{530})^{0.8}}\right] - \left[1 - e^{-(\frac{1,200}{530})^{0.8}}\right],$$
$$= e^{-(\frac{1,200}{530})^{0.8}} - e^{-(\frac{1,800}{530})^{0.8}},$$

or

$$Q_1(\Delta T_3) = 0.076219,$$

and similarly

$$Q_2(\Delta T_3) = Q_2(T_3) - Q_2(T_2),$$
$$= Q_2(1,800) - Q_2(1,200),$$
$$= \left[1 - e^{-(\frac{1,800}{5,500})^{1.35}}\right] - \left[1 - e^{-(\frac{1,200}{5,500})^{1.35}}\right],$$
$$= e^{-(\frac{1,200}{5,500})^{1.35}} - e^{-(\frac{1,800}{5,500})^{1.35}},$$

or

$$Q_2(\Delta T_3) = 0.078387.$$

Then, from Eq. (5.22),

$$P_3(n_3|j) = C_{n_3}^j \, Q_1^j(\Delta T_3) \, Q_2^{n_3-j}(\Delta T_3), \text{ for } j = 0, 1, 2, ..., n_3,$$

or

$$P_3(n_3 = 2|j) = C_2^j \, Q_1^j(\Delta T_3) \, Q_2^{2-j}(\Delta T_3), \text{ for } j = 0, 1, 2.$$

Substitution of $Q_1(\Delta T_3) = 0.076219$ and $Q_2(\Delta T_3) = 0.078387$ into this equation yields

$$P_3(n_3 = 2|j = 0) = C_2^0 \, (0.076219)^0 \, (0.078387)^{2-0} = 0.006147,$$

$$P_3(n_3 = 2|j = 1) = C_2^1 \, (0.076219)^1 \, (0.078387)^{2-1} = 0.011949,$$

$$P_3(n_3 = 2|j = 2) = C_2^2 \, (0.076219)^2 \, (0.078387)^{2-2} = 0.005809,$$

and

$$\sum_{j=0}^{2} P_3(n_3 = 2|j) = 0.006147 + 0.011949 + 0.005809 = 0.023903.$$

Substituting these values into Eq. (5.23) yields

$$P_3(j = 0|n_3 = 2) = \frac{P_3(n_3 = 2|j = 0)}{\sum_{j=0}^{2} P_3(n_3 = 2|j)},$$

$$= \frac{0.006147}{0.023903},$$

or

$$P_3(j = 0|n_3 = 2) = 0.257061,$$

$$P_3(j = 1|n_3 = 2) = \frac{P_3(n_3 = 2|j = 1)}{\sum_{j=0}^{2} P_3(n_3 = 2|j)},$$

$$= \frac{0.011949}{0.023903},$$

or

$$P_3(j = 1 | n_3 = 2) = 0.499902,$$

and

$$P_3(j = 2 | n_3 = 2) = \frac{P_3(n_3 = 2 | j = 2)}{\sum_{j=0}^{2} P_3(n_3 = 2 | j)},$$

$$= \frac{0.005809}{0.023903},$$

or

$$P_3(j = 2 | n_3 = 2) = 0.243038.$$

Finally, from Eq. (5.24), the expected number of failures belonging to the first (weak) subpopulation in (T_2, T_3) is

$$n_{31} = \sum_{j=0}^{n_3} j \, P_3(j | n_3),$$

$$= \sum_{j=0}^{2} j \, P_3(j | n_3 = 2),$$

$$= 0 \times P_3(j = 0 | n_3 = 2) + 1 \times P_3(j = 1 | n_3 = 2)$$

$$+ 2 \times P_3(j = 2 | n_3 = 2),$$

$$= 0 + 0.49902 + 2 \times 0.243038,$$

or

$$n_{31} = 0.98598.$$

The estimate of the posterior proportion of the weak subpopulation, $\hat{p}$, can be calculated by substituting the n_{i1} given in Table 5.9 into Eq. (5.25) with sample size $N = 90$. The result is

$$\hat{p} = 36.6124\%,$$

which turns out to be almost the same as that obtained by Jensen's graphical method.

From this, the sample sizes for the weak subsample and the strong subsample can be determined, respectively; which are,

$$N_1 = N \times \hat{p} = 90 \times 0.366124 = 32.95 \cong 33,$$

Table 5.9– Time intervals and the corresponding expected failure number belonging to the weak subpopulation, for Example 5–3.

Time interval, i	Time interval, $(T_{i-1}; T_i)$, hr	Failure number, n_i	Expected failure number belonging to the weak subpopulation, n_{i1}
1	(0; 600)	23	21.42957
2	(600; 1,200)	9	6.50100
3	(1,200; 1,800)	2	0.98598
4	(1,800; 2,400)	3	0.91037
5	(2,400; 3,000)	11	1.95299
6	(3,000; 3,600)	6	0.61617
7	(3,600; 4,200)	3	0.17991
8	(4,200; 4,800)	6	0.21403
9	(4,800; 5,400)	4	0.08670
10	(5,400; 6,000)	3	0.04036
11	(6,000; 6,600)	4	0.03409

PARAMETER ESTIMATION FOR THE BIMODAL CASE

and

$$N_2 = N - N_1 = 90 - 33 = 57.$$

Now the fractional ranks and the corresponding median ranks at the time interval ends, for both subpopulations, can be calculated using the expected failure numbers in Table 5.9, as given in Tables 5.10 and 5.11. Plotting the T_i versus the corresponding MR_{1i} and MR_{2i} values in Tables 5.10 and 5.11 on the WPP yields two straight lines as shown in Fig. 5.6. Then, the shape and scale parameters for the two subpopulations are read off by the standard Weibull plotting technique, and are found to be

$$\hat{\beta}_1 = 0.57,$$
$$\hat{\eta}_1 = 530 \text{ hr},$$
$$\hat{\beta}_2 = 1.82,$$

and

$$\hat{\eta}_2 = 5,900 \text{ hr}.$$

If the conventional separation plotting method is used, the whole sample can be divided at $T_3 = 1,800$ hr into two subsamples according to Fig. 5.5. A visual observation of Fig. 5.5 indicates that this separation point, which corresponds to the time interval (1,200 hr, 1,800 hr), may belong to either Subpopulation 1 or Subpopulation 2. Since there are two failures in this interval, it is decided that one of them is assigned to Subpopulation 1, which fails before the middle of the interval, or during (1,200 hr, 1,500 hr), and another one of the two is assigned to Subpopulation 2, which fails by the end of the interval, or by 1,800 hr. Therefore, two subsamples emerge with $N_1 = n_1 + n_2 + n_3/2 = 23+9+2/2=33$ and $N_2 = N - N_1 = 90-33=57$. Tables 5.12 and 5.13 list separately the calculated median ranks for these two subsamples. For example, at the end of the second interval of Subsample 1, $T_2 = 1,200$ hr, the cumulative number of failures is $r_2 = 32$. Then, the corresponding median rank is

$$MR_2 = \frac{r_2 - 0.3}{N_1 + 0.4},$$
$$= \frac{32 - 0.3}{33 + 0.4},$$

or

$$MR_2 = 94.91\%.$$

Table 5.10– Failure data, ranks and $MR's$ for the weak subsample with $N_1 = 33$, using the fractional rank method, for Example 5–3.

Time interval, i	Time at end of interval, T_i, hr	Cumulative failures, r_i	Weak failure rank, r_{1i}	Weak median rank, MR_{1i}, %
1	600	23	21.42957	63.26
2	1,200	32	27.93057	82.73
3	1,800	34	28.91655	85.68
4	2,400	37	29.82692	88.40
5	3,000	48	31.77991	94.25
6	3,600	54	32.39608	96.10
7	4,200	57	32.57599	96.63
8	4,800	63	32.79002	97.28
9	5,400	67	32.87672	97.54
10	6,000	70	32.91708	97.66
11	6,600	74	32.95117	97.76

As another example, at the end of the sixth interval of Subsample 2, $T_6 = 4,800$ hr, the cumulative number of failures is $r_6 = 30$. Then, the corresponding median rank is

$$MR_6 = \frac{r_6 - 0.3}{N_2 + 0.4},$$
$$= \frac{30 - 0.3}{57 + 0.4},$$

or

$$MR_6 = 51.74\%.$$

Plotting the results in Tables 5.12 and 5.13 on WPP, respectively, as shown in Fig. 5.7, yields the following parameter estimates:

$$\hat{\beta}_1 = 1.36,$$
$$\hat{\eta}_1 = 550 \text{ hr},$$
$$\hat{\beta}_2 = 4.0,$$

and

$$\hat{\eta}_2 = 5,000 \text{ hr}.$$

Table 5.11 – Failure data, ranks and $MR's$ for the strong subsample with $N_2 = 57$, using the fractional rank method, for Example 5-3.

Time interval, i	Time at end of interval, T_i, hr	Cumulative failures, r_i	Strong failure rank, r_{2i}	Strong median rank, MR_{2i}
1	600	23	1.57043	2.21
2	1,200	32	4.06943	6.57
3	1,800	34	5.08345	8.33
4	2,400	37	7.17308	11.97
5	3,000	48	16.22009	27.74
6	3,600	54	21.60392	37.11
7	4,200	57	24.42401	42.03
8	4,800	63	30.20998	52.11
9	5,400	67	34.12328	58.93
10	6,000	70	37.08292	64.08
11	6,600	74	41.04883	70.99

Table 5.12 – Failure data and $MR's$ for the weak subsample with $N_1 = 33$, using the separation method, for Example 5-3.

Time interval, i	Time at end of interval, T_i, hr	Cumulative failures, r_i	Median rank, MR_i, %
1	600	23	67.96
2	1,200	32	94.91
3	1,500	33	97.90

Table 5.13– Failure data and $MR's$ for the strong subsample with $N_2 = 57$, using the separation method, for Example 5–3.

Time interval, i	Time at end of interval, T_i, hr	Cumulative failures, r_i	Median rank, $MR_i, \%$
1	1,800	1	1.22
2	2,400	4	6.45
3	3,000	15	25.61
4	3,600	21	36.06
5	4,200	24	41.29
6	4,800	30	51.74
7	5,400	34	58.71
8	6,000	37	63.94
9	6,600	41	70.91

Table 5.14– Kolmogorov-Smirnov goodness-of-fit test on the parameter estimates, using Jensen's graphical method, for Example 5–3.

Time interval, i	Time at end of interval, T_i, hr	Cumulative failures, r_i	$Q_O(T_i)$*	$Q_E(T_i)$**	D***
1	600	23	0.255556	0.278237	0.022681
2	1,200	32	0.355556	0.391629	0.036073
3	1,800	34	0.377778	0.469214	0.091436
4	2,400	37	0.411111	0.532449	0.121338
5	3,000	48	0.533333	0.587976	0.054642
6	3,600	54	0.600000	0.638079	0.038079
7	4,200	57	0.633333	0.683574	0.050241
8	4,800	63	0.700000	0.724782	0.024782
9	5,400	67	0.744444	0.761883	0.017439
10	6,000	70	0.777778	0.795047	0.017269
11	6,600	74	0.822222	0.824437	0.002215
$T_{\max} = 2,400$ hr, $D_{\max} = 0.121338$					

* $Q_O(T_i)$ = Observed probability of failure or unreliability,
** $Q_E(T_i)$ = Expected probability of failure or unreliability,
$Q_E(T_i) = \hat{p}Q_1(T_i) + (1-\hat{p})Q_2(T_i)$,
*** D = $|Q_O(T_i) - Q_E(T_i)|$.

PARAMETER ESTIMATION FOR THE BIMODAL CASE 151

Table 5.15– Kolmogorov-Smirnov goodness-of-fit test on the parameter estimates, using the Bayesian approach and the separation plotting method, for Example 5-3.

Time interval, i	Time at end of interval, T_i, hr	Cumulative failures, r_i	$Q_O(T_i)$	$Q_E(T_i)$	D
1	600	23	0.255556	0.247466	0.008090
2	1,200	32	0.355556	0.347862	0.007694
3	1,800	34	0.377778	0.374252	0.003526
4	2,400	37	0.411111	0.398675	0.012436
5	3,000	48	0.533333	0.443158	0.090175
6	3,600	54	0.600000	0.515500	0.084500
7	4,200	57	0.633333	0.614715	0.018618
8	4,800	63	0.700000	0.728895	0.028895
9	5,400	67	0.744444	0.837388	0.092944
10	6,000	70	0.777778	0.920301	0.142523
11	6,600	74	0.822222	0.969556	0.147334
	$T_{max} = 6,600$ hr,		$D_{max} = 0.147334$		

Table 5.16– Kolmogorov-Smirnov goodness-of-fit test on the parameter estimates, using the Bayesian approach with fractional rank plotting, for Example 5-3.

Time interval, i	Time at end of interval, T_i, hr	Cumulative failures, r_i	$Q_O(T_i)$	$Q_E(T_i)$	D
1	600	23	0.255556	0.250766	0.004790
2	1,200	32	0.355556	0.325690	0.029866
3	1,800	34	0.377778	0.385950	0.008172
4	2,400	37	0.411111	0.443803	0.032692
5	3,000	48	0.533333	0.501700	0.031633
6	3,600	54	0.600000	0.559443	0.040557
7	4,200	57	0.633333	0.615997	0.017336
8	4,800	63	0.700000	0.670152	0.029848
9	5,400	67	0.744444	0.720817	0.023627
10	6,000	70	0.777778	0.767152	0.010626
11	6,600	74	0.822222	0.808613	0.013609
	$T_{max} = 3,600$ hr,		$D_{max} = 0.040557$		

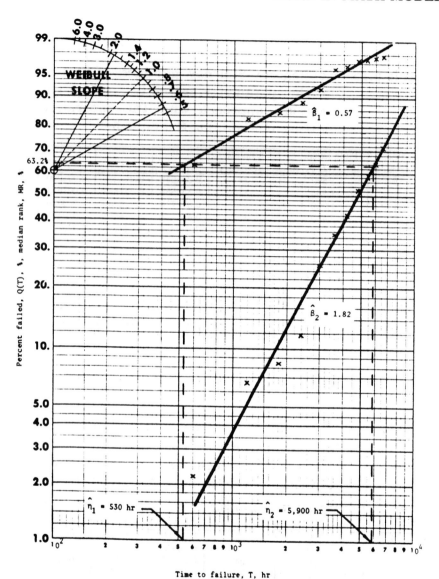

Fig. 5.6– Plot of the data in Tables 5.10 and 5.11 on WPP for the weak and strong subpopulations using the fractional rank plotting method, for Example 5-3.

PARAMETER ESTIMATION FOR THE BIMODAL CASE

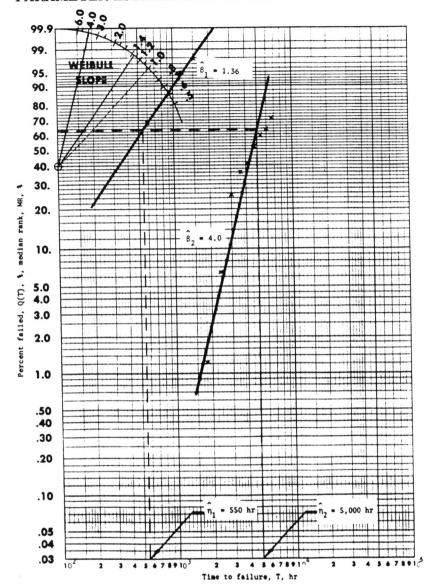

Fig. 5.7– Plot of the data in Tables 5.12 and 5.13 on WPP for the weak and strong subpopulations using the conventional separation plotting method, for Example 5–3.

Table 5.17– Comparison of the parameter estimates obtained using Jensen's graphical method and the Bayesian approach, with the separation plotting method (I) and with the fractional rank plotting method (II), for Example 5–3.

Parameter	$\hat{p}$, %	$\hat{\beta}_1$	$\hat{\eta}_1$, hr	$\hat{\beta}_2$	$\hat{\eta}_2$, hr	D_{max}
Jensen's graphical estimates	37.0000	0.8	530	1.35	5,500	0.121338
Bayesian estimates (I)	36.6124	1.36	550	4.00	5,000	0.147334
Bayesian estimates (II)	36.6124	0.57	530	1.82	5,900	0.040557

A comparison is made by conducting the Kolmogorov-Smirnov goodness-of-fit test on the parameter values obtained by Jensen's graphical method, by the Bayesian approach with the conventional *separation plotting method* and by the Bayesian approach with the *fractional rank plotting method*, as given in Tables 5.14, 5.15, 5.16 and 5.17, respectively. It may be seen that the Bayesian approach with the *fractional rank plotting method* yields the lowest D_{max}, Jensen's graphical method yields the middle D_{max}, and the Bayesian approach with the *separation plotting method* yields the highest D_{max}. The D_{max} values may be compared with the allowable values given in Appendix E of [1]. For $N = 90$ and level of significance, α, of 0.05, the allowable critical D_{max} value is 0.14117. On this basis and assuming that the values in this appendix apply to the two-subpopulation Weibull case, then the Bayesian estimates using the *fractional rank plotting method* and Jensen's graphical method estimates are accepted by the K-S test; however, the Bayesian estimates using the *separation plotting method* are rejected. As the Bayesian estimates using the *fractional rank plotting method* yield the lowest D_{max} value, and the K-S test accepts this method, it may be concluded that this is the method to be recommended. This proves, quantitatively, the following:

1. The Bayesian approach with the *fractional rank plotting method*

PARAMETER ESTIMATION FOR THE BIMODAL CASE

is more accurate than that with the *separation plotting method* and Jensen's graphical method.

2. The *fractional rank plotting method* is more reasonable and more logical than the *separation plotting method.*

3. For engineering purposes Jensen's graphical method provides good enough results for the *well-separated case*. This method does not require too much calculation; therefore, it is very easy to carry out and very quick to get the parameter estimates.

5.4.3 EML ALGORITHM FOR BOTH THE WELL-SEPARATED AND THE WELL-MIXED CASES

If the two subpopulations are not well separated; i.e., they are well-mixed, it will be very difficult to find $\hat{p}$ visually by Jensen's method. In this case, an Estimate and Maximum Likelihood (EML) approach is applicable to get the parameters. This approach can deal with both the well-mixed and the well-separated cases. Here only non-postmortem data for ungrouped and grouped cases are considered. For the postmortem data case see [7; 8]. Note that a postmortem data refers to the case where failure analysis is available after the occurrence of a failure so that its failure mode can be identified accordingly. On the contrary, a non-postmortem data refers to the case where failure analysis is not available after the occurrence of a failure and therefore its failure mode is not known.

5.4.3.1 THE MLE FOR UNGROUPED NON-POSTMORTEM DATA

Without loss of generality only the time-terminated sample is considered. The time to terminate the test, say t_T, is replaced by the time when the prefixed number of failures, say t_r, occur if the sample is failure terminated. Assume n units participate in a reliability life test, and the test is terminated at a fixed time t_T. Then, the data will consist of r times to failure out of a total of n units tested; i.e.,

$$t_1 \leq t_2 \leq \ldots \leq t_r \leq t_T,$$

and of $(n-r)$ survivals at time t_T. If the times to failure are independently and identically distributed (i.i.d.) and drawn from a finite, mixed-Weibull population, then the likelihood function is

$$\mathcal{L}(t; \Psi) = \frac{n!}{(n-r)!} [\prod_{j=1}^{r} f(t_j|\Psi)][R(t_T|\Psi)]^{n-r}, \qquad (5.26)$$

where

$$\Psi = (p_1, p_2, \eta_1, \eta_2, \beta_1, \beta_2),$$
$$f(t_j|\Psi) = p_1 f_1(t_j|\eta_1, \beta_1) + p_2 f_2(t_j|\eta_2, \beta_2),$$
$$R(t_T|\Psi) = p_1 R_1(t_T|\eta_1, \beta_1) + p_2 R_2(t_T|\eta_2, \beta_2),$$
$$f_i(t_j|\eta_i, \beta_i) = \frac{\beta_i}{\eta_i}\left(\frac{t_j}{\eta_i}\right)^{\beta_i-1} e^{-(t_j/\eta_i)^{\beta_i}}, \quad i=1,2,$$
$$R_i(t_T|\eta_i, \beta_i) = e^{-(t_T/\eta_i)^{\beta_i}}, \quad i=1,2,$$

and

$$\beta_1 > 0, \ \beta_2 > 0, \ \eta_1 > 0, \ \eta_2 > 0,$$
$$p_1 > 0, \ p_2 > 0, \ \text{and } p_1 + p_2 = 1.$$

Since the logarithm is a monotonic function, maximizing $\mathcal{L}(t; \Psi)$ is equivalent to maximizing $\log_e[\mathcal{L}(t, \Psi)]$.

The logarithmic likelihood function is

$$L(t; \Psi) = \log_e \frac{n!}{(n-r)!}$$
$$+ \sum_{j=1}^{r} \log_e[f(t_j|\Psi)] + (n-r)\log_e[R(t_T|\Psi)]. \quad (5.27)$$

Taking the derivative of Eq. (5.27) with respect to p_i, η_i and β_i, and setting them equal to zero can not yield analytic solutions. Therefore, a computer program was developed in FORTRAN language to solve for these parameters, and is listed in Appendix 5B.

5.4.3.2 THE MLE FOR GROUPED NON-POSTMORTEM DATA

Let r_j be the number of failures observed in the time interval $[t_{j-1}, t_j)$, $j = 1, 2, ..., k$, $t_0 = 0$, and r_{k+1} be the number of survivals at time t_k. So we have $k+1$ observations in $k+1$ time intervals which are

$$[0, t_1), [t_1, t_2), ..., [t_{k-1}, t_k), [t_k, \infty).$$

For the grouped, non-postmortem data, the likelihood function is

$$\mathcal{L}(t, r, ; \Psi) = \frac{n!}{\prod_{j=1}^{k+1} r_j!} \prod_{j=1}^{k+1} P_j^{r_j}, \quad (5.28)$$

PARAMETER ESTIMATION FOR THE BIMODAL CASE

where

P_j = likelihood for an observation in $[t_{j-1}, t_j)$,
$P_j^{r_j}$ = likelihood for all r_j observations in $[t_{j-1}, t_j)$,

$$P_1 = \int_0^{t_1} f(\tau|\Psi)d\tau = p_1 P_{11} + p_2 P_{12},$$

$$P_j = \int_{t_{j-1}}^{t_j} f(\tau|\Psi)d\tau = p_1 P_{j1} + p_2 P_{j2}, \quad j = 2, 3, ..., k,$$

$$P_{k+1} = \int_{t_k}^{\infty} f(\tau|\Psi)d\tau = p_1 P_{k+1,1} + p_2 P_{k+2,2},$$

$$P_{1i} = \int_0^{t_1} f_i(\tau|\eta_i, \beta_i)d\tau, \quad i = 1, 2,$$

$$P_{ji} = \int_{t_{j-1}}^{t_j} f_i(\tau|\eta_i, \beta_i)d\tau, \quad i = 1, 2,$$

and

$$P_{k+1,i} = \int_{t_k}^{\infty} f_i(\tau|\eta_i, \beta_i)d\tau, \quad i = 1, 2.$$

The logarithmic likelihood function is

$$L(t, r; \Psi) = \log_e\left(\frac{n!}{\prod_{j=1}^{k+1} r_j!}\right) + \sum_{j=1}^{k+1} r_j \log_e\left(\sum_{i=1}^{2} p_i P_{ji}\right). \quad (5.29)$$

Similarly, taking the derivative of Eq. (5.29) with respect to p_i, η_i and β_i, and setting them equal to zero can not yield analytic solutions. A computer program was developed in FORTRAN language to solve for these parameters, and is listed in Appendix 5C.

Three examples, given next, illustrate the use of the EML algorithm to find the parameters for grouped non-postmortem data.

EXAMPLE 5-4

Using the life test data given in Table 5.7, determine the mixed population's parameters using the EML algorithm.

SOLUTION TO EXAMPLE 5-4

After inputting the given failure data into the computer program, and taking the convergence precision, EPS, and the initial parameter estimates as

$$EPS = 5 \times 10^{-4},$$
$$p^0 = 0.400000,$$
$$\eta_1^0 = 10.000000 \text{ hr},$$
$$\beta_1^0 = 0.500000,$$
$$\eta_2^0 = 100.000000 \text{ hr},$$

and

$$\beta_2^0 = 1.000000,$$

the following final EML estimates are obtained after 645 iterations:

$$\hat{p} = 0.307302,$$
$$\hat{\eta}_1 = 37.577539 \text{ hr},$$
$$\hat{\beta}_1 = 0.531744,$$
$$\hat{\eta}_2 = 124.100548 \text{ hr},$$

and

$$\hat{\beta}_2 = 0.675260.$$

A comparison is made in Table 5.18 among the parameters obtained by the EML algorithm, and those by Jensen's graphical method and the Bayesian approach. It may be seen that there are significant differences in the p, γ, η and β values. The EML algorithm and the Bayesian method are recommended because they are theoretically accurate, especially for well-mixed cases.

EXAMPLE 5-5

Using the life test data given in Table 5.19, determine the mixed population's parameters using the EML algorithm.

SOLUTION TO EXAMPLE 5-5

After inputting the given failure data into the computer program, and taking the convergence precision, EPS, and the initial parameter estimates as

PARAMETER ESTIMATION FOR THE BIMODAL CASE 159

Table 5.18– Parameter estimates obtained using Jensen's graphical method, the Bayesian approach and the EML algorithm, for Example 5–4.

Parameter	$\hat{p}$, %	$\hat{\gamma}_1$, hr	$\hat{\beta}_1$	$\hat{\eta}_1$, hr	$\hat{\gamma}_2$, hr	$\hat{\beta}_2$	$\hat{\eta}_2$, hr
Jensen's estimate	61.0000	0.0	0.60	22	0.0	0.73	170
Bayesian estimate	46.1857	-40.0	3.90	13	30.0	0.97	135
EML algorithm	30.7302	0.0	0.53	38	0.0	0.68	124

Table 5.19– Raw failure data from life tests with a sample size of $N = 1,000$, for Example 5–5.

Time interval number, i	Time interval, (T_{i-1}, T_i), hr	Failure number, N_i
1	(0, 10)	101
2	(10, 20)	25
3	(20, 30)	14
4	(30, 40)	10
5	(40, 50)	7
6	(50, 60)	5
7	(60, 70)	4
8	(70, 80)	3
9	(80, 90)	3
10	(90, 100)	2
11	(100, 110)	2
12	(110, 120)	1
13	(120, 130)	1
14	(130, 140)	1
15	(140, 150)	1
16	(150, 160)	1
17	(160, 170)	1

$$EPS = 5 \times 10^{-4},$$
$$p^0 = 0.400000,$$
$$\eta_1^0 = 5.000000 \text{ hr},$$
$$\beta_1^0 = 0.600000,$$
$$\eta_2^0 = 10,000.000000 \text{ hr},$$

and

$$\beta_2^0 = 1.200000,$$

the following final EML estimates are obtained after $1,205$ iterations:

$$\hat{p} = 0.179997,$$
$$\hat{\eta}_1 = 14.661794 \text{ hr},$$
$$\hat{\beta}_1 = 0.543502,$$
$$\hat{\eta}_2 = 154607.840062 \text{ hr},$$

and

$$\hat{\beta}_2 = 0.719392.$$

EXAMPLE 5-6

Using the life test data given in Table 5.20, determine the mixed population's parameters using the EML algorithm.

SOLUTION TO EXAMPLE 5-6

After inputting the given failure data into the computer program, and taking the convergence precision, EPS, and the initial parameter estimates as

$$EPS = 5 \times 10^{-4},$$
$$p^0 = 0.001000,$$
$$\eta_1^0 = 10.000000 \text{ hr},$$
$$\beta_1^0 = 0.500000,$$
$$\eta_2^0 = 100,000.000000 \text{ hr},$$

Table 5.20 – Raw failure data from life test with a sample size of $N = 1,000,000$, for Example 5-6.

Time interval number, i	Time interval, (T_{i-1}, T_i), hr	Failure number, N_i
1	(0, 10)	68
2	(10, 20)	17
3	(20, 30)	12
4	(30, 40)	9
5	(40, 50)	8
6	(50, 60)	7
7	(60, 70)	7
8	(70, 80)	6
9	(80, 90)	6
10	(90, 100)	6
11	(100, 110)	6
12	(110, 120)	5
13	(120, 130)	5
14	(130, 140)	5
15	(140, 150)	5
16	(150, 160)	5
17	(160, 170)	5
18	(170, 180)	5
19	(180, 190)	5
20	(190, 200)	5

and

$$\beta_2^0 = 1.200000,$$

the following final EML estimates are obtained after 87 iterations:

$\hat{p} = 0.000127,$
$\hat{\eta}_1 = 21.589043$ hr,
$\hat{\beta}_1 = 0.431432,$
$\hat{\eta}_2 = 1,077,828.368873$ hr,

and

$$\hat{\beta}_2 = 1.098967.$$

5.4.4 THE LEAST-SQUARES METHOD FOR THE BATHTUB MODEL

In Section 5.2, four bathtub curve models were reviewed and Model 1 was recommended to represent the bimodal, mixed-Weibull distribution. The parameters of this model can be found using the least-squares method.

The failure rate function is

$$\lambda(T) = (\frac{\beta}{\alpha^\beta})T^{\beta-1}e^{(T/\alpha)^\beta}, \quad T \geq 0, \quad \alpha, \beta > 0, \qquad (5.30)$$

and the reliability function is

$$R(T) = e^{-[1-e^{(T/\alpha)^\beta}]}. \qquad (5.31)$$

Let

$Y = \log_e \log_e[1 - \log_e(R(T))],$
$X = \log_e T,$

and

$$a = -\beta \log_e \alpha,$$

then Eq. (5.31) becomes a linear equation of the form

$$Y = \beta X + a. \qquad (5.32)$$

PARAMETER ESTIMATION FOR THE BIMODAL CASE

Consider n pairs of failure data $[T_i, N_f(T_i)]$, $i = 1, 2, ..., n$, from life tests based on a sample of size N. Then, the average reliability at each time, T_i, can be calculated from

$$\hat{\bar{R}}(T_i) = 1 - \frac{N_f(T_i) - 0.3}{N + 0.4},$$

or

$$\hat{\bar{R}}(T_i) = \frac{N_s(T_i) + 0.7}{N + 0.4}, \tag{5.33}$$

where $N_s(T_i)$ is the number of units surviving at time T_i. Therefore, the estimates for β, a and α are [1, Vol. 1, pp. 601-603]

$$\hat{\beta} = \frac{L_{xy}}{L_{xx}} = \frac{\sum_{i=1}^{m}(X_i - \bar{X})(Y_i - \bar{Y})}{\sum_{i=1}^{n}(X_i - \bar{X})^2}, \tag{5.34}$$

$$\hat{a} = \bar{Y} - \hat{\beta}\bar{X}, \tag{5.35}$$

and

$$\hat{\alpha} = e^{-\hat{a}/\hat{\beta}}, \tag{5.36}$$

where

$$X_i = \log_e T_i,$$
$$Y_i = \log_e \log_e[1 - \log_e(R(T_i))],$$
$$\bar{X} = \frac{1}{n}\sum_{i=1}^{n} X_i,$$

and

$$\bar{Y} = \frac{1}{n}\sum_{i=1}^{n} Y_i.$$

The correlation coefficient, R_{xy}, is [1, Vol. 1, p. 601]

$$R_{xy} = \frac{L_{xy}}{\sqrt{L_{xx}L_{yy}}},$$

or

$$R_{xy} = \frac{\sum_{i=1}^{n}(X_i - \bar{X})(Y_i - \bar{Y})}{\sqrt{[\sum_{i=1}^{n}(X_i - \bar{X})^2][\sum_{i=1}^{n}(Y_i - \bar{Y})^2]}}. \tag{5.37}$$

A computer program named "LSQ," as listed in Appendix 5D, was developed to get $\hat{\beta}, \hat{\alpha}$ and R_{xy} from the raw test data $[T_i, N_f(T_i)]$, and is illustrated next.

EXAMPLE 5–7

Using the life test data given in Table 5.7, determine the mixed population's parameters for Bathtub Model 1 using the least-squares method.

SOLUTION TO EXAMPLE 5–7

From Table 5.7 the sample size, N, is 90, the time interval number is 11, and the time interval width is 600 hr. Inputting the given failure data information into the computer program "LSQ" yields the following parameter estimates:

$$\hat{\beta} = 0.578164,$$
$$\hat{\alpha} = 7.56028 \times 10^3 \text{ hr},$$

and

$$R_{xy} = 0.979237.$$

It may be seen that the Bathtub Model 1 fits the test data very well, with a high correlation coefficient of 0.979237.

EXAMPLE 5–8

Using the life test data given in Table 5.19, determine the mixed population's parameters for Bathtub Model 1 using the least-squares method.

SOLUTION TO EXAMPLE 5–8

From Table 5.19 the sample size, N, is 1,000, the time interval number is 17, and the time interval width is 10 hr. Inputting the given failure data information into the computer program "LSQ" yields the following parameter estimates:

$$\hat{\beta} = 0.193706,$$
$$\hat{\alpha} = 8.64919 \times 10^5 \text{ hr},$$

and

$$R_{xy} = 0.972924.$$

It may be seen that the Bathtub Model 1 fits the test data very well, with a high correlation coefficient of 0.972924.

EXAMPLE 5–9

Using the life test data given in Table 5.20, determine the mixed population's parameters for Bathtub Model 1 using the least-squares method.

SOLUTION TO EXAMPLE 5–9

From Table 5.20 the sample size, N, is 1,000,000, the number of time intervals is 20, and the time interval width is 10 hr. Inputting the given failure data information into the computer program "LSQ" yields the following parameter estimates:

$$\hat{\beta} = 0.355715,$$
$$\hat{\alpha} = 5.78176 \times 10^{12} \text{ hr},$$

and

$$R_{xy} = 0.997639.$$

It may be seen that the Bathtub Model 1 fits the test data very well, with a high correlation coefficient of 0.997639.

REFERENCES

1. Kececioglu, Dimitri B., *Reliability Engineering Handbook*, Prentice Hall, Inc., Upper Saddle River, NJ, Vol. 1, 720 pp. and Vol. 2, 568 pp., 1991, Fifth Printing 1995.

2. Yuan, J. and Shih, S. W., "Mixed Hazard Models and Their Applications," *Reliability Engineering and System Safety*, Vol. 33, pp. 115-129, 1991.

3. Jiang, Siyuan and Kececioglu, Dimitri B., "Graphical Representation of Two Mixed Weibull Distributions," *IEEE Transactions on Reliability*, Vol. 41, No. 2, pp. 241-247, 1992.

4. Jensen, Finn and Peterson, N. E., *Burn-in*, John Wiley & Sons, New York, 167 pp., 1982.

5. Kamath, A. R. R., Keller, A. Z. and Moss, T. R., "An Analysis of Transistor Failure Data," *5th Symposium on Reliability Technology*, Bradford, September 1978.

6. Kececioglu, Dimitri B. and Sun, Feng-Bin, "Mixed-Weibull Parameter Estimation for Burn-in Data Using the Bayesian Approach," *Microelectronics and Reliability*, Vol. 34, No. 10, pp. 1657-1679, 1994.

7. Jiang, Siyuan and Kececioglu, Dimitri B., "Maximum Likelihood Estimates from Censored Data for Mixed-Weibull Distributions," *IEEE Transactions on Reliability*, Vol. 41, No. 2, pp. 248-255, 1992.

8. Woodruff, B. W., Moore, A. H. and Cortes, R., "A Modified Kolmogorov-Smirnov Test for Weibull Distributions with Unknown Location and Scale Parameters," *IEEE Transactions on Reliability*, Vol. R-32, No. 2, pp. 209-213, June 1983.

ADDITIONAL REFERENCES

1. Jiang, R. and Murthy, D. N. P., "Modeling Failure Data by Mixture of Two Weibull Distributions: A Graphical Approach," *IEEE Transactions on Reliability*, Vol. 44, No. 3, pp. 479-488, 1995.

2. Jiang, R. and Murthy, D. N. P., "Reliability Modeling Involving Two Weibull Distributions," *Reliability Engineering and System Safety*, Vol. 47, No. 3, pp. 187-198, 1995.

APPENDIX 5A
DERIVATION OF THE THEORETICAL BASIS FOR JENSEN'S GRAPHICAL METHOD

Jensen's graphical method for the estimation of the parameters of the bimodal Weibull mixtures is mainly based on the three important features of the *cdf* curves on WPP, *when the two subpopulations are well separated*. These three characteristics are presented and proved next.

CHARACTERISTIC 1 — *If the slope of the plotted mixture cdf tends to zero in a certain region, then there exists an essentially horizontal portion on the cdf curve, and its Y-axis reading is asymptotically equal to the mixing weight p_1, as shown in Fig. 5.1.*

Proof

Let $R(t)$ and $Q(t)$ be the reliability and unreliability functions of a bimodal Weibull mixture, and $R_1(t)$ and $R_2(t)$ be the reliability functions of Subpopulations 1 and 2, respectively. Then,

$$R(t) = p_1 R_1(t) + (1-p_1) R_2(t) = 1 - Q(t). \tag{5A.1}$$

Let

$$x = \log_e t,$$

or

$$t = e^x,$$

and

$$Y = \log_e\{-\log_e[1 - Q(t)]\} = \log_e\{-\log_e[1 - Q(e^x)]\}. \tag{5A.2}$$

Taking the derivative of Eq. (5A.2) with respect to t, and using Eq. (5A.1), yields

$$\frac{dY}{dt} = \frac{-1}{\log_e[p_1 R_1(t) + (1-p_1) R_2(t)]} \cdot \frac{p_1 f_1(t) + (1-p_1) f_2(t)}{p_1 R_1(t) + (1-p_1) R_2(t)}.$$

Now it should be proved that a certain region does exist where the slope is zero for two well separated subpopulations. Consider the effect of changing the values of η_1 and η_2:

First fix η_2, β_1, β_2 and t, and let $\eta_1 \to 0$; then,

$$p_1 R_1(t) = p_1 e^{-(t/\eta_1)^{\beta_1}} \to 0,$$

and

$$p_1 f_1(t) = p_1 \frac{\beta_1}{\eta_1}(\frac{t}{\eta_1})^{\beta_1-1} e^{-(t/\eta_1)^{\beta_1}} \to 0.$$

Therefore,

$$\frac{dY}{dt} = \frac{-1}{\log_e[(1-p_1)R_2(t)]} \cdot \frac{(1-p_1)f_2(t)}{(1-p_1)R_2(t)}.$$

Second fix β_1, β_2 and t, and let $\eta_2 \to \infty$; then,

$$R_2(t) = e^{-(t/\eta_2)^{\beta_2}} \to 1,$$

and

$$f_2(t) = \frac{\beta_2}{\eta_2}(\frac{t}{\eta_2})^{\beta_2-1} e^{-(t/\eta_2)^{\beta_2}} \to 0.$$

These lead to

$$\frac{dY}{dt} \to 0.$$

Now rescale the abscissa, and let $x = \log_e t$, then $t = e^x$. Taking the derivative of Y with respect to x yields

$$\frac{dY}{dx} = \frac{dY}{dt}\frac{dt}{dx} = \frac{dY}{dt}e^x.$$

Therefore,

$$\frac{dY}{dx} \to 0 \text{ when } \frac{dY}{dt} \to 0.$$

It may also be seen that

$$R(t) = p_1 R_1(t) + (1-p_1)R_2(t) \to (1-p_1),$$

or

$$Q(t) = 1 - R(t) \to p_1,$$

as $\eta_1 \to 0$ and $\eta_2 \to \infty$, because then from above, $R_1(t) \to 0$ and $R_2(t) \to 1$.

APPENDIX 5A

CHARACTERISTIC 2– *The unreliability for $t = \eta_1 \ll \eta_2$, is*

$$Q(\eta_1) \to 0.632\, p_1,$$

and for $t = \eta_2$

$$Q(\eta_2) \to p_1 + 0.632\,(1 - p_1).$$

Proof

From Characteristic 1

$$Q(\eta_1) = 1 - [p_1 e^{-(\eta_1/\eta_1)^{\beta_1}} + (1 - p_1)e^{-(\eta_1/\eta_2)^{\beta_2}}].$$

Since when $\eta_1 \ll \eta_2$, $e^{-(\eta_1/\eta_2)^{\beta_2}} \to 1$, then

$$\begin{aligned}Q(\eta_1) &= 1 - [p_1 e^{-1} + (1 - p_1)], \\ &= p_1(1 - e^{-1}),\end{aligned}$$

or

$$Q(\eta_1) = 0.632\, p_1.$$

Also

$$Q(\eta_2) = 1 - [p_1 e^{-(\eta_2/\eta_1)^{\beta_1}} + (1 - p_1)e^{-(\eta_2/\eta_2)^{\beta_2}}],$$

and since when $\eta_1 \ll \eta_2$, $e^{-(\eta_2/\eta_1)^{\beta_1}} \to 0$, then

$$\begin{aligned}Q(\eta_2) &= 1 - [0 + (1 - p_1)e^{-1}], \\ &= p_1 + (1 - p_1) - (1 - p_1)e^{-1}, \\ &= p_1 + (1 - p_1)(1 - e^{-1}),\end{aligned}$$

or

$$Q(\eta_2) = p_1 + 0.632\,(1 - p_1).$$

CHARACTERISTIC 3– *The tangent lines drawn at the two ends of the cdf curves are asymptotically parallel to the straight lines which represent the two individual subpopulations, respectively. That is, for* $\eta_1 \ll \eta_2$

$$\frac{dY}{dx} \to \beta_1, \quad \text{as } x \to -\infty, \quad \text{or as } t \to 0,$$

and

$$\frac{dY}{dx} \to \beta_2, \quad \text{as } x \to +\infty, \quad \text{or as } t \to +\infty.$$

Proof

For fixed t it can be shown that $R_2(t) \to 1$ and $f_2(t) \to 0$ as $\eta_2 \to \infty$, and that $R_1(t) \to 0$ and $f_1(t) \to 0$ as $\eta_1 \to 0$. From Characteristic 2

$$\frac{dY}{dx} = \frac{dY}{dt}e^x = \frac{dY}{dt}t.$$

First fix the Subpopulation 1 and move the Subpopulation 2 far away to the right; i.e., let $\eta_2 \to \infty$, then $R_2(t) \to 1$ and $f_2(t) \to 0$, which yields

$$\frac{dY}{dx} = \frac{-1}{\log_e[p_1 R_1(t) + (1-p_1)]} \cdot \frac{p_1 f_1(t)}{p_1 R_1(t) + (1-p_1)} t.$$

Now

$$\lim_{t \to 0} R_1(t) = 1,$$
$$\lim_{t \to 0} f_1(t) = 0,$$
$$\lim_{t \to 0}[p_1 R_1(t) + (1-p_1)] = p_1 + (1-p_1) = 1,$$

and

$$\lim_{t \to 0} \log_e[p_1 R_1(t) + (1-p_1)] = \log_e 1 = 0.$$

Then,

$$\lim_{t \to 0} \frac{dY}{dx} = \left(\frac{-1}{0}\right)\left(\frac{0}{1}\right).$$

APPENDIX 5A

Using L'Hospital's rule yields

$$\lim_{t \to 0} \frac{dY}{dx} = (-p_1) \lim_{t \to 0} \frac{t f_1(t)}{\log_e[p_1 R_1(t) + (1-p_1)]},$$

$$= (-p_1) \lim_{t \to 0} \frac{[f_1(t) + t f_1'(t)]}{\left[\frac{1}{p_1 R_1(t)+(1-p_1)}\right][-p_1 f_1(t)]},$$

$$= \lim_{t \to 0} \left[1 + \frac{t f_1'(t)}{f(t)}\right].$$

Now

$$f_1(t) = \frac{\beta_1}{\eta_1} \left(\frac{t}{\eta_1}\right)^{\beta_1 - 1} e^{-\left(\frac{t}{\eta_1}\right)^{\beta_1}},$$

and

$$f_1'(t) = \frac{\beta_1}{\eta_1} \left\{ (\beta_1 - 1) \left(\frac{t}{\eta_1}\right)^{\beta_1 - 2} \left(\frac{1}{\eta_1}\right) e^{-\left(\frac{t}{\eta_1}\right)^{\beta_1}} \right.$$

$$\left. + \left(\frac{t}{\eta_1}\right)^{\beta_1 - 1} e^{-\left(\frac{t}{\eta_1}\right)^{\beta_1}} \left[-\beta_1 \left(\frac{t}{\eta_1}\right)^{\beta_1 - 1} \left(\frac{1}{\eta_1}\right)\right]\right\},$$

$$= \left(\frac{\beta_1 - 1}{t}\right) \left[\frac{\beta_1}{\eta_1} \left(\frac{t}{\eta_1}\right)^{\beta_1 - 1} e^{-\left(\frac{t}{\eta_1}\right)^{\beta_1}}\right]$$

$$- \frac{\beta_1}{\eta_1} \left(\frac{t}{\eta_1}\right)^{\beta_1 - 1} \left[\frac{\beta_1}{\eta_1} \left(\frac{t}{\eta_1}\right)^{\beta_1 - 1} e^{-\left(\frac{t}{\eta_1}\right)^{\beta_1}}\right],$$

$$= \left[\left(\frac{\beta_1 - 1}{t}\right) - \frac{\beta_1}{\eta_1} \left(\frac{t}{\eta_1}\right)^{\beta_1 - 1}\right] f_1(t).$$

Then,

$$\frac{t f_1'(t)}{f_1(t)} = t \left[\left(\frac{\beta_1 - 1}{t}\right) - \frac{\beta_1}{\eta_1} \left(\frac{t}{\eta_1}\right)^{\beta_1 - 1}\right],$$

$$= (\beta_1 - 1) - \beta_1 \left(\frac{t}{\eta_1}\right)^{\beta_1}.$$

Therefore,

$$\lim_{t \to 0} \frac{dY}{dx} = \lim_{t \to 0} \left[1 + (\beta_1 - 1) - \beta_1 \left(\frac{t}{\eta_1}\right)^{\beta_1}\right] = \beta_1.$$

Similarly, fix the Subpopulation 2 and move the Subpopulation 1 to the far left; i.e., let $\eta_1 \to 0$, then $R_1(t) \to 0$ and $f_1(t) \to 0$, which yields

$$\frac{dY}{dx} = \frac{-1}{\log_e[(1-p_1)R_2(t)]} \cdot \frac{(1-p_1)f_2(t)}{(1-p_1)R_2(t)} t,$$

or

$$\frac{dY}{dx} = \frac{-\beta_2 t^{\beta_2}}{\log_e[(1-p_1)R_2(t)]\eta_2^{\beta_2}}.$$

Since $\lim_{t \to \infty} R_2(t) = 0$, then

$$\lim_{t \to \infty} \frac{dY}{dx} = \frac{-\infty}{-\infty}.$$

Using L'Hospital's rule yields

$$\lim_{t \to \infty} \frac{dY}{dx} = \lim_{t \to \infty} \frac{-\beta_2^2 t^{\beta_2-1}}{\left[\frac{-(1-p_1)f_2(t)}{(1-p_1)R_2(t)}\right]\eta_2^{\beta_2}},$$

$$= \lim_{t \to \infty} \frac{\beta_2^2 t^{\beta_2-1}}{\left(\frac{\beta_2}{\eta_2}\right)\left(\frac{t}{\eta_2}\right)^{\beta_2-1}\eta_2^{\beta_2}},$$

or

$$\lim_{t \to \infty} \frac{dY}{dx} = \beta_2.$$

APPENDIX 5B
COMPUTER PROGRAM OF EML ALGORITHM FOR UNGROUPED NON-POSTMORTEM DATA[1]

```
C     $$$$$$$$$$$$$$$$$$$$$$$$$$$$$$$$$$$$$$$$$$$$$$$$$$$$$$$
C     $    MAXIMUM LIKELIHOOD ESTIMATE (EM-ALGORITHM) FOR    $
C     $      MIXED-WEIBULL DISTRIBUTION WITH UP TO FIVE      $
C     $      SUBPOPULATIONS. THE DATA ARE GIVEN IN THE       $
C     $       FORM OF INDIVIDUAL (COMPLETE OR CENSORED).     $
C     $$$$$$$$$$$$$$$$$$$$$$$$$$$$$$$$$$$$$$$$$$$$$$$$$$$$$$$
      IMPLICIT DOUBLE PRECISION (A-H,O-Z)
      REAL*8 PI(5),ETI(5),BEI(5)
C
      REAL*8 PO(5),ETO(5),BEO(5)
      INTEGER*4 M,N,R
      COMMON /XMAIN/ X(1000)
      COMMON /PRO/ PRO(1000,5)
      COMMON /PAR/ P(5),ET(5),BE(5)
      COMMON /INTE/ M,N,R
      COMMON /TIMET/ TT
C
      OPEN(UNIT=15,FILE='EMWI.IN',STATUS='OLD')
      OPEN(UNIT=14,FILE='EMWI.DAT',STATUS='OLD')
      OPEN(UNIT=16,FILE='EMWI.OUT',STATUS='UNKNOWN')
      OPEN(UNIT=1,FILE='PFIT.IN',STATUS='UNKNOWN')
C
      WRITE(*,*)' IF THE INPUT IS DATA ENTER 1: '
      READ(*,*)IDATA
      IF(IDATA.EQ.1)THEN
      WRITE(*,*)' ENTER THE SAMPLE SIZE: '
      READ(*,*)N
      WRITE(*,*)' ENTER THE NUMBER OF FAILURES: '
      READ(*,*)R
      WRITE(*,*)' ENTER THE NUMBER OF SUBPOPULATIONS: '
      READ(*,*)M
      WRITE(*,*)' ENTER THE ACCURACY: '
      READ(*,*)EPS
      WRITE(*,*)' ENTER THE TIME TO TERMINATE TEST (ENTER 0.0 '
      WRITE(*,*)' IF THE TEST IS NOT A TIME TERMINATED TEST): '
      READ(*,*)TT
```

---

[1] This computer program was developed by Dr. Siyuan Jiang presently at Ford Motor Company.

```
              WRITE(*,*)'DO YOU NEED TO FACTORIZE THE TIME?'
              WRITE(*,*)'ENTER THE FACTOR, IF NO ENTER 1: '
              READ(*,*)FACTOR
              DO 21 I=1,R
              READ(14,*)X(I)
21              CONTINUE
              EPSN=EPS
              WRITE(16,*)' EPS=',EPS,'   TT=',TT
              WRITE(16,*)'   N=',N,'    AND R=',R
              GOTO 222
              ENDIF
C
              READ(15,*)INI
              WRITE(16,111)
111             FORMAT(/,/,/)
              WRITE(16,*)' THE SEED =',INI
C              WRITE(*,*)' ENTER N,R,EPS'
              READ(15,*)N,R,EPS
              WRITE(*,*)' ENTER THE NUMBER OF THE SUBS.: '
              READ(*,*)M
              DO 2 J=1,M
              READ(15,*)PI(J),ETI(J),BEI(J)
2               CONTINUE
C              WRITE(*,*)'ENTER THE TIME OF TERMINATION OF THE TEST',
C       *        ' IF THE TEST IS NOT A TIME TERMINATED ENTER 0.0
              READ(15,*)TT
              WRITE(*,*)'DO YOU NEED TO FACTORIZE THE TIME?'
              WRITE(*,*)'ENTER THE FACTOR, IF NO ENTER 1: '
              READ(*,*)FACTOR
              DO 33 J=1,M
              ETI(J)=ETI(J)*FACTOR
33              CONTINUE
C
              RAM=RANO(INI)
              CALL WEII(M,PI,BEI,ETI,N,TT,R,X)
C
              WRITE(16,*)'   N=',N,'    AND R=',R
              WRITE(16,*)'---SUB.------PRO-------ETA------BETA----'
              DO 3 J=1,M
              WRITE(16,110)J,PI(J),ETI(J),BEI(J)
3               CONTINUE
              WRITE(16,*)' EPS=',EPS,'   TT=',TT
              EPSN=EPS
              RAM=RANO(INI)
```

APPENDIX 5B

```
C
222         WRITE(*,*)' IF THE INITIAL IS PARAMETER SET ENTER 1.'
            READ(*,*)ICON
            IF(ICON.EQ.1) GOTO 77
            DO 4 J=1,M
            DO 4 I=1,N
            PRO(I,J)=0.0
4           CONTINUE
            NSSUB=0
            DO 50 J=1,M
            WRITE(*,*)' ENTER THE INITIAL GUESS OF THE NUMBER BELONG TO'
            WRITE(*,*)' SUB.',J,' : '
            READ(*,*)ISUB
            WRITE(16,*)'NSUB',J,'==',ISUB
            NLSUB=NSSUB+1
            NUSUB=NSSUB+ISUB
            DO 30 I=NLSUB,NUSUB
            PRO(I,J)=1.
30          CONTINUE
            NSSUB=ISUB+NSSUB
50          CONTINUE
C
            CALL PROPOR
            NITER=1
            DO 36 I=1,M
            PO(I)=1./REAL(M)
            ETO(I)=1.0
            BEO(I)=1.0
36          CONTINUE
            GOTO 40
77          DO 37 I=1,M
            PO(I)=1./REAL(M)
            ETO(I)=1.0
            BEO(I)=1.0
37          CONTINUE
C
            WRITE(*,*)'ENTER ---P---'
            READ(*,*)P
            WRITE(*,*)'ENTER ---ET---'
            READ(*,*)ET
            WRITE(*,*)'ENTER ---BE---'
            READ(*,*)BE
            WRITE(*,*)'ENTER NITER'
            READ(*,*)NITER
```

```
            WRITE(16,*)'---SUB--------P0-------ETA0-------BETA0-------'
            DO 39 J=1,M
            WRITE(16,110)J,P(J),ET(J),BE(J)
39            CONTINUE
            DO 38 J=1,M
            ET(J)=ET(J)*FACTOR
38            CONTINUE
            CALL PROBI
            GOTO 66
40            DO 60 J=1,M
            CALL NEWBEI(ET(J),BE(J),J,EPSN)
60            CONTINUE
            CALL PROPOR
            WRITE(16,*)'---SUB--------P0-------ETA0-------BETA0-------'
            DO 65 J=1,M
            WRITE(16,110)J,P(J),ET(J),BE(J)
65            CONTINUE
C
            N10=1
66            DO 68 J=1,M
            CALL NEWBEI(ET(J),BE(J),J,EPSN)
68            CONTINUE
            CALL PROPOR
C
C       CALCULATE LN(L)
C
            CALL LIKHOD(F)
            IF(N10.GE.1) THEN
            WRITE(*,*)'P=',P
            WRITE(*,*)'ET=',ET
            WRITE(*,*)'BE=',BE
            WRITE(*,*)'F=',F
            N10=0
            END IF
            N10=N10+1
            DO 46 J=1,M
            IF( (ABS((P0(J)-P(J)) / P0(J)).GT.EPS)
     *        .OR. (ABS((ET0(J)-ET(J)) / ET0(J)).GT.EPS)
     *        .OR. (ABS((BE0(J)-BE(J)) / BE0(J)).GT.EPS))THEN
            CALL PROBI
            DO 45 I=1,M
            P0(I)=P(I)
            ET0(I)=ET(I)
            BE0(I)=BE(I)
```

```
45        CONTINUE
          NITER=NITER+1
          GOTO 66
          END IF
46        CONTINUE
          DO 70 J=1,M
          ET(J)=ET(J)/FACTOR
70        CONTINUE
          DO 75 I=1,R
          X(I)=X(I)/FACTOR
75        CONTINUE
          DO 151 I=1,R
          WRITE(1,*)X(I)
151       CONTINUE
          WRITE(16,*)'DURING THE CALCULATION X IS BEEN FACTORIZED BY'
       *      ,FACTOR
          WRITE(16,*)'TOTAL NO. OF ITERATION=',NITER
          WRITE(16,*)' LIKELIHOOD FUNCTION =',F
          WRITE(16,*)'---SUB--------P--------ETA--------BETA--------'
          DO 90 J=1,M
          WRITE(16,110)J,P(J),ET(J),BE(J)
90        CONTINUE
110        FORMAT(5X,I3,3F13.6)
          STOP
          END
C     ---------------------------------------------------------------
          SUBROUTINE PROBI
C     ---------------------------------------------------------------
C         CALCULATE THE PROBABILITIES OF THE MEMBERSHIPS OF THE
C         OBSERVATIONS WHICH ARE GIVEN IN THE FORM OF INDIVIDUAL.
C
          IMPLICIT DOUBLE PRECISION (A-H,O-Z)
          REAL*8 PDQ(1000,5),PDQQ(1000)
C
          REAL*8 PRO(1000,5),PO(5),
       *        ETO(5),BEO(5)
          INTEGER*4 M,N,R
          COMMON /XMAIN/ X(1000)
          COMMON /PRO/ PRO
          COMMON /PAR/ P(5),ET(5),BE(5)
          COMMON /INTE/ M,N,R
          COMMON /TIMET/ TT
C
          DO 20 I=1,R
```

```
              PDQQ(I)=0.0
              DO 20 J=1,M
              PDQ(I,J)=P(J)*PDF(X(I),J)
              PDQQ(I)=PDQQ(I)+PDQ(I,J)
       20     CONTINUE
              DO 30 I=1,R
              DO 30 J=1,M
              PRO(I,J)= PDQ(I,J)/PDQQ(I)
       30     CONTINUE
       C
              IF(R.GE.N) RETURN
       C
              IF(TT.EQ.0.0) THEN
              XT=X(R)
              ELSE
              XT=TT
              END IF
              PDQQ(R+1)=0.0
              DO 40 J=1,M
              PDQ(R+1,J)=P(J)*REL(XT,J)
              PDQQ(R+1)=PDQQ(R+1)+PDQ(R+1,J)
       40     CONTINUE
              DO 45 J=1,M
              PRO(R+1,J)=PDQ(R+1,J)/PDQQ(R+1)
       45     CONTINUE
              DO 50 I=R+2,N
              DO 50 J=1,M
              PRO(I,J)=PRO(R+1,J)
       50     CONTINUE
              RETURN
              END
       C      -----------------------------------------------
              SUBROUTINE PROPOR
       C      -----------------------------------------------
       C      ESTIMATE THE PROPORTIONS OF EACH SUBPOPULATION.
       C
              IMPLICIT DOUBLE PRECISION (A-H,O-Z)
              REAL*8 PRO(1000,5),PO(5),
             *       ETO(5),BEO(5)
              INTEGER*4 M,N,R
              COMMON /XMAIN/ X(1000)
              COMMON /PRO/ PRO
              COMMON /PAR/ P(5),ET(5),BE(5)
              COMMON /INTE/ M,N,R
```

```
              COMMON /TIMET/ TT
C
              DO 5 I=1,M
              P(I)=0.0
5             CONTINUE
              DO 20 J=1,M
              DO 10 I=1,N
              P(J)=P(J)+PRO(I,J)
10            CONTINUE
              P(J)=P(J)/DREAL(N)
20            CONTINUE
              RETURN
              END
C     ----------------------------------------------------------------
              SUBROUTINE NEWBEI(ETA,BETA,K,EPS,IER)
C     ----------------------------------------------------------------
C     ESTIMATE ETA(K) AND BETA(K), GIVEN P(K|T_J) AND T_J, USING
C     NEWTON-ITERATION METHOD.
C
              IMPLICIT DOUBLE PRECISION (A-H,O-Z)
              REAL*8 ETA,BETA,EPS
              INTEGER*4 K,IER
C
              REAL*8 PRO(1000,5),PO(5),
     *               ETO(5),BEO(5)
              INTEGER*4 M,N,R
              COMMON /XMAIN/ X(1000)
              COMMON /PRO/ PRO
              COMMON /PAR/ P(5),ET(5),BE(5)
              COMMON /INTE/ M,N,R
              COMMON /TIMET/ TT
C
              MM=0
              RR=0.0
              AF=0.0
              DO 2 I=1,R
              RR=RR+PRO(I,K)
2             CONTINUE
              PERL=0.1673*RR
              PERU=0.9737*RR
              RRO=0.0
              DO 5 I=1,N
              RRO=RRO+PRO(I,K)
              IF(RRO.GE.PERL) THEN
```

```
            IRO=I
            GOTO 6
            END IF
5           CONTINUE
6           DO 7 I=IRO+1,N
            RRO=RRO+PRO(I,K)
            IF(RRO.GE.PERU) THEN
            IR1=I
            GOTO 8
            END IF
7           CONTINUE
            IR1=R

8           IF((IR1.LE.0.0).OR.(IRO.LE.0.0)) GOTO 9
            TEM1=2.99/(DLOG(X(IR1))-DLOG(X(IRO)))
            IF(TEM1.LE.0.0) GOTO 9
            GOTO 10
C             WRITE(16,*)'BETAO=',TEM1,'  LOW   UP',IRO,IR1
9             TEM1=1.0
10          BETA=TEM1
            IF(BETA.GT.100.0)THEN
            IER=1
            RETURN
            END IF
            A=0.0
            B=0.0
            C=0.0
            E=0.0
            DO 20 I=1,R
            AL=DLOG(X(I))
            EX=X(I)**BETA
            A1=PRO(I,K)*EX
            B1=PRO(I,K)*AL
            A=A+A1*AL
            B=B+A1
            C=C+B1*EX*AL
            E=E+B1
20          CONTINUE
            A=A+(N-R)*PRO(R+1,K)*EX*AL
            B=B+(N-R)*PRO(R+1,K)*EX
            C=C+(N-R)*PRO(R+1,K)*EX*AL*AL
            IF((BETA.EQ.0.0).OR.(RR.EQ.0.0).OR.(B.EQ.0.0)) THEN
            CALL CHECK
            WRITE(16,*)'-B-',B
```

APPENDIX 5B

```
              WRITE(16,*)'-RR-',RR
              WRITE(16,*)'-MM-',MM
              GOTO 24
              END IF
              TEM1=1./BETA+E/RR-A/B
              TEM1=TEM1/(1./BETA/BETA+(B*C-A*A)/B/B)+BETA
              MM=MM+1
              IF(TEM1.LE.0.0) THEN
24            AF=AF+0.1
              TEM1=0.5+AF
              IF(AF.GT.50) THEN
              WRITE(*,*)' A NEGATIVE BETA WAS FOUND: BETA=', TEM1
              END IF
              GOTO 10
              END IF
              IF((ABS((BETA-TEM1)/BETA).GE.EPS).OR.(MM.LE.2)) GOTO 10
              BETA=TEM1
              ETA=(B/RR)**(1./BETA)
25            CONTINUE
              RETURN
              END
C     ----------------------------------------------------------
              SUBROUTINE CHECK
C     ----------------------------------------------------------
              IMPLICIT DOUBLE PRECISION (A-H,O-Z)
              REAL*8 PRO(1000,5),PO(5),
           *        ETO(5),BEO(5)
              INTEGER*4 M,N,R
              COMMON /XMAIN/ X(1000)
              COMMON /PRO/ PRO
              COMMON /PAR/ P(5),ET(5),BE(5)
              COMMON /INTE/ M,N,R
              COMMON /TIMET/ TT
C
              WRITE(16,*)'BETA',BE
              WRITE(16,*)'ETA',ET
              WRITE(16,*)'-P-',P
              RETURN
              END
C     ----------------------------------------------------------
              DOUBLE PRECISION FUNCTION RAN(ISEED)
C     ----------------------------------------------------------
              IMPLICIT DOUBLE PRECISION (A-H,O-Z)
              INTEGER*4 IA,IC,IM,ISEED
```

```fortran
      PARAMETER(IA=7141,IC=54773,IM=259200)
      ISEED=MOD(ISEED*IA+IC,IM)
      RAN=DREAL(ISEED)/DREAL(IM)
      RETURN
      END
C     -----------------------------------------------------
      DOUBLE PRECISION FUNCTION RANO(IDUM)
C     -----------------------------------------------------
      IMPLICIT DOUBLE PRECISION (A-H,O-Z)
      INTEGER*4 ISEED
      DIMENSION V(97)
      DATA IFF /0/
      IF(IDUM.LT.0.OR.IFF.EQ.0)THEN
      IFF=1
      ISEED=ABS(IDUM)
      IDUM=1
      DO 11 J=1,97
      DUM=RAN(ISEED)
11    CONTINUE
      DO 12 J=1,97
      V(J)=RAN(ISEED)
12    CONTINUE
      Y=RAN(ISEED)
      ENDIF
      J=1+INT(97.*Y)
      IF(J.GT.97.OR.J.LT.1)PAUSE
      Y=V(J)
      RANO=Y
      V(J)=RAN(ISEED)
      RETURN
      END
C     -----------------------------------------------------
      SUBROUTINE ORDER(X,N)
C     -----------------------------------------------------
      IMPLICIT DOUBLE PRECISION (A-H,O-Z)
      INTEGER*4 N
      DIMENSION X(N)
C
      DO 30 J=1,N
      DO 20 I=J+1,N
      IF(X(J).GE.X(I)) THEN
      Q=X(J)
      X(J)=X(I)
      X(I)=Q
```

APPENDIX 5B

```
              END IF
20            CONTINUE
30            CONTINUE
              RETURN
              END
C     ------------------------------------------------------
              SUBROUTINE LIKHOD(F)
C     ------------------------------------------------------
              IMPLICIT DOUBLE PRECISION (A-H,O-Z)
              REAL*8 F
C
              REAL*8 PRO(1000,5),PO(5),
     *               ETO(5),BEO(5)
              INTEGER*4 M,N,R
              COMMON /XMAIN/ X(1000)
              COMMON /PRO/ PRO
              COMMON /PAR/ P(5),ET(5),BE(5)
              COMMON /INTE/ M,N,R
              COMMON /TIMET/ TT
C
              Q=0.0
              DO 10 I=1,R
              QQ=0.0
              DO 5 J=1,M
              QQ=QQ+P(J)*PDF(X(I),J)
5             CONTINUE
              IF(QQ.LE.0.0) THEN
              CALL CHECK
              WRITE(16,*)'--I--',I
              WRITE(16,*)'-QQPDF-',QQ
              END IF
              Q=Q+DLOG(QQ)
10            CONTINUE
              IF(R.GE.N) THEN
              F=Q
              RETURN
              END IF
              IF(TT.LE.0.0) THEN
              XT=X(R)
              ELSE
              XT=TT
              END IF
              QQ=0.0
              DO 15 J=1,M
```

```
              QQ=QQ+P(J)*REL(XT,J)
15            CONTINUE
              IF(QQ.LE.0.0) THEN
              CALL CHECK
              WRITE(16,*)'--I--',I
              WRITE(16,*)'-QQREL-',QQ
              END IF
              F=Q+DREAL(N-R)*DLOG(QQ)
              RETURN
              END
C     ------------------------------------------------------
              DOUBLE PRECISION FUNCTION PDF(Y,K)
C     ------------------------------------------------------
              IMPLICIT DOUBLE PRECISION (A-H,O-Z)
              REAL*8 Y,QE
              INTEGER*4 K
C
              REAL*8 PRO(1000,5),PO(5),
       *            ETO(5),BEO(5)
              INTEGER*4 M,N,R
              COMMON /XMAIN/ X(1000)
              COMMON /PRO/ PRO
              COMMON /PAR/ P(5),ET(5),BE(5)
              COMMON /INTE/ M,N,R
              COMMON /TIMET/ TT
C
              QE=BE(K)*DLOG(Y/ET(K))
              IF(QE.GE.6.56) THEN
              PDF=0.0
              RETURN
              END IF
              PDF=BE(K)/ET(K)*(Y/ET(K))**(BE(K)-1.)*REL(Y,K)
              RETURN
              END
C     ------------------------------------------------------
              DOUBLE PRECISION FUNCTION REL(Y,K)
C     ------------------------------------------------------
              IMPLICIT DOUBLE PRECISION (A-H,O-Z)
              REAL*8 Y,QE
              INTEGER*4 K
C
              REAL*8 PRO(1000,5),PO(5),
       *            ETO(5),BEO(5)
              INTEGER*4 M,N,R
```

```
          COMMON /XMAIN/ X(1000)
          COMMON /PRO/ PRO
          COMMON /PAR/ P(5),ET(5),BE(5)
          COMMON /INTE/ M,N,R
          COMMON /TIMET/ TT
C
          QE=BE(K)*DLOG(Y/ET(K))
          IF(QE.GE.6.56) THEN
          REL=0.0
          RETURN
          END IF
          QE=DEXP(QE)
          REL=DEXP(-QE)
          RETURN
          END
C         ----------------------------------------------------------
          SUBROUTINE WEII(M,PI,BEI,ETI,N,TT,R,T)
C         ----------------------------------------------------------
C         TO GENERATE MIXED WEIBULL DISTRIBUTION OF INDIVIDUAL
C         TIME-TO-FAILURE, COMPLETE, TIME-TERMINATED OR FAILUE-
C         TERMINATED. PERFOME WITH FUNCTIONS RANO AND RAN, AND
C         SUBROUTINE ORDER.
C         >>>>>>>>>>>>>>>>>>>>>>>>>>>>>>>>>>>>>>>>>>>>>>>>>>>>>>
C         PARAMETERS:
C         M=NUMBER OF SUBPOPULATIONS.
C         PI(M)=ARRAY, MIXING WEIGHT.
C         BEI(M)=ARRAY, M SHAPE PARAMETERS OF M SUBPOPULATIONS.
C         ETI(M)=ARRAY, M SCALE PARAMETERS OF M SUBPOPULATIONS.
C         N=TOTAL NUMBER OF INDIVIDUALS IN THE LIFE TEST.
C         TT=TIME OF TERMINATING THE TEST, IF INPUT TT=0, IMPLIES
C         THAT THE TEST IS COMPLETE OR FAILURE TERMINATED.
C         R=INTEGER, NUMBER OF FAILED INDIVIDUAL, IF INPUT R=0,
C         IMPLIES THAT THE TEST IS COMPLETE OR TIME TERMINATED.
C         T(N)=ARRAY, N TIME-TO-FAILURE, IN THE VALUE OF INCREASING
C         ORDER.
C         ** INITIALIZE RANO BEFORE CALLING THIS SUBROUTINE.
C         >>>>>>>>>>>>>>>>>>>>>>>>>>>>>>>>>>>>>>>>>>>>>>>>>>>>>>
          IMPLICIT DOUBLE PRECISION (A-H,O-Z)
          INTEGER*4 M,N,R,RO
          REAL*8 PI(M),BEI(M),ETI(M),T(N),TT
          REAL*8 PP(10)
C
C         WRITE(*,*)M,N,R,TT
C         WRITE(*,*)' PI= ',PI(1),PI(2),PI(3)
```

```
C       WRITE(*,*)' ETI=',ETI(1),ETI(2),ETI(3)
C       WRITE(*,*)' BEI=',BEI(1),BEI(2),BEI(3)
        PP(1)=PI(1)
        DO 5 I=2,M
        PP(I)=PP(I-1)+PI(I)
5       CONTINUE
        DO 20 I=1,N
        U1=RAN0(INI)
        DO 10 J=1,M
        IF(U1.LE.PP(J)) THEN
6       U2=RAN0(INI)
        IF((U2.LE.1.0D-708).OR.(U2.GE.1.0D708)) GOTO 6
        T(I)=ETI(J)*(-DLOG(U2))**(1./BEI(J))
        GOTO 20
        END IF
10      CONTINUE
20      CONTINUE
        CALL ORDER(T,N)
        IF((TT.LT.1.0D-9)) RETURN
        R=N
        DO 30 I=1,N
        IF(T(I).GT.TT) THEN
        R=I-1
        RETURN
        END IF
30      CONTINUE
        R=N
        RETURN
        END
```

APPENDIX 5C
COMPUTER PROGRAM OF EML ALGORITHM
FOR GROUPED NON-POSTMORTEM DATA[2]

```
C     $$$$$$$$$$$$$$$$$$$$$$$$$$$$$$$$$$$$$$$$$$$$$$$$$$$$$$$$$$$$$$$$
C     $ MAXIMUM LIKELIHOOD ESTIMATE(EM-ALGORITHM) FOR MIXED-WEIBULL  $
C     $ DISTRIBUTION WITH UP TO FIVE SUBPOPULATIONS FOR GROUPED DATA.$
C     $$$$$$$$$$$$$$$$$$$$$$$$$$$$$$$$$$$$$$$$$$$$$$$$$$$$$$$$$$$$$$$$
      IMPLICIT DOUBLE PRECISION (A-H, O-Z)
      REAL*8 PI(5),ETI(5),BEI(5),XMID(200),X(2000),PARKK(2)
      COMMON /XMID/XMID
      COMMON /FISHER/ ALPHA(2,2),BETA(2),Y(2,2),C(2,2)
C
      REAL*8 XEND(200),PRO(200,5),
     *    P(5),ET(5),BE(5),P0(5),ET0(5),BE0(5)
      INTEGER NSUB,N,NK,NFAIL(200),KK
      COMMON /MAIN/ XEND,PRO
      COMMON /INTE/ NSUB,N,NK,NFAIL
      COMMON /KK/KK
      COMMON /PAR/P,ET,BE
C
      OPEN(UNIT=15,FILE='NEWG.IN',STATUS='OLD')
      OPEN(UNIT=16,FILE='NEWG.OUT',STATUS='UNKNOWN')
      OPEN(UNIT=4,FILE='NEWG.DAT',STATUS='OLD')
C
      WRITE(*,*) 'IF INPUT IS YOUR OWN DATA ENTER 1: '
      READ(*,*)IDATA
      IF(IDATA.EQ.1) THEN
      WRITE(*,*)' ENTER THE NUMBER OF THE SUBS.: '
      READ(*,*)NSUB
      WRITE(*,*)'ENTER THE SAMPLE SIZE N: '
      READ(*,*)N
      WRITE(*,*)' ENTER THE NUMBER OF THE GROUPS, EXCLUDING THE'
      WRITE(*,*)' GROUP OF SURVIVALS: '
      READ(*,*)NK
      WRITE(*,*)'DO YOU NEED TO FACTORIZE THE TIME?'
      WRITE(*,*)'ENTER THE FACTOR, IF NO ENTER 1: '
      READ(*,*)FACTOR
      WRITE(*,*)'ENTER THE EPS: '
      READ(*,*)EPS
```

---

[2]This computer program was developed by Dr. Siyuan Jiang presently at Ford Motor Company.

```
              DO 1000 I=1,NK
              READ(4,*)XEND(I)
              XEND(I)=XEND(I)*FACTOR
1000          CONTINUE
              DO 1001 I=1,NK+1
              READ(4,*)NFAIL(I)
1001          CONTINUE
              GOTO 1100
              END IF
C
C      ENTER THE INITIAL
C
              READ(15,*)INI
              WRITE(16,111)
111           FORMAT(/,/,/)
              WRITE(16,*)' THE SEED =',INI
C
C      ENTER TOTAL NUMBER OF DATA, N, AND EPS.
C
              READ(15,*)N,EPS
              WRITE(*,*)' ENTER THE NUMBER OF THE SUBS.'
              READ(*,*)NSUB
              DO 2 J=1,NSUB
              READ(15,*)PI(J),ETI(J),BEI(J)
2             CONTINUE
              WRITE(*,*)' ENTER THE NUMBER OF THE GROUPS, EXCLUSIVE THE'
              WRITE(*,*)' GROUP OF SURVIVALS.'
              READ(*,*)NK
C
C   'ENTER THE TIME OF TERMINATION OF THE TEST',
C
              READ(15,*)TTO
              WRITE(*,*)'DO YOU NEED TO FACTORIZE THE TIME?'
              WRITE(*,*)'ENTER THE FACTOR, IF NO ENTER 1'
              READ(*,*)FACTOR
              DO 33 J=1,NSUB
              ETI(J)=ETI(J)*FACTOR
33            CONTINUE
              TT=TTO*FACTOR
C
              WIDTH=(TT/REAL(NK))
              DO 41 I=1,NK
              XEND(I)=I*WIDTH
41            CONTINUE
```

APPENDIX 5C

```fortran
      RAM=RANO(INI)
      CALL WEII(NSUB,PI,BEI,ETI,N,TT,R,X)
      CALL WEIG(X,XEND,NFAIL,N,NK)
C
      WRITE(16,*)'---SUB.------PRO-------ETA------BETA----'
      DO 3 J=1,NSUB
      WRITE(16,110)J,PI(J),ETI(J),BEI(J)
3     CONTINUE
1100  XMID(1)=XEND(1)/2.
      DO 34 I=1,NK
      XMID(I)=(XEND(I)+XEND(I-1))/2.
34    CONTINUE
      WRITE(16,*)'----------------------------------------'
      WRITE(16,*)'   N=',N,'    NK=',NK,'    EPS=',EPS
      WRITE(16,*)'----------------------------------------'
      DO 1120 I=1,NK+1
      WRITE(16,*)I,NFAIL(I),XEND(I)
      WRITE(*,*)I,NFAIL(I),XEND(I)
1120  CONTINUE
      WRITE(16,*)'----------------------------------------'
      EPSN=EPS
C
      WRITE(*,*)' IF THE INITIAL IS PARAMETER SET ENTER 1.'
      READ(*,*)ICON
      IF(ICON.EQ.1) GOTO 77
      DO 4 J=1,NSUB
      DO 4 I=1,NK+1
      PRO(I,J)=0.0
4     CONTINUE
      NSSUB=0
      DO 50 J=1,NSUB
      WRITE(*,*)' ENTER THE INITIAL GUESS OF THE NUMBER BELONG TO'
      WRITE(*,*)'   SUB.',J
      READ(*,*)ISUB
      WRITE(16,*)'NSUB',J,'==',ISUB
      NLSUB=NSSUB+1
      NUSUB=NSSUB+ISUB
      DO 30 I=NLSUB,NUSUB
      PRO(I,J)=1.
30    CONTINUE
      NSSUB=ISUB+NSSUB
50    CONTINUE
C
      CALL PROPG
```

```
          NITER=1
          DO 36 I=1,NSUB
          PO(I)=1./REAL(NSUB)
          ETO(I)=1.0
          BEO(I)=1.0
36        CONTINUE
          GOTO 40
77        DO 37 I=1,NSUB
          PO(I)=1./REAL(NSUB)
          ETO(I)=1.0
          BEO(I)=1.0
37        CONTINUE
C
          do 888 j=1,5
          p(j)=0.0
          et(j)=0.0
          be(j)=0.0
888       continue
          WRITE(*,*)'ENTER P1, P2, .....'
          READ(*,*) (P(I),I=1,NSUB)
          WRITE(*,*)'ENTER ETA1, ETA2, .....'
          READ(*,*) (ET(I),I=1,NSUB)
          WRITE(*,*)'ENTER BETA1, BETA2, .....'
          READ(*,*) (BE(I),I=1,NSUB)
          WRITE(*,*)'ENTER NITER'
          READ(*,*)NITER
          WRITE(16,*)'---SUB--------PO-------ETAO-------BETAO-
          DO 39 J=1,NSUB
          WRITE(16,110)J,P(J),ET(J),BE(J)
39        CONTINUE
          DO 38 J=1,NSUB
          ET(J)=ET(J)*FACTOR
38        CONTINUE
          CALL PROBG
          GOTO 66
40        DO 60 KK=1,NSUB
          CALL NEWBEG(PARKK(1),PARKK(2),KK,EPS)
          WRITE(16,*)'--PARKK---',PARKK
          ET(KK)=PARKK(1)
          BE(KK)=PARKK(2)
          CALL MNEWT(20,PARKK,2,0.00001,0.00001)
          ET(KK)=PARKK(1)
          BE(KK)=PARKK(2)
60        CONTINUE
```

APPENDIX 5C

```
          CALL PROPG
          WRITE(16,*)'---SUB--------PO-------ETAO-------BETAO-------'
          DO 65 J=1,NSUB
          WRITE(16,110)J,P(J),ET(J),BE(J)
 65       CONTINUE
C
          N10=0
 66       DO 68 KK=1,NSUB
          CALL NEWBEG(PARKK(1),PARKK(2),KK,EPS)
          CALL MNEWT(20,PARKK,2,0.00001,0.00001)
          ET(KK)=PARKK(1)
          BE(KK)=PARKK(2)
 68       CONTINUE
          CALL PROPG
C
C     CALCULATE Ln(L)
C
          WRITE(*,*)'P=',P
          WRITE(*,*)'ET=',ET
          WRITE(*,*)'BE=',BE
          CALL LIKHOD(F)
          WRITE(*,*)'F=',F,'NITER=',NITER
          DO 46 J=1,NSUB
          IF( (ABS((PO(J)-P(J)) / PO(J)).GT.EPS)
     *        .OR. (ABS((ETO(J)-ET(J)) / ETO(J)).GT.EPS)
     *        .OR. (ABS((BEO(J)-BE(J)) / BEO(J)).GT.EPS))THEN
          CALL PROBG
          DO 45 I=1,NSUB
          PO(I)=P(I)
          ETO(I)=ET(I)
          BEO(I)=BE(I)
 45       CONTINUE
          NITER=NITER+1
          GOTO 66
          END IF
 46       CONTINUE
          DO 70 J=1,NSUB
          ET(J)=ET(J)/FACTOR
 70       CONTINUE
          DO 75 I=1,NK
          XEND(I)=XEND(I)/FACTOR
 75       CONTINUE
          WRITE(16,*)'DURING THE CALCULATION X IS BEEN FACTORIZED BY'
     *              ,FACTOR
```

```
      WRITE(16,*)'TOTAL NO. OF ITERATION=',NITER
      WRITE(16,*)' LIKELIHOOD FUNCTION =',F
      WRITE(16,*)'---SUB--------P--------ETA--------BETA-
      DO 90 J=1,NSUB
      WRITE(16,110)J,P(J),ET(J),BE(J)
90    CONTINUE
110   FORMAT(5X,I3,3F20.6)
      DO 200 J=1,NSUB
      WRITE(1,210)P(J),ET(J),BE(J)
200   CONTINUE
210   FORMAT(3F13.7)
      STOP
      END
C     ----------------------------------------------------
      SUBROUTINE WEIG(X,XEND,NFAIL,N,NK)
C     ----------------------------------------------------
C     GROUP THE WEIBULL DATA OBTAINED FROM WEII IN TO NK
C     GROUPS.
C
      implicit double precision (a-h, o-z)
      INTEGER N,NK,NFAIL(NK+1)
      REAL*8 X(N),XEND(NK)
C
      DO 15 I=1,NK
      NFAIL(I)=0
15    CONTINUE
      DO 16 I=1,N
      IF((X(I).LE.XEND(1)).AND.(X(I).GT.0.0)) THEN
      NFAIL(1)=NFAIL(1)+1
      GOTO 16
      END IF
      DO 17 J=2,NK
      IF((X(I).LE.XEND(J)).AND.(X(I).GT.XEND(J-1))) THEN
      NFAIL(J)=NFAIL(J)+1
      GOTO 16
      END IF
17    CONTINUE
      IF(X(I).GT.XEND(NK)) THEN
      NFAIL(J)=NFAIL(J)+1
      END IF
16    CONTINUE
      RETURN
      END
C     ----------------------------------------------------
```

APPENDIX 5C

```fortran
      SUBROUTINE PROBG
C     ---------------------------------------------------------------
C     CALCULATE THE PROBABILITIES OF THE MEMBERSHIPS OF THE OBSERVATION
C
      implicit double precision (a-h, o-z)
      REAL*8 AA1(5),AA2(5),Q1(5)
C
      REAL*8 XEND(200),PRO(200,5),
     *       P(5),ET(5),BE(5),P0(5),ET0(5),BE0(5)
      INTEGER NSUB,N,NK,NFAIL(200),KK
      COMMON /MAIN/ XEND,PRO
      COMMON /INTE/ NSUB,N,NK,NFAIL
      COMMON /KK/KK
      COMMON /PAR/P,ET,BE
C
      QQ=0.0
      DO 5 J=1,NSUB
      AA1(J)=1.
      AA2(J)=REL(XEND(1),J)
      Q1(J)=P(J)*(1.-AA2(J))
      QQ=QQ+P(J)*(AA1(J)-AA2(J))
5     CONTINUE
C
      DO 10 J=1,NSUB
      PRO(1,J)=Q1(J)/QQ
10    CONTINUE
C
      DO 20 I=2,NK
      DO 15 J=1,NSUB
      AA1(J)=AA2(J)
      AA2(J)=REL(XEND(I),J)
15    CONTINUE
      QQ=0.0
      DO 16 J=1,NSUB
      Q1(J)=P(J)*(AA1(J)-AA2(J))
      QQ=QQ+Q1(J)
16    CONTINUE
      DO 17 J=1,NSUB
      PRO(I,J)=Q1(J)/QQ
17    CONTINUE
20    CONTINUE
C
      QQ=0.0
      DO 30 J=1,NSUB
```

```fortran
              Q1(J)=P(J)*REL(XEND(NK),J)
              QQ=QQ+Q1(J)
30            CONTINUE
              DO 35 J=1,NSUB
              PRO(NK+1,J)=Q1(J)/QQ
35            CONTINUE
              RETURN
              END
C     ----------------------------------------------------------
              SUBROUTINE PROPG
C     ----------------------------------------------------------
C     ESTIMATE THE PROPORTIONS OF EACH SUB. POPULATIONS.
C
              implicit double precision (a-h, o-z)
              REAL*8 XEND(200),PRO(200,5),
     *             P(5),ET(5),BE(5),PO(5),ETO(5),BEO(5)
              INTEGER NSUB,N,NK,NFAIL(200),KK
              COMMON /MAIN/ XEND,PRO
              COMMON /INTE/ NSUB,N,NK,NFAIL
              COMMON /KK/KK
              COMMON /PAR/P,ET,BE
C
              DO 5 J=1,NSUB
              P(J)=0.0
5             CONTINUE
              DO 20 J=1,NSUB
              DO 10 I=1,NK+1
              P(J)=P(J)+NFAIL(I)*PRO(I,J)
10            CONTINUE
              P(J)=P(J)/REAL(N)
20            CONTINUE
              RETURN
              END
C     ----------------------------------------------------------
              SUBROUTINE LIKHOD(F)
C     ----------------------------------------------------------
              implicit double precision (a-h, o-z)
              REAL*8 F
C
              REAL*8 XEND(200),PRO(200,5),
     *             P(5),ET(5),BE(5),PO(5),ETO(5),BEO(5)
              INTEGER NSUB,N,NK,NFAIL(200),KK
              COMMON /MAIN/ XEND,PRO
              COMMON /INTE/ NSUB,N,NK,NFAIL
```

APPENDIX 5C

```
              COMMON /KK/KK
              COMMON /PAR/P,ET,BE
C
              Q=0.0
              QQ=0.0
              DO 5 J=1,NSUB
C             QQ=QQ+P(J)*DEXP(-(XEND(1)/ET(J))**BE(J))
              QQ=QQ+P(J)*REL(XEND(1),J)
5             CONTINUE
              QQ=1.-QQ
              IF(QQ.LE.0.0) THEN
              F=10.**10
              WRITE(*,*)'I=1'
              RETURN
              END IF
              Q=DLOG(QQ)*NFAIL(1)
              DO 10 I=2,NK
              QQ=0.0
              DO 8 J=1,NSUB
C             QQ=QQ+P(J)*(DEXP(-(XEND(I-1)/ET(J))**BE(J))-
C        *            DEXP(-(XEND(I)/ET(J))**BE(J)))
              QQ=QQ+P(J)*(REL(XEND(I-1),J)-REL(XEND(I),J))
8             CONTINUE
C
              IF(QQ.LE.0.0) THEN
              F=10.**10
              WRITE(*,*)'I=',I
              RETURN
              END IF
              Q=Q+DLOG(QQ)*NFAIL(I)
10            CONTINUE
              QQ=0.0
              DO 16 J=1,NSUB
C              QQ=QQ+P(J)*DEXP(-(XEND(NK)/ET(J))**BE(J))
              QQ=QQ+P(J)*REL(XEND(NK),J)
16            CONTINUE
              IF(QQ.LE.0.0) THEN
              F=10.**10
              WRITE(*,*)'I=NK+1'
              RETURN
              END IF
              F=Q+DLOG(QQ)*NFAIL(NK+1)
              RETURN
              END
```

```
C       ----------------------------------------------------------
        SUBROUTINE CHECK
C       ----------------------------------------------------------
        implicit double precision (a-h, o-z)
        REAL*8 XEND(200),PRO(200,5),
     *         P(5),ET(5),BE(5),PO(5),ETO(5),BEO(5)
        INTEGER NSUB,N,NK,NFAIL(200),KK
        COMMON /MAIN/ XEND,PRO
        COMMON /INTE/ NSUB,N,NK,NFAIL
        COMMON /KK/KK
        COMMON /PAR/P,ET,BE

        WRITE(6,*)'BETA',BE
        WRITE(6,*)'ETA',ET
        WRITE(6,*)'-P-',P
        RETURN
        END
C       ----------------------------------------------------------
        FUNCTION RAN(ISEED)
C       ----------------------------------------------------------
        PARAMETER(IA=7141,IC=54773,IM=259200)
        ISEED=MOD(ISEED*IA+IC,IM)
        RAN=FLOAT(ISEED)/FLOAT(IM)
        RETURN
        END
C       ----------------------------------------------------------
        FUNCTION RANO(IDUM)
C       ----------------------------------------------------------
        DIMENSION V(97)
        DATA IFF /0/
        IF(IDUM.LT.0.OR.IFF.EQ.0)THEN
        IFF=1
        ISEED=ABS(IDUM)
        IDUM=1
        DO 11 J=1,97
          DUM=RAN(ISEED)
11      CONTINUE
        DO 12 J=1,97
          V(J)=RAN(ISEED)
12      CONTINUE
        Y=RAN(ISEED)
        ENDIF
        J=1+INT(97.*Y)
        IF(J.GT.97.OR.J.LT.1)PAUSE
```

APPENDIX 5C

```
      Y=V(J)
      RANO=Y
      V(J)=RAN(ISEED)
      RETURN
      END
C     ------------------------------------------------------
      SUBROUTINE ORDER(X,N)
C     ------------------------------------------------------
      implicit double precision (a-h, o-z)
      INTEGER N
      DIMENSION X(N)
C
      DO 30 J=1,N
      DO 20 I=J+1,N
      IF(X(J).GE.X(I)) THEN
      Q=X(J)
      X(J)=X(I)
      X(I)=Q
      END IF
20    CONTINUE
30    CONTINUE
      RETURN
      END
C     ------------------------------------------------------
      DOUBLE PRECISION FUNCTION PDF(Y,K)
C     ------------------------------------------------------
      implicit double precision (a-h, o-z)
      REAL*8 Y
      INTEGER K
C
      REAL*8 XEND(200),PRO(200,5),
     *     P(5),ET(5),BE(5),PO(5),ETO(5),BEO(5)
      INTEGER NSUB,N,NK,NFAIL(200),KK
      COMMON /MAIN/ XEND,PRO
      COMMON /INTE/ NSUB,N,NK,NFAIL
      COMMON /KK/KK
      COMMON /PAR/P,ET,BE
C
      QE=BE(K)*DLOG(Y/ET(K))
      IF(QE.GE.6.56) THEN
      PDF=0.0D00
      RETURN
      END IF
      PDF=BE(K)/ET(K)*(Y/ET(K))**(BE(K)-1.)*REL(Y,K)
```

```
      RETURN
      END
C     ----------------------------------------------------------------
      DOUBLE PRECISION FUNCTION REL(Y,K)
C     ----------------------------------------------------------------
      implicit double precision (a-h, o-z)
      REAL*8 Y
      INTEGER K
C
      REAL*8 XEND(200),PRO(200,5),
     *       P(5),ET(5),BE(5),P0(5),ET0(5),BE0(5)
      INTEGER NSUB,N,NK,NFAIL(200),KK
      COMMON /MAIN/ XEND,PRO
      COMMON /INTE/ NSUB,N,NK,NFAIL
      COMMON /KK/KK
      COMMON /PAR/P,ET,BE
C
      QE=BE(K)*DLOG(Y/ET(K))
      IF(QE.GE.6.56) THEN
      REL=0.0D00
      RETURN
      END IF
      QE=DEXP(QE)
      REL=DEXP(-QE)
      RETURN
      END
C     ----------------------------------------------------------
      SUBROUTINE WEII(M,PI,BEI,ETI,N,TT,R,T)
C     ----------------------------------------------------------
C     TO GENERATE MIXED WEIBULL DISTRIBUTION OF INDIVIDUAL
C     TIME-TO-FAILURE, COMPLETE, TIME-TERMINATED OR FAILUE-
C     TERMINATED. PERFOME WITH FUNCTIONS RANO AND RAN, AND
C     SUBROUTINE ORDER.
C     >>>>>>>>>>>>>>>>>>>>>>>>>>>>>>>>>>>>>>>>>>>>>>>>>>>>>>
C     PARAMETERS:
C     M=NUMBER OF SUBPOPULATIONS.
C     PI(M)=ARRAY, MIXING WEIGHT.
C     BEI(M)=ARRAY, M SHAPE PARAMETERS OF M SUBPOPULATIONS.
C     ETI(M)=ARRAY, M SCALE PARAMETERS OF M SUBPOPULATIONS.
C     N=TOTAL NUMBER OF INDIVIDUALS IN THE LIFE TEST.
C     TT=TIME OF TERMINATING THE TEST, IF INPUT TT=0, IMPLIES
C         THAT THE TEST IS COMPLETE OR FAILURE TERMINATED.
C     R=INTEGER, NUMBER OF FAILED INDIVIDUAL, IF INPUT R=0,
C         IMPLIES THAT THE TEST IS COMPLETE OR TIME TERMINATED.
```

APPENDIX 5C

```
C     T(N)=ARRAY, N TIME-TO-FAILURE, IN THE VALUE OF INCREASING
C          ORDER.
C     ** INITIALIZE RANO BEFORE CALLING THIS SUBROUTINE.
C     >>>>>>>>>>>>>>>>>>>>>>>>>>>>>>>>>>>>>>>>>>>>>>>>>>>>>>>>>
      implicit double precision (a-h, o-z)
      INTEGER M,N,R
      REAL*8 PI(M),BEI(M),ETI(M),T(N),TT
      REAL*8 PP(10)
C
      WRITE(*,*)M,N,R,TT
      WRITE(*,*)PI,BEI,ETI
      PP(1)=PI(1)
      DO 5 I=2,M
      PP(I)=PP(I-1)+PI(I)
5     CONTINUE
      DO 20 I=1,N
      U1=RANO(INI)
      DO 10 J=1,M
      IF(U1.LE.PP(J)) THEN
      T(I)=ETI(J)*(-DLOG(RANO(INI)))**(1./BEI(J))
      GOTO 20
      END IF
10    CONTINUE
20    CONTINUE
      CALL ORDER(T,N)
      R=N
      IF((TT.EQ.0.0)) RETURN
      DO 30 I=1,N
      IF(T(I).GT.TT) THEN
      R=I-1
      RETURN
      END IF
30    CONTINUE
      R=N
      RETURN
      END
C     ----------------------------------------
      SUBROUTINE USRFUN(XX,NOVR,NP,ALPHA,BETA)
C     ----------------------------------------
C
      implicit double precision (a-h, o-z)
      INTEGER NOVR,NP,NFAIL(200)
      DIMENSION XX(NP),ALPHA(NP,NP),BETA(NP),BEET(5),RJ(200)
      DIMENSION TOE(200),EXTEB(200), ELNTE(200),
```

```
     *          PP1E(200),PP1B(200),PP2E(200),PP2B(200),
     *          PP2EB(200),PJI(200)
      COMMON /PAR/P(5),ET(5),BE(5)
      COMMON /INTE/ NSUB,N,NK,NFAIL
      COMMON /KK/KK
      COMMON /MAIN/ XEND(200),PRO(200,5)
C
c     write(*,*)' be=',be(1),be(2)
c     write(*,*)' et=',et(1),et(2)
c     write(*,*)' p=',p(1),p(2)
      DO 10 J=1,NK
      TOE(J)=XEND(J)/ET(KK)
      EXTEB(J)=TOE(J)**BE(KK)
      ELNTE(J)=DLOG(TOE(J))
      IF(EXTEB(J).GE.709.78D00)THEN
      RJ(J)=0.0D00
      ELSE
      RJ(J)=DEXP(-EXTEB(J))
      END IF
10    CONTINUE
C
      BEET(KK)=BE(KK)/ET(KK)
C
C     CALCULATE PJI
C
      PJI(1)=1.-RJ(1)
      DO 20 J=2,NK
      PJI(J)=RJ(J-1)-RJ(J)
20    CONTINUE
      PJI(NK+1)=RJ(NK)
C     WRITE(6,*)'-----PJI-----'
C     DO 21 J=1,NK+1
C     WRITE(6,*)J,PJI(J)
C21   CONTINUE
C
C     CALCULATE THE DERIVATIVES
C
      DO 25 J=1,NK
      TOE(J)=XEND(J)/ET(KK)
      EXTEB(J)=TOE(J)**BE(KK)
c     write(*,*)'toe=',toe(j)
      ELNTE(J)=DLOG(TOE(J))
      IF(EXTEB(J).GE.709.78)THEN
      RJ(J)=0.0D00
```

APPENDIX 5C

```
            ELSE
            RJ(J)=DEXP(-EXTEB(J))
c           write(*,*)'LLLLLLLLLLLLLLGGGGGGGGGGGGGGGGGGGG'
            END IF
25       CONTINUE
         PP1E(1)=-BEET(KK)*EXTEB(1)*RJ(1)
         PP2E(1)=-EXTEB(1)*RJ(1)*(BEET(KK)**2*EXTEB(1)
     *          -BE(KK)*(BE(KK)+1.)/ET(KK)**2)
         PP1B(1)=EXTEB(1)*ELNTE(1)*RJ(1)
         PP2B(1)=-ELNTE(1)**2*EXTEB(1)*(EXTEB(1)-1.)*RJ(1)
         PP2EB(1)=-BEET(KK)*RJ(1)*EXTEB(1)*(-EXTEB(1)*ELNTE(1)
     *          +ELNTE(1)+1./BE(KK))
C
         PP1E(NK+1)=BEET(KK)*EXTEB(NK)*RJ(NK)
         PP2E(NK+1)=EXTEB(NK)*RJ(NK)*(BEET(KK)**2*EXTEB(NK)
     *          -BE(KK)*(BE(KK)+1.)/ET(KK)**2)
         PP1B(NK+1)=-EXTEB(NK)*ELNTE(NK)*RJ(NK)
         PP2B(NK+1)=ELNTE(NK)**2*EXTEB(NK)*(EXTEB(NK)-1.)*RJ(NK)
         PP2EB(NK+1)=BEET(KK)*RJ(NK)*EXTEB(NK)*(-EXTEB(NK)*ELNTE(NK)
     *          +ELNTE(NK)+1./BE(KK))
         DO 30 J=2,NK
         PP1E(J)=BEET(KK)*(EXTEB(J-1)*RJ(J-1)-EXTEB(J)*RJ(J))
         PP2E(J)=EXTEB(J-1)*RJ(J-1)*(BEET(KK)**2*EXTEB(J-1)-
     *          BE(KK)*(BE(KK)+1.)/ET(KK)**2)-
     *          EXTEB(J)*RJ(J)*(BEET(KK)**2*EXTEB(J)-
     *          BE(KK)*(BE(KK)+1.)/ET(KK)**2)
         PP1B(J)=-EXTEB(J-1)*ELNTE(J-1)*RJ(J-1)+
     *          EXTEB(J)*ELNTE(J)*RJ(J)
         PP2B(J)=ELNTE(J-1)**2*EXTEB(J-1)*RJ(J-1)*(EXTEB(J-1)-1.)-
     *          ELNTE(J)**2*EXTEB(J)*RJ(J)*(EXTEB(J)-1.)
         PP2EB(J)=BEET(KK)*RJ(J-1)*EXTEB(J-1)*(-EXTEB(J-1)*ELNTE(J-1)
     *          +ELNTE(J-1)+1./BE(KK))-
     *          BEET(KK)*RJ(J)*EXTEB(J)*(-EXTEB(J)*ELNTE(J)
     *          +ELNTE(J)+1./BE(KK))
30       CONTINUE
C
C     CALCULATE ALPHA MATRIX
C
         ALPHA(1,1)=0.0
         ALPHA(1,2)=0.0
         ALPHA(2,2)=0.0
         DO 40 J=1,NK+1
         IF((PJI(J).LE.1.0E-30).OR.(PRO(J,KK).LE.0.0))GOTO 40
         ALPHA(1,1)=ALPHA(1,1)+NFAIL(J)*PRO(J,KK)*
```

```
     *                (PP2E(J)/PJI(J)-(PP1E(J)/PJI(J))**2)
          ALPHA(1,2)=ALPHA(1,2)+NFAIL(J)*PRO(J,KK)*
     *                (PP2EB(J)/PJI(J)-PP1E(J)/PJI(J)*PP1B(J)/PJI(J))
          ALPHA(2,2)=ALPHA(2,2)+NFAIL(J)*PRO(J,KK)*
     *                (PP2B(J)/PJI(J)-(PP1B(J)/PJI(J))**2)
40        CONTINUE
          ALPHA(2,1)=ALPHA(1,2)
C
C         CALCULATE THE VECTOR
C
          BETA(1)=0.0
          BETA(2)=0.0
          DO 50 J=1,NK+1
          IF((PJI(J).LE.1.0E-30).OR.(PRO(J,KK).LE.0.0))GOTO 50
          BETA(1)=BETA(1)-NFAIL(J)*PRO(J,KK)*PP1E(J)/PJI(J)
          BETA(2)=BETA(2)-NFAIL(J)*PRO(J,KK)*PP1B(J)/PJI(J)
50        CONTINUE
c         WRITE(6,*)'------BETA0 ----------'
c         WRITE(6,*)BETA
          RETURN
          END
C         ----------------------------------------
          SUBROUTINE MNEWT(NTRIAL,X,N,TOLX,TOLF)
C         ----------------------------------------
          PARAMETER (NP=2)
          implicit double precision (a-h, o-z)
          DIMENSION X(NP),ALPHA(NP,NP),BETA(NP),INDX(NP),
     *              Y(NP,NP),C(NP,NP)
          COMMON /PAR/ P(5),ET(5),BE(5)
          COMMON /KK/KK
          COMMON /FISHER/ALPHA,BETA,Y,C
          DO 13 K=1,NTRIAL
            CALL USRFUN(X,N,NP,ALPHA,BETA)
            ERRF=0.
            DO 11 I=1,N
              ERRF=ERRF+ABS(BETA(I))
11          CONTINUE
            IF(ERRF.LE.TOLF)GOTO 100
            CALL LUDCMP(ALPHA,N,NP,INDX,D)
            CALL LUBKSB(ALPHA,N,NP,INDX,BETA)
            ERRX=0.
            DO 12 I=1,N
              ERRX=ERRX+ABS(BETA(I))
              X(I)=X(I)+BETA(I)
```

APPENDIX 5C

```fortran
12        CONTINUE
          ET(KK)=X(1)
          BE(KK)=X(2)
16        CONTINUE
C          WRITE(6,*)'---X---',X
C          WRITE(6,*)'P=',P
C          WRITE(6,*)'ET=',ET
C          WRITE(6,*)'BE=',BE
          IF(ERRX.LE.TOLX)GOTO 100
13        CONTINUE
C
C100      WRITE(6,*)' THE NUMBER OF NEWTON ITERATIONS=',K
C
100       RETURN
          END
C         ------------------------------------
          SUBROUTINE LUBKSB(A,N,NP,INDX,B)
C         ------------------------------------
          implicit double precision (a-h, o-z)
          DIMENSION A(NP,NP),INDX(N),B(N)
          II=0
          DO 12 I=1,N
           LL=INDX(I)
           SUM=B(LL)
           B(LL)=B(I)
           IF (II.NE.0)THEN
             DO 11 J=II,I-1
               SUM=SUM-A(I,J)*B(J)
11           CONTINUE
           ELSE IF (SUM.NE.0.) THEN
             II=I
           ENDIF
           B(I)=SUM
12        CONTINUE
          DO 14 I=N,1,-1
           SUM=B(I)
           IF(I.LT.N)THEN
             DO 13 J=I+1,N
               SUM=SUM-A(I,J)*B(J)
13           CONTINUE
           ENDIF
           B(I)=SUM/A(I,I)
14        CONTINUE
          RETURN
```

```
            END
C     ---------------------------------------------
      SUBROUTINE LUDCMP(A,N,NP,INDX,D)
C     ---------------------------------------------
      implicit double precision (a-h, o-z)
      PARAMETER (NMAX=100,TINY=1.0E-20)
      DIMENSION A(NP,NP),INDX(N),VV(NMAX)
c
      D=1.
      DO 12 I=1,N
        AAMAX=0.
        DO 11 J=1,N
          IF (ABS(A(I,J)).GT.AAMAX) AAMAX=ABS(A(I,J))
11      CONTINUE
        IF (AAMAX.EQ.0.) PAUSE 'Singular matrix.'
        VV(I)=1./AAMAX
12    CONTINUE
      DO 19 J=1,N
        IF (J.GT.1) THEN
          DO 14 I=1,J-1
            SUM=A(I,J)
            IF (I.GT.1)THEN
              DO 13 K=1,I-1
                SUM=SUM-A(I,K)*A(K,J)
13            CONTINUE
              A(I,J)=SUM
            ENDIF
14        CONTINUE
        ENDIF
        AAMAX=0.
        DO 16 I=J,N
          SUM=A(I,J)
          IF (J.GT.1)THEN
            DO 15 K=1,J-1
              SUM=SUM-A(I,K)*A(K,J)
15          CONTINUE
            A(I,J)=SUM
          ENDIF
          DUM=VV(I)*ABS(SUM)
          IF (DUM.GE.AAMAX) THEN
            IMAX=I
            AAMAX=DUM
          ENDIF
16      CONTINUE
```

```
            IF (J.NE.IMAX)THEN
              DO 17 K=1,N
                DUM=A(IMAX,K)
                A(IMAX,K)=A(J,K)
                A(J,K)=DUM
17            CONTINUE
              D=-D
              VV(IMAX)=VV(J)
            ENDIF
            INDX(J)=IMAX
            IF(J.NE.N)THEN
              IF(A(J,J).EQ.0.)A(J,J)=TINY
              DUM=1./A(J,J)
              DO 18 I=J+1,N
                A(I,J)=A(I,J)*DUM
18          CONTINUE
            ENDIF
19      CONTINUE
        IF(A(N,N).EQ.0.)A(N,N)=TINY
c
        RETURN
        END

c     ---------------------------------------------------------------
        SUBROUTINE NEWBEG(ETA,BETA,K,EPS)
c     ---------------------------------------------------------------
c     ESTIMATE ETA(K) AND BETA(K), GIVEN P(K|T_J) AND T_J, USING
c     NEWTON-ITERATION METHOD.
c
        implicit double precision (a-h, o-z)
        REAL*8 ETA,BETA,EPS,XMID(200)
        INTEGER K
c
        REAL*8 XEND(200),PRO(200,5),
     *       P(5),ET(5),BE(5),P0(5),ET0(5),BE0(5)
        INTEGER NSUB,N,NK,NFAIL(200),KK
        COMMON /MAIN/ XEND,PRO
        COMMON /INTE/ NSUB,N,NK,NFAIL
        COMMON /KK/KK
        COMMON /PAR/P,ET,BE
        COMMON /XMID/ XMID
c
        MM=0
        RR=0.0
```

```
              DO 2 I=1,NK+1
              RR=RR+PRO(I,K)*NFAIL(I)
2             CONTINUE
              PERL=0.1673*RR
              PERU=0.9737*RR
              RRO=0.0
              DO 5 I=1,NK
              RRO=RRO+PRO(I,K)*NFAIL(I)
              IF(RRO.GE.PERL) THEN
              IRO=I
              GOTO 6
              END IF
5             CONTINUE
6             DO 7 I=IRO+1,NK
              RRO=RRO+PRO(I,K)*NFAIL(I)
              IF(RRO.GE.PERU) THEN
              IR1=I
              GOTO 8
              END IF
7             CONTINUE
              IR1=NK
c             WRITE(*,*)'-----NEWBEG------'
C     WRITE(*,*)'RR',RR,'IR1',IR1,'IRO',IRO
C     WRITE(*,*)'XMID(IR1)',XMID(IR1),'XMID(IRO)',XMID(IRO)
8             TEM1=2.99/(DLOG(XMID(IR1))-DLOG(XMID(IRO)))
C     WRITE(*,*)'BETAO=',TEM1,'  LOW   UP',IRO,IR1
C     WRITE(6,*)'BETAO=',TEM1,'  LOW   UP',IRO,IR1
              IF(TEM1.GT.10.) THEN
              WRITE(*,*)' BETAO > 10.'
              TEM1=10.
              END IF
10            BETA=TEM1
              A=0.0
              B=0.0
              C=0.0
              D=0.0
              E=0.0
              F=0.0
              BB=0.0
              CC=0.0
              DO 20 I=1,NK
              EX=XMID(I)**BETA
c             write(*,*) 'xmid=====================',xmid(i)
              AL=DLOG(XMID(I))
```

```
              A1=(PRO(I,K)*NFAIL(I))*EX
              B1=(PRO(I,K)*NFAIL(I))*AL
              A=A+PRO(I,K)*NFAIL(I)
              B=B+A1*AL
              D=D+A1
              F=F+B1
              BB=BB+A1*AL*AL
 20           CONTINUE
              EX=XEND(NK)**(BETA)
 c            write(*,*)'xend=============',xend(nk)
              AL=DLOG(XEND(NK))
              C=NFAIL(NK+1)*PRO(NK+1,K)*EX*AL
              E=NFAIL(NK+1)*PRO(NK+1,K)*EX
              CC=NFAIL(NK+1)*PRO(NK+1,K)*EX*AL*AL
              IF((BETA.EQ.0.0).OR.(RR.EQ.0.0).OR.(B.EQ.0.0)) THEN
              CALL CHECK
              WRITE(6,*)'-B-',B
              WRITE(6,*)'-RR-',RR
              WRITE(6,*)'-MM-',MM
              END IF
              TEM1=-A*(B+C)/(D+E)+F+A/BETA
              TEM2=(((D+E)*(BB+CC)-(B+C)*(B+C))/((D+E)*(D+E))+1./BETA**2)
              TEM2=-A*TEM2
 C      WRITE(*,*)'TEM1 AND TEM2',TEM1,TEM2,'BETA',BETA
              TEM1=BETA-(TEM1/TEM2)
              MM=MM+1
              IF((ABS((BETA-TEM1)/BETA).GE.EPS).OR.(MM.LE.2)) GOTO 10
              BETA=TEM1
 C            WRITE(*,*)'D E A',D,E,A,'BETA=',BETA
              ETA=((D+E)/A)**(1./BETA)
 25           CONTINUE
 C            WRITE(*,*)'BETA AND ETA',BETA,ETA
              RETURN
              END
```

APPENDIX 5D
COMPUTER PROGRAM OF "LSQ.FOR"

```fortran
c $$$$$$$$$$$$$$$$$$$$$$$$$$$$$$$$$$$$$$$$$$$$$$$$$$$$$$$$
c $      PARAMETERS' ESTIMATION FOR BATHTUB MODEL 1     $
c $              USING LEAST SQUARES METHOD             $
c $$$$$$$$$$$$$$$$$$$$$$$$$$$$$$$$$$$$$$$$$$$$$$$$$$$$$$$$
        implicit double precision (a-h, o-z)
        real almt
        dimension t(100),nf(100),r(100),x(100),y(100)
        common /div/dt,nd,n
        open(5,file='ilsq.dat',status='unknown')
        open(6,file='olsq.dat',status='new')
c
c       PARAMETER ESTIMATION BY LEAST SQUARES METHOD
c
5       nnf=0
        xa=0.0
        ya=0.0
        xx=0.0
        yy=0.0
        xy=0.0
        read(5,*) n,nd,dt
        write(6,8)
8       format(1x,///)
        write(6,10) n,nd,dt
10      format(15x, 'Sample size:      ',i20/15x,
     *  'Time interval number: ', i10
     *  /15x,'Interval width: ',f20.6)
c       write(6,12)
12      format(///5x,'No.',10x,'Ending time Ti',7x,'NF(DTi)',
     *  7x,'NS(Ti)',8x,'R(Ti)')
c       write(6,*)('=',i=1,78)
        do 20 i=1,nd
        read(5,*) t(i), nf(i)
        nnf=nnf+nf(i)
        ns=n-nnf
        r(i)=float(ns)/float(n)
c       write(6,14) i,t(i),nf(i),ns,r(i)
14      format(1x,i7,7x,f12.6,5x,i10,5x,i10,5x,f12.8)
        x(i)=dlog(t(i))
        y(i)=dlog(dlog(1-dlog(r(i))))
        xa=xa+x(i)
```

APPENDIX 5D

```
              ya=ya+y(i)
20            continue
c             write(6,*)('=',i=1,78)
              call rate(nf)
              write(6,8)
              xa=xa/float(nd)
              ya=ya/float(nd)
              do 30 i=1,nd
              xx=(x(i)-xa)*(x(i)-xa)+xx
              yy=(y(i)-ya)*(y(i)-ya)+yy
              xy=(x(i)-xa)*(y(i)-ya)+xy
30            continue
              beta=xy/xx
              a=ya-beta*xa
              alpha=dexp(-a/beta)
              cor=xy/sqrt(xx*yy)
              write(6,8)
              write(6,*)'======== Parameter Estimates ========'
              write(6,40) beta,alpha,cor
40            format(////10x,'beta=',E30.6/10x, 'alpha=',E30.6/10x,
     *        'Rxy=',f10.6///)
              write(6,*)('=',i=1,78)
              stop
              end
```

Chapter 6

BURN-IN TIME DETERMINATION USING A QUICK CALCULATION APPROACH

To determine the burn-in time, data like those given in Table 6.1 need to be obtained. This table is based, as an example, on 900 devices subjected to burn-in, and their performance is monitored in 10-hr intervals to determine the number of devices that fail in each interval. Next, the failure rate for each time interval is calculated. In Row 2 of Table 6.1 the equations for calculating the failure rate of these devices at each time interval are given. The first equation is

$$\hat{\bar{\lambda}}(T) = \frac{N_F(\Delta T)}{N_{BP}(T) \times \Delta T}, \qquad (6.1)$$

where

$\hat{\bar{\lambda}}(T)$ = average estimate of the failure rate in each ΔT time interval,

$N_F(\Delta T)$ = number of devices failing in ΔT time interval, which in Table 6.1 is 10 hr,

$N_{BP}(T)$ = number of devices at the beginning of each ΔT period involved,

and

ΔT = time interval for which the $\hat{\bar{\lambda}}(T)$ is calculated.

It may be seen that the denominator of Eq. (6.1) is the total operating device-hours during the ΔT interval. But in fact there are only $[N_{BP}(T) - N_F(\Delta T)]$ devices surviving this whole ΔT period, while $N_F(\Delta T)$ devices fail during each interval. Therefore, the actual total device hours of operation by all devices being burned-in is less than $N_{BP}(T) \times \Delta T$. To calculate the failure rate in each time interval more accurately, the following formula may be used:

$$\hat{\hat{\lambda}}(T) = \frac{N_F(\Delta T)}{[N_{BP}(T) - N_F(\Delta T)] \times \Delta T + \frac{1}{2} \times N_F(\Delta T) \times \Delta T}.$$

(6.2)

This equation is based on the assumption that the $N_F(\Delta T)$ failures occur uniformly during the ΔT interval. This formula gives a higher failure rate value than Eq. (6.1), but it approaches the true failure rate, $\lambda(T)$, more closely.

In the first time interval of 10 hr in Table 6.1, the number of failures observed is 33 out of the 900 devices put to burn-in. Consequently, the failure rate for the first time interval, if Eq. (6.1) is used, is

$$\hat{\hat{\lambda}}_1(\Delta T = 10 \text{ hr}) = \frac{33}{900 \times 10},$$

or

$$\hat{\hat{\lambda}}_1(\Delta T = 10 \text{ hr}) = 0.003667 \text{ fr/hr}.$$

If Eq. (6.2) is used, then

$$\hat{\hat{\lambda}}_1(\Delta T = 10 \text{ hr}) = \frac{33}{(900 - 33) \times 10 + \frac{1}{2} \times 33 \times 10},$$

or

$$\hat{\hat{\lambda}}_1(\Delta T = 10 \text{ hr}) = 0.003735 \text{ fr/hr} > 0.003667 \text{ fr/hr}.$$

It may be seen that Eq. (6.2) yields higher failure rate values. This procedure is used for the remaining time intervals as given in Table 6.1.

Next, the failure number in each time interval versus the burn-in hours accumulated up to the end of each ΔT period is plotted in histogram form as shown in Fig. 6.1. A smooth curve is then drawn favoring the midpoints of the tops of each histogram bar. As may be seen the number of failures drops relatively sharply, which is indicative of early, burn-in failures.

TABLE 6.1 – Field data for a population exhibiting early plus chance failure characteristics.

1	2	3	4	
Life, hr	Number of failures, N_F	No. of units surviving period, N_{se}	Failure rate	
			$\bar{\lambda} = \dfrac{N_F(\Delta T)}{N_{BP}(T) \times \Delta T}$, fr/hr	$\bar{\lambda} = \dfrac{N_F(\Delta T)}{[N_{BP}(T) - N_F(\Delta T)]\Delta T + \frac{1}{2}[N_F(\Delta T)](\Delta T)}$, fr/hr
0 - 10	33	900-33=867	$\dfrac{33}{900 \times 10} = 0.003667$	$\dfrac{33}{(900-33) \times 10 + \frac{1}{2}(33 \times 10)} = 0.003735$
10 - 20	31	867-31=836	$\dfrac{31}{867 \times 10} = 0.003576$	$\dfrac{31}{(867-31) \times 10 + \frac{1}{2}(31 \times 10)} = 0.003641$
20 - 30	28	836-28=808	$\dfrac{28}{836 \times 10} = 0.003349$	$\dfrac{28}{(836-28) \times 10 + \frac{1}{2}(28 \times 10)} = 0.003406$
30 - 40	25	808-25=783	$\dfrac{25}{808 \times 10} = 0.003094$	$\dfrac{25}{(808-25) \times 10 + \frac{1}{2}(25 \times 10)} = 0.003143$
40 - 50	23	783-23=760	$\dfrac{23}{783 \times 10} = 0.002937$	$\dfrac{23}{(783-23) \times 10 + \frac{1}{2}(23 \times 10)} = 0.002981$
50 - 60	21	760-21=739	$\dfrac{21}{760 \times 10} = 0.002763$	$\dfrac{21}{(760-21) \times 10 + \frac{1}{2}(21 \times 10)} = 0.002802$
60 - 70	20	739-20=719	$\dfrac{20}{739 \times 10} = 0.002706$	$\dfrac{20}{(739-20) \times 10 + \frac{1}{2}(20 \times 10)} = 0.002743$
70 - 80	19	719-19=700	$\dfrac{19}{719 \times 10} = 0.002643$	$\dfrac{19}{(719-19) \times 10 + \frac{1}{2}(19 \times 10)} = 0.002678$
80 - 90	18	700-18=682	$\dfrac{18}{700 \times 10} = 0.002571$	$\dfrac{18}{(700-18) \times 10 + \frac{1}{2}(18 \times 10)} = 0.002605$
90 - 100	18	682-18=664	$\dfrac{18}{682 \times 10} = 0.002639$	$\dfrac{18}{(682-18) \times 10 + \frac{1}{2}(18 \times 10)} = 0.002675$
Over 100	664	664-664=0	—	—
	N=900			

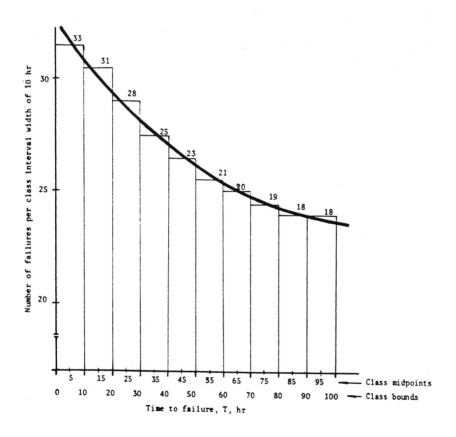

Fig. 6.1– Number of failures versus time-to-failure relationship during burn-in.

Next the failure rates which have been calculated are plotted, as shown in Fig. 6.2, versus the burn-in time, and two smooth curves are drawn favoring the midpoints of the tops of each histogram bar. These are the curves to be used to determine the burn-in time, which should be that time at which the failure rate stabilizes to a constant level, or becomes horizontal in disposition. In Fig. 6.2 the indicated burn-in time is approximately 95 hr, if Eq. (6.1) is used, and 97 hr if Eq. (6.2) is used. There is not much difference between these two burn-in times, though Curve 2 is higher than Curve 1. However, Curve 2 is recommended because it is more accurate.

In specific cases the time interval ΔT may need to be shorter or longer than 10 hr. This ΔT value may be established by studying past data on such devices. It is desirable also to monitor all components under burn-in, which may be accomplished by a microcomputer, which can check each component being burned-in, find out whether or not it has failed, and determine the time interval during which it failed.

Table 6.2 and Figs. 6.3 and 6.4 have also been included here to give an example of how it can be verified that the components have indeed been burned-in sufficiently to reach their useful life, during which the failure rate is essentially constant, as shown in Fig. 6.4.

REFERENCES

1. Kececioglu, Dimitri B., *Reliability Engineering Handbook*, Prentice Hall, Inc., Upper Saddle River, NJ, Vol. 1, 720 pp. and Vol. 2, 568 pp., 1991, Fifth Printing, 1995.

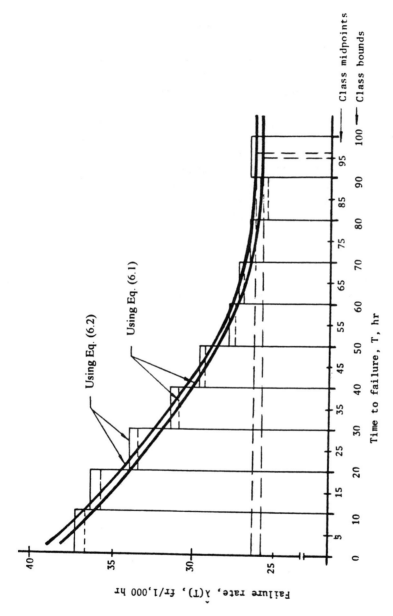

Fig. 6.2— Failure rate versus time-to-failure relationship during burn-in.

TABLE 6.2 – Field data for 900 identical units exhibiting chance failure characteristics.

1	2	3	4	
Life, hr, starting at 100 hr	Number of failures, N_F	No. of units surviving period, N_{se}	Failure rate	
			$\hat{\lambda} = \dfrac{N_F(\Delta T)}{N_{BP}(T) \times \Delta T}$, fr/hr	$\hat{\lambda} = \dfrac{N_F(\Delta T)}{[N_{BP}(T) - N_F(\Delta T)]\Delta T + \frac{1}{2}[N_F(\Delta T)](\Delta T)}$, fr/hr
0 - 20	47	900−47=853	$\dfrac{47}{900 \times 20} = 0.002611$	$\dfrac{47}{(900-47) \times 20 + \frac{1}{2}(47 \times 20)} = 0.002681$
20 - 40	44	853−44=809	$\dfrac{44}{853 \times 20} = 0.002579$	$\dfrac{44}{(853-44) \times 20 + \frac{1}{2}(44 \times 20)} = 0.002647$
40 - 60	42	809−42=767	$\dfrac{42}{809 \times 20} = 0.002596$	$\dfrac{42}{(809-42) \times 20 + \frac{1}{2}(42 \times 20)} = 0.002665$
60 - 80	39	767−39=728	$\dfrac{39}{767 \times 20} = 0.002542$	$\dfrac{39}{(767-39) \times 20 + \frac{1}{2}(39 \times 20)} = 0.002609$
80 - 100	38	728−38=690	$\dfrac{38}{728 \times 20} = 0.002610$	$\dfrac{38}{(728-38) \times 20 + \frac{1}{2}(38 \times 20)} = 0.002680$
100 - 120	35	690−35=655	$\dfrac{35}{690 \times 20} = 0.002536$	$\dfrac{35}{(690-35) \times 20 + \frac{1}{2}(35 \times 20)} = 0.002602$
Over 120	655	655−655=0	—	—
	N=900		$\hat{\lambda} = 0.002579$ fr/hr	$\hat{\lambda} = 0.002647$ fr/hr

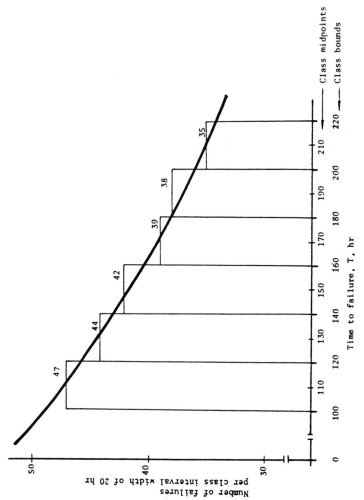

Fig. 6.3– Number of failures versus times-to-failure relationship for chance failures after burn-in.

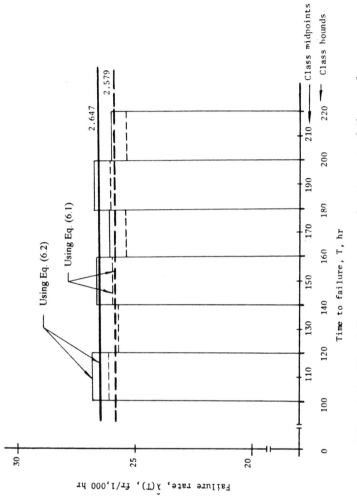

Fig. 6.4- The failure rate versus the times-to-failure relationship for chance failures after burn-in.

Chapter 7

BURN-IN TIME DETERMINATION BASED ON THE BIMODAL TIMES-TO-FAILURE DISTRIBUTION

It is often desired to determine the optimum burn-in time, T_b, such that performing a burn-in for time T_b brings the components into their useful life or the constant failure rate period of the components; or yields the components meeting specified reliability, failure rate or MTBF goals. Several approaches are presented here for different cases.

7.1 THE SUBPOPULATION TRUNCATION APPROACH

Consider the bimodal Weibull distribution with a *cdf* of

$$F(T) = pF_1(T) + (1-p)F_2(T),$$

where

$$F_1(T) = 1 - e^{-(T/\eta_1)^{\beta_1}},$$

and

$$F_2(T) = 1 - e^{-(T/\eta_2)^{\beta_2}}.$$

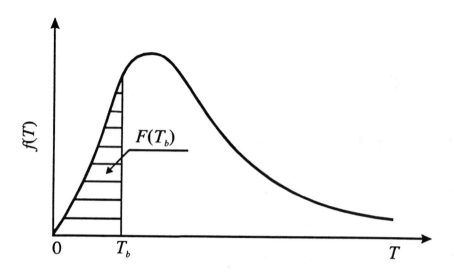

Fig. 7.1– Early part of the *pdf* of the mixed-Weibull *pdf* with burn-in continuing for time T_b [1].

Then,

$$F(T) = p\,(1 - e^{-(T/\eta_1)^{\beta_1}}) + (1-p)\,(1 - e^{-(T/\eta_2)^{\beta_2}}),$$

or

$$F(T) = 1 - p\,e^{-(T/\eta_1)^{\beta_1}} - (1-p)\,e^{-(T/\eta_2)^{\beta_2}}. \tag{7.1}$$

Figure 7.1 shows the probability density function of the times to failure, where burn-in continues for time T_b.

The new cumulative distribution function, $F_n(T)$, describing the remaining part of the distribution is

$$F_n(T) = \frac{F(T) - F(T_b)}{1 - F(T_b)}, \quad T_b \leq T < \infty. \tag{7.2}$$

Substituting Eq. (7.1) into Eq. (7.2) yields

$$F_n(T) = \frac{[1 - F(T_b)] - [p\,e^{-(T/\eta_1)^{\beta_1}} + (1-p)\,e^{-(T/\eta_2)^{\beta_2}}]}{1 - F(T_b)},$$

or

$$F_n(T) = 1 - \frac{p\, e^{-(T/\eta_1)^{\beta_1}} + (1-p)\, e^{-(T/\eta_2)^{\beta_2}}}{p\, e^{-(T_b/\eta_1)^{\beta_1}} + (1-p)\, e^{-(T_b/\eta_2)^{\beta_2}}}, \quad T_b \leq T < \infty. \tag{7.3}$$

Figure 7.2 gives a graphical comparison between the *pdf's* and *cdf's* of the original and the truncated life distributions.

If the burn-in for T_b hours has been carried out so that the early subpopulation has been more or less eliminated, then

$$e^{-(T/\eta_1)^{\beta_1}} \ll e^{-(T/\eta_2)^{\beta_2}}, \quad T_b \leq T < \infty,$$

and Eq. (7.3) becomes

$$F_n(T) \cong 1 - e^{-[(T/\eta_2)^{\beta_2} - (T_b/\eta_2)^{\beta_2}]}, \quad T_b \leq T < \infty. \tag{7.4}$$

For $T = T_b, F_n(T) \cong 0$; and for $T \gg T_b, F_n(T) \cong 1 - e^{-(T/\eta_2)^{\beta_2}}$. If $\beta_2 = 1$, then Eq. (7.4) becomes

$$F_n(T) \cong 1 - e^{-(T - T_b/\eta_2)^1}, \quad T_b \leq T < \infty. \tag{7.5}$$

From this it can be concluded that the plotting of $F_n(T)$ on Weibull probability paper, by plotting MR_i versus $(T_i - T_b)$, will yield a straight line, if β_2 has truly a value of 1 [1].

Therefore, based on the above discussion, to determine the burn-in time, or to evaluate the efficiency of burn-in, the following steps may be followed:

1. Arrange the times-to-failure data in ascending order and determine the corresponding median ranks.

2. Plot the median ranks versus the times to failure on Weibull probability paper to get an "S" shaped, or an approximately "S" shaped, curve which indicates a bimodal failure pattern.

3. Choose a time value near the flat portion, or near the knee, of the just plotted curve as the first estimate of the burn-in time, T_b.

4. Separate the remaining part of the sample after the truncation point T_b, and determine the new median ranks.

224 STATISTICAL BURN-IN TIME DETERMINATION

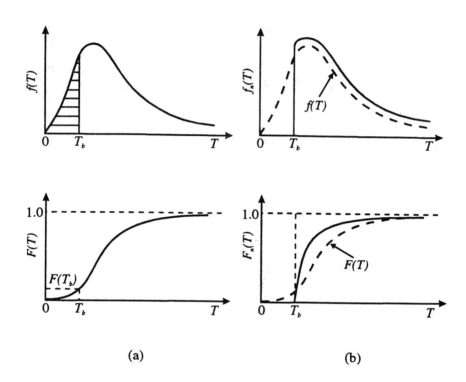

Fig. 7.2– (a) Original distribution. (b) Truncated distribution [1].

SUBPOPULATION TRUNCATION APPROACH

5. Replot the new median ranks, MR_j, versus the $(T_j - T_b)$ on Weibull probability paper. If replotting yields a straight line, it can be said that T_b hours of burn-in brings the components into their useful life period. If replotting does not yield a straight line, but an "S" shaped curve, indications are that the time of burn-in is not long enough. Then, adjust the T_b to another, more favorable, value (larger) and repeat Steps 4 and 5 until the best T_b is found.

6. It is recommended that, after getting a straight line, the line's slope be checked to see if β_2 is acceptably equal to 1.

It is seen that this approach is very easy to implement, and is independent of the behavior of the early failures. No parameter estimation is involved during the process, except at the last step where β_2 needs to be checked. Of course, the parameter p, estimated by the methods presented in Section 5.4, can also be used to quickly locate the proper truncation point. It should be pointed out that this truncation approach is true only for $\beta_2 = 1$, but in practice it may be taken to be valid for the $\beta_2 \neq 1$ case, within reasonable limits.

This approach is illustrated next by a numerical example.

EXAMPLE 7-1

Using the life test data given in Table 7.1, determine the burn-in time using the subpopulation truncation approach and Eq. (7.3).

SOLUTION TO EXAMPLE 7-1

The total failure number, and the corresponding median ranks at the end of each time interval, are given in Columns 4 and 5, respectively, of Table 7.1. The plotting of $MR(T_i)$ versus T_i on WPP yields an approximately "S" shaped curve composed of two "straight lines" crossing at $T = 70$ hr, which indicates a bimodal failure pattern as shown in Fig. 7.3.

Now choose the crossing time value as the truncation point; i.e., $T_b = 70$ hr. This T_b can be regarded as the first estimate of the burn-in time. Separate the remaining part of the sample after the truncation point $T_b = 70$ hr. The new median ranks are given in Table 7.2. Replotting the new median ranks, MR_j, versus the $(T_j - T_b)$ on Weibull probability paper yields a straight line as shown in Fig. 7.3. It may be seen that the Weibull slope of this line is $\hat{\beta} = 0.98$ which is almost 1. Therefore, it can be concluded that $T_b = 70$ hr of burn-in brings the components into their useful life period; i.e, the appropriate burn-in time for these components is 70 hr.

Table 7.1– Raw failure data from life testing with a sample size of $N = 900$, for Example 7–1.

Time interval number, i	Time interval, (T_{i-1}, T_i), hr	Failure number, N_i	Total failure number, $N_f(T_i)$	Median rank, $MR(T_i)$, %
1	(0, 10)	236	236	26.18
2	(10, 20)	95	331	36.73
3	(20, 30)	68	399	44.28
4	(30, 40)	54	453	50.28
5	(40, 50)	46	499	55.39
6	(50, 60)	39	538	59.72
7	(60, 70)	34	572	63.49
8	(70, 80)	30	602	66.83
9	(80, 90)	27	629	69.82
10	(90, 100)	24	653	72.49
11	(100, 110)	22	675	74.93
12	(110, 120)	20	695	77.15
13	(120, 130)	18	713	79.15
14	(130, 140)	16	729	80.93
15	(140, 150)	14	743	82.49
16	(150, 160)	13	756	83.93
17	(160, 170)	12	768	85.26
18	(170, 180)	11	779	86.48
19	(180, 190)	10	789	87.59
20	(190, 200)	9	798	88.59

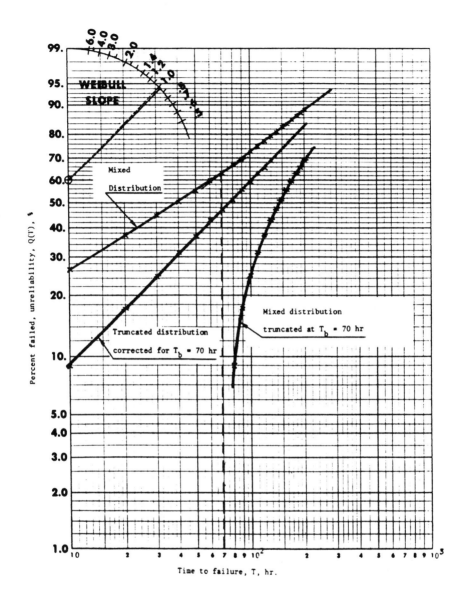

Fig. 7.3– Weibull plot of a bimodal mixture before and after truncation at T_b, for Example 7–1.

Table 7.2– New median rank calculation after truncation at $T_b = 70$ hr, for Example 7–1. Note that the sample size is decreased to $N = 900 - N(70 \text{ hr}) = 900 - 572 = 328$.

Time interval number, i	Time interval, (T_{i-1}, T_i), hr	Failure number, N_i	Total failure number, $N_f(T_i)$	Median rank, $MR(T_i)$, %
1	(70, 80)	30	30	9.04
2	(80, 90)	27	57	17.27
3	(90, 100)	24	81	24.57
4	(100, 110)	22	103	31.27
5	(110, 120)	20	123	37.37
6	(120, 130)	18	141	42.84
7	(130, 140)	16	157	47.72
8	(140, 150)	14	171	51.98
9	(150, 160)	13	184	55.94
10	(160, 170)	12	196	59.59
11	(170, 180)	11	207	62.94
12	(180, 190)	10	217	65.99
13	(190, 200)	9	226	68.73

7.2 THE BURN-IN TIME CORRESPONDING TO A ZERO SLOPE POINT OF THE FAILURE RATE CURVE

In the preceding section the failure rate function based on the bathtub curve Model 1 and the raw data $(T_i, N_f(T_i)), i = 1, 2, ..., n$, was obtained, which is

$$\hat{\lambda}(T) = (\frac{\hat{\beta}}{\hat{\alpha}^{\hat{\beta}}}) T^{\hat{\beta}-1} e^{(T/\hat{\alpha})^{\hat{\beta}}}.$$

The differentiation of $\lambda(T)$ yields

$$\frac{d[\hat{\lambda}(T)]}{dT} = \frac{\hat{\beta}}{\hat{\alpha}^{\hat{\beta}}} [(\hat{\beta}-1) T^{\hat{\beta}-2} e^{(T/\hat{\alpha})^{\beta}} + T^{\hat{\beta}-1} e^{(T/\hat{\alpha})^{\beta}} \hat{\beta} (\frac{T}{\hat{\alpha}})^{\hat{\beta}-1} \frac{1}{\hat{\alpha}}]. \quad (7.6)$$

Let

$$\frac{d[\hat{\lambda}(T)]}{dT} = 0,$$

then,

$$T_b = \hat{\alpha} \left(\frac{1}{\hat{\beta}} - 1\right)^{1/\hat{\beta}}. \quad (7.7)$$

This is the time at which the failure rate function has a zero slope, or it is the theoretical burn-in time.

7.3 THE BURN-IN TIME FOR A SPECIFIED ERROR ON THE ZERO FAILURE RATE CURVE SLOPE

Let

$$\frac{d[\lambda(T)]}{dT} = \varepsilon,$$

where $|\varepsilon|$ is larger than 0 but much less than 1. For the bimodal failure pattern ε is negative. Then,

$$\frac{\hat{\beta}}{\hat{\alpha}^{\hat{\beta}}} [(\hat{\beta}-1) T^{\hat{\beta}-2} e^{(T/\hat{\alpha})^{\beta}} + T^{\hat{\beta}-1} e^{(T/\hat{\alpha})^{\beta}} \hat{\beta} (\frac{T}{\hat{\alpha}})^{\hat{\beta}-1} \frac{1}{\hat{\alpha}}] = \varepsilon. \quad (7.8)$$

Rearranging this equation yields the iterative equation

$$T = \left[\frac{\varepsilon \times e^{-(T/\hat{\alpha})^{\hat{\beta}}}}{(\hat{\beta} - 1) + \hat{\beta} \times (T/\hat{\alpha})^{\hat{\beta}}}\right]^{1/\hat{\beta}-2}. \tag{7.9}$$

Then, the burn-in time corresponding to the failure rate function's slope of ε can be found, by putting an initial T_0 into Eq. (7.9) first and iterating step by step until the absolute difference between two adjacent T's are small enough, say less than 10^{-6}. It may be seen that the smaller the absolute value of ε is, the longer the T_b will be.

A computer program named "LSQ" has been developed, which covers the estimation of the two parameters β and α by the least-squares method, the calculation of the theoretical burn-in time corresponding to $\frac{d[\lambda(T)]}{dT} = 0$, and the calculation of the approximate burn-in time corresponding to $\frac{d[\lambda(T)]}{dT} = \varepsilon$. Appendix 7A is the code listing of this program. The numerical examples that follow illustrate this approach.

EXAMPLE 7-2

Ninety (90) identical units are put to a life test. The life test data are given in Table 5.7. Find the theoretical and the practical burn-in time with $\varepsilon = -0.01$, -0.001, -0.0001, -0.00001, -0.000001, and -0.0000001, using the Bathtub Model 1 to fit the data.

SOLUTION TO EXAMPLE 7-2

From Example 5-7, the least-squares estimates of the parameters of the Bathtub Model 1 are

$$\hat{\beta} = 0.578164,$$
$$\hat{\alpha} = 7.56028 \times 10^3 \text{ hr},$$

and

$$R_{xy} = 0.979237.$$

It may be seen that the Bathtub Model 1 fits the test data very well, with a high correlation coefficient of 0.979237.

Inputting the given failure data information into the computer program "LSQ" given in Appendix 7A yields the burn-in times given in Table 7.3 corresponding to different values of ε.

It may be seen that the lower the absolute error on the zero slope, the longer the burn-in duration. The theoretical burn-in time (with zero failure rate slope), of 4,382.69 hr, is much longer than the practical burn-in times with different errors on the zero slope.

Table 7.3– Burn-in times corresponding to different values of ε, for Example 7–2.

ε	T_b, hr
0	4,382.69
-0.01	0.25
-0.001	1.26
-0.0001	6.36
-0.00001	31.85
-0.000001	157.07
-0.0000001	725.49

EXAMPLE 7–3

One thousand (1,000) identical units are put to a life test. The life test data are given in Table 5.19. Find the theoretical and the practical burn-in times with errors on the zero failure rate slope of $\varepsilon = -0.01$, -0.001, -0.0001, -0.00001, -0.000001, and -0.0000001, using the Bathtub Model 1 to fit the data.

SOLUTION TO EXAMPLE 7–3

From Example 5–8, the least-squares estimates of the parameters of the Bathtub Model 1 are

$\hat{\beta} = 1.93706 \times 10^{-1}$,
$\hat{\alpha} = 8.64919 \times 10^5$ hr,

and

$R_{xy} = 0.972924$.

It may be seen that the Bathtub Model 1 fits the test data very well, with a high correlation coefficient of 0.972924.

Inputting the given failure data information into the computer program "LSQ" given in Appendix 7A yields the burn-in times given in Table 7.4 corresponding to different values of ε.

It may be seen that the lower the absolute error on the zero slope, the longer the burn-in duration. The theoretical burn-in time (with zero failure rate slope), of 1.3625×10^9 hr, is much longer than the practical burn-in times with different errors on the zero slope.

Table 7.4– Burn-in times corresponding to different values of ε, for Example 7–3.

ε	T_b, hr
0	1.3625×10^9
-0.01	1.09
-0.001	3.93
-0.0001	14.22
-0.00001	51.59
-0.000001	187.93
-0.0000001	688.07

EXAMPLE 7–4

One million (1,000,000) identical units are put to a life test. The life test data are given in Table 5.20. Find the theoretical and the practical burn-in time with errors on the zero failure rate slope of $\varepsilon = -0.01$, $-0.001, -0.0001, -0.00001, -0.000001, -0.0000001, -0.00000001$, and -0.000000001, using the Bathtub Model 1 to fit the data.

SOLUTION TO EXAMPLE 7–4

From Example 5–9, the least-squares estimates of the parameters of the Bathtub Model 1 are

$\hat{\beta} = 0.355715,$
$\hat{\alpha} = 5.78176 \times 10^{12}$ hr,

and

$R_{xy} = 0.997639.$

It may be seen that the Bathtub Model 1 fits the test data very well, with a high correlation coefficient of 0.997639.

Inputting the given failure data information into the computer program "LSQ" given in Appendix 7A yields the burn-in times given in Table 7.5 corresponding to different values of ε.

It may be seen that the lower the absolute error on the zero slope, the longer the burn-in duration. The theoretical burn-in time (with zero failure rate slope), of 3.0711×10^{13} hr, is much longer than the practical burn-in times with different errors on the zero slope.

Table 7.5 – Burn-in times corresponding to different values of ε, for Example 7-4.

ε	T_b, hr
0	3.0711×10^{13}
-0.01	0.012
-0.001	0.047
-0.0001	0.192
-0.00001	0.778
-0.000001	3.155
-0.0000001	12.799
-0.00000001	51.923
-0.000000001	210.639

7.4 THE BURN-IN TIME CORRESPONDING TO A SPECIFIED FAILURE RATE GOAL

If it is required that the failure rate of the units after T_b hours of burn-in should be equal to or less than a specified value, λ_G, the burn-in time can be decided by the following method:

The estimated failure rate function of the bathtub Model 1 is

$$\hat{\lambda}(T) = (\frac{\hat{\beta}}{\hat{\alpha}^{\hat{\beta}}}) T^{\hat{\beta}-1} e^{(T/\hat{\alpha})^{\hat{\beta}}}.$$

Let

$$\hat{\lambda}(T_b) = \lambda_G,$$

or

$$(\frac{\hat{\beta}}{\hat{\alpha}^{\hat{\beta}}}) T_b^{\hat{\beta}-1} e^{(T_b/\hat{\alpha})^{\hat{\beta}}} = \lambda_G.$$

Rearranging this equation yields the iterative equation

$$T_b = \left[\frac{\lambda_G \hat{\alpha}^{\hat{\beta}}}{\hat{\beta}} e^{-(T_b/\hat{\alpha})^{\hat{\beta}}} \right]^{1/(\hat{\beta}-1)}, \tag{7.10}$$

or

$$T_b^{i+1} = \left[\frac{\lambda_G \hat{\alpha}^{\hat{\beta}}}{\hat{\beta}} e^{-(T_b^i/\hat{\alpha})^{\hat{\beta}}} \right]^{1/(\hat{\beta}-1)}, \quad i = 0, 1, 2, ... \tag{7.11}$$

Table 7.6 – Burn-in times corresponding to different values of λ_G, for Example 7–5.

λ_G, fr/hr	T_b, hr
10^{-4}	6.0595×10^2
10^{-5}	1.3537×10^4
10^{-6}	3.9179×10^5
10^{-7}	2.5736×10^7

Then, the burn-in time, T_b, corresponding to the failure rate goal λ_G can be found by putting an initial T_b^0 into Eq. (7.11) first, and iterating step by step until the absolute difference between two adjacent $T_b's$ is small enough, say less than 10^{-6}.

This process can be achieved by the computer program included in "LSQ." The following examples illustrate this method.

EXAMPLE 7–5

Rework Example 7–3 by letting the failure rate goal λ_G in fr/hr be 10^{-4}, 10^{-5}, 10^{-6} and 10^{-7}.

SOLUTION TO EXAMPLE 7–5

Inputting the given failure data information into the computer program "LSQ" given in Appendix 7A yields the burn-in times given in Table 7.6 corresponding to different values of λ_G with a convergence precision of 10^{-7}.

It may be seen from Table 7.6 that the smaller the failure rate goal the longer the required burn-in time. Too low a failure rate goal may require an economically unreasonable burn-in time, such as $T_b = 2.5736 \times 10^7$ hr for a failure rate goal of $\lambda_G = 10^{-7}$ fr/hr. Therefore, cost is an important factor in the process of optimum burn-in time determination which will be discussed in detail in Chapter 9.

EXAMPLE 7–6

Rework Example 7–4 by letting the failure rate goal λ_G in fr/hr be 10^{-4}, 10^{-5}, 10^{-6}, 10^{-7} and 10^{-8}.

SOLUTION TO EXAMPLE 7–6

Inputting the given failure data information into the computer program "LSQ" given in Appendix 7A yields the burn-in times given in

Table 7.7– Burn-in times corresponding to different values of λ_G, for Example 7–6.

λ_G, fr/hr	T_b, hr
10^{-4}	0.03
10^{-5}	1.04
10^{-6}	37.14
10^{-7}	1,323.54
10^{-8}	47,218.10

Table 7.7 corresponding to different values of λ_G with a convergence precision of 10^{-7}.

It may be seen from Table 7.7 that the smaller the failure rate goal the longer the required burn-in time. Too low a failure rate goal may require an economically unreasonable burn-in time, such as $T_b = 47,218.10$ hr for a failure rate goal of $\lambda_G = 10^{-8}$ fr/hr. Therefore, cost is an important factor in the process of optimum burn-in time determination which will be discussed in detail in Chapter 9.

7.5 THE BURN-IN TIME CORRESPONDING TO A SPECIFIED RELIABILITY GOAL

It is sometimes required that burning-in for T_b^* hours should result in a specified high reliability goal during the useful life period, as shown in Fig. 7.4. Then, the product's reliability, for a mission time t after T_b hours of burn-in, is given by

$$R(t|T_b) = \frac{R(t+T_b)}{R(T_b)}.$$

Let

$$\frac{R(t+T_b^*)}{R(T_b^*)} = R_G; \qquad (7.12)$$

then, the burn-in time, T_b^*, can be found by solving Eq. (7.12), as illustrated in Examples 7–7 through 7–9.

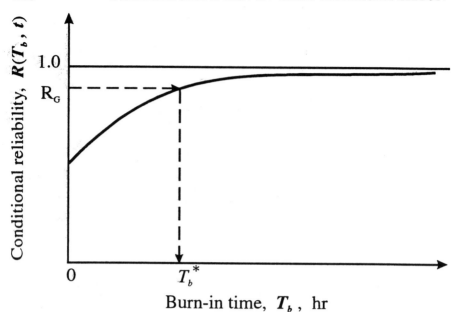

Fig. 7.4 – Burn-in time, T_b^*, for a specified reliability goal, R_G, for a mission duration t after burn-in.

7.5.1 APPLICATION OF EQ. (7.12) USING THE BATHTUB MODEL 1

If the bathtub Model 1 is used, Eq. (7.12) becomes

$$\frac{e^{1-e^{[(t+T_b^*)/\hat{\alpha}]^{\hat{\beta}}}}}{e^{1-e^{(T_b^*/\hat{\alpha})^{\hat{\beta}}}}} = R_G,$$

or

$$e^{e^{(T_b^*/\hat{\alpha})^{\hat{\beta}}} - e^{[(t+T_b^*)/\hat{\alpha}]^{\hat{\beta}}}} = R_G.$$

Taking the logarithm of both sides yields

$$e^{(T_b^*/\hat{\alpha})^{\hat{\beta}}} - e^{[(t+T_b^*)/\hat{\alpha}]^{\hat{\beta}}} = \log_e R_G,$$

or

$$e^{(T_b^*/\hat{\alpha})^{\hat{\beta}}} = e^{[(t+T_b^*)/\hat{\alpha}]^{\hat{\beta}}} + \log_e R_G.$$

Retaking the logarithm of both sides and rearranging yields the iteration equation

$$T_b^* = \hat{\alpha} \left\{ \log_e \left\{ e^{[(t+T_b^*)/\hat{\alpha}]^{\hat{\beta}}} + \log_e R_G \right\} \right\}^{1/\hat{\beta}}. \tag{7.13}$$

This can be solved by the computer program included in "LSQ."

EXAMPLE 7-7

Rework Example 7-3 by letting the reliability goal, R_G, be equal to 0.90 for a mission time of 100 hr.

SOLUTION TO EXAMPLE 7-7

Inputting the given failure data information into the computer program "LSQ" given in Appendix 7A yields the burn-in time as

$T_b^* = 554.95$ hr.

EXAMPLE 7-8

Rework Example 7-4 by letting the reliability goal R_G be equal to 0.9999 for a mission time of 1,000 hr.

SOLUTION TO EXAMPLE 7-8

Inputting the given failure data information into the computer program "LSQ" given in Appendix 7A yields the burn-in time as

$T_b^* = 885.86$ hr.

7.5.2 APPLICATION OF EQ. (7.12) USING THE BIMODAL MIXED-WEIBULL MODEL

If the bimodal, mixed-Weibull model of Eq. (5.4) is used, then Eq. (7.12) becomes

$$\frac{\hat{p}\, e^{-[(t+T_b^*)/\hat{\eta}_1]^{\hat{\beta}_1}} + (1-\hat{p})\, e^{-[(t+T_b^*)/\hat{\eta}_2]^{\hat{\beta}_2}}}{\hat{p}\, e^{-(T_b^*/\hat{\eta}_1)^{\hat{\beta}_1}} + (1-\hat{p})\, e^{-(T_b^*/\hat{\eta}_2)^{\hat{\beta}_2}}} = R_G, \tag{7.14}$$

and T_b^* can be solved by trial-and-error using the computer program "BIR" given in Appendix 7B.

EXAMPLE 7-9

Determine the burn-in time for the components with the life test data of Table 5.7, using Eq. (7.14) and the parameter estimates by the Bayesian approach with the fractional rank plotting method, as obtained in Example 7-3. Assume that the reliability goal is 0.99 and the mission duration is 50 hr.

SOLUTION TO EXAMPLE 7-9

From Example 5-3, the parameter estimates for the life test data of Table 5.7, using the Bayesian approach with the fractional rank plotting method, are

$\hat{p} = 0.366124,$
$\hat{\beta}_1 = 0.57,$
$\hat{\eta}_1 = 530 \text{ hr},$
$\hat{\beta}_2 = 1.82,$

and

$\hat{\eta}_2 = 5,900 \text{ hr}.$

Inputting these parameter estimates, and the given reliability goal and mission time, using Eq. (7.14), into the computer program "BIR" given in Appendix 7B, yields the burn-in time as

$T_b^* = 646.85 \text{ hr};$

i.e., burning-in the components, which are governed by the given life distribution parameters as estimated using the Bayesian approach, will guarantee a reliability goal of 0.99 during the specified 50-hr mission.

7.6 GRAPHICAL DETERMINATION OF THE BURN-IN PERIOD FROM THE FAILURE RATE AND RELIABILITY FUNCTIONS

METHOD 1 – DETERMINE THE BURN-IN PERIOD FROM THE FAILURE RATE FUNCTION

If the failure rate goal is given, then the burn-in period can be found from the plot of the failure rate function [2, Vol. 1, p. 559]; i.e., draw a horizontal line at the failure rate goal level, then find the intersection with the failure rate curve, drop vertically at the intersection, and read

GRAPHICAL APPROACH

off the burn-in time from the time axis. This burn-in time guarantees that the population will have a failure rate that is at least equal to or lower than the goal after the burn-in period.

If the failure rate goal is not given, then the minimum value of the failure rate function, with a reasonable accuracy, can be taken to be the failure rate goal. The minimum value of the failure rate function can be determined by taking the derivative of $\lambda(T)$ with respect to T, setting it equal to zero, finding T_{min}, then substituting back T_{min} in $\lambda(T)$, and obtaining $\lambda_{min} = \lambda(T_{min})$. Practically, λ_{min} can be determined directly from the plot of the failure rate function. When λ_{min} is determined, the failure rate goal can be obtained by adding the accuracy to λ_{min}; then the same method given in the previous paragraph may be used to find the burn-in period. This method is illustrated in Example 7-10.

METHOD 2 – DETERMINE THE BURN-IN PERIOD FROM THE RELIABILITY FUNCTION

If the reliability goal for a mission is given, then the burn-in period may be determined by plotting the conditional reliability function given by Eq. (7.12) versus the burn-in time [2, Vol. 1, p. 559]; i.e., draw a horizontal line at the post-burn-in reliability goal level, then find the intersection with the conditional reliability curve, drop vertically at the intersection, and read off the burn-in time from the time axis. This burn-in time guarantees that the population will have a reliability for a specified mission duration that is at least equal to or higher than the specified goal after the burn-in period. This method is also illustrated in Example 7-10.

EXAMPLE 7–10

A mixed sample of substandard and standard parts of a specific make is being used in a product. Five percent of the sample has been established to be substandard from past quality control results.

A plot of test results on Weibull paper provided the following early life *pdf* parameters:

$$\hat{\beta} = 0.6, \quad \hat{\eta} = 100 \text{ hr, and } \hat{\gamma} = 0 \text{ hr.}$$

Such a plot also revealed another subpopulation with the following characteristics:

$$\hat{\beta} = 1.0, \quad \hat{\eta} = 500 \text{ hr (or } \hat{\lambda} = \frac{1}{\hat{\eta}} = 2 \text{ fr}/10^3 \text{ hr)}, \text{ and } \hat{\gamma} = 0 \text{ hr.}$$

Do the following:

STATISTICAL BURN-IN TIME DETERMINATION

1. Write down the reliability function for this mixed sample.

2. Write down the *pdf* for this mixed sample.

3. Write down the failure rate function for this mixed sample.

4. Find the reliability of a component obtained from this sample for a mission of 30 minutes, starting the mission at age zero.

5. Find the reliability of a component obtained from this sample for an additional mission of 30 minutes.

6. Plot the reliability function.

7. Plot the *pdf*.

8. Plot the failure rate function.

9. If a failure rate of 2 fr/10^3 hr is desired with an error of 5%, find the burn-in period.

10. If a minimum reliability goal of 0.90 is desired for a mission duration of 50 hr after burn-in, find the burn-in period.

SOLUTIONS TO EXAMPLE 7–10

1. The substandard parts, which comprise 5% of the sample, correspond to the components which follow an early life Weibull *pdf*; i.e., $\hat{\beta} < 1.0$. Also,

$$\hat{\beta}_1 = 0.6, \quad \hat{\eta}_1 = 100 \text{ hr}, \quad \hat{\gamma}_1 = 0 \text{ hr},$$

and

$$\hat{\beta}_2 = 1.0, \quad \hat{\eta}_2 = 500 \text{ hr}, \quad \hat{\gamma}_2 = 0 \text{ hr}.$$

The reliability of this mixed sample is then given by

$$R(T) = \hat{p}\, R_1(T) + (1 - \hat{p})\, R_2(T),$$

or

$$R(T) = \hat{p}\, e^{-(\frac{T-\hat{\gamma}_1}{\hat{\eta}_1})^{\hat{\beta}_1}} + (1 - \hat{p})\, e^{-(\frac{T-\hat{\gamma}_2}{\hat{\eta}_2})^{\hat{\beta}_2}},$$

where
$$\hat{p} = 0.05,$$
and
$$1 - \hat{p} = 0.95.$$
Therefore,
$$R(T) = (0.05)e^{-(\frac{T}{100})^{0.6}} + (0.95)e^{-(\frac{T}{500})}.$$

2. The *pdf* for this mixed sample is given by
$$f(T) = \hat{p}\, f_1(T) + (1 - \hat{p})\, f_2(T),$$
where
$$f_1(T) = \frac{\hat{\beta}_1}{\hat{\eta}_1}(\frac{T - \hat{\gamma}_1}{\hat{\eta}_1})^{\hat{\beta}_1 - 1} e^{-(\frac{T - \hat{\gamma}_1}{\hat{\eta}_1})^{\hat{\beta}_1}},$$
and
$$f_2(T) = \frac{\hat{\beta}_2}{\hat{\eta}_2}(\frac{T - \hat{\gamma}_2}{\hat{\eta}_2})^{\hat{\beta}_2 - 1} e^{-(\frac{T - \hat{\gamma}_2}{\hat{\eta}_2})^{\hat{\beta}_2}}.$$
Therefore,
$$f(T) = (0.05)(\frac{0.6}{100})(\frac{T}{100})^{-0.4} e^{-(\frac{T}{100})^{0.6}} + (0.95)(\frac{1}{500}) e^{-(\frac{T}{500})}.$$

3. The failure rate function is given by
$$\lambda(T) = \frac{f(T)}{R(T)},$$
where $f(T)$ and $R(T)$ were found earlier.

4. The mission duration is 0.5 hr. Thus the reliability for a mission of T duration, starting the mission at age zero, is
$$R(T = 0.5 \text{ hr}) = (0.05)e^{-(\frac{0.5}{100})^{0.6}} + (0.95)e^{-(\frac{0.5}{500})},$$
or
$$R(T = 0.5 \text{ hr}) = 0.9970.$$

5. The reliability for another mission is

$$R(T,t) = \frac{R(T+t)}{R(T)},$$

where $T = 0.5$ hr and $t = 0.5$ hr; then

$$R(0.5 \text{ hr} + 0.5 \text{ hr}) = R(1 \text{ hr}),$$
$$= (0.05)e^{-(\frac{1}{100})^{0.6}} + (0.95)e^{-(\frac{1}{500})},$$

or

$$R(0.5 \text{ hr} + 0.5 \text{ hr}) = 0.9950,$$

and

$$R(T = 0.5 \text{ hr}) = 0.9970.$$

Therefore,

$$R(T = 0.5 \text{ hr}, t = 0.5 \text{ hr}) = \frac{R(0.5 \text{ hr} + 0.5 \text{ hr})}{R(0.5 \text{ hr})},$$
$$= \frac{0.9950}{0.9970},$$

or

$$R(T = 0.5 \text{ hr}, t = 0.5 \text{ hr}) = 0.9981.$$

6. The plot of the reliability function is given in Fig. 7.5. Table 7.8 gives the values used to plot Fig. 7.5.

7. The plot of the *pdf* is given in Fig. 7.5.

8. The plot of the failure rate function is given in Fig. 7.5.

9. The desired maximum failure rate is

$$\lambda = (2)(1 + 0.05) = 2.1 \text{ fr}/10^3 \text{ hr}.$$

From Fig. 7.5, the burn-in period corresponding to the failure rate of 2.1 fr/10^3 hr is 92 hr.

GRAPHICAL APPROACH

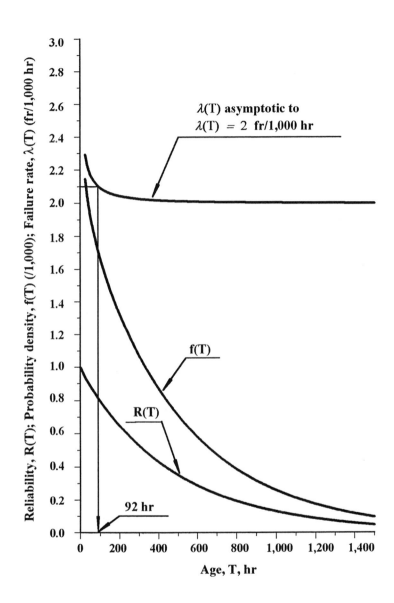

Fig. 7.5 – Plot of reliability, *pdf*, and failure rate versus age for Example 7–10.

TABLE 7.8 – Data for the reliability, *pdf*, and failure rate plots of Example 7–10.

Mission duration, T, hr	Reliability, $R(T)$	Probability density, $f(T) \times 10^3$	Failure rate, $\lambda(T)$, fr/10^3 hr
0	1.000000	∞	∞
25	0.936022	2.14533	2.29196
50	0.885445	1.92384	2.17274
100	0.796188	1.66595	2.09241
150	0.717743	1.47880	2.06035
200	0.647787	1.32355	2.04318
300	0.528605	1.07071	2.02554
400	0.431889	0.87105	2.01683
500	0.355102	0.71037	2.01180
600	0.288804	0.58009	2.00860
700	0.236277	0.47407	2.00642
800	0.193339	0.38762	2.00486
900	0.158225	0.31704	2.00370
1,000	0.129502	0.25937	2.00280
1,100	0.106001	0.21222	2.00208
1,200	0.085771	0.17367	2.00150
1,300	0.071033	0.14214	2.00100
1,400	0.058153	0.11634	2.00058
1,500	0.047610	0.09523	2.00020

10. The conditional reliability for a mission duration of 50 hr, given that the unit does not fail during the burn-in period, is

$$R(T_b, 50 \text{ hr}) = \frac{R(T_b + 50)}{R(T_b)},$$

or

$$R(T_b, 50 \text{ hr}) = \frac{(0.05)e^{-(\frac{T_b+50}{100})^{0.6}} + (0.95)e^{-(\frac{T_b+50}{500})}}{(0.05)e^{-(\frac{T_b}{100})^{0.6}} + (0.95)e^{-(\frac{T_b}{500})}}.$$

It may be seen that the conditional reliability varies with T_b. The values of $R(T_b, 50 \text{ hr})$ and T_b are given in Table 7.9, and the plot of $R(T_b, 50 \text{ hr})$ versus T_b is given in Fig. 7.6. From Fig. 7.6 the burn-in period corresponding to the reliability goal of 0.90 is 63 hr, i.e. the reliability goal of 0.90 for a 50-hr mission can be achieved after 63 hours of burn-in. For subsequent missions of 50-hr duration the conditional reliability will be higher than the goal, because there is no wearout failure mode in this population.

REFERENCES

1. Jensen, F. and Petersen, N. E., *Burn-in, An Engineering Approach to the Design and Analysis of Burn-in Procedures*, John Wiley & Sons, New York, 167 pp., 1982.

2. Kececioglu, Dimitri B., *Reliability Engineering Handbook*, Prentice Hall, Inc., Upper Saddle River, NJ, Vol. 1, 720 pp. and Vol. 2, 568 pp., 1991, Fifth Printing 1995.

TABLE 7.9 – The conditional reliability for a mission of 50 hr after T_b hours of burn-in.

Burn-in period, T_b, hr	Conditional reliability, $R(T_b, 50)$
0	0.88544
10	0.89325
20	0.89574
30	0.89729
40	0.89837
50	0.89920
60	0.89984
70	0.90036
80	0.90080
90	0.90116
100	0.90147

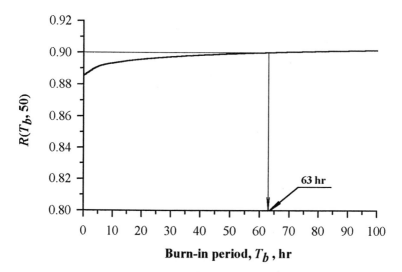

Fig. 7.6 – Conditional reliability function for a mission duration of 50 hr, given that the unit does not fail during the burn-in period, T_b.

APPENDIX 7A
COMPUTER PROGRAM OF "LSQ.FOR"

```
c     $$$$$$$$$$$$$$$$$$$$$$$$$$$$$$$$$$$$$$$$$$$$$$$$$$$$$$$$$$$$$
c     $       BURN-IN TIME DETERMINATION BY BATHTUB MODEL 1     $
c     $                  USING LEAST SQUARES METHOD             $
c     $$$$$$$$$$$$$$$$$$$$$$$$$$$$$$$$$$$$$$$$$$$$$$$$$$$$$$$$$$$$$
      implicit double precision (a-h, o-z)
      real almt
      dimension t(100),nf(100),r(100),x(100),y(100)
      common /div/dt,nd,n
      open(5,file='ilsq.dat',status='unknown')
      open(6,file='olsq.dat',status='new')
c
c     PARAMETER ESTIMATION BY LEAST SQUARES METHOD
c
      num=0
5     nnf=0
      xa=0.0
      ya=0.0
      xx=0.0
      yy=0.0
      xy=0.0
      read(5,*) n,nd,dt
      num=num+1
      write(6,8)
      write(6,*) '-----         CASE    ',num,'        -----'
      write(6,8)
8     format(1x,///)
      write(6,10) n,nd,dt
10    format(15x,'Sample size: ',i20/15x,'Time interval
     *  number: ', i10 /15x,'Interval width: ',f20.6)
c     write(6,12)
12    format(///5x,'No.',10x,'Ending time Ti',7x,'NF(DTi)',
     *  7x,'NS(Ti)',8x,'R(Ti)')
c     write(6,*)('=',i=1,78)
      do 20 i=1,nd
      read(5,*) t(i), nf(i)
      nnf=nnf+nf(i)
      ns=n-nnf
      r(i)=float(ns)/float(n)
c     write(6,14) i,t(i),nf(i),ns,r(i)
14    format(1x,i7,7x,f12.6,5x,i10,5x,i10,5x,f12.8)
```

```
              x(i)=dlog(t(i))
              y(i)=dlog(dlog(1-dlog(r(i))))
              xa=xa+x(i)
              ya=ya+y(i)
20            continue
c             write(6,*)('=',i=1,78)
              call rate(nf)
              write(6,8)
              xa=xa/float(nd)
              ya=ya/float(nd)
              do 30 i=1,nd
              xx=(x(i)-xa)*(x(i)-xa)+xx
              yy=(y(i)-ya)*(y(i)-ya)+yy
              xy=(x(i)-xa)*(y(i)-ya)+xy
30            continue
              beta=xy/xx
              a=ya-beta*xa
              alpha=dexp(-a/beta)
              cor=xy/sqrt(xx*yy)
              write(6,8)
              write(6,*)'========  Parameter Estimates   ========'
              write(6,40) beta,alpha,cor
40            format(////10x,'beta=',E30.6/10x, 'alpha=',E30.6
     *        /10x,'Rxy=',f10.6///)
              write(6,*)'      No.','           Time point',
     *        '            Failure rate'
              write(6,*)('=',i=1,78)
              Do 150 i=1,nd
              ti=0.5*dt*float(2*i-1)
              rt=beta/alpha**beta*ti**(beta-1.0)
     1        *dexp((ti/alpha)**beta)
150           write(6,160) i,ti,rt
160           format(5x,i4,6x,f20.6,7x,f12.8)
              write(6,*)('=',i=1,78)
              write(6,8)
C
C     Burn-in Time With Different Errors On the Zero Slope
C
              write(6,*)'========  Burn-in Time With Different
     *        Errors  ========'
              bt=1.0/beta
              tbi=alpha*(bt-1.0)**bt
              write(6,100)
100           format(///10x,' Burn-in Time',10x,
```

APPENDIX 7A

```
      *   'Error on zero slope'/)
          write(6,*)('=',i=1,78)
          write(6,120) tbi,0.0
55        write(*,*) 'Please input the convergence precision!'
          read(*,*) er
          e=-0.1
          do 167 ii=1,10
          e=e*0.1
          ee=e*alpha**beta/beta
          ncycle=0
          t0=0.5
60        qq=-(t0/alpha)**beta
          write(*,*) 'qq=====', qq
          tt=((ee*dexp(qq))/((beta-1.0)+beta*(t0/alpha)**beta))
      1   **(1.0/(beta-2.0))
          ncycle=ncycle+1
c         write(*,*) '   tt =',tt
          yt=abs(tt-t0)
          if(yt-er) 90,90,70
70        if(ncycle-500) 80,90,90
80        t0=tt
          goto 60
90        tbi=0.5*(tt+t0)
          write(6,120) tbi,e
120       format(2x,e20.6,10x,e20.6)
          almt=0.001
          do 245 ii=1,5
          almt=almt/10.0
          ncycle=0
          t0=0.5
200       pp=-(t0/alpha)**beta
          write(*,*) 'pp=====', pp
          tt=(almt*alpha**beta*dexp(pp)/beta)**(1.0/(beta-1.0))
          ncycle=ncycle+1
c         write(*,*) '   tt =',tt
          yt=abs(tt-t0)
          if(yt-er) 230,230,210
210       if(ncycle-500) 220,230,230
220       t0=tt
          goto 200
230       tbi=0.5*(tt+t0)
          write(6,240) almt,tbi
240       format(2x,'failure rate goal==',e15.9,
      *   'burn-in time is', e20.6)
```

```
245       continue
          write(*,*) 'Another case ? Yes=1/No=0'
          read(*,*) ian
          if(ian.eq.1) goto 5
C
C   Burn-in time corresponding to a specified RELIABILITY
C   goal, Rg, using the Bathtub Model 1
C
          write(6,*) ' Burn-in time corresponding to a
        * specified RELIABILITY goal, Rg'
          write(*,*) 'Give the RELIABILITY goal, Rg, and the
        * corresponding mission time, tm, please!'
          read(*,*) Rg, tm
          write(6,*)'Reliability Goal ==',Rg,'Mission Time==',tm
          write(*,*)'Reliability Goal ==',Rg,'Mission Time==',tm
          tt=0.00001
          kk=0
          do 260   i=1,1000
          tt=tt*1.5
250       p=dexp((tt/alpha)**beta)-dexp(((tm+tt)/alpha)**beta)
          R0=exp(p)
          write(*,*) 'Tb===',tt, 'R0===',R0
          if(R0.le.Rg) goto 252
          if(kk.ne.1) goto 252
          x2=tt
          y2=R0
          goto 280
252       if(R0-Rg) 255,270,300
255       x1=tt
          y1=R0
          kk=1
260       continue
270       tbi=tt
          goto 285
280       tbi=(Rg-y1)*(x2-x1)/(y2-y1)+x1
285       write(6,290) tbi
290       format(//2x,'The corresponding burn-in time is',
        * e20.6)
          write(*,*)'The corresponding burn-in time is',tbi
          goto 310
300       write(*,*)'Your Tb is less than 0.00002 hr!'
310       write(*,*) 'Another case ? Yes=1/No=0'
          read(*,*) ian
          if(ian.eq.1) goto 5
```

APPENDIX 7A

```
            stop
            end
C
            subroutine rate(nf)
c           implicit double precision (a-h, o-z)
            real lmd1,lmd2
            dimension nf(100)
            common /para/p,gamma1,beta1,eta1,gamma2,beta2,
          * eta2,nd,dt,nn
            cnf=0.0
            ns=nn
            write(6,10)
10          format(//5x,'Time Interval',10x,'Failure Number',
          * 8x,'Lmd1', 15x,'lmd2')
            write(6,*) ('*',i=1,78)
            do 50 i=1,nd
            ii=i-1
            cnf=cnf+nf(i)
            lmd1=nf(i)/(ns*dt)
            lmd2=nf(i)/((ns-nf(i))+0.5*nf(i))/dt
            write(6,30) ii*dt,i*dt,nf(i),lmd1,lmd2
30          format(1x,'(',f10.4,',',f10.4,')',5x,i6,10x,f12.9,
          * 5x,f12.9)
            ns=nn-cnf
50          continue
            write(6,*) ('*',i=1,78)
            return
            end
```

APPENDIX 7B
COMPUTER PROGRAM OF "BIR.FOR"

```fortran
C
C     $$$$$$$$$$$$$$$$$$$$$$$$$$$$$$$$$$$$$$$$$$$$$$$$$$$$$$$
C     $    BURN-IN TIME FOR A SPECIFIED RELIABILITY GOAL    $
C     $        USING THE BIMODAL, MIXED-WEIBULL  MODEL      $
C     $               AND THE BATHTUB MODEL 1               $
C     $$$$$$$$$$$$$$$$$$$$$$$$$$$$$$$$$$$$$$$$$$$$$$$$$$$$$$$
C
      implicit double precision (a-h, o-z)
      open(5,file='ibir.dat',status='unknown')
      open(6,file='obir.dat',status='new')
c
c  Burn-in time corresponding to a specified RELIABILITY
c  goal, Rg, using the Bimodal, Mixed-Weibull Model
c
5     write(*,*) 'Give parameters of mixed Weibull
     1 distribution: p, beta1,eta1, beta2, eta2 !!'
      read(5,*) p, beta1,eta1, beta2, eta2
      write(6,*) ' Burn-in time corresponding to a
     1 specified RELIABILITY goal, Rg'
      write(*,*) 'Give the RELIABILITY goal, Rg, and the
     1 corresponding mission time, tm, please!'
      read(*,*) Rg, tm
      write(6,*)'Reliability Goal ==',Rg,'Mission Time==',tm
      write(*,*)'Reliability Goal ==',Rg,'Mission Time==',tm
      tt=0.00001
      kk=0
      do 260  i=1,1000
      tt=tt*1.5
      rb1=exp(-((tm+tt)/eta1)**beta1)
      rb2=exp(-((tm+tt)/eta2)**beta2)
      r01=exp(-((tt)/eta1)**beta1)
      r02=exp(-((tt)/eta2)**beta2)
250   R0=(p*rb1+(1.0-p)*rb2)/(p*r01+(1.0-p)*r02)
      write(*,*) 'Tb===',tt, 'R0===',R0
      if(R0.le.Rg) goto 252
      if(kk.ne.1) goto 252
      x2=tt
      y2=R0
      goto 280
252   if(R0-Rg) 255,270,300
```

APPENDIX 7B

```
255     x1=tt
        y1=R0
        kk=1
260     continue
270     tbi=tt
        goto 285
280     tbi=(Rg-y1)*(x2-x1)/(y2-y1)+x1
285     write(6,290) tbi
290     format(//2x,'The corresponding burn-in time is',
      * e20.6)
        write(*,*)'The corresponding burn-in time is',tbi
        goto 310
300     write(*,*)'Your Tb is less than 0.00002 hr!'
310     write(*,*) 'Another case ? Yes=1/No=0'
        read(*,*) ian
        if(ian.eq.1) goto 5
        pause
        end
C
        FUNCTION FUNC(Z)
        FUNC=EXP(1-EXP((Z*ALPHA)**(-BETA)))/(Z*Z)
        RETURN
        END
        EXTERNAL FUNC
        A=0.0
        B=1.0/TT
        CALL QGAUS(FUNC,A,B,SS)
        WRITE(*,*) 'SS=',SS
        STOP
        END
C
        SUBROUTINE QGAUS(FUNC,A,B,SS)
        DIMENSION X(5),W(5)
        DATA X/0.1488743389,0.4333953941,0.6794095682,
      1 0.8650633666,0.9739065285/
        DATA W/0.2955242247,0.2692667193,0.2190863625,
      1 0.1494513491,0.0666713443/
        XM=0.5*(B+A)
        XR=0.5*(B-A)
        SS=0.0
        DO 11 J=1,5
        DX=XR*X(J)
        SS=SS+W(J)*(FUNC(XM+DX)+FUNC(XM-DX))
11      CONTINUE
```

```
SS=XR*SS
RETURN
END
```

Chapter 8

MEAN RESIDUAL LIFE (MRL) CONCEPT AND ITS APPLICATIONS TO BURN-IN TIME DETERMINATION

8.1 INTRODUCTION

The concept of mean residual life (MRL) has been used as far back as the third century A.D. [1]. In the last three decades, reliabilitists, statisticians, and others have shown intensified interest in the MRL and developed many useful results.

Given that an item is of age T, the remaining life after age T which is a random variable, is called the residual life. The expected value of this random residual life is called the mean residual life at age T. In general, MRL is a function of age T. Therefore, it is also referred to as the MRL function.

MRL has been recommended by many authors to be used as a very constructive tool in model building [2; 3; 4]. This is so because for a life distribution with a finite mean, the MRL acts like the probability density function (pdf), the moment generating function, or the characteristic function: it completely determines the distribution via an inverse transformation. The MRL is used not only for parametric modeling, but also for non-parametric modeling.

Since it was first developed, the MRL concept has had a tremendous range of applications, such as

1. optimum burn-in time determination [5; 6],

2. setting rates and benefits analysis of a life insurance plan [7; 8],

3. survival analysis in biomedical studies [7; 8],

4. life-length prediction of wars and strikes in social sciences [9],

5. optimum preventive replacement decision-making and renewal theory [10], and

6. dynamic programming and branching processes [10].

The applications of MRL concept in burn-in is the main focus of this chapter.

8.2 MATHEMATICAL DEFINITION OF MRL AND ITS RELATIONSHIP TO THE FAILURE RATE FUNCTION

Let T be the random life with a finite mean value and cdf, $F(T)$, for $T \geq 0$. Let $R(T)$ be the reliability function; i.e., $R(T) = 1 - F(T)$. Then, the residual life at any age t is given by $(T - t)$ and the mean residual life at age t, $m(t)$, is defined as the following conditional mean:

$$m(t) = E[T - t | T > t],$$
$$= \int_0^\infty R(x|t)\, dx,$$
$$= \int_0^\infty \frac{R(x+t)}{R(t)}\, dx,$$

or

$$m(t) = \frac{1}{R(t)} \int_t^\infty R(u)\, du, \qquad (8.1)$$

where $u = x + t$.

If a pdf exists for T, or $F(T)$ is differentiable; i.e.,

$$\frac{dF(T)}{dT} = -\frac{dR(T)}{dT} = f(T), \qquad (8.2)$$

then $m(t)$ in Eq. (8.1) can be rewritten as follows using the integration-by-parts procedure by noting that $\int v\, du = v\, u - \int u\, dv$ where $v = R(u)$ in this case:

$$m(t) = \frac{1}{R(t)} \left[u R(u) \Big|_t^\infty - \int_t^\infty u\, dR(u) \right],$$

MATHEMATICAL DEFINITION

from Eq. (8.2)

$$= \frac{1}{R(t)} \left[0 - t\,R(t) + \int_t^\infty u\,f(u)\,du \right],$$

or

$$m(t) = \frac{1}{R(t)} \int_t^\infty u\,f(u)\,du - t. \qquad (8.3)$$

Recall that the failure rate function at age t is defined as the following conditional probability:

$$\lambda(t) = \lim_{\Delta t \to 0} \left[\frac{F(t+\Delta t) - F(t)}{R(t)\,\Delta t} \right] = \frac{f(t)}{R(t)}, \qquad (8.4)$$

Though both the failure rate function at age t, $\lambda(t)$, and the MRL function at age t, are conditional on the survival of an item to age t, $m(t)$ provides the information about the *whole interval* after t (all after t), or for interval (t, ∞), while $\lambda(t)$ provides information about a *small interval* after age t (just after t), or for interval $(t, t+\Delta t)$. In addition, the existence of the MRL function, $m(t)$, does not necessarily imply the existence of a failure rate function, $\lambda(t)$, and the existence of $\lambda(t)$ does not necessarily imply the existence of $m(t)$.

When both $\lambda(t)$ and $m(t)$ exist, $\lambda(t)$ can be expressed in terms of $m(t)$ as follows:

$$\lambda(t) = \frac{1 + m'(t)}{m(t)}, \qquad (8.5)$$

which can be derived by taking the first derivative on both sides of Eq. (8.1) and rearranging; i.e.,

$$m'(t) = \frac{d}{dt}\left[\frac{1}{R(t)}\right] \int_t^\infty R(u)\,du + \left[\frac{1}{R(t)}\right] \frac{d}{dt}\left[\int_t^\infty R(u)\,du\right],$$

$$= \frac{f(t)}{[R(t)]^2} \int_t^\infty R(u)\,du + \frac{1}{R(t)}[0 - R(t)],$$

$$= \left[\frac{f(t)}{R(t)}\right] \left[\frac{\int_t^\infty R(u)\,du}{R(t)}\right] - 1,$$

$$= \lambda(t)\,m(t) - 1,$$

or

$$\lambda(t) = \frac{1 + m'(t)}{m(t)}.$$

Note from Eq. (8.5) that a contant MRL, $m(t) = m_0$, implies a constant failure rate, because then $\lambda(t) = \lambda_0 = \frac{1}{m_0}$, and this implies that the life distribution is exponential. But a constant failure rate, $\lambda(t) = \lambda_0$, dose not imply a constant MRL, because in this case, solving Eq. (8.5) for $m(t)$ yields

$$m(t) = a\, e^{\lambda_0\, t} + \frac{1}{\lambda_0},$$

which is a function of t, where a is a constant. However, a constant failure rate does imply an exponential distribution.

8.3 MRL COMPLETELY DETERMINES A LIFE DISTRIBUTION

The relationship between the MRL function and the failure rate function is given by Eq. (8.5). Now let's see how the reliability function, the unreliability function and the mean life are determined by MRL.

Equation (8.4) can be rewritten, using Eq. (8.2), as

$$\lambda(t) = -\frac{d[\log_e R(t)]}{dt}. \tag{8.6}$$

Substituting Eq. (8.6) into Eq. (8.5) yields

$$-\frac{d[\log_e R(t)]}{dt} = \frac{1 + m'(t)}{m(t)}.$$

Integrating both sides over the interval $(0, t)$ yields

$$\int_0^t d[\log_e R(x)] = -\int_0^t \left[\frac{1 + m'(x)}{m(x)}\right] dx,$$

or

$$\log_e R(t) - \log_e R(0) = -\int_0^t \frac{1}{m(x)} dx - \int_0^t \frac{m'(x)}{m(x)} dx. \tag{8.7}$$

Note that the second integral on the right side of Eq. (8.7) is

$$\int_0^t \frac{m'(x)}{m(x)} dx = \int_0^t \frac{1}{m(x)} d[m(x)],$$
$$= \left[\log_e m(x)\right]\Big|_0^t,$$
$$= \log_e m(t) - \log_e m(0),$$

or
$$\int_0^t \frac{m'(x)}{m(x)} dx = \log_e \frac{m(t)}{m(0)}.$$

Therefore, Eq. (8.7) becomes

$$\log_e R(t) - \log_e R(0) = -\int_0^t \frac{1}{m(x)} dx - \log_e \frac{m(t)}{m(0)},$$

$$\log_e R(t) = \log_e \left[\frac{m(0)}{m(t)} R(0)\right] - \int_0^t \frac{1}{m(x)} dx,$$

or

$$R(t) = \frac{m(0)}{m(t)} R(0) \, e^{-\int_0^t \frac{1}{m(x)} dx}. \tag{8.8}$$

Assume $R(0) \doteq 1$, then the reliability function is determined by $m(t)$ as follows:

$$R(t) = \frac{m(0)}{m(t)} \, e^{-\int_0^t \frac{1}{m(x)} dx}. \tag{8.9}$$

Correspondingly, the unreliability function, or the *cdf* of a life distribution is

$$F(t) = 1 - R(t) = 1 - \frac{m(0)}{m(t)} \, e^{-\int_0^t \frac{1}{m(x)} dx}. \tag{8.10}$$

The relationship between mean life, $\mu = E(T)$, and the MRL, $m(t)$, can be obtained using Eq. (8.3) and setting $t = 0$; i.e.,

$$m(0) = \frac{\int_0^\infty u \, f(u) \, du}{R(0)} - 0 = \frac{\mu}{R(0)},$$

or

$$\mu = m(0) \, R(0) \begin{cases} = m(0), & \text{if } R(0) = 1, \\ < m(0), & \text{if } 0 < R(0) < 1. \end{cases}$$

Assume $R(0) = 1$, then

$$\mu = m(0). \tag{8.11}$$

Various nonparametric classes of life distributions can be defined using the MRL concept, such as the following:

1. Decreasing mean residual life ($DMRL$) – A life distribution whose MRL is a decreasing function.

2. Increasing mean residual life ($IMRL$) – A life distribution whose MRL is an increasing function.

3. New better than used in expectation ($NBUE$) – A life distribution with $m(0) \geq m(t)$ for all $t \geq 0$.

4. New worse than used in expectation ($NWUE$) – A life distribution with $m(0) \leq m(t)$ for all $t \geq 0$.

5. Increasing then decreasing mean residual life ($IDMRL$) – A life distribution whose MRL is increasing on $[0, \tau)$ and decreasing on $[\tau, \infty)$ where $0 < \tau < \infty$.

6. Decreasing then increasing mean residual life ($DIMRL$) – A life distribution whose MRL is decreasing on $[0, \tau)$ and increasing on $[\tau, \infty)$ where $0 < \tau < \infty$.

Note that $DMRL$ implies $NBUE$ and $IMRL$ implies $NWUE$. A burn-in procedure is justified for $IMRL$ units, or equivalently, $NWUE$ units, and for $IDMRL$ units, because these units exhibit a decreasing failure rate during their early life period and burn-in before delivery will eliminate the defective units from the population and *increase the residual life of the delivered units.*

8.4 EMPIRICAL MRL FUNCTIONS

The empirical MRL function is a helpful addition to other life data techniques, such as total time on test (TTT) plots, empirical failure rate functions, etc. The MRL plot detects certain aspects of the distribution more readily than other techniques. A very simple equation is presented here for computing the empirical MRL at any age t, $\hat{m}(t)$, from a set of observed data, which is defined as [1; 11]

$$\hat{m}(t) = \frac{\text{Total time on test observed after } t}{\text{Total number of failures observed after } t}. \qquad (8.12)$$

The expressions of $\hat{m}(t)$ for grouped and ungrouped data sets with and without ties, are given next according to this definition. Numerical examples will be given in Section 8.8 where the empirical MRL functions and their plots are used to determine the optimum burn-in times directly.

8.4.1 FOR A COMPLETE, OR NONCENSORED, RANDOM SAMPLE

CASE 1 – Ungrouped, or individual, data set

First, a data set without ties is considered. Let $t_1 < t_2 < t_3 < \cdots < t_N$ be a complete, or noncensored, random sample from a life test of N components. Then, for any specified age t between t_k and t_{k+1}, or $t_k \leq t < t_{k+1}$, where $k = 0, 1, \cdots, (N-1)$ and $t_0 = 0$, the empirical MRL function is

$$\hat{m}_N(t) = \begin{cases} \dfrac{\sum_{i=k+1}^{N}(t_i - t)}{N - k}, & \text{for } t < t_N, \\ 0, & \text{for } t \geq t_N, \end{cases} \tag{8.13}$$

according to Eq. (8.12). Note that at $t = 0$ which corresponds to $k = 0$, substituting $t = 0$ and $k = 0$ into Eq. (8.13) yields

$$\hat{m}_N(0) = \dfrac{\sum_{i=1}^{N} t_i}{N},$$

which is the point estimate of the components' mean life.

Evaluating $\hat{m}_N(t)$, at $t = 0, t_1, t_2, \cdots, t_{N-1}, t_N$, yields $(N+1)$ points; i.e., $[0, \hat{m}_N(0)]$, $[t_1, \hat{m}_N(t_1)]$, $[t_2, \hat{m}_N(t_2)]$, $\cdots$, $[t_{N-1}, \hat{m}_N(t_{N-1})]$, $[t_N, 0]$. Connecting these points consecutively by straight lines yields an approximate empirical MRL function for the given life data set.

Second, a data set with ties is considered. In this case, N components are life tested until all fail. A total of n distinct failure times are observed, at $t_1 < t_2 < t_3 < \cdots < t_n$, respectively, where $n < N$. Let N_{fi} be the number of failures which occurred at t_i where $i = 1, 2, \cdots, n$. Therefore,

$$N = \sum_{i=1}^{n} N_{fi}.$$

Then, for any specified age t between t_k and t_{k+1}, or $t_k \leq t < t_{k+1}$, where $k = 0, 1, 2, \cdots, (n-1)$ and $t_0 = 0$, the empirical MRL function is

$$\hat{m}_N(t) = \begin{cases} \dfrac{\sum_{i=k+1}^{n} N_{fi}(t_i - t)}{\sum_{i=k+1}^{n} N_{fi}}, & \text{for } t < t_n, \\ 0, & \text{for } t \geq t_n, \end{cases} \tag{8.14}$$

according to Eq. (8.12). Note that at $t = 0$ which corresponds to $k = 0$, substituting $t = 0$ and $k = 0$ into Eq. (8.14) yields

$$\hat{m}_N(0) = \frac{\sum_{i=1}^{n} N_{fi}\, t_i}{\sum_{i=1}^{n} N_{fi}},$$

which is the point estimate of the components' mean life. Evaluating $\hat{m}_N(t)$, at $t = 0, t_1, t_2, \cdots, t_n$, yields $(n+1)$ points; i.e., $[0, \hat{m}_N(0)]$, $[t_1, \hat{m}_N(t_1)]$, $[t_2, \hat{m}_N(t_2)], \cdots, [t_n, 0]$. Connecting these points consecutively by straight lines yields an approximate empirical MRL function for the given life test data set.

CASE 2 – Grouped data set

In this case, N components are life tested until all fail. The number of failures are observed and recorded at the end of n time intervals as follows:

$$[(t_1, t_2); N_{f1}], [(t_2, t_3); N_{f2}], \cdots, [(t_n, t_{n+1}); N_{fn}].$$

Then, the empirical MRL's at the starting and ending points of each time interval can be written as:

$$\hat{m}_N(t_j) = \begin{cases} \dfrac{\sum_{i=j}^{n} N_{fi}\left[\frac{1}{2}(t_i + t_{i+1}) - t_j\right]}{\sum_{i=j}^{n} N_{fi}}, & \text{for } j = 0, 1, \cdots, n, \\ 0, & \text{for } j \geq n+1, \end{cases} \quad (8.15)$$

where

$n =$ total number of groups, or time intervals,

$t_i =$ starting point of the ith interval, (t_i, t_{i+1}), $i = 0, 1, 2, \cdots, n$,

$t_{i+1} =$ ending point of the ith interval, (t_i, t_{i+1}), $i = 0, 1, 2, \cdots, n$,

$N_{fi} =$ number of failures in the ith interval, (t_i, t_{i+1}), $i = 0, 1, 2, \cdots, n$,

$t_0 = 0$,

and

$N_{f0} = 0$.

Note that Eq. (8.15) is based on the assumption that N_{fi} failures occured uniformly over the interval (t_i, t_{i+1}); i.e., each with a mean life of $\frac{1}{2}(t_i + t_{i+1})$. The sample size in this case is

$$N = \sum_{i=1}^{n} N_{fi},$$

which corresponds to the total number of components involved in the test. Also note that for $j = 0$

$$\hat{m}_N(t_0) = \hat{m}_N(0) = \frac{\sum_{i=0}^{n} N_{fi} [\frac{1}{2}(t_i + t_{i+1})]}{N},$$

which is the point estimate of the components' mean life.

Connecting these points, $[0, \hat{m}_N(0)]$, $[t_1, \hat{m}_N(t_1)]$, $\cdots$, $[t_n, \hat{m}_N(t_n)]$, $(t_{n+1}, 0)$, consecutively by straight lines yields an approximate empirical MRL function for the given grouped data set.

8.4.2 FOR A CENSORED RANDOM SAMPLE

CASE 1 – Ungrouped, or individual, data sets

First, a data set without ties is considered. In this case, N components are life tested and the test is terminated at t_d. A total of n failures are observed at $t_1 < t_2 < t_3 < \cdots < t_n$, respectively, where $t_n \leq t_d$. Then, for any specified age t between t_k and t_{k+1}, or $t_k \leq t < t_{k+1}$, where $k = 0, 1, 2, \cdots, n-1$ and $t_0 = 0$, the empirical MRL function is

$$\hat{m}_{(N,n,t_d)}(t) = \begin{cases} \frac{\sum_{i=k+1}^{n}(t_i - t) + N_s(t_d - t)}{n - k}, & \text{for } t < t_n, \\ \text{unknown from the test,} & \text{for } t \geq t_n, \end{cases} \quad (8.16)$$

according to Eq. (8.12), where

N_s = total number of components surviving the test,

and

$$N_s = N - n.$$

Note that for $t = 0$ which corresponds to $k = 0$, substituting $t = 0$ and $k = 0$ into Eq. (8.16) yields

$$\hat{m}_{(N,n,t_d)}(0) = \frac{\sum_{i=1}^{n} t_i + N_s t_d}{n},$$

which is the point estimate of the components' mean life.

Evaluating $\hat{m}_{(N,n,t_d)}(t)$, at $t = 0$, t_1, t_2, $\cdots$, t_{n-1}, yields n points; i.e., $[0, \hat{m}_{(N,n,t_d)}(0)]$, $[t_1, \hat{m}_{(N,n,t_d)}(t_1)]$, $\cdots$, $[t_{n-1}, \hat{m}_{(N,n,t_d)}(t_{n-1})]$. Connecting these points consecutively by straight lines yields an approximate MRL function for the given life data set.

Second, a data set with ties is considered. In this case, N components are life tested and the test is terminated at t_d. A total of n failures are observed at L distinct times, $t_1 < t_2 < \cdots < t_L$, respectively, where $L < n < N$ and $t_L \leq t_d$. Also let N_{fi} be the number of failures which occurred at t_i for $i = 1, 2, \cdots, L$. Therefore,

$$n = \sum_{i=1}^{L} N_{fi}.$$

Then, for any specified age t between t_k and t_{k+1}, or $t_k \leq t < t_{k+1}$, where $k = 0, 1, 2, \cdots, (L-1)$ and $t_0 = 0$, the empirical MRL function is

$$\hat{m}_{(N,n,t_d)}(t) = \begin{cases} \dfrac{\sum_{i=k+1}^{L} N_{fi}(t_i - t) + N_s(t_d - t)}{\sum_{i=k+1}^{L} N_{fi}}, & \text{for } t < t_L, \\ \text{unknown from the test,} & \text{for } t \geq t_L, \end{cases} \quad (8.17)$$

according to Eq. (8.12), where

N_s = total number of components surviving the test,

and

$$N_s = N - n = N - \sum_{i=1}^{L} N_{fi}.$$

Note that for $t = 0$ which corresponds to $k = 0$, substituting $t = 0$ and $k = 0$ into Eq. (8.17) yields

$$\hat{m}_{(N,n,t_d)}(0) = \frac{\sum_{i=1}^{L} N_{fi} t_i + N_s t_d}{\sum_{i=1}^{L} N_{fi}},$$

which is the point estimate of the components' mean life.

Evaluating $\hat{m}_{(N,n,t_d)}(t)$ at $t = 0, t_1, t_2, \cdots, t_{L-1}$ yields L points; i.e., $[0, \hat{m}_{(N,n,t_d)}(0)]$, $[t_1, \hat{m}_{(N,n,t_d)}(t_1)]$, $\cdots$, $[t_{L-1}, \hat{m}_{(N,n,t_d)}(t_{L-1})]$. Connecting these points consecutively by straight lines yields an approximate MRL function for the given life data set.

CASE 2 – Grouped data sets

In this case, N components are life tested and the test is terminated at t_d. A total of r failures are observed and recorded during n time intervals as follows:

$$[(t_1, t_2); N_{f1}], \quad [(t_2, t_3); N_{f2}], \quad \cdots, \quad [(t_n, t_{n+1}); N_{fn}],$$

where

N_{fi} = number of failures in interval (t_i, t_{i+1}),

$$r = \sum_{i=1}^{n} N_{fi},$$

and

$$t_{n+1} \leq t_d.$$

Then, the empirical MRL's at the starting and ending points of each time interval can be written as

$$\hat{m}_{(N,r,t_d)}(t_j) = \begin{cases} \dfrac{\sum_{i=j}^{n} N_{fi} \left[\frac{1}{2}(t_i + t_{i+1}) - t_j\right] + N_s (t_d - t_j)}{\sum_{i=j}^{n} N_{fi}}, & \text{for } j = 0, 1, 2, \cdots, n, \\ \text{unknown from the test}, & \text{for } j \geq n+1, \end{cases}$$

(8.18)

where

n = total number of groups, or time intervals,

t_i = starting point of the ith interval, (t_i, t_{i+1}),
    for $i = 0, 1, 2, \cdots, n$,

t_{i+1} = ending point of the ith interval, (t_i, t_{i+1}),
    for $i = 0, 1, 2, \cdots, n$,

$t_0 = 0$,

and

$N_{f0} = 0.$

Note that Eq. (8.18) is based on the assumption that N_{fi} failures occured uniformly over the interval (t_i, t_{i+1}); i.e., each with a mean life of $\frac{1}{2}(t_i + t_{i+1})$. Also note that for $j = 0$, $t_j = t_0 = 0$, then Eq. (8.18) becomes

$$\hat{m}_{(N,r,t_d)}(0) = \frac{\sum\limits_{i=0}^{n} N_{fi} \left[\frac{1}{2}(t_i + t_{i+1})\right] + N_s\, t_d}{\sum\limits_{i=0}^{n} N_{fi}},$$

or

$$\hat{m}_{(N,r,t_d)}(0) = \frac{\sum\limits_{i=1}^{n} N_{fi} \left[\frac{1}{2}(t_i + t_{i+1})\right] + N_s\, t_d}{\sum\limits_{i=1}^{n} N_{fi}}, \qquad (8.19)$$

due to $N_{f0} = 0$. Equation (8.19) gives the point estimate of the components' mean life.

8.5 MEAN RESIDUAL LIFETIMES FOR SOME FREQUENTLY USED LIFE DISTRIBUTIONS

8.5.1 THE WEIBULL DISTRIBUTION

The *pdf* of the Weibull life distribution is

$$f_W(T) = \frac{\beta}{\eta}\left(\frac{T}{\eta}\right)^{\beta-1} e^{-\left(\frac{T}{\eta}\right)^{\beta}}, \text{ for } T \geq 0, \beta > 0, \eta > 0. \qquad (8.20)$$

The reliability function is

$$R_W(T) = \int_T^{\infty} f_W(u)\, du = e^{-\left(\frac{T}{\eta}\right)^{\beta}}. \qquad (8.21)$$

Then, the mean residual life at any age t can be obtained by substituting Eqs. (8.20) and (8.21) into Eq. (8.3), or

$$m_W(t) = \frac{1}{R_W(t)} \int_t^{\infty} u\, f_W(u)\, du - t,$$

$$= e^{\left(\frac{t}{\eta}\right)^{\beta}} \int_t^{\infty} u\, \frac{\beta}{\eta}\left(\frac{u}{\eta}\right)^{\beta-1} e^{-\left(\frac{u}{\eta}\right)^{\beta}}\, du - t,$$

or

$$m_W(t) = e^{\left(\frac{t}{\eta}\right)^{\beta}} \int_t^{\infty} \beta\left(\frac{u}{\eta}\right)^{\beta} e^{-\left(\frac{u}{\eta}\right)^{\beta}}\, du - t. \qquad (8.22)$$

Let $\left(\frac{u}{\eta}\right)^\beta = x$, then $u = \eta\, x^{\frac{1}{\beta}}$ and $du = \frac{\eta}{\beta} x^{\frac{1}{\beta}-1}\, dx$. In addition, let

$$\Gamma(a; b) = \int_b^\infty y^{a-1}\, e^{-y}\, dy, \tag{8.23}$$

which is the so-called Incomplete Gamma Function. Substituting these into Eq. (8.22) yields

$$m_W(t) = e^{\left(\frac{t}{\eta}\right)^\beta} \int_{\left(\frac{t}{\eta}\right)^\beta}^\infty (\beta\, x\, e^{-x}) \left(\frac{\eta}{\beta}\right) x^{\frac{1}{\beta}-1}\, dx - t,$$

$$= \eta\, e^{\left(\frac{t}{\eta}\right)^\beta} \int_{\left(\frac{t}{\eta}\right)^\beta}^\infty x^{\frac{1}{\beta}}\, e^{-x}\, dx - t,$$

or [11] from Eq. (8.23)

$$m_W(t) = \eta\, e^{\left(\frac{t}{\eta}\right)^\beta} \Gamma\left[\frac{1}{\beta} + 1; \left(\frac{t}{\eta}\right)^\beta\right] - t. \tag{8.24}$$

8.5.2 THE GAMMA DISTRIBUTION

The *pdf* of the Gamma life distribution is

$$f_G(T) = \frac{1}{\eta\, \Gamma(\beta)} \left(\frac{T}{\eta}\right)^{\beta-1} e^{-\left(\frac{T}{\eta}\right)}, \quad \text{for } T \geq 0, \beta > 0, \eta > 0. \tag{8.25}$$

The reliability function is

$$R_G(T) = \int_T^\infty f_G(u)\, du = \int_T^\infty \frac{1}{\eta\, \Gamma(\beta)} \left(\frac{u}{\eta}\right)^{\beta-1} e^{-\left(\frac{u}{\eta}\right)} du. \tag{8.26}$$

Then, the mean residual life at any age t can be obtained by substituting Eqs. (8.25) and (8.26) into Eq. (8.3), or

$$m_G(t) = \frac{1}{R_G(t)} \int_t^\infty u\, f_G(u)\, du - t,$$

$$= \frac{\eta\, \Gamma(\beta)}{\int_t^\infty \left(\frac{u}{\eta}\right)^{\beta-1} e^{-\left(\frac{u}{\eta}\right)} du}$$

$$\cdot \int_t^\infty u\, \frac{1}{\eta\, \Gamma(\beta)} \left(\frac{u}{\eta}\right)^{\beta-1} e^{-\left(\frac{u}{\eta}\right)} du - t,$$

or [11]

$$m_G(t) = \frac{\int_t^\infty u^\beta\, e^{-\left(\frac{u}{\eta}\right)} du}{\int_t^\infty u^{\beta-1}\, e^{-\left(\frac{u}{\eta}\right)} du} - t. \tag{8.27}$$

Let $y = \frac{u}{\eta}$. Then, $u = \eta y$ and $du = \eta\, dy$. The numerator in Eq. (8.27) becomes

$$\int_t^\infty u^\beta e^{-\left(\frac{u}{\eta}\right)} du = \int_{\frac{t}{\eta}}^\infty \eta^\beta y^\beta e^{-y} \eta\, dy,$$

$$= \eta^{\beta+1} \int_{\frac{t}{\eta}}^\infty y^{(\beta+1)-1} e^{-y}\, dy,$$

$$= \eta^{\beta+1} \Gamma(\beta+1; \frac{t}{\eta}),$$

according to Eq. (8.23). Similarly, the denominator in Eq. (8.27) becomes

$$\int_t^\infty u^{\beta-1} e^{-\left(\frac{u}{\eta}\right)} du = \int_{\frac{t}{\eta}}^\infty \eta^{\beta-1} y^{\beta-1} e^{-y} \eta\, dy,$$

$$= \eta^\beta \int_{\frac{t}{\eta}}^\infty y^{\beta-1} e^{-y}\, dy,$$

$$= \eta^\beta \Gamma(\beta; \frac{t}{\eta}).$$

Therefore, Eq (8.27) becomes

$$m_G(t) = \frac{\eta^{\beta+1} \Gamma(\beta+1; \frac{t}{\eta})}{\eta^\beta \Gamma(\beta; \frac{t}{\eta})} - t,$$

or

$$m_G(t) = \frac{\Gamma(\beta+1; \frac{t}{\eta})}{\Gamma(\beta; \frac{t}{\eta})} \eta - t. \tag{8.28}$$

8.5.3 THE EXPONENTIAL DISTRIBUTION

The exponential distribution is a special case of both the Weibull and the Gamma distribution when $\beta = 1$. The *pdf* of the exponential life distribution is

$$f_E(T) = \lambda e^{-\lambda T}, \text{ for } T \geq 0 \text{ and } \lambda > 0. \tag{8.29}$$

The reliability function is

$$R_E(T) = \int_T^\infty f_E(u)\, du = e^{-\lambda T}. \tag{8.30}$$

Then, the mean residual life at any age t can be obtained by substituting Eqs. (8.30) into Eq. (8.1), or

$$m_E(t) = \frac{1}{R_E(t)} \int_t^\infty R_E(u)\,du,$$

$$= e^{\lambda t} \int_t^\infty e^{-\lambda u}\,du,$$

$$= e^{\lambda t} \left(\frac{1}{\lambda} e^{-\lambda t} \right),$$

or

$$m_E(t) = \frac{1}{\lambda} = \text{constant}. \tag{8.31}$$

It may be seen that the mean residual life of the exponential distribution at any age t is a constant and equal to its mean life. This is another indication of the memorilessness of exponential life distribution; i.e., the remaining life of an exponential component at any age t is independent of its operation history, or age.

8.5.4 THE TRUNCATED NORMAL DISTRIBUTION

The *pdf* of the truncated normal life distribution is

$$f_N(T) = \frac{\psi}{\sqrt{2\pi}\,\sigma} e^{-\frac{1}{2}\left(\frac{T-\mu}{\sigma}\right)^2}, \text{ for } T \geq 0, \mu > 0, \sigma > 0, \tag{8.32}$$

where

$$\psi = \Phi\left(\frac{\mu}{\sigma}\right),$$

and

$\Phi(\ ) = cdf$ of the standardized normal distribution, $N(0,1)$.

The reliability function is

$$R_N(T) = \int_T^\infty f_N(u)\,du = \psi\,\Phi\left(\frac{\mu - T}{\sigma}\right). \tag{8.33}$$

Then, the mean residual life at any age t can be obtained by substituting Eqs. (8.32) and (8.33) into Eq. (8.3), or

$$m_N(t) = \frac{1}{R_N(t)} \int_t^\infty u\,f_N(u)\,du - t,$$

$$= \frac{1}{\psi\,\Phi\left(\frac{\mu-t}{\sigma}\right)} \int_t^\infty u\,\frac{\psi}{\sqrt{2\pi}\,\sigma} e^{-\frac{1}{2}\left(\frac{u-\mu}{\sigma}\right)^2} du - t,$$

$$= \frac{1}{\sqrt{2\pi}\,\Phi\left(\frac{\mu-t}{\sigma}\right)} \int_t^\infty \frac{u}{\sigma} e^{-\frac{1}{2}\left(\frac{u-\mu}{\sigma}\right)^2} du - t,$$

or

$$m_N(t) = \frac{1}{\sqrt{2\pi}\,\Phi\left(\frac{\mu-t}{\sigma}\right)} I(t) - t, \tag{8.34}$$

where

$$I(t) = \int_t^\infty \frac{u}{\sigma} e^{-\frac{1}{2}\left(\frac{u-\mu}{\sigma}\right)^2} du, \tag{8.35}$$

which can be expressed in terms of the Incomplete Gamma Function as follows:

$$I(t) = \begin{cases} \sigma\, e^{-\frac{1}{2}\left(\frac{t-\mu}{\sigma}\right)^2} - \frac{\mu}{\sqrt{2}}\left\{2\pi - \Gamma\left[\frac{1}{2};\frac{1}{2}\left(\frac{t-\mu}{\sigma}\right)^2\right]\right\}, & \text{for } 0 < t < \mu, \\ \sigma\, e^{-\frac{1}{2}\left(\frac{t-\mu}{\sigma}\right)^2} - \frac{\mu}{\sqrt{2}}\,\Gamma\left[\frac{1}{2};\frac{1}{2}\left(\frac{t-\mu}{\sigma}\right)^2\right], & \text{for } t \geq \mu. \end{cases}$$

$$(8.36)$$

The derivation of Eq. (8.36) is given in Appendix 8A.

8.5.5 THE LOGNORMAL DISTRIBUTION

The *pdf* of the lognormal life distribution is

$$f_{LN}(T) = \frac{1}{\sqrt{2\pi}\,\sigma' T} e^{-\frac{1}{2}\left(\frac{\log_e T - \mu'}{\sigma'}\right)^2}, \quad \text{for } T \geq 0, \mu' > 0, \sigma' > 0.$$

$$(8.37)$$

The reliability function is

$$R_{LN}(T) = \int_T^\infty f_{LN}(u)\, du = \Phi\left(\frac{\mu' - \log_e T}{\sigma'}\right). \tag{8.38}$$

Then, the mean residual life at any age t can be obtained by substituting Eqs. (8.37) and (8.38) into Eq. (8.3), or

$$m_{LN}(t) = \frac{1}{R_{LN}(t)} \int_t^\infty u\, f_{LN}(u)\, du - t,$$

$$= \frac{1}{\Phi\left(\frac{\mu' - \log_e t}{\sigma'}\right)} \int_t^\infty u\, \frac{1}{\sqrt{2\pi}\,\sigma' u} e^{-\frac{1}{2}\left(\frac{\log_e u - \mu'}{\sigma'}\right)^2} du - t,$$

$$= \frac{1}{\Phi\left(\frac{\mu' - \log_e t}{\sigma'}\right)} \int_t^\infty \frac{1}{\sqrt{2\pi}\,\sigma'} e^{-\frac{1}{2}\left(\frac{\log_e u - \mu'}{\sigma'}\right)^2} du - t,$$

or

$$m_{LN}(t) = \frac{I'(t)}{\Phi\left(\frac{\mu' - \log_e t}{\sigma'}\right)} - t, \tag{8.39}$$

where

$$I'(t) = \int_t^\infty \frac{1}{\sqrt{2\pi}\,\sigma'} e^{-\frac{1}{2}\left(\frac{\log_e u - \mu'}{\sigma'}\right)^2} du, \tag{8.40}$$

which can be expressed in terms of $\Phi(\ )$, the *cdf* of the standardized normal distribution, $N(0,1)$, as

$$I'(t) = e^{\mu' + \frac{1}{2}(\sigma')^2}\,\Phi\left[\frac{(\sigma')^2 + \mu' - \log_e t}{\sigma'}\right]. \tag{8.41}$$

The derivation of Eq. (8.41) is given in Appendix 8B. Substituting Eq. (8.41) into Eq. (8.39) yields

$$m_{LN}(t) = \frac{\Phi\left[\frac{(\sigma')^2 + \mu' - \log_e t}{\sigma'}\right]}{\Phi\left(\frac{\mu' - \log_e t}{\sigma'}\right)} e^{\mu' + \frac{1}{2}(\sigma')^2} - t. \tag{8.42}$$

8.5.6 THE RAYLEIGH DISTRIBUTION

The Rayleigh life distribution is a special case of Weibull life distribution when $\beta = 2$. Its *pdf* is

$$f_R(T) = \frac{T}{\eta^2} e^{-\frac{1}{2}\left(\frac{T}{\eta}\right)^2}, \text{ for } T \geq 0, \eta > 0. \tag{8.43}$$

The reliability function is

$$R_R(T) = \int_T^\infty f_R(u)\,du = e^{-\frac{1}{2}\left(\frac{T}{\eta}\right)^2}. \tag{8.44}$$

Then, the mean residual life at any age t can be obtained by substituting Eqs. (8.44) into Eq. (8.1), or

$$\begin{aligned} m_R(t) &= \frac{1}{R_R(t)} \int_t^\infty R_R(u)\,du, \\ &= e^{\frac{1}{2}\left(\frac{t}{\eta}\right)^2} \int_t^\infty e^{-\frac{1}{2}\left(\frac{u}{\eta}\right)^2} du. \end{aligned}$$

Let $x = \frac{1}{2}\left(\frac{u}{\eta}\right)^2$. Then, $u = \eta\sqrt{2x}$ and $du = \eta\left(\frac{1}{2}\right)\frac{2\,dx}{\sqrt{2x}} = \frac{\eta}{\sqrt{2}} x^{-\frac{1}{2}}\,dx$.
Then,

$$m_R(t) = e^{\frac{1}{2}\left(\frac{t}{\eta}\right)^2} \int_{\frac{1}{2}\left(\frac{t}{\eta}\right)^2}^{\infty} e^{-x}\, \frac{\eta}{\sqrt{2}}\, x^{-\frac{1}{2}}\,dx,$$

$$= \frac{\eta}{\sqrt{2}}\, e^{\frac{1}{2}\left(\frac{t}{\eta}\right)^2} \int_{\frac{1}{2}\left(\frac{t}{\eta}\right)^2}^{\infty} x^{\frac{1}{2}-1}\, e^{-x}\,dx,$$

or

$$m_R(t) = \frac{\eta}{\sqrt{2}}\, e^{\frac{1}{2}\left(\frac{t}{\eta}\right)^2} \Gamma\left[\frac{1}{2}; \frac{1}{2}\left(\frac{t}{\eta}\right)^2\right], \tag{8.45}$$

according to Eq. (8.23).

8.5.7 THE MIXED-LIFE DISTRIBUTION

The *pdf* of a mixed-life distribution is

$$f_M(T) = \sum_{i=1}^{n} p_i\, f_i(T), \tag{8.46}$$

where

n = number of subpopulations,
$f_i(T)$ = *pdf* of subpopulation i,
p_i = proportion of subpopulation i,

and

$$\sum_{i=1}^{n} p_i = 1.$$

The reliability function is

$$R_M(T) = \int_T^{\infty} f_M(u)\,du = \sum_{i=1}^{n} p_i\, R_i(T), \tag{8.47}$$

where

$R_i(T)$ = reliability function of subpopulation i.

MRL FOR FREQUENTLY USED DISTRIBUTIONS

Then, the mean residual life at any age t can be obtained by substituting Eqs. (8.46) and (8.47) into Eq. (8.3), or

$$m_M(t) = \frac{1}{R_M(t)} \int_t^\infty u\, f_M(u)\, du - t,$$

$$= \frac{1}{\sum_{i=1}^n p_i R_i(t)} \int_t^\infty u \left[\sum_{i=1}^n p_i f_i(u)\right] du - t,$$

$$= \frac{1}{\sum_{i=1}^n p_i R_i(t)} \int_t^\infty \left[\sum_{i=1}^n p_i u f_i(u)\right] du - t,$$

or

$$m_M(t) = \frac{1}{\sum_{i=1}^n p_i R_i(t)} \sum_{i=1}^n \left[p_i \int_t^\infty u f_i(u)\, du\right] - t. \tag{8.48}$$

For example, the MRL for a mixed-Weibull distribution can be obtained by substituting Eqs. (8.20) and (8.21) into Eq. (8.48); i.e.,

$$m_{MW}(t) = \frac{1}{\sum_{i=1}^n p_i\, e^{-\left(\frac{t}{\eta_i}\right)^{\beta_i}}} \sum_{i=1}^n \left[p_i \int_t^\infty u\, \frac{\beta_i}{\eta_i}\left(\frac{u}{\eta_i}\right)^{\beta_i-1} e^{-\left(\frac{u}{\eta_i}\right)^{\beta_i}} du\right] - t.$$

Following the same procedure as that for deriving Eq. (8.24) which gives the MRL for a single Weibull life distribution, but substituting the mixed-Weibull pdf in place of the single Weibull pdf, yields

$$m_{MW}(t) = \frac{1}{\sum_{i=1}^n p_i\, e^{-\left(\frac{t}{\eta_i}\right)^{\beta_i}}} \sum_{i=1}^n \left\{p_i\, \eta_i\, \Gamma\left[\frac{1}{\beta_i}+1;\left(\frac{t}{\eta_i}\right)^{\beta_i}\right]\right\} - t. \tag{8.49}$$

As a second example, the MRL for mixed-Gamma distribution can be obtained by substituting Eqs. (8.25) and (8.26) into Eq. (8.48); i.e.,

$$m_{MG}(t) = \left(\frac{1}{\sum_{i=1}^n p_i \int_t^\infty \frac{1}{\eta_i \Gamma(\beta_i)}\left(\frac{u}{\eta_i}\right)^{\beta_i-1} e^{-\left(\frac{u}{\eta_i}\right)} du}\right)$$
$$\cdot \sum_{i=1}^n \left[p_i \int_t^\infty u\, \frac{1}{\eta_i \Gamma(\beta_i)}\left(\frac{u}{\eta_i}\right)^{\beta_i-1} e^{-\left(\frac{u}{\eta_i}\right)} du\right] - t,$$

$$= \left(\frac{1}{\sum_{i=1}^{n} \left[\frac{p_i}{\eta_i^{\beta_i} \Gamma(\beta_i)} \int_t^\infty u^{\beta_i-1} e^{-\left(\frac{u}{\eta_i}\right)} du \right]} \right)$$

$$\cdot \sum_{i=1}^{n} \left[\frac{p_i}{\eta_i^{\beta_i} \Gamma(\beta_i)} \int_t^\infty u^{\beta_i} e^{-\left(\frac{u}{\eta_i}\right)} du \right] - t.$$

Following the same procedure as that for deriving Eq. (8.28), yields

$$m_{MG}(t) = \left(\frac{1}{\sum_{i=1}^{n} \left[\frac{p_i \, \eta_i^{\beta_i}}{\eta_i^{\beta_i} \Gamma(\beta_i)} \Gamma\left(\beta_i; \frac{t}{\eta_i}\right) \right]} \right)$$

$$\cdot \sum_{i=1}^{n} \left[\frac{p_i \, \eta_i^{\beta_i+1}}{\eta_i^{\beta_i} \Gamma(\beta_i)} \Gamma\left(\beta_i + 1; \frac{t}{\eta_i}\right) \right] - t,$$

or

$$m_{MG}(t) = \frac{\sum_{i=1}^{n} \left[p_i \, \eta_i \frac{\Gamma\left(\beta_i+1; \frac{t}{\eta_i}\right)}{\Gamma(\beta_i)} \right]}{\sum_{i=1}^{n} \left[p_i \frac{\Gamma\left(\beta_i; \frac{t}{\eta_i}\right)}{\Gamma(\beta_i)} \right]} - t. \tag{8.50}$$

8.6 THE EFFECT OF BURN-IN ON THE MRL ASSUMING A BATHTUB-SHAPED FAILURE RATE CURVE

It has been discussed in Chapter 4 that for many non-screened components, the bathtub curve provides a very good description on their failure rate behavior. A general bathtub failure rate curve, such as the one shown in Fig. 4.1, is a continuous function and is composed of the following three sections:

1. Early failure period, $[0, T_1)$, during which $\lambda(T)$ is a decreasing function (DFR).

2. Chance failure period, $[T_1, T_2)$, during which $\lambda(T)$ is a constant (CFR).

3. Wearout period, $[T_2, \infty)$, during which $\lambda(T)$ is an increasing function (IFR).

Now let's take a look at the effect of burn-in on the MRL for the following two cases of the bathtub-shaped failure rate function [12; 13]:

1. Without wearout failure period, or $T_2 \to \infty$.

2. With wearout failure period, or $T_2 < \infty$.

According to Eq. (8.1), the MRL of the product after T_b hours of burn-in is

$$m(T_b) = \frac{1}{R(T_b)} \int_{T_b}^{\infty} R(u)\, du. \tag{8.51}$$

Then, taking the first derivative of both sides of Eq. (8.51) with respect to T_b yields

$$\begin{aligned}
m'(T_b) &= \frac{d}{dT_b}\left[\frac{1}{R(T_b)}\right] \int_{T_b}^{\infty} R(u)\, du + \left[\frac{1}{R(T_b)}\right] \frac{d}{dT_b}\left[\int_{T_b}^{\infty} R(u)\, du\right], \\
&= \frac{f(T_b)}{[R(T_b)]^2} \int_{T_b}^{\infty} R(u)\, du + \frac{1}{R(T_b)}[0 - R(T_b)], \\
&= \frac{1}{R(T_b)}\left[\frac{f(T_b)}{R(T_b)} \int_{T_b}^{\infty} R(u)\, du - R(T_b)\right], \\
&= \frac{1}{R(T_b)}\left[\lambda(T_b) \int_{T_b}^{\infty} R(u)\, du - R(T_b)\right],
\end{aligned}$$

or

$$m'(T_b) = N(T_b)/R(T_b), \tag{8.52}$$

where

$$N(T_b) = \lambda(T_b) \int_{T_b}^{\infty} R(u)\, du - R(T_b), \tag{8.53}$$

which is a continuously differentiable function, with its first derivative being

$$N'(T_b) = \lambda'(T_b) \int_{T_b}^{\infty} R(u)\, du + \lambda(T_b)[0 - R(T_b)] + f(T_b),$$

or

$$N'(T_b) = \lambda'(T_b) \int_{T_b}^{\infty} R(u)\, du, \tag{8.54}$$

and its limiting value being

$$N(\infty) = \lim_{T_b \to \infty} N(T_b),$$

$$= \lim_{T_b \to \infty} \left[\lambda(T_b) \int_{T_b}^{\infty} R(u)\,du - R(T_b)\right],$$

or

$$N(\infty) = 0. \tag{8.55}$$

CASE 1– For a bathtub-shaped failure rate curve without wearout, or $T_2 \to \infty$.

In this case, $\lambda(T_b)$ is decreasing in $[0, T_1)$ and constant in $[T_1, \infty)$. A constant $\lambda(T_b)$ in $[T_1, \infty)$ implies that $\lambda'(T_b) = 0$ and therefore $N'(T_b) = 0$ in $[T_1, \infty)$ according to Eq. (8.54). Consequently, $N(T_b)$ is constant in $[T_1, \infty)$. But according to Eq. (8.55), $N(\infty) = 0$. Then, $N(T_b) = 0$ in $[T_1, \infty)$ which implies $m'(T_b) = 0$ in $[T_1, \infty)$ according to Eq. (8.52), and which in turn implies that $m(T_b)$ is constant in $[T_1, \infty)$.

Now consider the interval $[0, T_1)$ where $\lambda(T_b)$ is a decreasing function which implies that $\lambda'(T_b) < 0$, and which in turn implies that $N'(T_b) < 0$ in $[0, T_1)$ according to Eq. (8.54). Therefore, $N(T_b)$ is a decreasing function in $[0, T_1)$. But $N(T_b) = 0$ in $[T_1, \infty)$ which implies that $N(T_1) = 0$. Then, $N(T_b) > 0$ in $[0, T_1)$ and in turn $m'(T_b) > 0$ in $[0, T_1)$ according to Eq. (8.52), which imply that $m(T_b)$ is an increasing function in $[0, T_1)$.

In summary, when there is no wearout failure period, the MRL, $m(T_b)$, is an increasing function in the early failure period, $[0, T_1)$, and constant in the chance failure period, $[T_1, \infty)$. This behavior is illustrated in Fig. 8.1.

CASE 2– For a bathtub-shaped failure rate curve with wearout failure period, or $T_2 < \infty$.

In this case, $\lambda(T_b)$ is a decreasing function in $[0, T_1)$, constant in $[T_1, T_2)$ and an increasing function in $[T_2, \infty)$. An increasing $\lambda(T_b)$ in $[T_2, \infty)$ implies that $\lambda'(T_b) > 0$ in $[T_2, \infty)$. Then, according to Eq. (8.54), $N'(T_b) > 0$ in $[T_2, \infty)$, which implies that $N(T_b)$ strictly increases in $[T_2, \infty)$. But $N(\infty) = 0$ according to Eq. (8.55). Then, $N(T_b) < 0$ and consequently $m'(T_b) < 0$ in $[T_2, \infty)$ according to Eq. (8.52), which implies that $m(T_b)$ is a strictly decreasing function in $[T_2, \infty)$.

Now consider the interval $[T_1, T_2)$ where $\lambda(T_b)$ is constant, which implies that $\lambda'(T_b) = 0$, which in turn implies that $N'(T_b) = 0$ in $[T_1, T_2)$ according to Eq. (8.54), or $N(T_b)$ is constant in $[T_1, T_2)$. But since $N(T_b) < 0$ in $[T_2, \infty)$, then $N(T_2) < 0$. Therefore, $N(T_b) =$

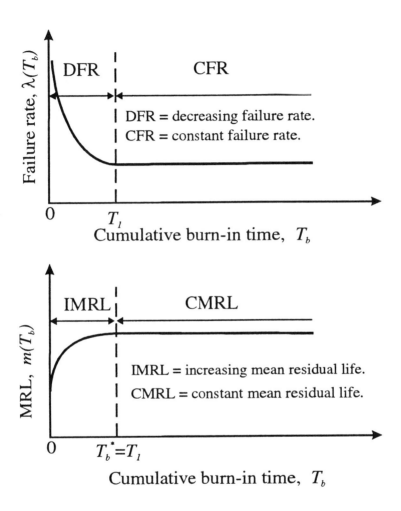

Fig. 8.1– Behavior of MRL versus burn-in time for a bathtub-shaped failure rate function without wearout failure period.

$N(T_2) < 0$ in $[T_1, T_2)$. Consequently, $m'(T_b) < 0$ in $[T_1, T_2)$ according to Eq. (8.52), which implies that $m(T_b)$ is also a strictly decreasing function in $[T_1, T_2)$.

Finally, consider the interval $[0, T_1)$ where $\lambda(T_b)$ is a decreasing function. According to the discussion in Case 1, $N(T_b)$ is a decreasing function in $[0, T_1)$. Now since $N(T_b) < 0$ in $[T_1, T_2)$, then $N(T_1) < 0$. Therefore, there are two possibilities for the value of $N(T_b)$ in $[0, T_1)$:

1. $N(T_1) < N(T_b) < 0$ in $[0, T_1)$.

2. $0 < N(T_b) < \infty$ in $[0, T_0)$ and $N(T_1) < N(T_b) \leq 0$ in $[T_0, T_1)$ where $0 < T_0 < T_1$.

For the first case, $N(T_b) < 0$ implies that $m'(T_b) < 0$ in $[0, T_1)$ according to Eq. (8.52), which implies that $m(T_b)$ is a decreasing function in $[0, T_1)$.

Following the same reasoning, for the second case, $m(T_b)$ will be an increasing function in $[0, T_0)$ but a decreasing function in $[T_0, T_1)$.

In summary, when the wearout failure period is present, the MRL, $m(T_b)$, may behave in the following two ways:

1. **Decreases throughout $[0, \infty)$.**

2. **Increases in $[0, T_0)$ and decreases in $[T_0, \infty)$ where $0 < T_0 < T_1$.**

This behavior is illustrated in Fig. 8.2. It may be seen from Figs. 8.1 and 8.2 that if the objective of a burn-in is to achieve the maximum MRL, then the optimum burn-in time T_b^* will be

1. $T_b^* = T_1$ for a bathtub-shaped failure rate function without wearout failure period.

2. $0 \leq T_b^* < T_1$ for a bathtub-shaped failure rate function with wearout failure period.

This conclusion will be demonstrated by examples in the next section.

8.7 APPLICATION OF THE MRL CONCEPT TO THE OPTIMUM BURN-IN TIME DETERMINATION

8.7.1 OPTIMUM BURN-IN TIME FOR A PRESPECIFIED MRL

It has been said that units with a decreasing failure rate (DFR) should be burned-in before they are delivered to the customers. Let T_b be the

APPLICATION IN BURN-IN

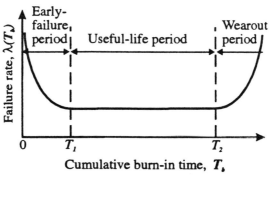

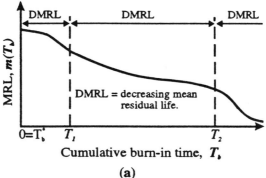

(a)

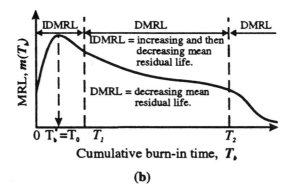

(b)

Fig. 8.2– Behavior of MRL versus burn-in time for a bathtub-shaped failure rate function with a wearout failure period. (a) First possibility: $m(T_b)$ decreases throughout the whole life. (b) Second possibility: $m(T_b)$ increases in $[0, T_0)$ and decreases in $[T_0, \infty)$.

burn-in time. Then, the mean residual life of those units surviving the burn-in of period T_b is given by Eq. (8.1), or

$$m(T_b) = \frac{1}{R(T_b)} \int_{T_b}^{\infty} R(u) \, du. \tag{8.56}$$

Often the product is expected to reach a specified mean residual life goal, m_G, after it is burned-in for a particular period of time, T_b^*. In this case, the MRL function of the product after burn-in, $m(T_b)$, should be determined first. Let it equal to m_G, and then solve for T_b^*.

Now let

$$m(T_b^*) = m_G.$$

Then, substituting Eq. (8.56) yields

$$\frac{1}{R(T_b^*)} \int_{T_b^*}^{\infty} R(u) \, du = m_G. \tag{8.57}$$

Solving Eq. (8.57) will yield the T_b^* corresponding to the specified m_G. Presented next are two applications of Eq. (8.57).

8.7.1.1 APPLICATION OF EQ. (8.57) USING THE BATHTUB MODEL 1

If the Bathtub Model 1 is used, Eq. (8.57) becomes

$$\left[\int_{T_b^*}^{\infty} e^{1-e^{(u/\hat{\alpha})^{\hat{\beta}}}} du \right] \Big/ e^{1-e^{(T_b^*/\hat{\alpha})^{\hat{\beta}}}} = m_G. \tag{8.58}$$

T_b^* can be solved by trial-and-error plus a routine integration in the infinity interval. A computer program named "BIM" has been developed for this and is listed in Appendix 8C.

EXAMPLE 8-1

Determine the burn-in time for an electronic component with the Bathtub Model 1 parameter estimates of

$$\hat{\beta} = 0.62,$$

and

$$\hat{\alpha} = 70,000 \text{ hr},$$

and using Eq. (8.58). Assume that the MRL goal is 39,000 hr.

SOLUTION TO EXAMPLE 8-1

Inputting the given parameter estimates, and the MRL goal, into the computer program "BIM" given in Appendix 8C yields the burn-in time as

$$T_b^* = 426.66 \text{ hr.}$$

EXAMPLE 8-2

Determine the burn-in time for an electronic component with the Bathtub Model 1 parameter estimates of

$$\hat{\beta} = 0.571,$$

and

$$\hat{\alpha} = 54,321 \text{ hr},$$

and using Eq. (8.58). Assume that the MRL goal is 30,000 hr.

SOLUTION TO EXAMPLE 8-2

Inputting the given parameter estimates, and the MRL goal, into the computer program "BIM" given in Appendix 8C yields the burn-in time of

$$T_b^* = 170.70 \text{ hr.}$$

8.7.1.2 APPLICATION OF EQ. (8.57) USING THE BIMODAL, MIXED-WEIBULL MODEL

If the bimodal, mixed-Weibull model is used, Eq. (8.57) becomes

$$\frac{\int_0^\infty \left[\hat{p}\, e^{-(\frac{u+T_b^*}{\hat{\eta}_1})^{\hat{\beta}_1}} + (1-\hat{p})\, e^{-(\frac{u+T_b^*}{\hat{\eta}_2})^{\hat{\beta}_2}} \right] du}{\hat{p}\, e^{-(T_b^*/\hat{\eta}_1)^{\hat{\beta}_1}} + (1-\hat{p})\, e^{-(T_b^*/\hat{\eta}_2)^{\hat{\beta}_2}}} = m_G,$$

or

$$\frac{\hat{p} \int_{T_b^*}^\infty e^{-(u/\hat{\eta}_1)^{\hat{\beta}_1}} du + (1-\hat{p}) \int_{T_b^*}^\infty e^{-(u/\hat{\eta}_2)^{\hat{\beta}_2}} du}{\hat{p}\, e^{-(T_b^*/\hat{\eta}_1)^{\hat{\beta}_1}} + (1-\hat{p})\, e^{-(T_b^*/\hat{\eta}_2)^{\hat{\beta}_2}}} = m_G. \quad (8.59)$$

The T_b^* can be solved by trial-and-error and a routine integration, which are included in the computer program "BIM" given in Appendix 8C.

EXAMPLE 8-3

Rework Example 7-9 by letting the MRL goal of the system be 4,000 hr, and using Eq. (8.59).

SOLUTION TO EXAMPLE 8-3

From Example 7-9, the system's parameter estimates for the Bimodal, Mixed-Weibull Model are

$$\hat{p} = 0.366124,$$
$$\hat{\beta}_1 = 0.57,$$
$$\hat{\eta}_1 = 530 \text{ hr},$$
$$\hat{\beta}_2 = 1.82,$$

and

$$\hat{\eta}_2 = 5,900 \text{ hr}.$$

Inputting these parameter estimates, and the given MRL goal, into the computer program "BIM" given in Appendix 8C yields the burn-in time as

$$T_b^* = 89.67 \text{ hr}.$$

8.7.2 OPTIMUM BURN-IN TIME FOR THE MAXIMUM MRL

When the burn-in goal is to achieve a maximum MRL, the optimum burn-in time, T_b^*, can be obtained by maximizing Eq. (8.56). The first derivative of $m(T_b)$ is given by Eq. (8.5), or

$$m'(T_b) = \lambda(T_b)\, m(T_b) - 1. \tag{8.60}$$

Then, the optimum burn-in time, T_b^*, can be obtained by letting the right side of Eq. (8.60) equal to zero and solving for T_b; i.e.,

$$\lambda(T_b^*)\, m(T_b^*) - 1 = 0,$$

or

$$m(T_b^*) = \frac{1}{\lambda(T_b^*)}. \tag{8.61}$$

8.7.2.1 APPLICATION OF EQ. (8.61) USING THE BATHTUB MODEL 1

If the Bathtub Model 1 is used, then substituting Eqs. (5.8) and (8.58) into Eq. (8.61) yields

$$\left[\int_{T_b^*}^{\infty} e^{1-e^{(u/\hat{\alpha})^{\hat{\beta}}}} du\right] \bigg/ e^{1-e^{(T_b^*/\hat{\alpha})^{\hat{\beta}}}} = \left(\frac{\hat{\alpha}^{\hat{\beta}}}{\hat{\beta}}\right)(T_b^*)^{1-\hat{\beta}} e^{-(T_b^*/\hat{\alpha})^{\hat{\beta}}},$$

(8.62)

or

$$\int_{T_b^*}^{\infty} e^{1-e^{(u/\hat{\alpha})^{\hat{\beta}}}} du = \left(\frac{\hat{\alpha}^{\hat{\beta}}}{\hat{\beta}}\right)(T_b^*)^{1-\hat{\beta}} e^{1-(T_b^*/\hat{\alpha})^{\hat{\beta}} - e^{(T_b^*/\hat{\alpha})^{\hat{\beta}}}}.$$

T_b^* can be solved by the following iteration relation:

$$T_b^* = \left[\left(\frac{\hat{\beta}}{\hat{\alpha}^{\hat{\beta}}}\right) e^{(T_b^*/\hat{\alpha})^{\hat{\beta}} + e^{(T_b^*/\hat{\alpha})^{\hat{\beta}}} - 1} \int_{T_b^*}^{\infty} e^{1-e^{(u/\hat{\alpha})^{\hat{\beta}}}} du\right]^{\frac{1}{1-\hat{\beta}}}.$$

(8.63)

The corresponding maximum MRL is the left side of Eq. (8.62), or

$$MRL^* = \left[\int_{T_b^*}^{\infty} e^{1-e^{(u/\hat{\alpha})^{\hat{\beta}}}} du\right] \bigg/ e^{1-e^{(T_b^*/\hat{\alpha})^{\hat{\beta}}}}. \qquad (8.64)$$

A computer program named "MRLMAX1" has been developed in FORTRAN language for this and is listed in Appendix 8D.

EXAMPLE 8-4

Rework Example 8-1 for the maximum MRL.

SOLUTION TO EXAMPLE 8-4

Inputting the given parameter estimates into the computer program "MRLMAX1" given in Appendix 8D yields the optimum burn-in time as

$$T_b^* = 13,299.09 \text{ hr.}$$

The corresponding maximum MRL is

$$MRL^* = 42,026.97 \text{ hr.}$$

Comparing $T_b^* = 13,299.09$ hr here for the maximum MRL, $MRL^* = 42,026.97$ hr, with $T_b^* = 426.66$ hr obtained in Example 8–1 for a specified MRL goal of $m_G = 39,000$ hr, it may be seen that to achieve the maximum MRL, the required burn-in time is much longer and even economically unjustifiable. That is why cost should also be considered in determining a practically optimum burn-in time, which will be covered and illustrated in the next chapter.

EXAMPLE 8–5

Rework Example 8–2 for the maximum MRL.

SOLUTION TO EXAMPLE 8–5

Inputting the given parameter estimates into the computer program "MRLMAX1" given in Appendix 8D yields the optimum burn-in time as

$$T_b^* = 16,130.81 \text{ hr.}$$

The corresponding maximum MRL is

$$MRL^* = 34,276.79 \text{ hr.}$$

Comparing $T_b^* = 16,130.81$ hr here for the maximum MRL, $MRL^* = 34,276.79$ hr, with $T_b^* = 170.70$ hr obtained in Example 8–2 for a specified MRL goal of $m_G = 30,000$ hr, it may be seen that to achieve the maximum MRL, the required burn-in time is much longer and even economically unjustifiable. That is why cost should also be considered in determining a practically optimum burn-in time, which will be covered and illustrated in the next chapter.

8.7.2.2 APPLICATION OF EQ. (8.61) USING THE BIMODAL, MIXED-WEIBULL MODEL

If the bimodal, mixed-Weibull model is used, Eq. (8.61) becomes, according to Eq. (8.59) and the failure rate function of the mixed-Weibull distribution,

$$\frac{\hat{p} \int_{T_b^*}^{\infty} e^{-(u/\hat{\eta}_1)^{\hat{\beta}_1}} du + (1-\hat{p}) \int_{T_b^*}^{\infty} e^{-(u/\hat{\eta}_2)^{\hat{\beta}_2}} du}{\hat{p}\, e^{-(T_b^*/\hat{\eta}_1)^{\hat{\beta}_1}} + (1-\hat{p})\, e^{-(T_b^*/\hat{\eta}_2)^{\hat{\beta}_2}}}$$

$$= \frac{\hat{p}\, e^{-(T_b^*/\hat{\eta}_1)^{\hat{\beta}_1}} + (1-\hat{p})\, e^{-(T_b^*/\hat{\eta}_2)^{\hat{\beta}_2}}}{\hat{p}\, \frac{\hat{\beta}_1}{\hat{\eta}_1} \left(\frac{T_b^*}{\hat{\eta}_1}\right)^{\hat{\beta}_1 - 1} e^{-(T_b^*/\hat{\eta}_1)^{\hat{\beta}_1}} + (1-\hat{p})\, \frac{\hat{\beta}_2}{\hat{\eta}_2} \left(\frac{T_b^*}{\hat{\eta}_2}\right)^{\hat{\beta}_2 - 1} e^{-(T_b^*/\hat{\eta}_2)^{\hat{\beta}_2}}}.$$

APPLICATION IN BURN-IN

$$\tag{8.65}$$

The corresponding maximum MRL is the left side of Eq. (8.65), or

$$MRL^* = \frac{\hat{p}\int_{T_b^*}^{\infty} e^{-(u/\hat{\eta}_1)^{\hat{\beta}_1}}\,du + (1-\hat{p})\int_{T_b^*}^{\infty} e^{-(u/\hat{\eta}_2)^{\hat{\beta}_2}}\,du}{\hat{p}\,e^{-(T_b^*/\hat{\eta}_1)^{\hat{\beta}_1}} + (1-\hat{p})\,e^{-(T_b^*/\hat{\eta}_2)^{\hat{\beta}_2}}}.$$

$$\tag{8.66}$$

T_b^* can be solved by trial-and-error plus a routine integration in the infinity interval which are included in the computer program "MRL-MAX2" given in Appendix 8E.

EXAMPLE 8–6

Rework Example 8–3 for the maximum MRL.

SOLUTION TO EXAMPLE 8–6

From Example 7–9, the system's parameter estimates for the Bi-modal, Mixed-Weibull Model are

$\hat{p} = 0.366124,$

$\hat{\beta}_1 = 0.57,$

$\hat{\eta}_1 = 530$ hr,

$\hat{\beta}_2 = 1.82,$

and

$\hat{\eta}_2 = 5,900$ hr.

Inputting these parameter estimates into the computer program "MRLMAX2" given in Appendix 8E yields the optimum burn-in time as

$T_b^* = 485.00$ hr.

The corresponding maximum MRL is

$MRL^* = 4,203.63$ hr.

Comparing $T_b^* = 485.00$ hr here for the maximum MRL, $MRL^* = 4,203.63$ hr, with $T_b^* = 89.67$ hr obtained in Example 8–3 for a specified MRL goal of $m_G = 4,000$ hr, it may be seen that to achieve the maximum MRL, the required burn-in time is much longer and even economically injustifiable. That is why cost should also be considered in determining a practically optimum burn-in time, which will be covered and illustrated in the next chapter.

8.8 DETERMINING THE OPTIMUM BURN-IN TIME DIRECTLY FROM THE EMPIRICAL MRL FUNCTION OR PLOT

In practical situations, often a set of life testing data or burn-in testing data are provided instead of giving directly a distribution model, such as the Bathtub model 1 or the bimodal mixed-Weibull model as is the case in Section 8.7. In this case, the empirical MRL function or plot is a very attractive and economical approach for the optimum burn-in time determination. This is so because the empirical MRL functions, as presented in Section 8.4, and their plots, offer a very straightforward and non-parametric description of the life behavior since these functions and plots are not based on any distribution assumption.

After evaluating the empirical MRL function values at the observed failure time points, using an appropriate MRL equation presented in Section 8.4, the empirical MRL plot can be obtained by connecting these points consecutively by *straight lines*. Then, the optimum burn-in time for a specified MRL goal, or for the maximum MRL, can be obtained from this plot which will be discussed next and illustrated by several numerical examples afterwards.

Two computer programs have been developed in FORTRAN language to facilitate the calculation of the empirical MRL functions which are presented in Section 8.4. These two computer programs, named "MRL-EMP1.FOR" and "MRL-EMP2.FOR," are listed in Appendices 8F and 8G, respectively. "MRL-EMP1.FOR" is designed for the ungrouped data set – complete or censored, with or without ties. "MRL-EMP2.FOR" is designed for the grouped data set – complete or censored.

8.8.1 OPTIMUM BURN-IN TIME FOR A SPECIFIED MRL GOAL

To determine the optimum burn-in time for a prespecified MRL goal, m_G, draw a horizontal line at $\hat{m}(t) = m_G$, then find the intersection with the empirical MRL plot, drop vertically at this intersection and read off the burn-in time from the time axis. This procedure yields the optimum burn-in time, T_b^*, between two adjacent failure times, say t_i and t_{i+1}, as shown in Fig. 8.3. The value of T_b^*, as read off from the time axis, can also be calculated accurately using the linear interpolation technique as follows:

$$\frac{T_b^* - t_i}{t_{i+1} - t_i} = \frac{m_G - \hat{m}_i}{\hat{m}_{i+1} - \hat{m}_i},$$

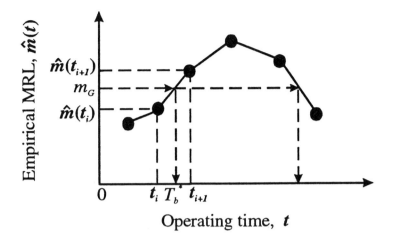

Fig. 8.3– Determination of the optimum burn-in time for a specified MRL goal using the empirical MRL plot.

or

$$T_b^* = t_i + \frac{m_G - \hat{m}_i}{\hat{m}_{i+1} - \hat{m}_i}(t_{i+1} - t_i).$$

Note that if the horizontal line, $\hat{m}(t) = m_G$, has two or more intersection points with the empirical MRL plot, as is the case in Fig. 8.3, then always pick the first intersection point which gives the shortest burn-in time for the same MRL goal.

8.8.2 OPTIMUM BURN-IN TIME FOR THE MAXIMUM MRL

From the empirical MRL plot, locate the point where the calculated empirical MRL is the maximum, say at t_i, such as the one shown in Fig. 8.4. The burn-in time at this point can be regarded as a rough estimate of the optimum burn-in time for the maximum MRL.

To be more precise, a *quadratic function* can be fitted using two points next to t_i, t_{i-1} and t_{i+1}, as shown in Fig. 8.4, plus t_i itself; i.e.,

$$\hat{m}(t) = a\,t^2 + b\,t + c, \qquad (8.67)$$

where a, b and c are three unknown constants to be determined from $[t_{i-1}, \hat{m}_{i-1}]$, $[t_i, \hat{m}_i]$ and $[t_{i+1}, \hat{m}_{i+1}]$. Note that the first constant, a, in

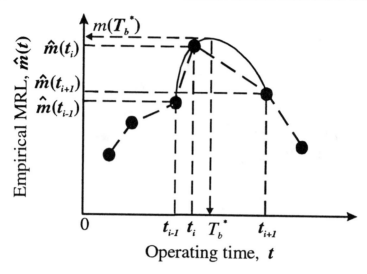

Fig. 8.4– Determination of the optimum burn-in time for the maximum MRL using the empirical MRL plot and a quadratic fitting function in the neighborhood of the maximum calculated empirical MRL point.

Eq. (8.67) should always be a negative value, or

$$a < 0,$$

because the MRL is concave in the neighborhood of t_i where the real maximum MRL occurs. Substituting $[t_{i-1}, \hat{m}_{i-1}]$, $[t_i, \hat{m}_i]$ and $[t_{i+1}, \hat{m}_{i+1}]$ into Eq, (8.67) yields

$$\begin{cases} \hat{m}_{i-1} = a\,t_{i-1}^2 + b\,t_{i-1} + c, \\ \hat{m}_i \;\;\;\, = a\,t_i^2 + b\,t_i + c, \\ \hat{m}_{i+1} = a\,t_{i+1}^2 + b\,t_{i+1} + c. \end{cases}$$

Solving these three equations simultaneously yields

$$\begin{cases} a = \frac{D_a}{D}, \\ b = \frac{D_b}{D}, \\ c = \frac{D_c}{D}, \end{cases} \tag{8.68}$$

where

$$D = t_{i-1}\,t_i\,(t_{i-1} - t_i) + t_i\,t_{i+1}\,(t_i - t_{i+1}) \\ + t_{i+1}\,t_{i-1}\,(t_{i+1} - t_{i-1}),$$

$$D_a = m_{i-1}(t_i - t_{i+1}) + m_i(t_{i+1} - t_{i-1}) + m_{i+1}(t_{i-1} - t_i),$$
$$D_b = t_{i-1}^2(m_i - m_{i+1}) + t_i^2(m_{i+1} - m_{i-1}) + t_{i+1}^2(m_{i-1} - m_i),$$

and

$$D_c = t_{i-1} t_i m_{i+1}(t_{i-1} - t_i) + t_i t_{i+1} m_{i-1}(t_i - t_{i+1})$$
$$+ t_{i+1} t_{i-1} m_i(t_{i+1} - t_{i-1}).$$

Then, the optimum burn-in time, T_b^*, can be determined by taking the first derivative of $\hat{m}(t)$, given by Eq. (8.67), with respect to t, let it equal to zero and solve for t; i.e.,

$$\frac{\partial \hat{m}(t)}{\partial t} = 2at + b = 0,$$

or

$$T_b^* = -\frac{b}{2a}, \quad (8.69)$$

which is the more precise estimate of the optimum burn-in time. It may be seen from Eq. (8.69) that constant b should always be a positive value to maintain a positive T_b^*; i.e.,

$$b > 0,$$

since a is always a negative value.

The corresponding maximum MRL is given by

$$MRL_{\max} = a(T_b^*)^2 + bT_b^* + c.$$

Substituting Eq. (8.69) into this equation yields

$$MRL_{\max} = a\left(-\frac{b}{2a}\right)^2 + b\left(-\frac{b}{2a}\right) + c,$$

or

$$MRL_{\max} = \frac{4ac - b^2}{4a}. \quad (8.70)$$

This procedure of determining the optimum burn-in time using a quadratic fitting function in the neighborhood of the "maximum" calculated empirical MRL point is also included in computer programs "MRL-EMP1.FOR" and "MRL-EMP2.FOR" to facilitate the calculation.

EXAMPLE 8-7

A life test on 1,000,000 electronic components yields Table 8.1 where a total of 197 failures are observed during 200 hr of operation. Determine the optimum burn-in time using the empirical MRL approach for the following requirements:

Table 8.1– Raw failure data from a life test of 200 hr on $N = 1,000,000$ electronic components, for Example 8–7.

Time interval number, i	Time interval, (t_i, t_{i+1}), hr	Failure number, N_{fi}
1	(0, 10)	68
2	(10, 20)	17
3	(20, 30)	12
4	(30, 40)	9
5	(40, 50)	8
6	(50, 60)	7
7	(60, 70)	7
8	(70, 80)	6
9	(80, 90)	6
10	(90, 100)	6
11	(100, 110)	6
12	(110, 120)	5
13	(120, 130)	5
14	(130, 140)	5
15	(140, 150)	5
16	(150, 160)	5
17	(160, 170)	5
18	(170, 180)	5
19	(180, 190)	5
20	(190, 200)	5

DETERMINE BURN-IN TIME FROM EMPIRICAL MRL

1. For a specified MRL goal of 1.72×10^6 hr.

2. For the maximum MRL.

SOLUTIONS TO EXAMPLE 8-7

1. Given in Table 8.1 is a censored and grouped data set. Then, Eq. (8.18) should be used to evaluate the empirical MRL values at the starting and ending points of each time interval; i.e.,

$$\hat{m}_{(N,r,t_d)}(t_j) = \begin{cases} \dfrac{\sum_{i=j}^{n} N_{fi}\left[\frac{1}{2}(t_i+t_{i+1})-t_j\right]+N_s(t_d-t_j)}{\sum_{i=j}^{n} N_{fi}}, \\ \text{for } j = 0, 1, 2, \cdots, n, \\ \text{unknown from the test, for } j \geq n+1, \end{cases}$$

or

$$\hat{m}_{(10^6;197;200)}(t_j) = \begin{cases} \dfrac{\sum_{i=j}^{20} N_{fi}\left[\frac{1}{2}(t_i+t_{i+1})-t_j\right]+999{,}803\,(200-t_j)}{\sum_{i=j}^{20} N_{fi}}, \\ \text{for } j = 0, 1, 2, \cdots, 20, \\ \text{unknown from the test, for } j \geq 21, \end{cases}$$

(8.71)

where

$$N = 10^6,$$
$$r = 197,$$
$$t_d = 200 \text{ hr},$$
$$n = \text{total number of groups, or time intervals,}$$
$$n = 20,$$
$$N_s = \text{total number of survivors at } t_d,$$

and

$$N_s = N - r = 10^6 - 197 = 999{,}803.$$

Table 8.2– Empirical MRL function values calculated for Table 8.1 using Eq. (8.71), for Example 8–7.

Time order, j	Failure time, t_j, hr	Empirical MRL, $\widehat{MRL}_j$, hr
1	0	1,015,087.000
2	10	1,472,655.875
3	20	1,606,904.000
4	30	1,699,742.625
5	40	1,757,969.000
6	50	1,806,943.125
7	60	1,841,809.500
8	70	1,883,749.250
9	80	1,904,445.500
10	90	1,929,498.750
11	100	1,960,447.125
12	110	1,999,651.000
13	120	1,999,646.000
14	130	1,999,641.000
15	140	1,999,636.000
16	150	1,999,631.000
17	160	1,999,626.000
18	170	1,999,621.000
19	180	1,999,616.000
20	190	1,999,611.000

Equation (8.71) is evaluated using the computer program "MRL-EMP2.FOR" in Appendix 8G and summarized in Table 8.2. A sample calculation is given here. For example, at $t_j = t_{19} = 180$ hr,

$$\hat{m}_{(10^6;197;200)}(180)$$

$$= \frac{\sum\limits_{i=19}^{20} N_{fi}\left[\frac{1}{2}(t_i + t_{i+1}) - 180\right] + 999,803\,(200 - 180)}{\sum\limits_{i=19}^{20} N_{fi}},$$

$$= \left\{ N_{f19}\left[\frac{1}{2}(t_{19} + t_{20}) - 180\right]\right.$$

$$\left. + N_{f20}\left[\frac{1}{2}(t_{20} + t_{21}) - 180\right] + 19,996,060 \right\}$$

$$\Big/ (N_{f19} + N_{f20}),$$

$$= \left\{ 5\left[\frac{1}{2}(180 + 190) - 180\right]\right.$$

$$\left. + 5\left[\frac{1}{2}(190 + 200) - 180\right] + 19,996,060 \right\}$$

$$\Big/ (5 + 5),$$

$$= 19,996,160\Big/10,$$

or

$$\hat{m}_{(10^6;197;200)}(180) = 1,999,616 \text{ hr.}$$

The results given in Table 8.2 are plotted in Fig. 8.5. From Fig. 8.5, it may be seen that the optimum burn-in time corresponding to the $MRL_{\text{goal}} = 1.72 \times 10^6$ hr is between $t_4 = 30$ hr and $t_5 = 40$ hr. Linear interpolation can be applied here; i.e.,

$$\frac{T_b^* - 30}{40 - 30} = \frac{1.72 \times 10^6 - 1,699,742.625}{1,757,969.000 - 1,699,742.625},$$

or

$$T_b^* = 33.479 \text{ hr, or } 34 \text{ hr.}$$

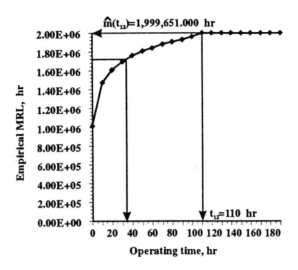

Fig. 8.5– Empirical MRL plot using the data in Table 8.2, for Example 8–7.

2. From Fig. 8.5, the maximum calculated MRL is achieved at $t_{12} = 110$ hr where $\widehat{MRL} = 1,999,651.000$ hr. Therefore, $t_{12} = 110$ hr can be regarded as a rough estimate of the optimum burn-in time.

To get the more precise maximum MRL point, fit the quadratic function given in Eq. (8.67) in the neighborhood of $t_{12} = 110$ hr using the following points: $[t_{11}, \hat{m}(t_{11})]=(100$ hr, $1,960,447.125$ hr), $[t_{12}, \hat{m}(t_{12})]=(110$ hr, $1,999,651.000$ hr) and $[t_{13}, \hat{m}(t_{13})]=(120$ hr, $1,999,646.000$ hr). Substituting these three points into Eq. (8.67) yields

$$\begin{cases} 100^2\, a + 100\, b + c = 1,960,447.125, \\ 110^2\, a + 110\, b + c = 1,999,651.000, \\ 120^2\, a + 120\, b + c = 1,999,646.000. \end{cases}$$

Solving these three equations simultaneously, or substituting $[t_{11}, \hat{m}(t_{11})] = (100$ hr, $1,960,447.125$ hr$)$, $[t_{12}, \hat{m}(t_{12})] = (110$ hr, $1,999,651.000$ hr$)$, and $[t_{13}, \hat{m}(t_{13})] = (120$ hr, $1,999,646.000$ hr$)$ into Eq. (8.68) directly, yields

$$\begin{cases} a = -196.044, \\ b = 45,089.706, \\ c = -588,079.750. \end{cases}$$

Then, the fitted quadratic function in the neighborhood of $t_{12} = 110$ hr is

$$\hat{m}(t) = -196.044\, t^2 + 45,089.706\, t - 588,079.750.$$

The optimum burn-in time is given by Eq. (8.69); i.e.,

$$T_b^* = -\frac{b}{2a},$$
$$= -\frac{45,089.706}{2\,(-196.044)},$$

or

$$T_b^* = 114.999 \text{ hr, or } 115 \text{ hr}.$$

The corresponding maximum MRL is given by Eq. (8.70); i.e.,

$$MRL_{\max} = \frac{4\,a\,c - b^2}{4\,a},$$
$$= \frac{4\,(-196.044)\,(-588,079.750) - 45,089.706^2}{4\,(-196.044)},$$

or

$$MRL_{\max} = 2,004,549.625 \text{ hr}.$$

It may be seen that the MRL at $T_b^* = 115$ hr is greater than that at $t_{12} = 110$ hr, or 2,004,549.625 hr $>$ 1,999,651.000 hr. Then, 2,004,549.625 hr is really the maximum MRL.

EXAMPLE 8-8

A life test on 150 electronic components yields the data in Table 8.3, where a total of 19 failures are observed at 12 distinctive failure times during 4,500 hr of operation. Determine the optimum burn-in time using the empirical MRL approach for the following requirements:

1. For a specified MRL goal of 35,000 hr.
2. For the maximum MRL.

Table 8.3– Raw failure data from a life test of 4,500 hr on $N = 150$ electronic components, for Example 8-8.

Time order, i	Failure time, t_i, hr	Failure number, N_{fi}
1	100	1
2	200	1
3	250	1
4	420	2
5	588	3
6	708	1
7	1,044	1
8	2,892	2
9	3,396	2
10	3,997	3
11	4,165	1
12	4,500	1

SOLUTIONS TO EXAMPLE 8-8

1. Given in Table 8.3 is a censored (and ungrouped) data set with ties. Then, Eq. (8.17) should be used to evaluate the empirical MRL value at any specified age t between t_k and t_{k+1}, or $t_k \leq t < t_{k+1}$, where $k = 0, 1, 2, \cdots, 11$ and $t_0 = 0$; i.e.,

$$\hat{m}_{(N,n,t_d)}(t) = \begin{cases} \dfrac{\sum\limits_{i=k+1}^{L} N_{fi}(t_i - t) + N_s(t_d - t)}{\sum\limits_{i=k+1}^{L} N_{fi}}, & \text{for } t < t_L, \\ \text{unknown from the test}, & \text{for } t \geq t_L, \end{cases}$$

or

$$\hat{m}_{(150;19;4,500)}(t) = \begin{cases} \dfrac{\sum\limits_{i=k+1}^{12} N_{fi}(t_i - t) + 131(4,500 - t)}{\sum\limits_{i=k+1}^{12} N_{fi}}, & \text{for } t < 4,500 \text{ hr}, \\ \text{unknown from the test}, & \text{for } t \geq 4,500 \text{ hr}, \end{cases}$$

(8.72)

where

$N = 150$,
$n = $ total number of failures in the test,
$n = 19$,
$t_d = 4,500$ hr,
$L = $ total number of distinctive failure times,
$L = 12$,
$N_s = $ total number of components surviving the test,

and

$$N_s = N - n = 150 - 19 = 131.$$

Equation (8.72) was evaluated using the computer program "MRL-EMP1.FOR" in Appendix 8F. The results are summarized in Table 8.4.

Table 8.4– Empirical MRL function values calculated for Table 8.3 using Eq. (8.72), for Example 8–8.

Time order, j	Failure time, t_j, hr	Empirical MRL, $\widehat{MRL}_j$, hr
0	0	33,033.578
1	100	34,035.445
2	200	35,161.059
3	250	36,896.125
4	420	40,382.000
5	588	49,180.727
6	708	52,394.801
7	1,044	52,952.445
8	2,892	31,121.715
9	3,396	29,660.000
10	3,997	33,282.000
11	4,165	44,220.000

A sample calculation is given here. For example, at $t = t_j = t_{10} = 3,997$ hr,

$$\hat{m}_{(150;19;4,500)}(3,997)$$

$$= \frac{\sum_{i=11}^{12} N_{fi}(t_i - 3,997) + 131(4,500 - 3,997)}{\sum_{i=11}^{12} N_{fi}},$$

$$= \frac{N_{f11}(t_{11} - 3,997) + N_{f12}(t_{12} - 3,997) + 65,893}{N_{f11} + N_{f12}},$$

$$= \frac{1(4,165 - 3,997) + 1(4,500 - 3,997) + 65,893}{1 + 1},$$

or

$$\hat{m}_{(150;19;4,500)}(3,997) = 33,282 \text{ hr}.$$

The results given in Table 8.4 are plotted in Fig. 8.6. From Fig. 8.6, it may be seen that the optimum burn-in time corresponding

DETERMINE BURN-IN TIME FROM EMPIRICAL MRL   299

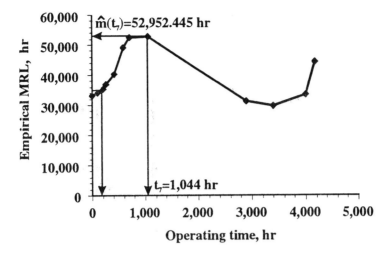

Fig. 8.6– Empirical MRL plot using data from Table 8.4, for Example 8–8.

to $MRL_{\text{goal}} = 35{,}000$ hr is between $t_1 = 100$ hr and $t_2 = 200$ hr. Linear interpolation can be applied here; i.e.,

$$\frac{T_b^* - 100}{200 - 100} = \frac{35{,}000 - 34{,}035.445}{35{,}161.059 - 34{,}035.445},$$

or

$$T_b^* = 185.691 \text{ hr, or } 186 \text{ hr}.$$

2. From Fig. 8.6, the maximum calculated MRL is achieved at $t_7 = 1{,}044$ hr where $\widehat{MRL} = 52{,}952.445$ hr. Therefore, $t_7 = 1{,}044$ hr can be regarded as a rough estimate of the optimum burn-in time.

To get a more precise maximum MRL point, fit the quadratic function given in Eq. (8.67) in the neighborhood of $t_7 = 1{,}044$ hr using the following points: $[t_6, \hat{m}(t_6)] = (708$ hr, $52{,}394.801$ hr$)$, $[t_7, \hat{m}(t_7)] = (1{,}044$ hr, $52{,}952.445$ hr$)$ and $[t_8, \hat{m}(t_8)] = (2{,}892$ hr, $31{,}121.715$ hr$)$. Substituting these three points into Eq. (8.67) yields

$$\begin{cases} 708^2\, a + 708\, b + c = 52{,}394.801, \\ 1{,}044^2\, a + 1{,}044\, b + c = 52{,}952.445, \\ 2{,}892^2\, a + 2{,}892\, b + c = 31{,}121.715. \end{cases}$$

Solving these three equations simultaneously, or substituting $[t_6, \hat{m}(t_6)] = (708 \text{ hr}, 52{,}394.801 \text{ hr})$, $[t_7, \hat{m}(t_7)] = (1{,}044 \text{ hr}, 52{,}952.445 \text{ hr})$ and $[t_8, \hat{m}(t_8)] = (2{,}892 \text{ hr}, 31{,}121.715 \text{ hr})$ into Eq. (8.68) directly, yields

$$\begin{cases} a = -0.006, \\ b = 12.468, \\ c = 46{,}660.028. \end{cases}$$

Then, the fitted quadratic function in the neighborhood of $t_7 = 1{,}044$ hr is

$$\hat{m}(t) = -0.006\, t^2 + 12.468\, t - 46{,}660.028.$$

The optimum burn-in time is given by Eq. (8.69); i.e.,

$$T_b^* = -\frac{b}{2a},$$
$$= -\frac{12.468}{2(-0.006)},$$

or

$$T_b^* = 1{,}010.519 \text{ hr or } 1{,}011 \text{ hr}.$$

The corresponding maximum MRL is given by Eq. (8.70); i.e.,

$$MRL_{\max} = \frac{4ac - b^2}{4a},$$
$$= \frac{4(-0.006)(46{,}660.028) - 12.468^2}{4(-0.006)},$$

or

$$MRL_{\max} = 52{,}959.359 \text{ hr}.$$

It may be seen that the MRL at $T_b^* = 1{,}011$ hr is greater than that at $t_7 = 1{,}044$ hr, or 52,959.359 hr $>$ 52,952.445 hr. Then, 52,959.359 hr is really the maximum MRL.

Table 8.5– Raw failure data from a life test of 6,600 hr on $N = 90$ electronic components, for Example 8-9.

Time interval number, i	Time interval, (t_i, t_{i+1}), hr	Failure number, N_{fi}
1	(0, 600)	23
2	(600, 1,200)	9
3	(1,200, 1,800)	2
4	(1,800, 2,400)	3
5	(2,400, 3,000)	11
6	(3,000, 3,600)	6
7	(3,600, 4,200)	3
8	(4,200, 4,800)	6
9	(4,800, 5,400)	4
10	(5,400, 6,000)	3
11	(6,000, 6,600)	4

EXAMPLE 8-9

A life test on 90 electronic components yields the data in Table 8.5 where a total of 74 failures are observed during 6,600 hr of operation. Determine the optimum burn-in time using the empirical MRL approach for the following requirements:

1. For a specified MRL goal of 4,500 hr.
2. For the maximum MRL.

SOLUTIONS TO EXAMPLE 8-9

1. Given in Table 8.5 is a censored and grouped data set. Then, Eq. (8.18) should be used to evaluate the empirical MRL values at the starting and ending points of each time interval; i.e.,

$$\hat{m}_{(N,r,t_d)}(t_j) = \begin{cases} \dfrac{\sum\limits_{i=j}^{n} N_{fi}\left[\frac{1}{2}(t_i+t_{i+1})-t_j\right]+N_s\,(t_d-t_j)}{\sum\limits_{i=j}^{n} N_{fi}}, \\ \quad \text{for } j = 0, 1, 2, \cdots, n, \\ \text{unknown from the test}, \quad \text{for } j \geq n+1, \end{cases}$$

or

$$\hat{m}_{(90;74;6,600)}(t_j) = \begin{cases} \dfrac{\sum\limits_{i=j}^{11} N_{fi}\left[\frac{1}{2}(t_i+t_{i+1})-t_j\right]+16\,(6,600-t_j)}{\sum\limits_{i=j}^{11} N_{fi}}, \\ \quad \text{for } j = 0, 1, 2, \cdots, 11, \\ \text{unknown from the test,} \quad \text{for } j \geq 12, \end{cases}$$

(8.73)

where

$N = 90,$
$r = 74,$
$t_d = 6,600 \text{ hr},$
$n = $ total number of groups, or time intervals,
$n = 11,$
$N_s = $ total number of survivors at t_d,

and

$$N_s = N - r = 90 - 74 = 16.$$

Equation (8.73) was evaluated using the computer program "MRL-EMP2.FOR" in Appendix 8G. The results are summarized in Table 8.6.

The sample calculation is ommitted here since it is similar to that given in Example 8–7. The results in Table 8.6 are plotted in Fig. 8.7. From Fig. 8.7, it may be seen that the optimum burn-in time corresponding to $MRL_{\text{goal}} = 4,500$ hr is between $t_1 = 0$ hr and $t_2 = 600$ hr. Linear interpolation can be applied here; i.e.,

$$\frac{T_b^* - 0}{600 - 0} = \frac{4,500 - 3,794.594}{4,582.353 - 3,794.594},$$

or

$$T_b^* = 537.275 \text{ hr, or } 538 \text{ hr}.$$

Table 8.6– Empirical MRL function values calculated for Table 8.5 using Eq. (8.73), for Example 8–9.

Time order, j	Failure time, t_j, hr	Empirical MRL, $\widehat{MRL}_j$, hr
1	0	3,794.594
2	600	4,582.353
3	1,200	4,671.429
4	1,800	4,050.000
5	2,400	3,494.594
6	3,000	3,876.923
7	3,600	3,870.000
8	4,200	3,335.294
9	4,800	3,518.182
10	5,400	3,385.714
11	6,000	2,700.000

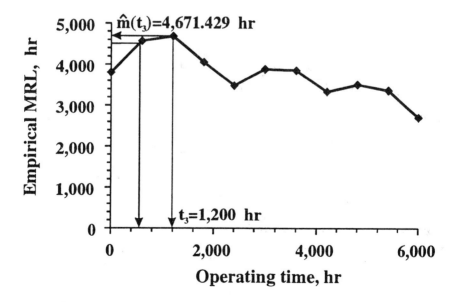

Fig. 8.7– Empirical MRL plot using data from Table 8.6, for Example 8–9.

2. From Fig. 8.7, the maximum calculated MRL is achieved at $t_3 = 1,200$ hr where $\widehat{MRL} = 4,671.429$ hr. Therefore, $t_3 = 1,200$ hr can be regarded as a rough estimate of the optimum burn-in time.

To get the more precise maximum MRL point, fit the quadratic function given in Eq. (8.67) in the neighborhood of $t_3 = 1,200$ hr using the following points: $[t_2, \hat{m}(t_2)] = (600$ hr, 4,582.353 hr$)$, $[t_3, \hat{m}(t_3)] = (1,200$ hr, 4,671.429 hr$)$ and $[t_4, \hat{m}(t_4)] = (1,800$ hr, 4,050.000 hr$)$. Substituting these three points into Eq. (8.67) yields

$$\begin{cases} 600^2 \, a + 600 \, b + c = 4,582.353, \\ 1,200^2 \, a + 1,200 \, b + c = 4,671.429, \\ 1,800^2 \, a + 1,800 \, b + c = 4,050.000. \end{cases}$$

Solving these three equations simultaneously, or substituting $[t_2, \hat{m}(t_2)] = (600$ hr, 4,582.353 hr$)$, $[t_3, \hat{m}(t_3)] = (1,200$ hr, 4,671.429 hr$)$ and $[t_4, \hat{m}(t_4)] = (1,800$ hr, 4,050.000 hr$)$. into Eq. (8.68) directly, yields

$$\begin{cases} a = -0.001, \\ b = 1.925, \\ c = 3,782.772. \end{cases}$$

Then, the fitted quadratic function in the neighborhood of $t_3 = 1,200$ hr is

$$\hat{m}(t) = -0.001 \, t^2 + 1.925 \, t + 3,782.772.$$

The optimum burn-in time is given by Eq. (8.69); i.e.,

$$\begin{aligned} T_b^* &= -\frac{b}{2a}, \\ &= -\frac{1.925}{2(-0.001)}, \end{aligned}$$

or

$$T_b^* = 975.222 \text{ hr, or } 975 \text{ hr.}$$

The corresponding maximum MRL is given by Eq. (8.70); i.e.,

$$MRL_{\max} = \frac{4ac - b^2}{4a},$$
$$= \frac{4(-0.001)(3,782.772) - 1.925^2}{4(-0.001)},$$

or

$$MRL_{\max} = 4,721.288 \text{ hr}.$$

It may be seen that the MRL at $T_b^* = 975$ hr is greater than that at $t_3 = 1,200$ hr, or 4,721.288 hr > 4,671.429 hr. Then, 4,721.288 hr is really the maximum MRL.

EXAMPLE 8–10

Table 8.7 is taken from [14, p. 537] where ninety (90) electronic components are life tested until all fail. The failures are counted and recorded at the end of seventeen (17) time intervals. It was found in [14, pp. 537-556] that a trimodal mixed-Weibull distribution with the following parameters fit these data very well:

$\beta_1 = 0.96, \ \eta_1 = 650 \text{ hr}, \quad \gamma_1 = 0 \text{ hr},$
$\beta_2 = 3.00, \ \eta_2 = 4,450 \text{ hr}, \ \gamma_2 = 0 \text{ hr},$
$\beta_3 = 4.10, \ \eta_3 = 8,000 \text{ hr}, \ \gamma_3 = 0 \text{ hr}.$

The fitted *pdf*, failure rate and reliability functions are plotted in Fig. 8.8 [14, p. 547].

Determine the optimum burn-in time using the empirical MRL approach for the following requirements:

1. For a specified MRL goal of 3,500 hr.

2. For the maximum MRL.

3. Compare the result from Case 2 with that obtained from Fig. 8.8 but for the minimum failure rate.

Table 8.7– Raw failure data from a life test on $N = 90$ electronic components, for Example 8–10.

Time interval number, i	Time interval, (t_i, t_{i+1}), hr	Failure number, N_{fi}
1	(0, 600)	23
2	(600, 1,200)	9
3	(1,200, 1,800)	2
4	(1,800, 2,400)	3
5	(2,400, 3,000)	11
6	(3,000, 3,600)	6
7	(3,600, 4,200)	3
8	(4,200, 4,800)	6
9	(4,800, 5,400)	4
10	(5,400, 6,000)	3
11	(6,000, 6,600)	4
12	(6,600, 7,200)	8
13	(7,200, 7,800)	3
14	(7,800, 8,400)	0
15	(8,400, 9,000)	1
16	(9,000, 9,600)	2
17	(9,600, 10,200)	2

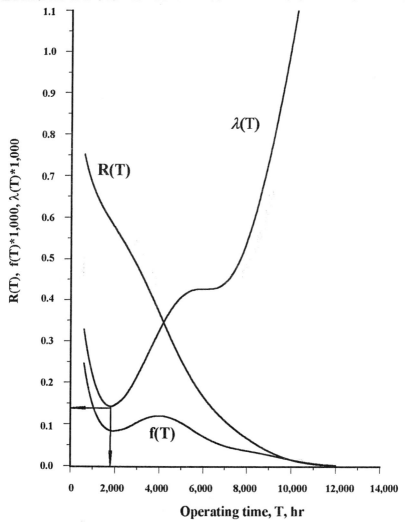

Fig. 8.8– Plots of the *pdf*, failure rate and reliability functions of the fitted trimodal mixed-Weibull distribution [14, p. 547] for the life testing data given in Table 8.5, for Example 8–10.

SOLUTIONS TO EXAMPLE 8–10

1. Given in Table 8.7 is a complete and grouped data set. Then, Eq. (8.15) should be used to evaluate the empirical MRL values at the starting and ending points of each time interval; i.e.,

$$\hat{m}_N(t_j) = \begin{cases} \dfrac{\sum_{i=j}^{n} N_{fi}\left[\frac{1}{2}(t_i+t_{i+1})-t_j\right]}{\sum_{i=j}^{n} N_{fi}}, & \text{for } j = 0, 1, \cdots, n, \\ 0, & \text{for } j \geq n+1, \end{cases}$$

or

$$\hat{m}_{90}(t_j) = \begin{cases} \dfrac{\sum_{i=j}^{17} N_{fi}\left[\frac{1}{2}(t_i+t_{i+1})-t_j\right]}{\sum_{i=j}^{17} N_{fi}}, & \text{for } j = 0, 1, \cdots, 17, \\ 0, & \text{for } j \geq 18, \end{cases} \qquad (8.74)$$

where

$N = 90,$

$n = $ total number of groups, or time intervals,

and

$n = 17.$

Equation (8.74) was evaluated using the computer program "MRL-EMP2.FOR" in Appendix 8G. The results are summarized in Table 8.8.

A sample calculation is given here. For example, at $t_j = t_{16} = 9,000$ hr,

$$\hat{m}_{90}(9,000 \text{ hr})$$

$$= \dfrac{\sum_{i=16}^{17} N_{fi}\left[\frac{1}{2}(t_i+t_{i+1}) - 9,000\right]}{\sum_{i=16}^{17} N_{fi}},$$

$$= \dfrac{N_{f16}\left[\frac{1}{2}(t_{16}+t_{17}) - 9,000\right] + N_{f17}\left[\frac{1}{2}(t_{17}+t_{18}) - 9,000\right]}{N_{f16} + N_{f17}},$$

Table 8.8– Empirical MRL function values calculated for Table 8.7 using Eq. (8.74), for Example 8–10.

Time order, j	Failure time, t_j, hr	Empirical MRL, $\widehat{MRL}_j$, hr
1	0	3,333.333
2	600	3,774.627
3	1,200	3,713.793
4	1,800	3,235.714
5	2,400	2,801.887
6	3,000	2,857.143
7	3,600	2,683.333
8	4,200	2,300.000
9	4,800	2,144.444
10	5,400	1,865.217
11	6,000	1,500.000
12	6,600	1,200.000
13	7,200	1,500.000
14	7,800	1,620.000
15	8,400	1,020.000
16	9,000	600.000
17	9,600	300.000
18	10,200	0.000

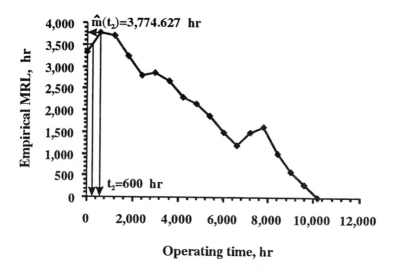

Fig. 8.9– Empirical *MRL* plot using data from Table 8.8, for Example 8–10.

$$= \frac{2\left[\frac{1}{2}(9,000 + 9,600) - 9,000\right] + 2\left[\frac{1}{2}(9,600 + 10,200) - 9,000\right]}{2 + 2}$$

$$= \frac{2,400}{4},$$

or

$$\hat{m}_{90}(9,000 \text{ hr}) = 600 \text{ hr}.$$

The results in Table 8.8 are plotted in Fig. 8.9. From Fig. 8.9, it may be seen that the optimum burn-in time corresponding to $MRL_{\text{goal}} = 3,500$ hr is between $t_1 = 0$ hr and $t_2 = 600$ hr. Linear interpolation can be applied here; i.e.,

$$\frac{T_b^* - 0}{600 - 0} = \frac{3,500 - 3333.333}{3,774.627 - 3,333.333},$$

or

$$T_b^* = 226.607 \text{ hr, or } 227 \text{ hr}.$$

2. From Fig. 8.9, the maximum calculated MRL is achieved at $t_2 = 600$ hr where $\widehat{MRL} = 3{,}774.627$ hr. Therefore, $t_2 = 600$ hr can be regarded as a rough estimate of the optimum burn-in time.

To get the more precise maximum MRL point, fit the quadratic function given in Eq. (8.67) in the neighborhood of $t_2 = 600$ hr using the following points: $[t_1, \hat{m}(t_1)] = (0 \text{ hr}, 3{,}333.333 \text{ hr})$, $[t_2, \hat{m}(t_2)] = (600 \text{ hr}, 3{,}774.627 \text{ hr})$ and $[t_3, \hat{m}(t_3)] = (1{,}200 \text{ hr}, 3{,}713.793 \text{ hr})$. Substituting these three points into Eq. (8.67) yields

$$\begin{cases} 0^2 a + 0\, b + c & = 3{,}333.333, \\ 600^2 a + 600\, b + c & = 3{,}774.627, \\ 1{,}200^2 a + 1{,}200\, b + c & = 3{,}713.793. \end{cases}$$

Solving these three equations simultaneously, or substituting $[t_1, \hat{m}(t_1)] = (0 \text{ hr}, 3{,}333.333 \text{ hr})$, $[t_2, \hat{m}(t_2)] = (600 \text{ hr}, 3{,}774.627 \text{ hr})$ and $[t_3, \hat{m}(t_3)] = (1{,}200 \text{ hr}, 3{,}713.793 \text{ hr})$ into Eq. (8.68) directly, yields

$$\begin{cases} a = -0.0007, \\ b = 1.154, \\ c = 3{,}333.333. \end{cases}$$

Then, the fitted quadratic function in the neighborhood of $t_2 = 600$ hr is

$$\hat{m}(t) = -0.0007\, t^2 + 1.154\, t + 3{,}333.333.$$

The optimum burn-in time is given by Eq. (8.69); i.e.,

$$\begin{aligned} T_b^* &= -\frac{b}{2\,a}, \\ &= -\frac{1.154}{2(-0.0007)}, \end{aligned}$$

or

$$T_b^* = 827.309 \text{ hr, or } 827 \text{ hr}.$$

The corresponding maximum MRL is given by Eq. (8.70); i.e.,

$$MRL_{\max} = \frac{4ac - b^2}{4a},$$
$$= \frac{4(-0.0007)(3,333.333) - 1.154^2}{4(-0.0007)},$$

or

$$MRL_{\max} = 3,810.661 \text{ hr}.$$

It may be seen that the MRL at $T_b^* = 827$ hr is greater than that at $t_2 = 600$ hr, or 3,810.661 hr $>$ 3,774.627 hr. Then, 3,810.661 hr is really the maximum MRL.

3. From Fig. 8.8, however, the minimum failure rate occurs at $T_b^* = 1,830$ hr, where

$$\lambda_{\min}(1,830 \text{ hr}) = 1.4235 \times 10^{-4} \text{ fr/hr}.$$

This burn-in time of $T_b^* = 1,830$ hr, which corresponds to the minimum failure rate, is much larger than that for the maximum MRL which is $T_b^* = 827$ hr.

As a matter of fact, this result is consistent with the conclusion which was obtained in Section 8.6 and illustrated in Fig. 8.2, about the effect of burn-in on the MRL when a Bathtub-shaped failure curve with wearour period is assumed. That is, when the wearout period is present on a Bathtub-shaped failure rate curve, the maximum MRL point is between zero (0) and the minimum-failure-rate point.

8.9 FURTHER COMMENTS

As being stated at the begining of this chapter, the mean residual life concept has a tremendous range of applications in addition to being used in determining the optimum burn-in time as presented in this chapter. For more characterizations and other applications of the mean residual life concept, the readers are referred to the ADDITIONAL REFERENCES as listed at the end of this chapter, which give a comprehensive summary of the literature on MRL published since 1988.

REFERENCES

1. Guess, F. and Proschan, F., "Mean Residual Life: Theory and Applications," Krishnaiah, P. R. and Rao, C. R. eds., *Handbook of Statistics*, Vol. 7, Elsevier Science Publishers B.V., pp. 215-224, 1988.

2. Cox, D. R., *Renewal Theory*, Methuen, London, 142 pp., 1962.

3. Kotz, S. and Shanbhag, D. N., "Some New Approaches to Probability Distributions," *Advances in Applied Probability*, Vol. 12, pp. 903-921, 1980.

4. Hall, W. J. and Wellner, J. A., "Mean Residual Life," Csorgo, M., Dawson, D. A., Rao, J. N. K. and Saleh, A. K. Md. E. eds., *Statistics and Related Topics*, North-Holland, Amsterdam, pp. 169-184, 1981.

5. Watson, G. S. and Wells, W. T., "On the Possibility of Improving the Mean Useful Life of Items by Eliminating Those with Short Lives," *Technometrics*, Vol. 3, No. 2, pp. 281-298, 1961.

6. Kuo, W., "Reliability Enhancement through Optimum Burn-in," *IEEE Transactions on Reliability*, Vol. 33, pp. 145-156, 1984.

7. Gross, A. J. and Clark, V. A., *Survival Distributions: Reliability Applications in the Biomedical Science*, John Wiley & Sons, New York, 1975.

8. Elandt-Johnson, R. C. and Johnson, N. J., *Survival Models and Data Analysis*, John Wiley & Sons, New York, 1980.

9. Morrison, D. G., "On Linearly Increasing Mean Residual Lifetimes," *Journal of Applied Probability*, Vol. 15, pp. 617-620, 1978.

10. Bhattacharjee, M. C., "The Class of Mean Residual Lives and Some Consequences," *SIAM Journal of Algebraic Discrete Methods*, Vol. 3, pp. 56-65, 1982.

11. Siddiqui, M. M. and Caglar, M., "Residual Lifetime Distribution and Its Applications," *Microelectronics and Reliability*, Vol. 34, No. 2, pp. 211-227, 1994.

12. Park. K. S., "Effect of Burn-in on Mean Residual Life," *IEEE Transactions on Reliability*, Vol. R-34, No. 5, pp. 522-523, 1985.

13. Mi, J., "Bathtub Failure Rate and Upside-Down Bathtub Mean Residual Life," *IEEE Transactions on Reliability*, Vol. 44, No. 3, pp. 388-391, 1995.

14. Kececioglu, D., *Reliability Engineering Handbook*, Prentice Hall, Inc., Upper Saddle River, NJ, Vol. 1, 720 pp., 1991, Fifth Printing, 1995.

ADDITIONAL REFERENCES

1. Abouammoh, A. M., et al., "On Some Aspects of Variance Remaining Life Distributions," *Microelectronics and Reliability*, Vol. 30, No. 4, pp. 751-760, 1990

2. Abouammoh, A. M. and Ahmed, A. N., "On Renewal Failure Rate Classes of Life Distributions," *Statistics and Probability Letters*, Vol. 14, No. 3, pp. 211-217, 1992.

3. Ahmad, I. A., "A New Test for Mean Residual Life Times," *Biometrika*, Vol. 79, No. 2, pp. 416-419, 1992.

4. Ahmed, A. H. and Alzaid, A. A., "Systems with Exponential Life Using Components with HNBUE Lives," *IEEE Transactions on Reliability*, Vol. 37, No. 4, pp. 424-426, 1988.

5. Alam, K. and Kulasekera, K. B., "Estimation of the Quantile Function of Residual Life Time Distribution," *Journal of Statistics*, Planning and Inference, Vol. 37, No. 3, pp. 327-337, 1993.

6. Aly, E. E. A. A., "On Some Confidence Bands for Percentile Residual Life Functions," *Journal of Nonparametric Statistics*, Vol. 2, No. 1, pp. 59-70, 1992.

7. Aly, E. E. A. A., "On Testing for Decreasing Mean Residual Life Ordering," *Naval Research Logistics, An International Journal*, Vol. 40, No. 5, pp. 633-642, 1993.

8. Alzaid, A. A., "Aging Concepts for Items of Unknown Age," *Communications in Statistics*, Vol. 10, No. 3, pp. 649-659, 1994.

9. Chang, D. S., "Critical Time of the Lognormal Distribution," Vol. 34, No. 2, pp. 261-266, 1994.

10. Csorgo, S. and Viharos, L., "Confidence Bands for Percentile Residual Lifetimes," *Journal of Statistical Planning and Inference*, Vol. 30, No. 3, pp. 327-337, 1992.

11. Dagpunar, J. S., "Some Necessary and Sufficient Conditions for Age Replacement with Non-zero Downtimes," *Journal of the Operational Research Society*, Vol. 45, No. 2, pp. 225-229, 1994.

12. Dickson, T. L. and Simonen, F. A., "Application of Probabilistic Fracture Analysis to Residual Life Evaluation of Embrittled Reactor Vessels," *Proceedings of the Winter Annual Meeting of the American Society of Mechanical Engineering*, Anaheim, CA, *Reliability Technology – 1992 American Society of Mechanical Engineers, Aerospace Division*, Vol. 28, pp. 43-55, 1992.

13. Ebrahimi, N., "On Estimating the Change Point in a Mean Residual Life Function," *The Indian Journal of Statistics*, Series A, Vol. 53, No. 2, pp. 206-219, 1991.

14. Fagiuoli, E. and Pellerey, F., "Mean Residual Life and Increasing Convex Comparison of Shock Models," *Statistics and Probability Letters*, Vol. 20, No. 5, pp. 337-345, 1994.

15. Galambos, J. and Xu, Y., "Regularly Varying Expected Residual Life and Domains of Attraction of Extreme Value Distributions," *Annales of Universities Scientiarum Budapestinensis de Rolando Eotvos Nominatae*, Sectio Mathematica, Vol. 33, pp. 105-108, 1991.

16. Galambos, J. and Hagwood, C., "The Characterization of a Distribution Function by the Second Moment of the Residual Life," *Communications in Statistics*, Theory and Methods, Vol. 21, No. 5, pp. 1463-1468, 1992.

17. Galambos, J. and Hagwood, C., "An Unreliable Server Characterization of the Exponential Distribution," *Journal of Applied Probability*, Vol. 31, No. 1, pp. 274-279, 1994.

18. Guess, F. M. and Kitchin, J. F., "Mean Time to System Recovery," *Quality and Reliability Engineering International*, Vol. 7, No. 1, pp. 5-6, 1991.

19. Guess, F. M. and Park, D. H., "Modeling Discrete Bathtub and Upside-Down Bathtub Mean Residual-Life Functions," *IEEE Transactions on Reliability*, Vol. 37, No. 5, pp. 545-549, 1988.

20. Guess, F. M. and Park, D. H., "Nonparametric Confidence Bounds, Using Censored Data, on the Mean Residual Life," *IEEE Transactions on Reliability*, Vol. 40, No. 1, pp. 78-80, 1991.

21. Gupta, P. L. and Gupta, R. D., "Relative Errors in Reliability Measures," *Collection: Topics in Statistical Dependence*, pp. 251-256, 1987, published by Institution of Mathematical Statistics, Hayward, CA, 1990.

22. Hansen, B. G. and Frenk, J. B. G., "Some Monotonicity Properties of the Delayed Renewal Function," *Journal of Applied Probability*, Vol. 28, No. 4, pp. 811-821, 1991.

23. Hawkins, D. L., Kochar, S. and Loader, C., "Testing Exponentiality Against IDMRL Distributions with Unknown Change Point," *The Annals of Statistics*, Vol. 20, No. 1, pp. 280-290, 1992.

24. Jaisingh, L. R., "Characterization of Inverse Gaussian and Inverted Gamma Distributions," *Microelectronics and Reliability*, Vol. 30, No. 4, pp. 775-780, 1990.

25. Kalpakam, S., "On the Quasi-stationary Distribution of the Residual Lifetime," *IEEE Transactions on Reliability*, Vol. 42, No. 4, pp. 623-624, 1993.

26. Karpinski, J., "Distribution of Residual System Life After Partial Failures," *IEEE Transactions on Reliability*, Vol. 37, No. 5, pp. 539-544, 1988.

27. Kopnov, V. A., "Residual Life, Linear Fatigue Damage Accumulation and Optimal Stopping Time," *Reliability Engineering and System Safety*, Vol. 40, No. 3, pp. 319-325, 1993.

28. Korwar, R. M., "A Characterization of the Family of Distributions with a Linear Mean Residual Life Function," *The Indian Journal of Statistics*, Series B, Vol. 54, No. 2, pp. 257-260, 1992.

29. Kulasekera, K. B., "Smooth Nonparametric Estimation of Mean Residual Life," *Microelectronics and Reliability*, Vol. 31, No. 1, pp. 97-108, 1991.

30. Launer, R. L., "Graphical Techniques for Analyzing Failure Data with the Percentile Residual Life Function," *IEEE Transactions on Reliability*, Vol. 42, No. 1, pp. 71-75, 1993.

31. Li, W. and Cao, J. H., "Limiting Distribution of the Residual Lifetimes of Several Repairable Systems," *Microelectronics and Reliability*, Vol. 33, No. 8, pp. 1069-1072, 1993.

32. Li, W. and Cao, J. H., "Limiting Distribution of the Residual Lifetime of a Markov Repairable System," *Reliability Engineering and System Safety*, Vol. 41, No. 2, pp. 103-105, 1993.

33. Lim, J. H. and Park, D. H., "Test for DMRL Using Censored Data," *Journal of Nonparametric Statistics*, Vol. 3, No. 2, pp. 167-173, 1993.

34. Maguluri, G. and Zhang, C. H., "Estimation in the Mean Residual Life Regression Model," *Journal of the Royal Statistical Society*, Series B, Methodological, Vol. 56, No. 3, pp. 477-489, 1994.

35. Mi, J., "Discrete Bathtub Failure Rate and Upside-down Bathtub Mean Residual Life," *Naval Research Logistics, An International Journal*, Vol. 40, No. 3, pp. 361-371, 1993.

36. Mi, J., "Estimation Related to Mean Residual Life," *Journal of Nonparametric Statistics*, Vol. 4, No. 2, pp. 179-190, 1994.

37. Mukherjee, S. P. and Chatterjee, A., "Closure Under Convolution of Dominance Relations," *Calcutta Statistical Association Bulletin*, Vol. 42, No. 167-168, pp. 251-254, 1992.

38. Osaki, S. and Li, X. X., "Characterization of Gamma and Negative Binomial Distributions," *IEEE Transactions on Reliability*, Vol. 37, No. 4, pp. 379-382, 1988.

39. Park, B. G. and Sohn, J. K., "Nonparametric Estimation of Mean Residual Life Function Under Random Censorship," *Journal of the Korean Statistical Society*, Vol. 22, No. 2, pp. 147-157, 1993.

40. Park, K. S. and Yoo, Y. K., "(τ, k) Block Replacement Policy with Idle Count," *IEEE Transactions on Reliability*, Vol. 42, No. 4, pp. 561-565, 1993.

41. Pellerey, F., "Partial Ordering Under Cumulative Damage Shock Models," *Advances in Applied Probability*, Vol. 25, No. 4, pp. 939-946, 1993.

42. Raja, R. B., et al., "Setting the Clock Back to Zero Property of a Class of Bivariate Life Distributions," *Communications in Statistics*, Theory and Methods, Vol. 22, No. 7, pp. 2067-2080, 1993.

43. Rao, B. R., Damaraju, C. V. and Alhumoud, J. M., "Covariate Effect on the Life Expectancy and Percentile Residual Life Functions Under the Proportional Hazards and the Accelerated Life Models," *Communications in Statistics*, Theory and Methods, Vol. 22, No. 1, pp. 257-281, 1993.

44. Roy, D., "Characterizations Through Failure Rate and Mean Residual Life Transforms," *Microelectronics and Reliability*, Vol. 33, No. 2, pp. 141-142, 1993.

45. Ruiz, J. M. and Nararro, J., "Characterization of Distributions by Relationships between Failure Rate and Mean Residual Life," *IEEE Transactions on Reliability*, Vol. 43, No. 4, pp. 640-644, 1994.

46. Sakon, T., et al., "Residual Life Assessment of Pump Casing Considering Thermal Fatigue Crack Propagation," *Proceedings of the Symposium on Thermo-mechanical Fatigue Behavior of Materials*, San Diego, CA, ASTM Special Technical Publication, No. 1186, pp. 239-252, 1993.

47. Sankaran, P. G. and Nair, N. U., "Bivariate Variance Residual Life," *Journal of the Indian Association for Productivity*, Quality and Reliability, Vol. 18, No. 2, pp. 1-6, 1993.

48. Sankaran, P. G. and Unnikrishnan-Nair, N., "Characterizations by Properties of Residual Life Distributions," *Journal of Theoretical and Applied Statistics*, Vol. 24, No. 3, pp. 245-251, 1993.

49. Sen, K. and Jain, M. B., "Tests for Bivariate Mean Residual Life," *Communications in Statistics*, Theory and Methods, Vol. 20, No. 8, pp. 2549-2558, 1991.

50. Shaked, M. and Shanthikumar, J. G., "Multivariate Conditional Hazard Rate and Mean Residual Life Functions and Their Applications," *Proceedings of the Conference on Reliability and Decision Making*, The University of Siena, Siena, Edited by Barlow R. E., etc., pp. 137-155, October 15-26, 1990.

51. Shaked, M. and Shanthikumar, J. G., "Dynamic Multivariate Aging Notions in Reliability Theory," *Stochastic Processes and Their Applications*, Vol. 38, No. 1, pp. 85-97, 1991.

52. Shaked, M. and Shanthikumar, J. G., "Dynamic Conditional Marginal Distributions in Reliability Theory," *Journal of Applied Probability*, Vol. 30, No. 2, pp. 421-428, 1993.

53. Tsang, A. H. C. and Jardine, A. K. S., "Estimators of 2-parameter Weibull Distributions From Incomplete Data with Residual Lifetimes," *IEEE Transactions on Reliability*, Vol. 42, No. 2, pp. 291-298, 1993.

APPENDIX 8A
DERIVATION OF EQ. (8.36)

From Eq. (8.35),

$$I(t) = \int_t^\infty \frac{(u-\mu)+\mu}{\sigma} e^{-\frac{1}{2}(\frac{u-\mu}{\sigma})^2} du,$$

$$= \int_t^\infty \frac{u-\mu}{\sigma} e^{-\frac{1}{2}(\frac{u-\mu}{\sigma})^2} du + \int_t^\infty \frac{\mu}{\sigma} e^{-\frac{1}{2}(\frac{u-\mu}{\sigma})^2} du,$$

$$= \sigma \int_t^\infty \frac{u-\mu}{\sigma} e^{-\frac{1}{2}(\frac{u-\mu}{\sigma})^2} d\left(\frac{u-\mu}{\sigma}\right)$$

$$+ \mu \int_t^\infty e^{-\frac{1}{2}(\frac{u-\mu}{\sigma})^2} d\left(\frac{u-\mu}{\sigma}\right),$$

or

$$I(t) = FT - ST, \tag{8.75}$$

where

$$FT = \sigma \int_t^\infty \frac{u-\mu}{\sigma} e^{-\frac{1}{2}(\frac{u-\mu}{\sigma})^2} d\left(\frac{u-\mu}{\sigma}\right),$$

and

$$ST = \mu \int_t^\infty e^{-\frac{1}{2}(\frac{u-\mu}{\sigma})^2} d\left(\frac{u-\mu}{\sigma}\right).$$

Now evaluate FT and ST, respectively. The first term, FT, is

$$FT = \sigma \int_t^\infty e^{-\frac{1}{2}(\frac{u-\mu}{\sigma})^2} d\left[\frac{1}{2}\left(\frac{u-\mu}{\sigma}\right)^2\right],$$

$$= \sigma \left[-e^{-\frac{1}{2}(\frac{u-\mu}{\sigma})^2}\right]\bigg|_t^\infty,$$

or

$$FT = \sigma \, e^{-\frac{1}{2}(\frac{t-\mu}{\sigma})^2}. \tag{8.76}$$

For the second term, the following two situations need to be considered:

CASE 1: $0 < t < \mu$.

CASE 2: $t \geq \mu$.

Now let $y = \frac{1}{2}\left(\frac{u-\mu}{\sigma}\right)^2$. Then, $\frac{u-\mu}{\sigma} = \pm\sqrt{2y}$ and $d\left(\frac{u-\mu}{\sigma}\right) = \pm\frac{1}{\sqrt{2}}y^{-\frac{1}{2}}dy$.

CASE 1: Since $t < \mu$, then for $t \leq u < \mu$, $\frac{u-\mu}{\sigma} = -\sqrt{2y}$ and $d\left(\frac{u-\mu}{\sigma}\right) = -\frac{1}{\sqrt{2}}y^{-\frac{1}{2}}dy$; and for $u \geq \mu$, $\frac{u-\mu}{\sigma} = \sqrt{2y}$ and $d\left(\frac{u-\mu}{\sigma}\right) = \frac{1}{\sqrt{2}}y^{-\frac{1}{2}}dy$. The second term, ST, becomes

$$ST = \mu\left[\int_t^\mu e^{-\frac{1}{2}\left(\frac{u-\mu}{\sigma}\right)^2} d\left(\frac{u-\mu}{\sigma}\right) + \int_\mu^\infty e^{-\frac{1}{2}\left(\frac{u-\mu}{\sigma}\right)^2} d\left(\frac{u-\mu}{\sigma}\right)\right],$$

$$= \mu\left[\int_{\frac{1}{2}\left(\frac{t-\mu}{\sigma}\right)^2}^{0} e^{-y}\left(-\frac{1}{\sqrt{2}}y^{-\frac{1}{2}}dy\right) + \int_0^\infty e^{-y}\left(-\frac{1}{\sqrt{2}}y^{-\frac{1}{2}}dy\right)\right],$$

$$= \frac{\mu}{\sqrt{2}}\left[\int_0^{\frac{1}{2}\left(\frac{t-\mu}{\sigma}\right)^2} y^{\frac{1}{2}-1}e^{-y}dy + \int_0^\infty y^{\frac{1}{2}-1}e^{-y}dy\right],$$

$$= \frac{\mu}{\sqrt{2}}\left[\left(\int_0^\infty y^{\frac{1}{2}-1}e^{-y}dy - \int_{\frac{1}{2}\left(\frac{t-\mu}{\sigma}\right)^2}^{\infty} y^{\frac{1}{2}-1}e^{-y}dy\right)\right.$$
$$\left.+ \int_0^\infty y^{\frac{1}{2}-1}e^{-y}dy\right],$$

$$= \frac{\mu}{\sqrt{2}}\left(2\int_0^\infty y^{\frac{1}{2}-1}e^{-y}dy - \int_{\frac{1}{2}\left(\frac{t-\mu}{\sigma}\right)^2}^{\infty} y^{\frac{1}{2}-1}e^{-y}dy\right),$$

$$= \frac{\mu}{\sqrt{2}}\left\{2\Gamma\left(\frac{1}{2}\right) - \Gamma\left[\frac{1}{2};\frac{1}{2}\left(\frac{t-\mu}{\sigma}\right)^2\right]\right\},$$

or

$$ST = \frac{\mu}{\sqrt{2}}\left\{2\pi - \Gamma\left[\frac{1}{2};\frac{1}{2}\left(\frac{t-\mu}{\sigma}\right)^2\right]\right\}. \tag{8.77}$$

CASE 2: Since $t \geq \mu$, then $\frac{u-\mu}{\sigma} = \sqrt{2y}$ and $d\left(\frac{u-\mu}{\sigma}\right) = \frac{1}{\sqrt{2}}y^{-\frac{1}{2}}dy$. The second term, ST, then becomes

$$ST = \mu\int_{\frac{1}{2}\left(\frac{t-\mu}{\sigma}\right)^2}^{\infty} e^{-y}\left(\frac{1}{\sqrt{2}}y^{-\frac{1}{2}}dy\right),$$

$$= \frac{\mu}{\sqrt{2}}\int_{\frac{1}{2}\left(\frac{t-\mu}{\sigma}\right)^2}^{\infty} y^{\frac{1}{2}-1}e^{-y}dy,$$

APPENDIX 8A

or

$$ST = \frac{\mu}{\sqrt{2}} \Gamma\left[\frac{1}{2};\frac{1}{2}\left(\frac{t-\mu}{\sigma}\right)^2\right]. \tag{8.78}$$

Therefore, substituting Eqs. (8.76), (8.77) and (8.78) into Eq. (8.75) yields

$$I(t) = FT - ST = FT = \sigma\, e^{-\frac{1}{2}\left(\frac{t-\mu}{\sigma}\right)^2} - ST,$$

or

$$I(t) = \begin{cases} \sigma\, e^{-\frac{1}{2}\left(\frac{t-\mu}{\sigma}\right)^2} - \frac{\mu}{\sqrt{2}}\left\{2\pi - \Gamma\left[\frac{1}{2};\frac{1}{2}\left(\frac{t-\mu}{\sigma}\right)^2\right]\right\}, & \text{for } 0 < t < \mu, \\ \sigma\, e^{-\frac{1}{2}\left(\frac{t-\mu}{\sigma}\right)^2} - \frac{\mu}{\sqrt{2}} \Gamma\left[\frac{1}{2};\frac{1}{2}\left(\frac{t-\mu}{\sigma}\right)^2\right], & \text{for } t \geq \mu. \end{cases}$$

$$\tag{8.79}$$

APPENDIX 8B
DERIVATION OF EQ. (8.41)

From Eq. (8.40),

$$I'(t) = \int_t^\infty \frac{1}{\sqrt{2\pi}\,\sigma'} e^{-\frac{1}{2}\left(\frac{\log_e u - \mu'}{\sigma'}\right)^2} du. \tag{8.80}$$

Now let $z = \frac{\log_e u - \mu'}{\sigma'}$, then $u = e^{\mu' + \sigma' z}$ and $du = \sigma' e^{\mu' + \sigma' z} dz$. Then,

$$I'(t) = \int_{\frac{\log_e t - \mu'}{\sigma'}}^\infty \frac{1}{\sqrt{2\pi}\,\sigma'} e^{-\frac{1}{2} z^2} \sigma' e^{\mu' + \sigma' z} dz,$$

$$= \int_{\frac{\log_e t - \mu'}{\sigma'}}^\infty \frac{1}{\sqrt{2\pi}} e^{-\frac{1}{2} z^2 + \sigma' z + \mu'} dz,$$

$$= \int_{\frac{\log_e t - \mu'}{\sigma'}}^\infty \frac{1}{\sqrt{2\pi}} e^{-\frac{1}{2}[z^2 - 2\sigma' z + (\sigma')^2] + [\mu' + \frac{1}{2}(\sigma')^2]} dz,$$

or

$$I'(t) = e^{\mu' + \frac{1}{2}(\sigma')^2} \int_{\frac{\log_e t - \mu'}{\sigma'}}^\infty \frac{1}{\sqrt{2\pi}} e^{-\frac{1}{2}\left(\frac{z - \sigma'}{1}\right)^2} dz. \tag{8.81}$$

Note that the integrand in Eq. (8.81) turns out to be the *pdf* of a normal distribution with mean σ' and standard deviation 1. Therefore, Eq. (8.81) becomes

$$I'(t) = e^{\mu' + \frac{1}{2}(\sigma')^2} \left[1 - \int_{-\infty}^{\frac{\log_e t - \mu'}{\sigma'}} \frac{1}{\sqrt{2\pi}} e^{-\frac{1}{2}\left(\frac{z - \sigma'}{1}\right)^2} dz \right],$$

$$= e^{\mu' + \frac{1}{2}(\sigma')^2} \left[1 - \Phi\left(\frac{\frac{\log_e t - \mu'}{\sigma'} - \sigma'}{1}\right) \right],$$

$$= e^{\mu' + \frac{1}{2}(\sigma')^2} \Phi\left(\sigma' - \frac{\log_e t - \mu'}{\sigma'}\right),$$

or

$$I'(t) = e^{\mu' + \frac{1}{2}(\sigma')^2} \Phi\left[\frac{(\sigma')^2 + \mu' - \log_e t}{\sigma'}\right]. \tag{8.82}$$

APPENDIX 8C
COMPUTER PROGRAM OF "BIM.FOR"

```
C $$$$$$$$$$$$$$$$$$$$$$$$$$$$$$$$$$$$$$$$$$$$$$$$$$$$$
C $      BURN-IN TIME FOR A SPECIFIED MRL GOAL      $
C $      USING THE BIMODAL, MIXED WEIBULL MODEL     $
C $            AND THE BATHTUB MODEL                $
C $$$$$$$$$$$$$$$$$$$$$$$$$$$$$$$$$$$$$$$$$$$$$$$$$$$$$
      IMPLICIT DOUBLE PRECISION (A-H, O-Z)
      REAL M1,M2,M0,MG
      EXTERNAL FUNC, FUNC0, FUNC1, FUNC2, DQDAGI
      COMMON /PAR1/ALPHA,/PAR2/BETA
      COMMON /PAR3/BETA1,/PAR4/ETA1,/PAR5/BETA2,
     1 /PAR6/ETA2
      OPEN(5,FILE='IBIM.DAT',STATUS='UNKNOWN')
      OPEN(6,FILE='OBIM.DAT',STATUS='NEW')
C
C  BURN-IN TIME CORRESPONDING TO A SPECIFIED MRL GOAL,
C  MG, USING THE BIMODAL, MIXED-WEIBULL MODEL.
C
5     WRITE(*,*) 'GIVE PARAMETERS OF MIXED WEIBULL
     1 DISTRIBUTION: P, BETA1,ETA1, BETA2, ETA2 !!'
      READ(5,*) P, BETA1,ETA1, BETA2, ETA2
      WRITE(5,*) P, BETA1,ETA1, BETA2, ETA2
      WRITE(6,*) ' BURN-IN TIME CORRESPONDING TO A
     1 SPECIFIED MRL GOAL, MG'
      WRITE(*,*) 'GIVE THE MRL GOAL, MG, PLEASE!'
      READ(*,*) MG
      WRITE(6,*)'MRL GOAL ==',MG
      WRITE(*,*)'MRL GOAL ==',MG
      TT=0.1
      KK=0
      DO 260  I=1,1000
      TT=TT*1.5
      CALL DQDAGI(FUNC1,TT,1,0.0,0.00001,M1,ERREST)
      CALL DQDAGI(FUNC2,TT,1,0.0,0.00001,M2,ERREST)
      RB1=EXP(-((TT)/ETA1)**BETA1)
      RB2=EXP(-((TT)/ETA2)**BETA2)
      M0=(P*M1+(1.0-P)*M2)/(P*RB1+(1.0-P)*RB2)
      WRITE(*,*) 'Tb===',TT, 'M0==',M0
260   CONTINUE
      IF(M0.LE.MG) GOTO 252
      IF(KK.NE.1) GOTO 252
```

```
              X2=TT
              Y2=MO
              GOTO 280
252           IF(MO-MG) 255,270,300
255           X1=TT
              Y1=MO
              KK=1
270           TBI=TT
              GOTO 285
280           TBI=(MG-Y1)*(X2-X1)/(Y2-Y1)+X1
285           WRITE(6,290) TBI
290           FORMAT(//2X,'THE CORRESPONDING BURN-IN TIME IS',
             1 E20.6)
              WRITE(*,*)'THE CORRESPONDING BURN-IN TIME IS',TBI
              GOTO 310
300           WRITE(*,*)'YOUR TB IS LESS THAN 0.00002 HR!'
310           WRITE(*,*) 'ANOTHER CASE ? YES=1/NO=0'
              READ(*,*) IAN
              IF(IAN.EQ.1) GOTO 5
C
C   BURN-IN TIME CORRESPONDING TO A SPECIFIED MRL GOAL, MG,
C   USING THE BATHTUB MODEL 1.
C
86            WRITE(*,*) 'GIVE PARAMETERS OF THE BATHTUB MODEL 1
             1 ALPHA AND BETA !!'
              READ(5,*) ALPHA, BETA
              WRITE(*,*) ALPHA, BETA
              WRITE(6,*) ' BURN-IN TIME CORRESPONDING TO A
             1 SPECIFIED MRL GOAL, MG'
              WRITE(*,*) 'GIVE THE MRL GOAL, MG, PLEASE!'
              READ(*,*) MG
              WRITE(6,*)'MRL GOAL ==',MG
              WRITE(*,*)'MRL GOAL ==',MG
              TT=0.1
              KK=0
              DO 860 I=1,10000
              TT=TT*1.1
C             WRITE(*,*)'TT=',TT
              CALL DQDAGI(FUNC,TT,1,0.0,0.00001,RESULT,ERREST)
C             WRITE(*,*) 'VAL====',RESULT
              RB=EXP(1-EXP((TT/ALPHA)**BETA))
              MO=RESULT/RB
              WRITE(*,*) 'TB===',TT, 'RB==',RB,'MO==',MO
              IF(MO.LE.MG) GOTO 852
```

```
              IF(KK.NE.1) GOTO 852
              X2=TT
              Y2=MO
              GOTO 880
852           IF(MO-MG) 855,870,800
855           X1=TT
              Y1=MO
              KK=1
860           CONTINUE
870           TBI=TT
              GOTO 885
880           TBI=(MG-Y1)*(X2-X1)/(Y2-Y1)+X1
885           WRITE(6,890) TBI
890           FORMAT(//2X,'THE CORRESPONDING BURN-IN TIME IS',
             1 E20.6)
              WRITE(*,*)'THE CORRESPONDING BURN-IN TIME IS',TBI
              GOTO 810
800           WRITE(*,*)'YOUR TB IS LESS THAN 0.00002 HR!'
810           WRITE(*,*) 'ANOTHER CASE ? YES=1/NO=0'
              READ(*,*) IAN
              IF(IAN.EQ.1) GOTO 86
              STOP
              END
C
              FUNCTION FUNC1(Z)
              COMMON /PAR3/BETA1,/PAR4/ETA1
C             WRITE(*,*) BETA1,ETA1
              FUNC1=EXP(-(Z/ETA1)**BETA1)
C             WRITE(*,*)'FUNC1====',FUNC1
              RETURN
              END
C
              FUNCTION FUNC2(Z)
              COMMON /PAR5/BETA2,/PAR6/ETA2
C             WRITE(*,*)BETA2,ETA2
              FUNC2=EXP(-(Z/ETA2)**BETA2)
              RETURN
              END
C
              FUNCTION FUNC(Z)
              COMMON /PAR1/ALPHA,/PAR2/BETA
              FUNC=EXP(1-EXP((Z/ALPHA)**BETA))
              RETURN
              END
```

APPENDIX 8D
COMPUTER PROGRAM OF "MRLMAX1.FOR"

```fortran
C
C $$$$$$$$$$$$$$$$$$$$$$$$$$$$$$$$$$$$$$$$$$$$$$$$$$$
C $      BURN-IN TIME FOR THE MAXIMUM MRL VALUE     $
C $            USING THE BATHTUB MODEL 1            $
C $$$$$$$$$$$$$$$$$$$$$$$$$$$$$$$$$$$$$$$$$$$$$$$$$$$
C
      IMPLICIT DOUBLE PRECISION (A-H, O-Z)
      EXTERNAL FUNC, FUNC1, FUNC2, DQDAGI
      REAL MRLMAX
      COMMON /PAR1/ALPHA,/PAR2/BETA
      COMMON /PAR3/BETA1,/PAR4/ETA1,/PAR5/BETA2,
     1 /PAR6/ETA2
      OPEN(5,FILE='IMRL.DAT',STATUS='UNKNOWN')
      OPEN(6,FILE='OMRL.DAT',STATUS='NEW')
C
C  BURN-IN TIME FOR THE MAXIMUM MEAN RESIDUAL LIFE
C  USING THE BATHTUB MODEL 1.
C
      WRITE(*,*) 'GIVE PARAMETERS: ALPHA AND BETA !!'
      READ(5,*) ALPHA, BETA
      WRITE(*,*) ALPHA, BETA
      WRITE(*,*)' GIVE THE CONVERGENCE PRECISION!'
      READ(*,*) EPSILON
      WRITE(*,*) EPSILON
      WRITE(6,*) ' BURN-IN TIME FOR THE MAXIMUM MRL'
      WRITE(6,*) '================================='
      WRITE(*,*)'ENTER AN INITIAL GUESS FOR TB!'
      READ(*,*) TT
260   CALL DQDAGI(FUNC,TT,1,0.0,0.00001,RESULT,ERREST)
      RB=(BETA/ALPHA**BETA)*EXP((TT/ALPHA)**BETA
     1 +EXP((TT/ALPHA)**BETA)-1.0)
      TK=(RB*RESULT)**(1.0/(1.0-BETA))
      WRITE(*,*) 'TT===',TT, 'TK==',TK
      ERROR=ABS((TT-TK)/TT)
      IF(ERROR.LT.EPSILON)GOTO 265
      TT=TK
      GOTO 260
265   TB=TK
      MRLMAX=RESULT/EXP(1.0-EXP((TB/ALPHA)**BETA))
      WRITE(*,*)'BURN-IN TIME IS',TB
```

APPENDIX 8D

```
      WRITE(*,*)'THE MAXIMUM MRL IS', MRLMAX
      STOP
      END
C
      FUNCTION FUNC1(Z)
      COMMON /PAR3/BETA1,/PAR4/ETA1
      FUNC1=EXP(-(Z/ETA1)**BETA1)
      RETURN
      END
C
      FUNCTION FUNC2(Z)
      COMMON /PAR5/BETA2,/PAR6/ETA2
      FUNC2=EXP(-(Z/ETA2)**BETA2)
      RETURN
      END
C
      FUNCTION FUNC(Z)
      COMMON /PAR1/ALPHA,/PAR2/BETA
      FUNC=EXP(1-EXP((Z/ALPHA)**BETA))
      RETURN
      END
```

APPENDIX 8E
COMPUTER PROGRAM OF "MRLMAX2.FOR"

```
C
C       ************************************************************
C       *    COMPUTER PROGRAM FOR THE OPTIMUM BURN-IN TIME          *
C       *    FOR THE MAXIMUM MEAN RESIDUAL LIFE (MRL) USING         *
C       *    BIMODAL, MIXED WEIBULL LIFE DISTRIBUTION AND           *
C       *         USING TRIAL-AND-ERROR APPROACH                    *
C       ************************************************************
C
        IMPLICIT DOUBLE PRECISION (A-H, O-Z)
        EXTERNAL FUNC1, FUNC2, DQDAGI,D
        COMMON /PAR2/P,/PAR3/BETA1,/PAR4/ETA1,/PAR5/BETA2,
      1 /PAR6/ETA2
        REAL M0,MRL
        OPEN(5,FILE='IMRL2.DAT',STATUS='UNKNOWN')
        OPEN(6,FILE='OMRL2.DAT',STATUS='NEW')
        WRITE(6,10)
10      FORMAT(1X, 'GIVE PARAMETERS OF MIXED WEIBULL
      1 DISTRIBUTION: P, BETA1,ETA1, BETA2, ETA2 !!')
        READ(5,*) P, BETA1,ETA1, BETA2, ETA2
        WRITE(5,*) P, BETA1,ETA1, BETA2, ETA2
        WRITE(6,*) ' BURN-IN TIME FOR MAXIMUM MRL'
        WRITE(6,*) '================================'
        WRITE(*,*)'ENTER THE LOWER LIMIT FOR TB!'
        READ(*,*)A1
        T0=A1
        M0=D(T0)
15      TT=T0+1.0
        MRL=D(TT)
        IF(MRL.LE.M0)GOTO 20
        T0=TT
        M0=MRL
        GOTO 15
20      TB=T0
        MRL=M0
        WRITE(6,*)'TB=',TB,'MRL=',MRL
        WRITE(*,*)'======================================'
        WRITE(*,*)'THE OPTIMUM BURN-IN TIME IS:'
        WRITE(*,*)'TB*=',TB
        WRITE(*,*)'======================================'
        WRITE(*,*)'THE MAXIMUM MRL IS:'
```

APPENDIX 8E

```
          WRITE(*,*)'MRL*=',MRL
          WRITE(*,*)'======================================'
          STOP
          END
C
C ***********************************************************
C *         SUBPROGRAM FOR THE OBJECTIVE FUNCTION           *
C *               --------THE MEAN RESIDUAL LIFE            *
C ***********************************************************
C
          FUNCTION D(TT)
          IMPLICIT DOUBLE PRECISION (A-H, O-Z)
          COMMON /PAR2/P,/PAR3/BETA1,/PAR4/ETA1,/PAR5/BETA2,
        1 /PAR6/ETA2
          EXTERNAL FUNC1,FUNC2,DQDAGI
          REAL M1,M2,MRL
          CALL DQDAGI(FUNC1,TT,1,0.0,0.00001,M1,ERREST)
          CALL DQDAGI(FUNC2,TT,1,0.0,0.00001,M2,ERREST)
          RB1=EXP(-((TT)/ETA1)**BETA1)
          RB2=EXP(-((TT)/ETA2)**BETA2)
          MRL=(P*M1+(1.0-P)*M2)/(P*RB1+(1.0-P)*RB2)
          D=MRL
          RETURN
          END
C
          FUNCTION FUNC1(Z)
          IMPLICIT DOUBLE PRECISION (A-H, O-Z)
          COMMON /PAR3/BETA1,/PAR4/ETA1
          FUNC1=EXP(-(Z/ETA1)**BETA1)
          RETURN
          END
C
          FUNCTION FUNC2(Z)
          IMPLICIT DOUBLE PRECISION (A-H, O-Z)
          COMMON /PAR5/BETA2,/PAR6/ETA2
          FUNC2=EXP(-(Z/ETA2)**BETA2)
          RETURN
          END
```

APPENDIX 8F
COMPUTER PROGRAM OF "MRL-EMP1.FOR"

```
C
C $$$$$$$$$$$$$$$$$$$$$$$$$$$$$$$$$$$$$$$$$$$$$$$$$$$
C $ COMPUTATION OF THE EMPIRICAL MEAN RESIDUAL LIFE  $
C $ FOR CENSORED, UNGROUPED BURN-IN TESTING DATA SET $
C $ AND OPTIMUM BURN-IN TIME FOR MAXIMUM MRL VALUE   $
C $$$$$$$$$$$$$$$$$$$$$$$$$$$$$$$$$$$$$$$$$$$$$$$$$$$
C
      IMPLICIT DOUBLE PRECISION (A-H, O-Z)
      DIMENSION NF(50)
      REAL T(50), MRL(50), MRL0, MRLMAX, M1, M2, M3
      OPEN(5,FILE='IMRL.DAT',STATUS='UNKNOWN')
      OPEN(6,FILE='OMRL.DAT',STATUS='NEW')
C
C  EMPIRICAL MEAN RESIDUAL LIFE CALCULATION
C
      WRITE(*,*) 'ENTER THE TOTAL NUMBER OF COMPONENTS
     1 IN THE TEST, NT, PLEASE!'
      WRITE(6,*) 'ENTER THE TOTAL NUMBER OF COMPONENTS
     1 IN THE TEST, NT, PLEASE!'
      READ(*,*) NT
      WRITE(6,*) 'NT=', NT
      WRITE(*,*) 'ENTER THE TOTAL NUMBER OF DISTINCT
     1 FAILURE TIMES, N, PLEASE!'
      WRITE(6,*) 'ENTER THE TOTAL NUMBER OF DISTINCT
     1 FAILURE TIMES, N, PLEASE!'
      READ(*,*) N
      WRITE(6,*) 'N=', N
      WRITE(*,*) 'ENTER THE TEST TERMINATION TIME,
     1 TD, PLEASE!'
      WRITE(6,*) 'ENTER THE TEST TERMINATION TIME,
     1 TD, PLEASE!'
      READ(*,*) TD
      WRITE(6,*) 'TD=', TD
      DO 10 I=1,N
      READ(5,*) T(I)
      WRITE(*,*) 'T(',I,')=', T(I)
      WRITE(6,*) 'T(',I,')=', T(I)
10    CONTINUE
      NFT=0
      DO 20 I=1,N
```

APPENDIX 8F

```
              READ(5,*) NF(I)
              WRITE(*,*) 'NF(',I,')=', NF(I)
              WRITE(6,*) 'NF(',I,')=', NF(I)
              NFT=NFT+NF(I)
20            CONTINUE
              NS=NT-NFT
              WRITE(*,*) 'TOTAL # OF FAILURES, NFT=', NFT
              WRITE(6,*) 'TOTAL # OF FAILURES, NFT=', NFT
              WRITE(*,*) 'TOTAL # OF SURVIVORS, NS=', NS
              WRITE(6,*) 'TOTAL # OF SURVIVORS, NS=', NS
              WRITE(6,*) '====================================='
              WRITE(6,*) '    EMPIRICAL MEAN RESIDUAL LIFE'
              WRITE(6,*) '====================================='
              IF(T(1).EQ.0.0)GOTO 25
              X=0.0
              NK=0
              DO 22 K=1,N
              X=X+NF(K)*T(K)
              NK=NK+NF(K)
22            CONTINUE
              ANK=NK
              MRLO=(X+NS*TD)/ANK
              WRITE(*,50) 0, 0.0, 0, MRLO
              WRITE(6,50) 0, 0.0, 0, MRLO
25            DO 40 J=1,N-1
              X=0.0
              NK=0.0
              DO 30 K=J+1,N
              X=X+ NF(K)*(T(K)-T(J))
              NK=NK+NF(K)
30            CONTINUE
              ANK=NK
              MRL(J)=(X+NS*(TD-T(J)))/ANK
              WRITE(*,50) J, T(J), J, MRL(J)
              WRITE(6,50) J, T(J), J, MRL(J)
40            CONTINUE
50            FORMAT (2X,'T(', I4, ')= ', F13.4,
             1 ' MRL(', I4, ')= ', F13.4)
C
C     OPTIMUM BURN-IN TIME FOR THE MAXIMUM MRL
C     USING QUADRATIC FUNCTION APPROXIMATION
C     FOR MRL FUNCTION AROUND MAXIMUM
C     EMPIRICAL MRL POINT
C
```

```
              IF(T(1).GT.0.0) GOTO 60
              IMAX=1
              MRLMAX=MRL(1)
              GOTO 70
60            IMAX=0
              MRLMAX=MRL0
70            DO 80 I=1,N-1
              IF(MRL(I).LT.MRLMAX) GOTO 80
              IMAX=I
              MRLMAX=MRL(I)
80            CONTINUE
              I1=IMAX-1
              I2=IMAX
              I3=IMAX+1
              T1=T(I1)
              T2=T(I2)
              T3=T(I3)
              M1=MRL(I1)
              M2=MRL(I2)
              M3=MRL(I3)
              D=T1*T2*(T1-T2)+T2*T3*(T2-T3)+T3*T1*(T3-T1)
              DA=M1*(T2-T3)+M2*(T3-T1)+M3*(T1-T2)
              DB=T1*T1*(M2-M3)+T2*T2*(M3-M1)+T3*T3*(M1-M2)
              DC=T1*T2*M3*(T1-T2)+T2*T3*M1*(T2-T3)+T3*T1*M2*(T3-T1)
              A=DA/D
              B=DB/D
              C=DC/D
              TB=-B/(2.0*A)
              MRLMAX=(4.0*A*C-B*B)/(4.0*A)
              WRITE(*,85) A, B, C
              WRITE(6,85) A, B, C
              WRITE(*,90) TB, MRLMAX
              WRITE(6,90) TB, MRLMAX
85            FORMAT(2X, 'A=', F13.4, ' B=', F13.4, ' C=', F13.4)
90            FORMAT(2X, 'TB=', F13.4, ' MRLMAX=', F13.4)
              STOP
              END
```

APPENDIX 8G
COMPUTER PROGRAM OF "MRL-EMP2.FOR"

```
C
C $$$$$$$$$$$$$$$$$$$$$$$$$$$$$$$$$$$$$$$$$$$$$$$$$$$$$$
C $ COMPUTATION OF THE EMPIRICAL MEAN RESIDUAL LIFE    $
C $ FOR CENSORED, UNGROUPED BURN-IN TESTING DATA SET   $
C $ AND OPTIMUM BURN-IN TIME FOR MAXIMUM MRL VALUE     $
C $$$$$$$$$$$$$$$$$$$$$$$$$$$$$$$$$$$$$$$$$$$$$$$$$$$$$$
C
      IMPLICIT DOUBLE PRECISION (A-H, O-Z)
      DIMENSION NF(50)
      REAL T(50), MRL(50), MRL0, MRLMAX, M1, M2, M3
      OPEN(5,FILE='IMRL.DAT',STATUS='UNKNOWN')
      OPEN(6,FILE='OMRL.DAT',STATUS='NEW')
C
C  EMPIRICAL MEAN RESIDUAL LIFE CALCULATION
C
      WRITE(*,*) 'ENTER THE TOTAL NUMBER OF COMPONENTS
     1 IN THE TEST, NT, PLEASE!'
      WRITE(6,*) 'ENTER THE TOTAL NUMBER OF COMPONENTS
     1 IN THE TEST, NT, PLEASE!'
      READ(*,*) NT
      WRITE(6,*) 'NT=', NT
      WRITE(*,*) 'ENTER THE TOTAL NUMBER OF INTERVALS,
     1 N, PLEASE!'
      WRITE(6,*) 'ENTER THE TOTAL NUMBER OF INTERVALS,
     1 N, PLEASE!'
      READ(*,*) N
      WRITE(6,*) 'N=', N
      WRITE(*,*) 'ENTER THE TEST TERMINATION TIME,
     1 TD, PLEASE!'
      WRITE(6,*) 'ENTER THE TEST TERMINATION TIME,
     1 TD, PLEASE!'
      READ(*,*) TD
      WRITE(6,*) 'TD=', TD
      DO 10 I=1,N+1
      READ(5,*) T(I)
      WRITE(*,*) 'T(',I,')=', T(I)
      WRITE(6,*) 'T(',I,')=', T(I)
10    CONTINUE
      NFT=0
      DO 20 I=1,N
```

```
              READ(5,*) NF(I)
              WRITE(*,*) 'NF(',I,')=', NF(I)
              WRITE(6,*) 'NF(',I,')=', NF(I)
              NFT=NFT+NF(I)
     20       CONTINUE
              NS=NT-NFT
              WRITE(*,*) 'TOTAL # OF FAILURES, NFT=', NFT
              WRITE(6,*) 'TOTAL # OF FAILURES, NFT=', NFT
              WRITE(*,*) 'TOTAL # OF SURVIVORS, NS=', NS
              WRITE(6,*) 'TOTAL # OF SURVIVORS, NS=', NS
              WRITE(6,*) '=================================='
              WRITE(6,*) '    EMPIRICAL MEAN RESIDUAL LIFE'
              WRITE(6,*) '=================================='
              IF(T(1).EQ.0.0)GOTO 25
              X=0.0
              NK=0
              DO 22 K=1,N
              X=X+NF(K)*(0.5*(T(K)+T(K+1)))
              NK=NK+NF(K)
     22       CONTINUE
              ANK=NK
              MRLO=(X+NS*TD)/ANK
              WRITE(*,50) 0, 0.0, 0, MRLO
              WRITE(6,50) 0, 0.0, 0, MRLO
     25       DO 40 J=1,N
              X=0.0
              NK=0
              DO 30 K=J,N
              X=X+NF(K)*(0.5*(T(K)+T(K+1))-T(J))
              NK=NK+NF(K)
     30       CONTINUE
              ANK=NK
              MRL(J)=(X+NS*(TD-T(J)))/ANK
              WRITE(*,50) J, T(J), J, MRL(J)
              WRITE(6,50) J, T(J), J, MRL(J)
     40       CONTINUE
     50       FORMAT(2X, 'T(', I4, ')= ', F13.4,
             1 ' MRL(', I4, ')= ', F13.4)
      C
      C  OPTIMUM BURN-IN TIME FOR THE MAXIMUM MRL
      C  USING QUADRATIC FUNCTION APPROXIMATION
      C  FOR MRL FUNCTION AROUND MAXIMUM
      C  EMPIRICAL MRL POINT
      C
```

APPENDIX 8G

```
          IF(T(1).GT.0.0) GOTO 60
          IMAX=1
          MRLMAX=MRL(1)
          GOTO 70
60        IMAX=0
          MRLMAX=MRL0
70        DO 80 I=1,N-1
          IF(MRL(I).LT.MRLMAX) GOTO 80
          IMAX=I
          MRLMAX=MRL(I)
80        CONTINUE
          I1=IMAX-1
          I2=IMAX
          I3=IMAX+1
          T1=T(I1)
          T2=T(I2)
          T3=T(I3)
          M1=MRL(I1)
          M2=MRL(I2)
          M3=MRL(I3)
          D=T1*T2*(T1-T2)+T2*T3*(T2-T3)+T3*T1*(T3-T1)
          DA=M1*(T2-T3)+M2*(T3-T1)+M3*(T1-T2)
          DB=T1*T1*(M2-M3)+T2*T2*(M3-M1)+T3*T3*(M1-M2)
          DC=T1*T2*M3*(T1-T2)+T2*T3*M1*(T2-T3)+T3*T1*M2*(T3-T1)
          A=DA/D
          B=DB/D
          C=DC/D
          TB=-B/(2.0*A)
          MRLMAX=A*TB*TB + B*TB + C
          WRITE(*,85) A, B, C
          WRITE(6,85) A, B, C
          WRITE(*,90) TB, MRLMAX
          WRITE(6,90) TB, MRLMAX
85        FORMAT(2X, 'A=', F13.4, ' B=', F13.4, ' C=', F13.4)
90        FORMAT(2X, 'TB=', F13.4, ' MRLMAX=', F13.4)
          STOP
          END
```

Chapter 9

BURN-IN TIME DETERMINATION FOR THE MINIMUM COST

In the preceding chapters the burn-in time was determined from the point of view of a good leveling off the failure rate signifying the end of the burn-in time (*Quick Calculation Approach*), of getting a Weibull shape parameter of 1 (*Subpopulation Truncation Approach*), of meeting a specified *failure rate goal*, of meeting a specified *reliability goal*, and of meeting a specified *MRL goal* or for the maximum *MRL*. These approaches focus mainly on maximizing the essential physical performance requirements, but ignore perhaps the most important consideration of all: cost. Very rarely it is true that "money is no object" in the business world. In the few cases where this may be true and where performance is all important, then and only then may we be justified in using a decision model containing only physical performance criteria. With a fixed amount of funds to spend, some constraint must be placed on performance. Hence, the situation ultimately comes down to fulfilling mission or performance requirements in the most economic manner possible.

If the burn-in period is chosen too short some substandard quality components will still be left. If these components fail after being put into operation they will, in some cases, result in very high costs. On the other hand, if the burn-in period is too long the burn-in cost will be too high. This is demonstrated in a typical example depicted in Fig. 9.1, portraying burn-in costs, field failure costs, and the total cost, as a function of burn-in time.

The burn-in cost curve is composed of two elements; i.e., fixed and

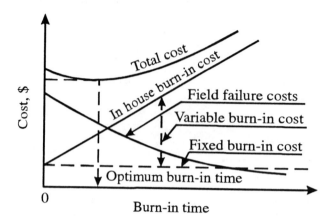

Fig. 9.1– Cost versus burn-in time analysis.

variable costs. Fixed costs include such items as investment in burn-in test facilities, generation of test procedures and software programs. Variable costs include test labor, utilities, and support functions; these costs may vary nearly linearly with burn-in time.

Total cost, the uppermost curve, is the sum of the burn-in cost plus the field failure cost. It will typically have a rather broad and shallow minimum cost region which increases the manufacturer's flexibility in planning the burn-in duration.

Generally, the optimum burn-in is defined as that which will be the most cost effective when considering burn-in cost, field failure savings, and the impact on system reliability. Different cost models can be developed according to different evaluation criteria and cost categories covered. In the following sections, several models are presented and illustrated with numerical examples. The readers are referred to the ADDITIONAL REFERENCES for the other models.

9.1 COST MODEL 1

This model was proposed by Jensen and Peterson 1982 [1, pp. 91-98] which classifies the burn-in costs into the following five categories:

1. *Burn-in constant costs*, abbreviated as *BICC*. This cost may be expressed as a cost per product, which includes the costs of installing the product in the burn-in chambers and removing them

again, and the costs associated with unpacking and repacking of the products.

2. *Burn-in failure costs*, abbreviated as *BIFC*. This cost may be expressed as the cost of repair per failure per product, which includes handling costs (removing and reinstalling the product in the burn-in chamber), and costs of all types of repairs.

3. *Burn-in time-dependent costs*, abbreviated as *BITC*. This cost may be expressed as cost per test equipment per day, which includes three parts. Firstly, the cost associated with the burn-in chamber itself, which may be expressed as a cost per test equipment per day (24 hours). Secondly, the losses caused by production delay due to the number of days or hours the products are on burn-in. And thirdly, the costs associated with product testing and failure monitoring during burn-in.

4. *Customer repair cost*, abbreviated as *CRC*. This cost may be expressed as a cost per failure per product while it is being used by the customer and is under warranty.

5. The costs per failure due to *loss of goodwill*, abbreviated as *LG*. Here only failures occurring during the warranty period are considered, and the cost due to the loss of goodwill is assumed to be a constant per failure. Furthermore, it is assumed that the probability of two failures occurring in any one product during the warranty period is negligible.

Then, the *total cost associated with burn-in*, abbreviated as *TCBI*, is

$$TCBI(T_b) = BICC + FDBI(T_b)\,BIFC + T_b\,BITC \\ + FABI(T_b)\,(CRC + LG), \tag{9.1}$$

where

$$T_b = \text{number of days of burn-in},$$
$$FDBI(T_b) = \text{number of failures during burn-in},$$

and

$$FABI(T_b) = \text{number of failures after burn-in}.$$

The *total cost* if the burn-in is entirely omitted, or the total warranty costs, per product, abbreviated as *TC*, is

$$TC = FWBI\,(CRC + LG), \tag{9.2}$$

where

$FWBI$ = number of failures without burn-in, or the number of failures in the hands of the customer during the warranty period.

If the cost difference $[TCBI(T_b) - TC]$ is denoted by C, C will be the function of the independent variable T_b. Therefore,

$$C(T_b) = BICC + FDBI(T_b)BIFC + T_b BITC$$
$$+ [FABI(T_b) - FWBI](CRC + LG). \qquad (9.3)$$

The optimum burn-in time T_b^* should satisfy the following conditions:

$$\begin{cases} C(T_b^*) & < 0, \\ \left.\frac{dC(T_b)}{dT_b}\right|_{T_b=T_b^*} & = 0, \\ \left.\frac{d^2C(T_b)}{dT_b^2}\right|_{T_b=T_b^*} & > 0. \end{cases} \qquad (9.4)$$

Equation (9.4) is the basic requirement that the total cost with burn-in be less that the total cost without burn-in.

In Eq. (9.3), $BICC$, $BIFC$, $BITC$, CRC and LG are constants and can be evaluated easily from past experience, but $FDBI(T_b)$ and $FABI(T_b)$ are not constants and are dependent on burn-in time duration, operating stress and environment. However, it can be reasonably assumed that the substandard components which do not fail during burn-in will fail in the consumer's hand within the warranty period. In other words, $FWBI$ is equal to the sum of $FDBI$ and $FABI$, or

$$FWBI = FDBI(T_b) + FABI(T_b). \qquad (9.5)$$

Due to the difference in the nature of the environment nature between burn-in and field use, it is often difficult to estimate the values of $FDBI(T_b)$ and $FABI(T_b)$. Jensen and Peterson [1] provided a good approximation method to compute these three quantities using the curve of the times-to-the-first-failure. It is summarized as follows:

If a burn-in test lasts long enough to show the leveling-off of the early failure rate pattern, or to show an "S" shaped trend on the Weibull curve of the times-to-the-first-failure, denote the time period for this experiment as U_{max}. Then, the mean values of $FDBI(T_b)$ and $FWBI$ can be calculated using the following equations which are based on the Poisson process:

$$F(T_b) = 1 - e^{-\int_0^{T_b} \lambda(T)dT} = 1 - e^{-FDBI(T_b)},$$

TABLE 9.1– Burn-in test results for Example 9–1.

Burn-in time, T_b, days	1	2	3	4	5	6	7
Percent failed, $F(T_b)$	0.060	0.101	0.120	0.138	0.145	0.150	0.151

or

$$FDBI(T_b) = -\log_e[1 - F(T_b)], \qquad (9.6)$$

and

$$FWBI = -\log_e[1 - F(U_{max})]. \qquad (9.7)$$

Substituting Eqs. (9.6) and (9.7) into Eq. (9.5) yields

$$FABI(T_b) = FWBI - FDBI(T_b) = \log_e\left[\frac{1 - F(T_b)}{1 - F(U_{max})}\right]. \qquad (9.8)$$

Following is an example which shows how this model works to determine the cost-optimized burn-in time.

EXAMPLE 9–1

A burn-in test ends with the results listed in Table 9.1. The associated cost parameters have been provided to be the following:

1. Burn-in constant costs, $BICC$ =$10/product being burned-in.

2. Burn-in failure costs, $BIFC$ =$15/failure/product being burned-in.

3. Time-dependent costs, $BITC$ =$3/product being burned-in/day.

4. Customer repair costs, CRC =$100/failure/product being burned-in.

5. Loss of goodwill, LG =$100/failure/product being burned-in.

Determine the cost-optimized burn-in time T_b^*.

TABLE 9.2– $FDBI$ and $FABI$ corresponding to different burn-in time durations, for Example 9–1.

T_b, days	$FDBI(T_b)$ $= -\log_e[1 - F(T_b)]$	$FABI(T_b)$ $= FWBI - FDBI(T_b)$
1	0.062	0.102
2	0.106	0.058
3	0.128	0.036
4	0.149	0.015
5	0.157	0.007
6	0.163	0.001
7	0.164	0.001

SOLUTION TO EXAMPLE 9–1

Another table, Table 9.2, is prepared to calculate the values of $FDBI(T_b)$ and $FABI(T_b)$. Here, from the predicted *cdf* curve of the times-to-failure Weibull graph, it would seem reasonable to set $U_{max} = 7$ days. Then,

$$FWBI = -\log_e[1 - F(7)] = -\log_e[1 - 0.151] = 0.164.$$

By substituting the given cost parameters and the information given in Table 9.1, the total cost associated with burn-in, $TCBI(T_b)$, the total warranty cost, TC, and their difference, $C(T_b)$, are obtained as listed in Table 9.3.

A sample calculation is given next for evaluating $FDBI(T_b)$, $FABI(T_b)$, $TCBI(T_b)$, TC and finally $C(T_b)$ when $T_b = 3$ days.

When $T_b = 3$ days, according to Table 9.1 the observed percentage of failures is

$$F(T_b) = F(3) = 0.120.$$

Then, from Eq. (9.6),

$$\begin{aligned} FDBI(T_b) &= FDBI(3), \\ &= -\log_e[1 - F(3)], \\ &= -\log_e(1 - 0.120), \end{aligned}$$

or

$$FDBI(3) = 0.128.$$

COST MODEL 1

TABLE 9.3– Total burn-in cost as a function of burn-in time, for Example 9–1.

T_b, days	$TCBI(T_b)$, $/product	TC, $/product	$C(T_b) = TCBI(T_b) - TC$, $/product
1	34	33	1
2	29	33	-4
3	28	33	-5
4	27	33	-6*
5	29	33	-4
6	31	33	-2
7	33	33	0

From Eq. (9.8),

$$FABI(T_b) = FABI(3),$$
$$= FWBI - FDBI(3),$$
$$= 0.164 - 0.128,$$

or

$$FABI(3) = 0.036.$$

Now substituting $T_b = 3$, $BICC = 10$, $FDBI(3) = 0.128$, $BIFC = 15$, $BITC = 3$, $FABI(3) = 0.036$, $CRC = 100$ and $LG = 100$ into Eq. (9.1) yields

$$TCBI(3) = 10 + (0.128)\, 15 + 3 \times 3 + 0.036\,(100 + 100),$$

or

$$TCBI(3) = \$28.12/\text{product, or } \$28/\text{product}.$$

Substituting $FWBI = 0.164$, $CRC = 100$ and $LG = 100$ into Eq. (9.2) yields

$$TC = 0.164\,(100 + 100) = \$32.8/\text{product, or } \$33/\text{product}.$$

Therefore, the difference between the total cost with burn-in and the total cost without burn-in for $T_b = 3$ hr is given by Eq. (9.3), or

$$C(3) = TCBI(3) - TC = 28 - 33 = -\$5/\text{product}.$$

From Table 9.3, it may be seen that burn-in is cost effective ($TCBI(T_b) < TC$) for all burn-in times between 2 and 6 days inclusive, with an optimum burn-in time at 4 days. This behavior is illustrated in Fig. 9.2.

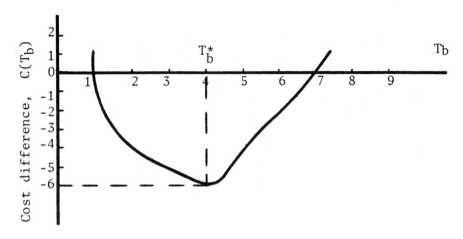

Fig. 9.2– The cost difference, $C(T_b)$, versus burn-in time, T_b, for Example 9–1.

9.2 COST MODEL 2

This model was proposed by Plesser and Field in 1977 [2] which permits determining the duration of cost-optimized burn-in and evaluating the resultant saving for repairable electronic systems. It is assumed that infant mortality failures occur according to a nonhomogeneous Poisson process with a decreasing failure rate, and repair actions restore the system to the failure rate it had just before failure. The expected cost associated with factory and field failures are traded off with the costs of implementing a burn-in program under the assumption that devices will be operated for a fixed useful life after burn-in ceases, regardless of the burn-in duration.

The expected value of the cost of operating a unit with N_b failures during burn-in and N_d failures during deployment is

$$C(T_b) = C_0 + C_b T_b + C_{rb} E\{N_b\} + C_{rd} E\{N_d\}, \qquad (9.9)$$

where

C_0 = fixed cost of burn-in per unit, $C_0 = 0$ if $T_b = 0$,
C_b = cost per unit per hour of burn-in,
C_{rb} = repair cost for a unit which fails during burn-in,
C_{rd} = repair cost for a unit which fails during deployment,

and

T_b = burn-in time duration.

COST MODEL 2

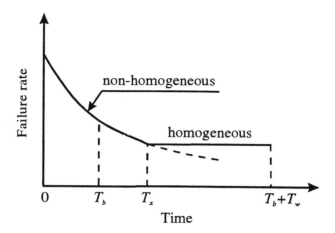

Fig. 9.3– Failure rate under a nonhomogeneous Poisson process in region $[0, T_x)$ and under a homogeneous Poisson process in region $[T_x, (T_b + T_w)]$.

Consider a class of devices in which the failure rate is monotonically decreasing until a time T_x, after which time the failure rate remains constant. Subsequent to burn-in, the device is operated for time T_w and is then retired. The failure rate of such devices is depicted in Fig. 9.3.

Since the expected number of failures in an interval is equal to the integral of the failure rate over that interval, Eq. (9.9) can be written as

$$C(T_b) = C_0 + C_b T_b + C_{rb} \int_0^{T_b} \lambda(T)\, dT + C_{rd} \int_{T_b}^{T_b+T_w} \lambda(T)\, dT.$$

(9.10)

Note that

$$\int_{T_b}^{T_b+T_w} \lambda(T)\, dT = \int_{T_b}^{T_x} \lambda(T)\, dT + \int_{T_x}^{T_b+T_w} \lambda(T)\, dT,$$

and

$$\lambda(T) = \lambda(T_x) = \text{constant during } [T_x, T_b + T_w].$$

Then,

$$\int_{T_x}^{T_b+T_w} \lambda(T)\, dT = \lambda(T_x)\left[(T_b + T_w) - T_x\right],$$

and
$$\int_{T_b}^{T_b+T_w} \lambda(T)\,dT = \int_{T_b}^{T_x} \lambda(T)\,dT + \lambda(T_x)[(T_b+T_w)-T_x]. \quad (9.11)$$

Substituting Eq. (9.11) into Eq. (9.10) yields
$$C(T_b) = C_0 + C_b T_b + C_{rb}\int_0^{T_b}\lambda(T)\,dT + C_{rd}\int_{T_b}^{T_x}\lambda(T)\,dT$$
$$+C_{rd}\lambda(T_x)[(T_b+T_w)-T_x]. \quad (9.12)$$

The optimum T_b corresponding to the minimum $C(T_b)$ will occur when
$$\frac{\partial C(T_b)}{\partial T_b} = 0,$$

or
$$0 + C_b + C_{rb}\frac{\partial}{\partial T_b}\left[\int_0^{T_b}\lambda(T)\,dT\right]$$
$$+C_{rd}\frac{\partial}{\partial T_b}\left[\int_{T_b}^{T_x}\lambda(T)\,dT\right] + C_{rd}\lambda(T_x) = 0.$$

Now,
$$\frac{\partial}{\partial T_b}\left[\int_0^{T_b}\lambda(T)\,dT\right] = \lambda(T_b),$$

and
$$\frac{\partial}{\partial T_b}\left[\int_{T_b}^{T_x}\lambda(T)\,dT\right] = -\lambda(T_b).$$

Then,
$$\frac{\partial C(T_b)}{\partial T_b} = C_b + (C_{rb}-C_{rd})\lambda(T_b) + C_{rd}\lambda(T_x) = 0,$$

or
$$\lambda(T_b) = [C_b + C_{rd}\lambda(T_x)]/(C_{rd}-C_{rb}), \quad (9.13)$$

and
$$\frac{\partial^2 C(T_b)}{\partial T_b^2} = (C_{rb}-C_{rd})\frac{\partial\lambda(T_b)}{\partial T_b} > 0. \quad (9.14)$$

COST MODEL 2

TABLE 9.4– Times-to-failure data for the countermeasure receiver of Example 9-2.

	Receiver group number								
	# 1	# 2	# 3	# 4	# 5	# 6	# 7	# 8	#9
Times	41	7	10	170	6	35	4	192	64
to	101	17	147			49			75
failure,	102	85	225						150
hr	175								

For each device that is burned-in until time T_b^*, the expected saving over "no burn-in whatsoever" is independent of T_w, and is

$$S(T_b^*) = [C_{rd} - C_{rb}] \left[\int_0^{T_b^*} \lambda(T)\, dT - T_b^* \lambda(T_b^*) \right] - C_0. \qquad (9.15)$$

Beyond the optimum burn-in point, a positive saving is realized up to a burn-in time T_0 such that

$$\lambda(T_b^*)\, T_0 = \int_0^{T_0} \lambda(T)\, dT, \qquad (9.16)$$

which is independent of T_w. Following is an example illustrating the application of this model.

EXAMPLE 9-2

Consider a group of nine (9) i.i.d. countermeasure receivers whose behavior was observed, with the times-to-failure data given in Table 9.4, during the first 225 hr of life.

Assume that the expected number of failures per component in the interval $(0, T)$ results from a nonhomogeneous Poisson process (with a nonconstant failure rate) having a mean value function of the form

$$m(T) = \int_0^T \lambda(x)\, dx = T^\beta / \eta, \qquad 0 < T < T_x, \qquad (9.17)$$

where η and β are positive constants.

The failure rate will then satisfy the decreasing assumption for $0 < \beta < 1$ and $\eta > 0$, and will be of the form of the Weibull failure rate, or

$$\lambda(T) = \begin{cases} (\beta/\eta) T^{\beta-1}, & 0 < T < T_x, \\ (\beta/\eta) T_x^{\beta-1}, & T_x < T < T_b + T_w. \end{cases} \qquad (9.18)$$

Sufficient field data have been accumulated to show a mature stable $MTBF$ of 200 hr for these countermeasure receivers. The associated cost values are given as C_{rb}=\$500, C_{rd}=\$5,000, C_0=\$200 and C_b=\$18/hr. Determine the cost-optimized burn-in time, using the Cost Model 2.

SOLUTION TO EXAMPLE 9-2

Estimates of η and β can be determined using the least-squares method by rewriting Eq. (9.17) in the form

$$\log_e m(T) = \beta \log_e T - \log_e \eta. \tag{9.19}$$

Let

$$X = \log_e T,$$
$$Y = \log_e m(T),$$
$$a = -\log_e \eta, \tag{9.20}$$

and

$$b = \beta. \tag{9.21}$$

Then, Eq. (9.19) can be rewritten as

$$Y = a + bX.$$

The least-squares estimators for a and b are given by

$$\begin{cases} \hat{b} = \frac{L_{XY}}{L_{XX}}, \\ \hat{a} = \overline{Y} - \hat{b}\overline{X}, \end{cases} \tag{9.22}$$

with a correlation coefficient of

$$\rho = \frac{L_{XY}}{\sqrt{L_{XX} L_{YY}}}, \tag{9.23}$$

where

$$L_{XY} = \sum_{i=1}^{n}(X_i - \overline{X})(Y_i - \overline{Y}),$$

$$L_{XX} = \sum_{i=1}^{n}(X_i - \overline{X})^2,$$

$$L_{YY} = \sum_{i=1}^{n}(Y_i - \overline{Y})^2,$$

$$X_i = \log_e T_i,$$

COST MODEL 2

$$Y_i = \log_e m(T_i),$$

$$\overline{X} = \frac{1}{n}\sum_{i=1}^{n} X_i,$$

$$\overline{Y} = \frac{1}{n}\sum_{i=1}^{n} Y_i,$$

and

$n =$ sample size.

Table 9.5 is prepared for the estimation of a and b based on $n = 19$ data points given in Table 9.4. Substituting the (X_i)'s and (Y_i)'s obtained in Table 9.5 into Eqs. (9.22) and (9.23) yields

$$\begin{cases} \hat{a} = -2.543, \\ \hat{b} = 0.616, \end{cases}$$

and

$$\rho = 0.979.$$

Then, from Eqs. (9.20) and (9.21)

$$\hat{\eta} = e^{-\hat{a}} = e^{2.543} = 12.718 \text{ hr},$$

and

$$\hat{\beta} = \hat{b} = 0.616.$$

Since sufficient field data have also been accumulated to show a mature stable $MTBF$ of 200 hr, the value of the constant T_x can be determined by substituting the inverse of this steady state $MTBF$ as $\lambda(T_x)$ into Eq. (9.18) and solving for $T = T_x$. This procedure yields

$$\frac{1}{\lambda(T_x)} = \frac{\hat{\eta}}{\hat{\beta}}(T_x)^{1-\hat{\beta}} = 200,$$

or

$$\frac{12.718}{0.616}T_x^{1-0.616} = 200,$$

or

$$T_x = 370.005 \text{ hr}.$$

TABLE 9.5– Calculations for the least-squares parameter estimates of η and β, for Example 9–2.

Failure order, i	Failure time, T_i, hr	$X_i = \log_e T_i$	Observed number of failures in $[0, T_i]$, $m(T_i) = i$	$Y_i = \log_e m(T_i)$
1	4	1.386	1	0.000
2	6	1.792	2	0.693
3	7	1.946	3	1.099
4	10	2.303	4	1.386
5	17	2.833	5	1.609
6	35	3.555	6	1.792
7	41	3.714	7	1.946
8	49	3.892	8	2.079
9	64	4.159	9	2.197
10	75	4.317	10	2.303
11	85	4.443	11	2.398
12	101	4.615	12	2.485
13	102	4.625	13	2.565
14	147	4.990	14	2.639
15	150	5.011	15	2.708
16	170	5.136	16	2.773
17	175	5.165	17	2.833
18	192	5.257	18	2.890
19	225	5.416	19	2.944
		$\overline{X} = 3.924$		$\overline{Y} = 2.071$

Substituting $\lambda(T_b)$ and $\lambda(T_x)$ given by Eq. (9.18) into Eq. (9.13), and solving for T_b^*, yields

$$\frac{\hat{\beta}}{\hat{\eta}}(T_b)^{\hat{\beta}-1} = \left[C_b + C_{rd}\frac{\hat{\beta}}{\hat{\eta}}(T_x)^{\hat{\beta}-1}\right]\Big/(C_{rd} - C_{rb}).$$

Multiplying both sides by $\left(\frac{\hat{\eta}}{\hat{\beta}}\right)$ yields

$$(T_b)^{\hat{\beta}-1} = \left[\frac{C_b\,\hat{\eta}}{(C_{rd} - C_{rb})\hat{\beta}} + \frac{C_{rd}}{C_{rd} - C_{rb}}T_x^{\hat{\beta}-1}\right].$$

Then,

$$T_b^* = \left[\frac{C_b\,\hat{\eta}}{(C_{rd} - C_{rb})\hat{\beta}} + \frac{C_{rd}}{(C_{rd} - C_{rb})}T_x^{\hat{\beta}-1}\right]^{\frac{1}{\hat{\beta}-1}}. \qquad (9.24)$$

Substituting Eqs. (9.18) and (9.17) into Eq. (9.15) yields

$$S(T_b^*) = (C_{rd} - C_{rb})\left\{(T_b^*)^{\hat{\beta}}\Big/\hat{\eta} - T_b^*\left[(\hat{\beta}/\hat{\eta})(T_b^*)^{\hat{\beta}-1}\right]\right\} - C_0,$$

or

$$S(T_b^*) = \left(\frac{C_{rd} - C_{rb}}{\hat{\eta}}\right)(1 - \hat{\beta})(T_b^*)^{\hat{\beta}} - C_0. \qquad (9.25)$$

Substituting Eqs. (9.18) and (9.17) into Eq. (9.16) yields

$$(\hat{\beta}/\hat{\eta})(T_b^*)^{\hat{\beta}-1}T_0 = \int_0^{T_0}(\hat{\beta}/\hat{\eta})\,T^{\hat{\beta}-1}\,dT = (T_0)^{\hat{\beta}}\Big/\hat{\eta},$$

or

$$(T_0)^{\hat{\beta}-1} = \hat{\beta}\,(T_b^*)^{\hat{\beta}-1}.$$

Then,

$$T_0 = \hat{\beta}^{\left(\frac{1}{\hat{\beta}-1}\right)}T_b^*. \qquad (9.26)$$

Substituting the given cost values, the least-squares estimates of the Poisson process parameters, and the T_x value found before, into Eq. (9.24) yields the cost-optimized burn-in duration as

$$T_b^* = \left[\frac{18\,(12.718)}{(5,000 - 500)\,0.616} + \frac{5,000}{(5,000 - 500)}(370.005)^{0.616-1}\right]^{\left(\frac{1}{0.616-1}\right)},$$

or

$$T_b^* = 68.500 \text{ hr, or } 69 \text{ hr.}$$

Substituting this T_b^*, the given cost values and the least-squares estimates of the Poisson process parameters into Eqs. (9.25) and (9.26) yields the net saving as

$$S(T_b^*) = S(68.500),$$
$$= \left(\frac{5,000 - 500}{12.718}\right)(1 - 0.616)(68.500)^{0.616} - 200,$$

or

$$S(T_b^*) = \$1,636.159/\text{product},$$

and the time T_0 as

$$T_0 = 0.616^{\left(\frac{1}{0.616-1}\right)}(68.500) = 241.912 \text{ hr,}$$

or

$$T_0 = 242 \text{ hr.}$$

This information says that the burn-in time can be longer than 68.5 hr, which is the economically optimum with a net saving of \$1,636/product. If 68.5 hr are not long enough to meet the specified reliability, failure rate or $MTBF$ goals, we still save money even if the burn-in time is as long as 242 hr.

9.3 COST MODEL 3

This model is a modified version of the cost model proposed by Washburn in 1989 [3] which is capable of determining the optimum burn-in time by combining the elements of physical performance with the economic facts of real life. The generalized gamma (GG) distribution, which includes, as special cases, such distributions as the normal, Rayleigh, Maxwell, χ^2, Weibull, exponential, ordinary gamma, etc., is used as the life model. The costs considered here are those due to burn-in operation, production and sales. The total cost model is

$$C(T_b) = C_1 T_b + C_2 N_b F(T_b) + C_3 N_m, \qquad (9.27)$$

COST MODEL 3

where

C_1 = operational cost of the burn-in facility per hour of burn-in time, which includes capital investment, interest, overhead maintenance, all related costs for the physical plant and equipment, the cost of all associated labor (both direct and indirect), and a reasonable rate of return on investment,

C_2 = cost per unit that failed in the burn-in process, which includes all the associated costs of producing a unit and placing it in the burn-in facility,

N_b = minimum number of units to be placed on burn-in such that the customer's requirements will be satisfied,

and

$F(T_b)$ = cumulative probability of failure of the unit up to burn-in time T_b,

$C_2 \, N_b \, F(T_b)$ = total cost of the units failed in the burn-in process,

C_3 = sale price of the units that survive the burn-in process,

N_m = required number of units to satisfy mission requirements with the given level of the unit's prevailing system effectiveness, or

$$N_m = K_3/P_E(T_b), \tag{9.28}$$

K_3 = minimum number of units required by the customer to satisfy his requirements,

and

$P_E(T_b)$ = system effectiveness of each burned-in unit.

Note that the number of serviceable units required is $N_m = K_3/P_E(T_b)$, and the number of unserviceable units (units failed in the burn-in process) is $N_b \, F(T_b)$. The number of units placed on burn-in, N_b, includes both the failed and unfailed units; i.e.,

$$N_b = K_3/P_E(T_b) + N_b \, F(T_b),$$

$$N_b - N_b F(T_b) = K_3/P_E(T_b),$$
$$N_b [1 - F(T_b)] = K_3/P_E(T_b),$$
$$N_b R(T_b) = K_3/P_E(T_b),$$

or

$$N_b = K_3/[P_E(T_b) R(T_b)], \qquad (9.29)$$

which is the minimum number of units to be placed on burn-in to fulfil the customer's needs, because K_3 is the minimum number of units required by the customer to satisfy his requirements.

Substituting Eqs. (9.28) and (9.29) into Eq. (9.27) yields

$$C(T_b) = C_1 T_b + C_2 \left[\frac{K_3}{P_E(T_b) R(T_b)}\right] F(T_b) + C_3 \left[\frac{K_3}{P_E(T_b)}\right]. \qquad (9.30)$$

It may be worth noting that though the costs included in C_2 are quite similar to those included in C_1, C_2 does not include a profit; but it does include material costs. C_3 is essentially the same as C_2, with the additional cost of any handling, testing, and shipping associated with delivering the end product to the customer. Of course, an additional cost to provide for a reasonable profit is also included in C_3.

Meanwhile, it should be pointed out that the burn-in process affects system effectiveness, $P_E(T_b)$, only in two distinct ways. First, it considers how the mission reliability is influenced by population truncation (infant mortality deletion) due to burn-in. Second, it takes into consideration the improvement in the expected time to failure from the untruncated to the truncated population. This improvement increases system availability and in turn the system's operational readiness. For the purpose of this investigation, $P_E(T_b)$ is composed only of the product's mission reliability and operational readiness. Since the design adequacy is not influenced by burn-in, it is considered to be unity.

Therefore, the system effectiveness of each unit is given by

$$P_E(T_b) = R(T_m|T_b) \cdot P_{OR}(T_b),$$
$$= R(T_m|T_b) \cdot [A(T_b)(1 - P_s) + K_1 P_s], \qquad (9.31)$$

where

T_m = mission duration, hours,
$R(T_m|T_b)$ = mission reliability; i.e., probability of a unit, which has been burned-in for T_b hours, surviving the mission of duration T_m,
$P_{OR}(T_b)$ = unit's operational readiness,
P_s = conditional probability that the unit is in nonuse,

or in storage, at any time point, provided that it
was operational when last used or placed in
storage,

$1 - P_s$ = probability that the unit is either in operation
or undergoing maintenance at any point in time,

$A(T_b)$ = unit's availability,

$$A(T_b) = \frac{MTBF(T_b)/K_2}{MTBF(T_b)/K_2 + MTTR},$$

or

$$A(T_b) = \{1 + [K_2 \cdot MTTR/MTTF(T_b)]\}^{-1}, \tag{9.32}$$

K_2 = unit's use coefficient, $0 < K_2 \leq 1$,

K_1 = probability of unit's survival during storage or nonuse,

$MTTR$ = mean time to repair the unit,

and

$MTTF(T_b)$ = mean time to failure of the units that survive the burn-in process lasting T_b hours.

The optimum burn-in time corresponding to the minimum $C(T_b)$ will exist when

$$\frac{dC(T_b)}{dT_b} = 0, \tag{9.33}$$

and

$$\frac{d^2C(T_b)}{dT_b^2} > 0. \tag{9.34}$$

Because of the complicated nature of the first and second derivatives of $P_E(T_b)$ with respect to T_b, a reasonable way is to solve Eqs. (9.33) and (9.34) graphically. But this is more tedious and less rewarding than graphing Eq. (9.30) directly and finding its minimum. The graph of Eq. (9.30) would reveal all of the relative extrema, inflection points, and other critical points that exist in the domain of T_b exercised. Hence, it can be seen at a glance that the selection of the minimum point depends not only on where it lies, but also on how critical it is.

The generalized gamma distribution is used here. The primary failure probability density function $f(T)$ is defined as follows:

$$f(T) = \begin{cases} \frac{\beta}{\eta}(\frac{T-\gamma}{\eta})^{\alpha-1} e^{-(\frac{T-\gamma}{\eta})^{\beta}} / \Gamma_{\infty}(\alpha/\beta), & \gamma < T < \infty, \\ 0, & \text{otherwise,} \end{cases} \quad (9.35)$$

where

β = shape parameter,
α = auxiliary shape parameter,
η = scale parameter,
γ = location parameter,

and

$$\Gamma_{\infty}(\alpha/\beta) = \int_0^{\infty} x^{\frac{\alpha}{\beta}-1} e^{-x} dx.$$

It may be seen that when α is equal to β, Eq. (9.35) yields the Weibull distribution; when β is equal to 1, Eq. (9.35) yields the gamma distribution.

The cumulative probability of failure of the unit up to burn-in time T_b is then given by

$$F(T_b) = 1 - R(T_b),$$
$$= 1 - \int_{T_b}^{\infty} f(T) dT,$$

or

$$F(T_b) = 1 - \left[\int_{T_b}^{\infty} \frac{\beta}{\eta} (\frac{T-\gamma}{\eta})^{\alpha-1} e^{-(\frac{T-\gamma}{\eta})^{\beta}} dT \right] / \Gamma_{\infty}(\alpha/\beta), \quad (9.36)$$

for $T_b > \gamma$. Let

$$X = (\frac{T-\gamma}{\eta})^{\beta},$$

or

$$\frac{T-\gamma}{\eta} = X^{\frac{1}{\beta}}. \quad (9.37)$$

Then,

$$dX = \frac{\beta}{\eta}(\frac{T-\gamma}{\eta})^{\beta-1}dT,$$

$$= \frac{\beta}{\eta}X^{\frac{\beta-1}{\beta}}dT,$$

or

$$dT = \frac{\eta}{\beta}X^{\frac{1}{\beta}-1}dX. \tag{9.38}$$

Substituting Eqs. (9.37) and (9.38) into Eq. (9.36) yields

$$F(T_b) = 1 - \left[\int_{(\frac{T_b-\gamma}{\beta})^\beta}^{\infty} \left(\frac{\beta}{\eta}X^{\frac{\alpha-1}{\beta}}e^{-X}\right)\left(\frac{\eta}{\beta}X^{\frac{1}{\beta}-1}dX\right)\right]/\Gamma_\infty(\alpha/\beta),$$

$$= 1 - \left[\int_{(\frac{T_b-\gamma}{\beta})^\beta}^{\infty} X^{\frac{\alpha}{\beta}-1}e^{-X}dX\right]/\Gamma_\infty(\alpha/\beta),$$

or

$$F(T_b) = 1 - \Gamma(\frac{\alpha}{\beta})\Big|_{X_b}^{\infty}/\Gamma_\infty(\alpha/\beta), \tag{9.39}$$

where

$$\Gamma(\frac{\alpha}{\beta})\Big|_{X_b}^{\infty} = \int_{X_b}^{\infty} T^{\frac{\alpha}{\beta}-1}e^{-T}dT,$$

and

$$X_b = (\frac{T_b-\gamma}{\eta})^\beta.$$

Therefore,

$$R(T_b) = 1 - F(T_b) = \Gamma(\frac{\alpha}{\beta})\Big|_{X_b}^{\infty}/\Gamma_\infty(\alpha/\beta). \tag{9.40}$$

The system effectiveness, modified for burn-in truncation and unity design adequacy, is then given by substituting Eqs. (9.32) and (9.40) into Eq. (9.31); i.e.,

$$P_E(T_b) = R(T_m|T_b)\,P_{OR}(T_b),$$

$$= \frac{R(T_m+T_b)}{R(T_b)}P_{OR}(T_b),$$

$$= \frac{\Gamma(\frac{\alpha}{\beta})\Big|_{Y}^{\infty}/\Gamma_\infty(\alpha/\beta)}{\Gamma(\frac{\alpha}{\beta})\Big|_{X_b}^{\infty}/\Gamma_\infty(\alpha/\beta)}\left[\frac{1-P_S}{1+\frac{K_2\cdot MTTR}{MTTF(T_b)}}+K_1 P_S\right],$$

or

$$P_E(T_b) = \frac{\Gamma(\frac{\alpha}{\beta})|_Y^\infty}{\Gamma(\frac{\alpha}{\beta})|_{X_b}^\infty} \left[\frac{1-P_S}{1 + \frac{K_2 \cdot MTTR}{MTTF(T_b)}} + K_1 P_S \right], \quad (9.41)$$

where

$$\Gamma(\frac{\alpha}{\beta})|_Y^\infty = \int_Y^\infty T^{\frac{\alpha}{\beta}-1} e^{-T} dT,$$

$$Y = (\frac{T_b + T_m - \gamma}{\eta})^\beta,$$

and

$$MTTF(T_b) = \int_{T_b}^\infty R(T) dT \Big/ R(T_b).$$

Integrating the right side of the $MTTF(T_b)$ equation by parts yields

$$MTTF(T_b) = \left\{ T\ R(T)\big|_{T_b}^\infty - \int_{T_b}^\infty T[-f(T)] dT \right\} \Big/ R(T_b),$$

$$= \left[-T_b R(T_b) + \int_{T_b}^\infty (T - \gamma + \gamma) f(T) dT \right] \Big/ R(T_b),$$

$$= -T_b + \left[\int_{T_b}^\infty (T - \gamma) f(T) dT + \gamma \int_{T_b}^\infty f(T) dT \right] \Big/ R(T_b),$$

$$= -T_b + \left[\int_{T_b}^\infty (T - \gamma) f(T) dT + R(T_b) \gamma \right] \Big/ R(T_b),$$

$$= \frac{\int_{T_b}^\infty (T - \gamma) f(T) dT}{\int_{T_b}^\infty f(T) dT} - T_b + \gamma,$$

$$= \frac{\eta \left[\int_{T_b}^\infty \frac{\beta}{\eta} (\frac{T-\gamma}{\eta})^\alpha e^{-(\frac{T-\gamma}{\eta})^\beta} dT \right] \Big/ \Gamma_\infty(\alpha/\beta)}{\left[\int_{T_b}^\infty \frac{\beta}{\eta} (\frac{T-\gamma}{\eta})^{\alpha-1} e^{-(\frac{T-\gamma}{\eta})^\beta} dT \right] \Big/ \Gamma_\infty(\alpha/\beta)} - T_b + \gamma,$$

or

$$MTTF(T_b) = \eta \left[\frac{\Gamma(\frac{\alpha+1}{\beta})|_{X_b}^\infty}{\Gamma(\frac{\alpha}{\beta})|_{X_b}^\infty} \right] - T_b + \gamma. \quad (9.42)$$

If the unit's life distribution parameters and other associated quantities are given, it will be possible to evaluate the total cost for different burn-in times, and then find the optimum burn-in time corresponding to the minimum cost. A numerical example is given next.

COST MODEL 3

EXAMPLE 9-3

The generalized gamma distribution was fitted to a given set of burn-in testing data with the estimated parameters given as follows:

$$\alpha = 0.6, \beta = 0.2, \gamma = 0, \text{ and } \eta = 50 \text{ hr}.$$

Use the Cost Model 3 to determine the optimum burn-in time, T_b^*, and the minimum number of units to be placed on burn-in, N_b, according to the following quantities:

$$C_1 = \$20/\text{hr}, \quad C_2 = \$1/\text{unit}, \quad C_3 = \$5/\text{unit},$$

$$K_1 = 0.98, \quad K_2 = 0.80, \quad K_3 = 10{,}000 \text{ unit},$$

$$MTTR = 3.5 \text{ hr}, \quad P_S = 0.1 \text{ and } T_m = 300 \text{ hr}.$$

SOLUTION TO EXAMPLE 9-3

Since the distribution parameters, α, β, γ and η are given, then $F(T_b)$ and $MTTF(T_b)$ can be evaluated by Eqs. (9.39) and (9.42) at different values of burn-in time T_b. With the given values of T_m, K_1, K_2, K_3, $MTTR$ and P_S, $P_E(T_b)$ and thereafter $C(T_b)$ can be evaluated by Eqs. (9.41) and (9.30) at different values of burn-in time T_b.

During this process the gamma function, $\Gamma_\infty(\frac{\alpha}{\beta})$, and the incomplete gamma functions, $\Gamma(\frac{\alpha}{\beta})\big|_Y^\infty$, $\Gamma(\frac{\alpha}{\beta})\big|_{X_b}^\infty$ and $\Gamma(\frac{\alpha+1}{\beta})\big|_{X_b}^\infty$, can be either found out by referring to the gamma and the incomplete gamma function tables or evaluated by numerical integration. A sample calculation is given next for evaluating the total cost when $T_b = 30$ hr.

For $T_b = 30$ hr, the total cost is given by Eq. (9.30), or

$$C(30) = 20\,(30) + 1 \left[\frac{10{,}000}{P_E(30)\,R(30)} \right] F(30) + 5 \left[\frac{10{,}000}{P_E(30)} \right], \quad (9.43)$$

which requires the evaluation of $F(30)$, $R(30)$ and $P_E(30)$. Now from Eq. (9.39),

$$F(T_b) = 1 - \Gamma(\frac{\alpha}{\beta})\big|_{X_b}^\infty / \Gamma_\infty(\alpha/\beta),$$

where

$$X_b = \left(\frac{T_b - \gamma}{\eta} \right)^\beta = \left(\frac{30 - 0}{50} \right)^{0.2} = 0.9029,$$

and

$$\alpha/\beta = 0.6/0.2 = 3.$$

Then,
$$F(30) = 1 - \Gamma(3)\Big|_{0.9029}^{\infty} \Big/ \Gamma_{\infty}(3),$$
where
$$\Gamma_{\infty}(3) = (3-1)! = 2,$$
and
$$\Gamma(3)\Big|_{0.9029}^{\infty} = \int_{0.9029}^{\infty} X^{3-1} e^{-X} dX.$$

Note that for any positive integer, n, the incomplete gamma function
$$\Gamma(n)\Big|_{a}^{\infty} = \int_{a}^{\infty} X^{n-1} e^{-X} dX \tag{9.44}$$
can be evaluated directly using the following equation:
$$\Gamma(n)\Big|_{a}^{\infty} = (n-1)! e^{-a} \sum_{k=0}^{n-1} \left(\frac{a^k}{k!}\right) \tag{9.45}$$
which can be derived from Eq. (9.44) using integration by parts $(n-1)$ times. Now for $n = 3$ and $a = 0.9029$,
$$\Gamma(3)\Big|_{0.9029}^{\infty} = (3-1)! e^{-0.9029} \sum_{k=0}^{3-1} \frac{0.9029^k}{k!},$$
$$= 2 e^{-0.9029} \left(1 + 0.9029 + \frac{0.9029^2}{2}\right),$$
or
$$\Gamma(3)\Big|_{0.9029}^{\infty} = 1.8733.$$

Then,
$$F(30) = 1 - 1.8733/2 = 0.0633,$$
and
$$R(30) = 1 - F(30) = 0.9367.$$

From Eq. (9.42),
$$MTTF(30) = 50 \left[\frac{\Gamma\left(\frac{0.6+1}{0.2}\right)\Big|_{0.9029}^{\infty}}{\Gamma\left(\frac{0.6}{0.2}\right)\Big|_{0.9029}^{\infty}}\right] - 30 + 0,$$
$$= 50 \left[\frac{\Gamma(8)\Big|_{0.9029}^{\infty}}{\Gamma(3)\Big|_{0.9029}^{\infty}}\right] - 30.$$

COST MODEL 3

Now $\Gamma(3)\big|_{0.9029}^{\infty} = 1.8733$ and from Eq. (9.45),

$$\Gamma(8)\big|_{0.9029}^{\infty} = (8-1)!\,e^{-0.9029} \sum_{k=0}^{8-1} \left(\frac{0.9029^k}{k!}\right),$$

$$= 5{,}040\,e^{-0.9029} \left(1 + \frac{0.9029^1}{1!} + \cdots + \frac{0.9029^7}{7!}\right),$$

or

$$\Gamma(8)\big|_{0.9029}^{\infty} = 5{,}039.9752.$$

Then,

$$MTTF(30) = 50\left(\frac{5{,}039.9752}{1.8733}\right) - 30 = 134{,}488.7278 \text{ hr}.$$

From Eq. (9.41),

$$P_E(T_b) = \frac{\Gamma(\frac{\alpha}{\beta})\big|_Y^{\infty}}{\Gamma(\frac{\alpha}{\beta})\big|_{X_b}^{\infty}} \left[\frac{1 - P_S}{1 + \frac{K_2 \cdot MTTR}{MTTF(T_b)}} + K_1 P_S\right], \qquad (9.46)$$

where

$$Y = \left(\frac{T_b + T_m - \gamma}{\eta}\right)^{\beta},$$

$$= \left(\frac{30 + 300 - 0}{50}\right)^{0.2},$$

or

$$Y = 1.4585,$$

and

$$\Gamma(\frac{\alpha}{\beta})\big|_Y^{\infty} = \Gamma\left(\frac{0.6}{0.2}\right)\big|_{1.4585}^{\infty},$$

$$= \Gamma(3)\big|_{1.4585}^{\infty},$$

$$= (3-1)!\,e^{-1.4585} \sum_{k=0}^{3-1} \frac{1.4585^k}{k!},$$

$$= 2\,e^{-1.4585}\left(1 + 1.4585 + \frac{1.4585^2}{2}\right),$$

or

$$\Gamma(\frac{\alpha}{\beta})\Big|_Y^\infty = 1.6384.$$

Substituting the given and the calculated quantities into Eq. (9.46) yields

$$P_E(30) = \frac{1.6384}{1.8733}\left[\frac{1-0.1}{1+\frac{0.8\,(3.5)}{134{,}488.7278}} + 0.98\,(0.1)\right],$$

or

$$P_E(30) = 0.8728.$$

Substituting $P_E(30) = 0.8728$, $F(30) = 0.0633$ and $R(30) = 0.9367$ into Eq. (9.43) yields

$$C(30) = 600 + \left[\frac{10{,}000}{0.8728\,(0.9367)}\right]0.0633 + 5\left(\frac{10{,}000}{0.8728}\right),$$

or

$$C(30) = \$58{,}660.8374.$$

Table 9.6 illustrates how the total costs, $C(T_b)$, together with reliability, $R(T_b)$, system effectiveness, $P_E(T_b)$, and mean life after burn-in, $MTTF(T_b)$, behave when the burn-in times vary in the range of $0 \sim 100$ hr. The computation for this table is accomplished using the computer software MATHCAD. From Table 9.6, it may be seen that the optimum burn-in time is in the neighborhood of $T_b = 50$ hr. To get a more accurate minimum-cost burn-in time, more calculations are made using MATHCAD on $R(T_b)$, $P_E(T_b)$, $MTTF(T_b)$ and $C(T_b)$ in the neighborhood of $T_b = 50$ hr, or for $T_b = 40, 41, \cdots, 59, 60$ hr. They are listed in another table, Table 9.7, and plotted in Fig. 9.4. From Table 9.7 and Fig. 9.4, it may be seen that the optimum burn-in time is

$$T_b^* = 54 \text{ hr},$$

which corresponds to the minimum cost of

$$C_{min} = C(T_b^*) = C(54 \text{ hr}) = \$58{,}529.0036.$$

Correspondingly, the minimum number of units to be placed on burn-in is given by Eq. (9.29); i.e.,

$$N_b = K_3/[P_E(T_b^*)\,R(T_b^*)],$$
$$= 10{,}000/[0.8861\,(0.9168)],$$

COST MODEL 4   363

TABLE 9.6– Total burn-in and associated costs, $C(T_b)$, reliability function, $R(T_b)$, system effectiveness, $P_E(T_b)$, and mean life after burn-in, $MTTF(T_b)$, for different T_b values, using the Cost Model 3, for Example 9–3.

T_b, hr	$C(T_b)$, $	$R(T_b)$	$P_E(T_b)$	$MTTF(T_b)$, hr
0	60,657.7868	1.0000	0.8243	126,000.0000
10	59,216.3004	0.9628	0.8538	130,861.6406
20	58,847.0155	0.9478	0.8649	132,921.4312
30	58,660.8374	0.9367	0.8728	134,488.7278
40	58,567.2891	0.9275	0.8791	135,800.8121
50	58,531.4566	0.9197	0.8843	136,950.0000
60	58,535.5631	0.9127	0.8887	137,983.6994
70	58,569.1419	0.9065	0.8926	138,930.0764
80	58,625.4411	0.9007	0.8961	139,807.4748
90	58,699.8214	0.8954	0.8993	140,628.6259
100	58,788.9428	0.8904	0.9021	141,402.7908

or

$N_b = 12,308.8466$ or $12,309$.

Note that Table 9.7 also provides information on the post-burn-in mean life, $MTTF(T_b)$, in the neighborhood of the optimum burn-in time, $T_b^* = 54$ hr. If

$$MTTF(T_b^*) = MTTF(54 \text{ hr}) = 137,375.4682 \text{ hr}$$

does not satisfy the prespecified mean life requirement, T_b^* can be further adjusted, or increased in this case, because the total cost function is quite shallow in the neighborhood of $T_b^* = 54$ hr as may be observed from both Table 9.7 and Fig. 9.4.

9.4 COST MODEL 4

This model is a modified version of the cost model proposed by Kuo in 1984 [4] and the cost model proposed by Yang in 1993 [5], which enables the determination of the optimum burn-in time such that the total cost

TABLE 9.7– More calculations on total costs, $C(T_b)$, reliability function, $R(T_b)$, system effectiveness, $P_E(T_b)$, and mean life after burn-in, $MTTF(T_b)$, for different T_b values in the range of [40 hr, 60 hr], for determining the more accurate optimum burn-in time, T_b^*, for Example 9–3.

T_b, hr	$C(T_b)$, $	$R(T_b)$	$P_E(T_b)$	$MTTF(T_b)$, hr
40	58,567.2891	0.9275	0.8791	135,800.8121
41	58,561.4867	0.9267	0.8796	135,922.0564
42	58,556.2273	0.9259	0.8802	136,041.7622
43	58,551.4903	0.9251	0.8807	136,159.9842
44	58,547.2562	0.9243	0.8812	136,276.7744
45	58,543.5068	0.9235	0.8818	136,392.1813
46	58,540.2248	0.9227	0.8823	136,506.2513
47	58,537.3938	0.9219	0.8828	136,619.0279
48	58,534.9984	0.9212	0.8833	136,730.5524
49	58,533.0240	0.9204	0.8838	136,840.8641
50	58,531.4566	0.9197	0.8843	136,950.0000
51	58,530.2831	0.9190	0.8847	137,057.9956
52	58,529.4910	0.9182	0.8852	137,164.8844
53	58,529.0683	0.9175	0.8857	137,270.6984
54	58,529.0036	0.9168	0.8861	137,375.4682
55	58,529.2863	0.9161	0.8866	137,479.2229
56	58,529.9058	0.9154	0.8870	137,581.9902
57	58,530.8525	0.9148	0.8875	137,683.7966
58	58,532.1169	0.9141	0.8879	137,784.6676
59	58,533.6900	0.9134	0.8883	137,884.6274
60	58,535.5631	0.9127	0.8887	137,983.6994

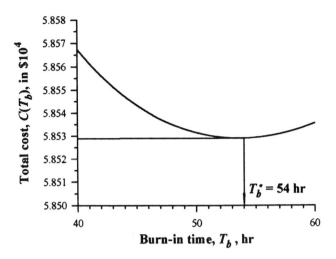

Fig. 9.4– The optimum burn-in time, under the generalized gamma distribution assumption, for Example 9–3.

during burn-in and warranty is minimized. The bimodal mixed two-parameter-Weibull distribution is used as the life model. It is further assumed that the probability of two or more failures occurring in any unit during the warranty period is negligible. Then, the total expected cost during burn-in and warranty is given by

$$C(T_b) = C_b T_b + C_{fb} N_b F(T_b) + C_{fw} (1 + \xi_L) N_s(T_b) F(T_w|T_b),$$

(9.47)

where

C_b = operational cost of the burn-in facility per hour of burn-in time, which includes capital investment, interest, overhead maintenance, all related costs for the physical plant and equipment, the cost of all associated labor (both direct and indirect), and a reasonable rate of return on investment,

C_{fb} = cost per unit that failed in the burn-in process, which includes all the associated costs of producing a unit and placing it in the burn-in facility,

C_{fw} = cost per unit that survived the burn-in but failed in the field operation during warranty of length T_w, including the costs of an unit, labor and production loss due to failure repair,

T_b = burn-in length,

N_b = number of units to be placed on burn-in,

$F(T_b)$ = cdf, or the probability of failure, of an unit during burn-in,

ξ_L = ratio of the cost per failure during warranty due to loss of reputation to the cost per failure during warranty due to repair, C_{fw},

$\xi_L C_{fw}$ = cost per failure during warranty due to loss of reputation,

$C_{fw}(1+\xi_L)$ = total cost per failure during warranty,

$N_s(T_b)$ = number of units out of N_b surviving the burn-in,

and

$F(T_w|T_b)$ = conditional cdf, or the probability of failure, of a burned-in unit during warranty.

Note

$$N_s(T_b) = N_b\, R(T_b), \tag{9.48}$$

and

$$\begin{aligned}F(T_w|T_b) &= 1 - R(T_w|T_b), \\ &= 1 - R(T_b + T_w)/R(T_b),\end{aligned}$$

or

$$F(T_w|T_b) = \frac{R(T_b) - R(T_b + T_w)}{R(T_b)}, \tag{9.49}$$

where

$$R(T_b) = 1 - F(T_b),$$

or

$$F(T_b) = 1 - R(T_b). \tag{9.50}$$

COST MODEL 4

Substituting Eqs. (9.48), (9.49) and (9.50) into Eq. (9.47) yields

$$C(T_b) = C_b T_b + C_{fb} N_b [1 - R(T_b)]$$
$$+ C_{fw}(1 + \xi_L)[N_b R(T_b)] \left[\frac{R(T_b) - R(T_b + T_w)}{R(T_b)}\right],$$

or

$$C(T_b) = C_b T_b + C_{fb} N_b [1 - R(T_b)]$$
$$+ C_{fw}(1 + \xi_L) N_b [R(T_b) - R(T_b + T_w)]. \quad (9.51)$$

For the bimodal mixed two-parameter-Weibull distribution,

$$R(T_b) = p_1 e^{-\left(\frac{T_b}{\eta_1}\right)^{\beta_1}} + (1 - p_1) e^{-\left(\frac{T_b}{\eta_2}\right)^{\beta_2}}, \quad (9.52)$$

and

$$R(T_b + T_w) = p_1 e^{-\left(\frac{T_b + T_w}{\eta_1}\right)^{\beta_1}} + (1 - p_1) e^{-\left(\frac{T_b + T_w}{\eta_2}\right)^{\beta_2}}, \quad (9.53)$$

where

p_1 = percentage of the substandard subpopulation,
$1 - p_1$ = percentage of the standard subpopulation,
β_1, η_1 = shape and scale parameters, respectively,
of the substandard subpopulation,

and

β_2, η_2 = shape and scale parameters, respectively,
of the standard subpopulation.

Given the number of units to be placed on burn-in, their life distribution parameters, the warranty period, and other associated costs, the total expected costs for various burn-in times can be evaluated and the optimum burn-in time corresponding to the minimum cost can be determined. This is illustrated next by a numerical example.

EXAMPLE 9-4

The bimodal, mixed, two-parameter Weibull distribution was fitted to a given set of burn-in testing data. The estimated parameters are given as follows:

$$p_1 = 0.05, \quad \beta_1 = 0.4, \quad \eta_1 = 50 \text{ hr}, \quad \beta_2 = 1.2, \text{ and } \eta_2 = 86,450 \text{ hr}.$$

A total of 10,000 such units are to be placed on burn-in and their associated costs are

$$C_b = \$50/\text{hr}, \quad C_{fb} = \$5/\text{unit}, \quad \text{and} \quad C_{fw} = \$80/\text{unit}.$$

Determine the optimum burn-in time, using the Cost Model 4, for a warranty period of $T_w = 10,000$ hr, and for $\xi_L = 1$ (low penalty), $\xi_L = 2$ (medium penalty), and $\xi_L = 3$ (high penalty), respectively.

SOLUTIONS TO EXAMPLE 9-4

Since the warranty period, T_w, and the life distribution parameters, p_1, β_1, η_1, β_2 and η_2, are given, then $R(T_b)$ and $R(T_b + T_w)$ can be evaluated using Eqs. (9.52) and (9.53) at various values of T_b. With the given values of N_b, C_b, C_{fb} and C_{fw}, the total expected cost, $C(T_b)$, can be evaluated at various values of T_b. A sample calculation is given next for evaluating the total expected cost when $T_b = 100$ hr and $\xi_L = 2$.

When $T_b = 100$ hr, the reliability functions $R(T_b)$ and $R(T_b + T_w)$ are given by Eqs. (9.52) and (9.53); i.e.,

$$R(T_b) = R(100 \text{ hr}) = 0.05\, e^{-\left(\frac{100}{50}\right)^{0.4}} + 0.95\, e^{-\left(\frac{100}{86,450}\right)^{1.2}},$$

or

$$R(100 \text{ hr}) = 0.9631,$$

and

$$R(T_b + T_w) = R(10,100 \text{ hr}),$$
$$= 0.05\, e^{-\left(\frac{10,100}{50}\right)^{0.4}} + 0.95\, e^{-\left(\frac{10,100}{86,450}\right)^{1.2}},$$

or

$$R(10,100 \text{ hr}) = 0.8804.$$

Substituting these and the other given information into Eq. (9.51) yields

$$C(T_b) = C(100 \text{ hr}),$$
$$= 50\,(100) + 5\,(10,000)\,(1 - 0.9631)$$
$$+ 80\,(1+2)\,10,000\,(0.9631 - 0.8804),$$

or

$$C(100 \text{ hr}) = \$2.0516 \times 10^5.$$

COST MODEL 4

TABLE 9.8– Total expected costs for different T_b values, using the Cost Model 4, under low penalty ($\xi_L = 1$), medium penalty ($\xi_L = 2$) and high penalty ($\xi_L = 3$), respectively, for Example 9–4.

Burn-in time, T_b, hr	Total expected cost, $C(T_b)$, in $\$10^5$			$R(T_w\|T_b)$
	$\xi_L = 1$	$\xi_L = 2$	$\xi_L = 3$	
100	1.3906	2.0516	2.7127	0.9142
110	1.3859	2.0420	2.6981	0.9148
120	1.3824	2.0340	2.6856	0.9153
130	1.3796	2.0272	2.6748	0.9158
140	1.3776	2.0216	2.6655	0.9162
150	1.3762	2.0169	2.6576	0.9166
160	1.3754	2.0130	2.6506	0.9169
170	<u>1.3750</u>	2.0099	2.6447	0.9173
180	1.3751	2.0074	2.6396	0.9176
190	1.3756	2.0054	2.6353	0.9179
200	1.3763	2.0040	2.6317	0.9181
210	1.3774	2.0031	2.6287	0.9183
220	1.7880	2.0025	2.6263	0.9186
230	1.3804	<u>2.0023</u>	2.6243	0.9188
240	1.3822	2.0025	2.6228	0.9190
250	1.3842	2.0030	2.6217	0.9191
260	1.3864	2.0037	2.6210	0.9193
270	1.3888	2.0048	2.6207	0.9195
280	1.3914	2.0060	<u>2.6206</u>	0.9196
290	1.3941	2.0075	2.6209	0.9198
300	1.3969	2.0092	2.6214	0.9199
310	1.3998	2.0110	2.6222	0.9200
320	1.4029	2.0131	2.6233	0.9201
330	1.4021	2.0153	2.6245	0.9202
340	1.4093	2.0176	2.6260	0.9203
350	1.4127	2.0202	2.6276	0.9204
360	1.4162	2.0228	2.6294	0.9205

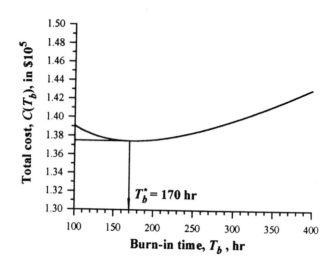

Fig. 9.5– The total expected cost, $C(T_b)$, versus burn-in time, T_b, under low penalty ($\xi_L = 1$), for Example 9–4.

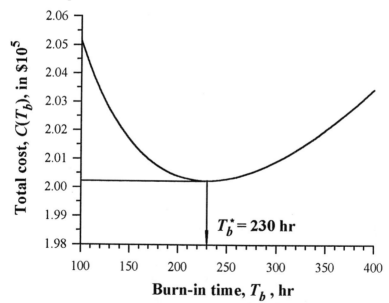

Fig. 9.6– The total expected cost, $C(T_b)$, versus burn-in time, T_b, under medium penalty ($\xi_L = 2$), for Example 9–4.

COST MODEL 4

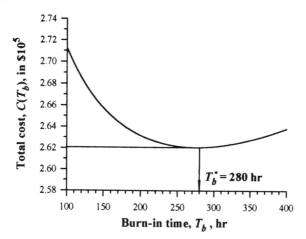

Fig. 9.7– The total expected cost, $C(T_b)$, versus burn-in time, T_b, under high penalty ($\xi_L = 3$), for Example 9–4.

Table 9.8 lists the total expected cost values corresponding to various T_b values in the range of 100 to 360 hr where the minimum costs occur for $\xi_L = 1$, $\xi_L = 2$ and $\xi_L = 3$, respectively, and which are plotted in Figs. 9.5, 9.6 and 9.7, respectively.

From Table 9.8 and Fig. 9.5, it may be seen that the optimum burn-in time when a low penalty is assumed for the loss of reputation of a failure during warranty, or when $\xi_L = 1$, is

$T_b^* = 170$ hr.

The corresponding minimum total expected cost is

$C(170) = \$1.3750 \times 10^5$.

From Table 9.8 and Fig. 9.6, it may be seen that the optimum burn-in time when a medium penalty is assumed for the loss of reputation of a failure during warranty, or when $\xi_L = 2$, is

$T_b^* = 230$ hr.

The corresponding minimum total expected cost is

$C(230) = \$2.0023 \times 10^5$.

From Table 9.8 and Fig. 9.7, it may be seen that the optimum burn-in time when a high penalty is assumed for the loss of reputation of a failure during warranty, or when $\xi_L = 3$, is

$T_b^* = 280$ hr.

The corresponding minimum total expected cost is

$C(280) = \$2.6206 \times 10^5.$

It may be seen that the higher penalty on the loss of reputation, due to failure during warranty, requires longer burn-in time.

Note that Table 9.8 also provides the information on the mission reliability of the burned-in units at the end of the warranty period,

$$R(T_w|T_b) = \frac{R(T_b + T_w)}{R(T_b)}.$$

If $R(T_w|T_b^*)$ does not satisfy the prespecified requirement, a further adjustment on T_b^* can be made, by increasing T_b^* in this case, because $R(T_w|T_b)$ is an increasing function of T_b and the total expected cost function is quite shallow in the neighborhood of T_b^*, as may be observed from Table 9.8 and Figs. 9.5, 9.6 and 9.7.

REFERENCES

1. Jensen, Finn and Peterson, N. E., *Burn-in*, John Wiley & Sons, New York, 167 pp., 1982.

2. Plesser, K. T. and Field, T. O., "Cost-optimized Burn-in Duration for Repairable Electronic Systems," *IEEE Transactions on Reliability*, Vol. R-26, No. 3, pp. 195-197, August 1977.

3. Washburn, L. A., "Determination of Optimum Burn-in Time : a Composite Criterion," *IEEE Transactions on Reliability*, Vol. 38, No. 2, pp. 193-198, June 1989.

4. Kuo, W., "Reliability Enhancement Through Optimal Burn-in," *IEEE Transactions on Reliability*, Vol. R-33, No. 2, pp. 145-156, June 1984.

5. Yang, G. B., "An Optimal Strategy for Capacitor Burn-in," *Proceedings of Capacitor & Resistor Technology Symposium*, Costa Mesa, CA, 1993.

ADDITIONAL REFERENCES

1. Buck C. N., Cartalano, L. J. and Shalvoy, C. E., "Justifying the Cost of Test During Burn-in," *Electronics Test*, pp. 31-35, April 1988.

2. Chandrasekaran, R., "Optimal Policies for Burn-in Procedures," *Operations Research*, Vol. 14, No. 3, pp. 149-160, 1977.

3. Chi, D. H. and Kuo, W., "Burn-in Optimization under Reliability and Capacity Restrictions," *IEEE Transactions on Reliability*, Vol. 38, No. 2, pp. 193-198, June 1989.

REFERENCES

4. Kuo, W. and Kuo, Y., "Facing the Headaches of Early Failures: A State-of-the-Art Review of Burn-in Decisions," *Proceedings of the IEEE*, Vol. 71, No. 11, pp. 1257-1266, 1983.

5. Marko, D. M. and Schoonmaker, T. D., "Optimizing Spare Module Burn-in," *Proceedings of Annual Reliability and Maintainability Symposium*, pp. 83-86, 1982.

6. Nguyen, D. G. and Murthy, D. N. P., "Optimal Burn-in Time to Minimize Cost for Products Sold Under Warranty," *IIE Transactions*, Vol. 14, No. 3, pp.167-172, 1982.

7. Shaw, M., "Recognizing the Optimum Burn-in Period," *Quality and Reliability Engineering International*, Vol. 3, No. 4, pp. 259-263, 1987.

8. Sultan, T. I., "Optimum Burn-in Time: Model and Application," *Microelectronics and Reliability*, Vol. 26, No. 5, pp. 909-916, 1986.

9. Vassilev, V., "Cost Effectiveness of Burn-in Procedures of Semiconductor Devices and Integrated Circuits," *Proceedings of 7th Symposium on Reliability in Electronics*, Vol. 11, pp. 543-553, 1988.

10. Wager, A. J., Thompson, D. L. and Forcier, A. C., "Implications of a Model for Optimum Burn-in," *Proceedings of Annual Reliability Physics Symposium*, pp. 286-291, 1983.

11. Weiss, G. H. and Dishon, M., "Some Economic Problems Related to Burn-in Programs," *IEEE Transactions on Reliability*, Vol. R-20, No. 3, pp. 190-195, August 1971.

Chapter 10

BURN-IN QUANTIFICATION AND OPTIMIZATION USING THE BIMODAL MIXED-EXPONENTIAL LIFE DISTRIBUTION

10.1 INTRODUCTION

In the previous chapters, burn-in testing has been quantified and optimized using various life distributions, such as the mixed Weibull, various bathtub models, etc. As a special case of the mixed-Weibull distribution, the mixed-exponential life distribution has its unique mathematical features. It is simple and easy to manipulate. More importantly, it always has a decreasing failure rate (DFR) function which makes it mathematically qualified to describe the times-to-failure behavior during burn-in.

10.2 THE MIXED-EXPONENTIAL LIFE DISTRIBUTION

The *pdf* of a general mixed-exponential life distribution is given by

$$f(T) = \sum_{i=1}^{n} p_i \, f_i(T) = \sum_{i=1}^{n} p_i \, \lambda_i \, e^{-\lambda_i \, T}, \quad T \geq 0, \tag{10.1}$$

where

n = total number of subpopulations, $n \geq 2$,
$f_i(T)$ = pdf of the ith subpopulation, $i = 1, 2, \cdots, n$,
$f_i(T) = \lambda_i\, e^{-\lambda_i T}$,
λ_i = failure rate of the ith subpopulation, $i = 1, 2, \cdots, n$,
$\lambda_i \neq \lambda_j$, for $i \neq j$, $i = 1, 2, \cdots, n$, and $j = 1, 2, \cdots, n$,
p_i = proportion of the ith subpopulation, $i = 1, 2, \cdots, n$,

and

$$\sum_{i=1}^{n} p_i = 1.$$

The *cdf* and the reliability functions are given by

$$F(T) = \sum_{i=1}^{n} p_i\, F_i(T) = \sum_{i=1}^{n} p_i\, (1 - e^{-\lambda_i T}), \qquad (10.2)$$

and

$$R(T) = \sum_{i=1}^{n} p_i\, R_i(T) = \sum_{i=1}^{n} p_i\, e^{-\lambda_i T}, \qquad (10.3)$$

respectively. The failure rate function is given by

$$\lambda(T) = \frac{f(T)}{R(T)} = \frac{\sum_{i=1}^{n} p_i\, f_i(T)}{\sum_{i=1}^{n} p_i\, R_i(T)},$$

or

$$\lambda(T) = \frac{\sum_{i=1}^{n} p_i\, \lambda_i\, e^{-\lambda_i T}}{\sum_{i=1}^{n} p_i\, e^{-\lambda_i T}}, \qquad (10.4)$$

which is always a decreasing function of T, as shown next.
The first derivative of Eq. (10.4) with respect to T is

$$\frac{d\lambda(T)}{dT} = \left[-(\sum_{i=1}^{n} p_i\, \lambda_i^2\, e^{-\lambda_i T})(\sum_{i=1}^{n} p_i\, e^{-\lambda_i T}) \right.$$
$$\left. + (\sum_{i=1}^{n} p_i\, \lambda_i\, e^{-\lambda_i T})(\sum_{i=1}^{n} p_i\, \lambda_i\, e^{-\lambda_i T}) \right]$$

$$\bigg/\bigg(\sum_{i=1}^{n} p_i\, e^{-\lambda_i T}\bigg)^2,$$

$$= \bigg\{-\bigg[\sum_{i=1}^{n} p_i^2\, \lambda_i^2\, e^{-2\lambda_i T} + \sum_{\substack{j=1\\i\neq j}}^{n}\sum_{i=1}^{n} p_i\, p_j\, (\lambda_i^2 + \lambda_j^2)\, e^{-(\lambda_i+\lambda_j) T}\bigg]$$

$$+ \bigg[\sum_{i=1}^{n} p_i^2\, \lambda_i^2\, e^{-2\lambda_i T} + 2\sum_{\substack{j=1\\i\neq j}}^{n}\sum_{i=1}^{n} p_i\, p_j\, \lambda_i\, \lambda_j\, e^{-(\lambda_i+\lambda_j) T}\bigg]\bigg\}$$

$$\bigg/\bigg(\sum_{i=1}^{n} p_i\, e^{-\lambda_i T}\bigg)^2,$$

$$= -\bigg[\sum_{\substack{j=1\\i\neq j}}^{n}\sum_{i=1}^{n} p_i\, p_j\, (\lambda_i^2 + \lambda_j^2)\, e^{-(\lambda_i+\lambda_j) T}$$

$$-2\sum_{\substack{j=1\\i\neq j}}^{n}\sum_{i=1}^{n} p_i\, p_j\, \lambda_i\, \lambda_j\, e^{-(\lambda_i+\lambda_j) T}\bigg]\bigg/\bigg(\sum_{i=1}^{n} p_i\, e^{-\lambda_i T}\bigg)^2,$$

$$= -\bigg[\sum_{\substack{j=1\\i\neq j}}^{n}\sum_{i=1}^{n} p_i\, p_j\, (\lambda_i^2 - 2\lambda_i\lambda_j + \lambda_j^2)\, e^{-(\lambda_i+\lambda_j) T}\bigg]$$

$$\bigg/\bigg(\sum_{i=1}^{n} p_i\, e^{-\lambda_i T}\bigg)^2,$$

or

$$\frac{d\lambda(T)}{dT} = -\frac{\sum_{\substack{j=1\\i\neq j}}^{n}\sum_{i=1}^{n} p_i\, p_j\, (\lambda_i - \lambda_j)^2\, e^{-(\lambda_i+\lambda_j) T}}{\big(\sum_{i=1}^{n} p_i\, e^{-\lambda_i T}\big)^2} < 0. \tag{10.5}$$

Equation (10.5) indicates that the first derivative of the failure rate function of the mixed-exponential life distribution is always negative, or $\lambda(T)$ is always a decreasing function of T.

10.3 THE BIMODAL MIXED-EXPONENTIAL LIFE DISTRIBUTION

The mathematics of the burn-in process shows that the residue of latent defects, left over after burn-in, is a direct function of the burn-in duration for any prescribed stress level [1]. Assume that in a lot there is a proportion, p_b, of substandard components with latent defects which will eventually cause early failures. Obviously, if the mean time to failure of good components in the lot is in the millions of hours, and of the bad components in the lot failing is in a few thousand, hundred or ten hours, then the latter are failing due to early causes. If in a lot of N components there are N_b weak components and N_g good components, the proportion of latent defectives is

$$p_b = N_b/N.$$

The proportion of good components in the lot is then

$$p_g = N_g/N,$$

so that

$$N_b + N_g = N,$$

and

$$p_b + p_g = 1.$$

The quantities p_b and p_g are also the probabilities of picking an early failing component or picking a good component, respectively, when drawing a component at random from the lot. Of course, one cannot know a priori whether the drawn component comes from the p_b or the p_g subpopulation in the lot.

Now let the weak components in the lot have a failure rate of λ_b and the good components in the lot a failure rate of λ_g, with the respective reliabilities of

$$R_b(T) = e^{-\lambda_b T},$$

and

$$R_g(T) = e^{-\lambda_g T}.$$

Since with a probability of p_b the component drawn at random from the lot will have a reliability of $R_b(T)$, and with a probability of p_g it will have a reliability of $R_g(T)$, any component drawn at random from the lot will have a statistical probability of survival (reliability) of

$$R(T) = p_b R_b(T) + p_g R_g(T),$$

or

$$R(T) = p_b e^{-\lambda_b T} + p_g e^{-\lambda_g T}. \tag{10.6}$$

BIMODAL MIXED-EXPONENTIAL LIFE DISTRIBUTION 379

Equation (10.6) is based on the principle of uncertainty because we do not know what component in the lot we have drawn.

The probability density function of such a randomly drawn component in the lot is then

$$f(T) = -d[R(T)]/dT,$$

which results in

$$f(T) = p_b \lambda_b e^{-\lambda_b T} + p_g \lambda_g e^{-\lambda_g T}, \tag{10.7}$$

and the component's failure rate is

$$\lambda(T) = f(T)/R(T). \tag{10.8}$$

Substituting Eqs. (10.6) and (10.7) into Eq. (10.8) yields the failure rate equation

$$\lambda(T) = \frac{p_b \lambda_b e^{-\lambda_b T} + p_g \lambda_g e^{-\lambda_g T}}{p_b e^{-\lambda_b T} + p_g e^{-\lambda_g T}}. \tag{10.9}$$

This function decreases with time because the first derivative of $\lambda(T)$ with respect to T is always negative; i.e.,

$$\begin{aligned}
\frac{d\lambda(T)}{dT} &= [(-p_b \lambda_b^2 e^{-\lambda_b T} - p_g \lambda_g^2 e^{-\lambda_g T})(p_b e^{-\lambda_b T} + p_g e^{-\lambda_g T}) \\
&\quad - (p_b \lambda_b e^{-\lambda_b T} + p_g \lambda_g e^{-\lambda_g T}) \\
&\quad \cdot (-p_b \lambda_b e^{-\lambda_b T} - p_g \lambda_g e^{-\lambda_g T})] \\
&\quad \Big/ (p_b e^{-\lambda_b T} + p_g e^{-\lambda_g T})^2, \\
&= \Big[-(p_b^2 \lambda_b^2 e^{-2\lambda_b T} + p_b p_g \lambda_b^2 e^{-(\lambda_b+\lambda_g)T} \\
&\quad + p_b p_g \lambda_g^2 e^{-(\lambda_b+\lambda_g)T} + p_g^2 \lambda_g^2 e^{-2\lambda_g T}) \\
&\quad + (p_b^2 \lambda_b^2 e^{-2\lambda_b T} + 2 p_b p_g \lambda_b \lambda_g e^{-(\lambda_b+\lambda_g)T} \\
&\quad + p_g^2 \lambda_g^2 e^{-2\lambda_g T}) \Big] \Big/ (p_b e^{-\lambda_b T} + p_g e^{-\lambda_g T})^2, \\
&= \frac{-p_b p_g (\lambda_b^2 - 2\lambda_b \lambda_g + \lambda_g^2) e^{-(\lambda_b+\lambda_g)T}}{(p_b e^{-\lambda_b T} + p_g e^{-\lambda_g T})^2},
\end{aligned}$$

or

$$\frac{d\lambda(T)}{dT} = -\frac{p_b p_g (\lambda_b - \lambda_g)^2 e^{-(\lambda_b+\lambda_g)T}}{(p_b e^{-\lambda_b T} + p_g e^{-\lambda_g T})^2} < 0, \tag{10.10}$$

which is consistent with Eq. (10.5).

The initial value of $\lambda(T)$, λ_i, at $T = 0$ is

$$\lambda_i = \lambda(0) = p_b \lambda_b + p_g \lambda_g, \tag{10.11}$$

while the limiting final value, λ_f, is that of a good component in the lot, λ_g, as T approaches infinity, and it thus becomes

$$\lambda_f = \lim_{T \to \infty} \lambda(T) = \lambda_g. \tag{10.12}$$

This may be proven if we rewrite Eq. (10.9) in the form

$$\lambda(T) = \lambda_g + \frac{p_b (\lambda_b - \lambda_g) e^{-(\lambda_b - \lambda_g) T}}{p_g + p_b e^{-(\lambda_b - \lambda_g) T}}, \tag{10.13}$$

and see that the numerator of the second term on the right vanishes as T goes to infinity.

Figure 10.1 [1] is a plot of the failure rate function given by Eq. (10.13) for an assumed components population which contains 1% of weak components ($p_b = 0.01$) and 99% of good components ($p_g = 0.99$). Let the good components in the lot, under a "ground benign" environment, have a failure rate of 10^{-7} failures per hour (MIL-HDBK-217) while the weak components in the lot have a failure rate of 10^{-3} failures per hour, under the same environment. During an accelerated burn-in let an acceleration factor of 100 be applied (without damaging components in the lot or subtracting from their life span). By failing the weak components in the lot during burn-in the initial failure rate of the component's population will drop from 10^{-5} failures per hour [see Eq. (10.11)] to a value very close to the failure rate of the good components in the lot (10^{-7} fr/hr). This process of elimination of the weak components in the lot may take thousands of hours under non-accelerated conditions while it will take much less time under accelerated burn-in. After burn-in we have a highly reliable components population, while in the first several thousand hours – without burn-in – the components in the lot would be very unreliable. This can be ascertained by looking at the ordinate of Fig. 10.1.

The proper physical interpretation of the decaying failure rate phenomenon is that the longer a component in the lot survives, the more likely it is that it came from the good population p_g. One should remember that no single physical component in the lot ever has the failure rate of Eq. (10.13). It is either good with a failure rate of λ_g, or it is weak with a failure rate of λ_b. Yet, from Eq. (10.13) these very important conclusions can be drawn.

BIMODAL MIXED-EXPONENTIAL LIFE DISTRIBUTION 381

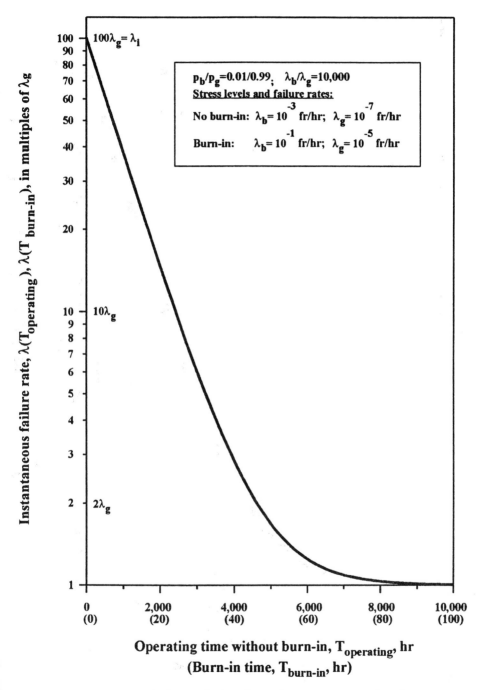

Fig. 10.1– Early failures during burn-in versus stress level.

10.4 THE OPTIMUM BURN-IN TIME FOR A SPECIFIED POST-BURN-IN MISSION RELIABILITY

The post-burn-in reliability of a component with a bimodal mixed-exponential life distribution is given by the following conditional reliability:

$$R(t|T_b) = \frac{R(T_b + t)}{R(T_b)} = \frac{p_b \, e^{-\lambda_b \, (T_b+t)} + p_g \, e^{-\lambda_g \, (T_b+t)}}{p_b \, e^{-\lambda_b \, T_b} + p_g \, e^{-\lambda_g \, T_b}}.$$

(10.14)

For a specified mission reliability goal, R_G, the optimum burn-in time can be determined by solving the following equation for T_b:

$$\frac{p_b \, e^{-\lambda_b \, (T_b+t)} + p_g \, e^{-\lambda_g \, (T_b+t)}}{p_b \, e^{-\lambda_b \, T_b} + p_g \, e^{-\lambda_g \, T_b}} = R_G.$$

(10.15)

Multiplying both sides of Eq. (10.15) by $(p_b \, e^{-\lambda_b \, T_b} + p_g \, e^{-\lambda_g \, T_b})$ yields

$$p_b \, e^{-\lambda_b \, (T_b+t)} + p_g \, e^{-\lambda_g \, (T_b+t)} = R_G \, (p_b \, e^{-\lambda_b \, T_b} + p_g \, e^{-\lambda_g \, T_b}).$$

(10.16)

Dividing both sides of Eq. (10.16) by $e^{-\lambda_g \, T_b}$ yields

$$p_b \, e^{-\lambda_b \, t} \, e^{-(\lambda_b - \lambda_g) \, T_b} + p_g \, e^{-\lambda_g \, t} = R_G \, [p_b \, e^{-(\lambda_b - \lambda_g) \, T_b} + p_g].$$

Rearranging and factoring yields

$$p_g \, (e^{-\lambda_g \, t} - R_G) = p_b \, (R_G - e^{-\lambda_b \, t}) \, e^{-(\lambda_b - \lambda_g) \, T_b},$$

or

$$T_b^* = \left(\frac{1}{\lambda_b - \lambda_g}\right) \log_e \frac{p_b \, (R_G - e^{-\lambda_b \, t})}{p_g \, (e^{-\lambda_g \, t} - R_G)},$$

(10.17)

which is the optimum burn-in time for the specified post-burn-in mission reliability, R_G. It may be seen from Eq. (10.17) that a feasible solution for T_b^* exists only for

$$e^{-\lambda_b \, t} < R_G < e^{-\lambda_g \, t},$$

(10.18)

and

$$\frac{p_b(R_G - e^{-\lambda_b t})}{p_g(e^{-\lambda_g t} - R_G)} > 1,$$

or

$$R_G > p_b e^{-\lambda_b t} + p_g e^{-\lambda_g t}. \tag{10.19}$$

Note that the right side of Eq. (10.19) is greater than $e^{-\lambda_b t}$ because

$$\left(p_b e^{-\lambda_b t} + p_g e^{-\lambda_g t}\right) - e^{-\lambda_b t}$$
$$= p_g e^{-\lambda_g t} - (1 - p_b) e^{-\lambda_b t},$$
$$= p_g e^{-\lambda_g t} - p_g e^{-\lambda_b t},$$
$$= p_g \left(e^{-\lambda_g t} - e^{-\lambda_b t}\right) > 0,$$

due to $\lambda_g < \lambda_b$. Combining Eqs. (10.18) and (10.19) yields the feasible domain for Eq. (10.17); i.e.,

$$(p_b e^{-\lambda_b t} + p_g e^{-\lambda_g t}) < R_G < e^{-\lambda_g t}. \tag{10.20}$$

EXAMPLE 10–1

The data analysis of burn-in test results of a certain electronic component yielded the following estimated parameters when the bimodal mixed-exponential life distribution was fitted:

$$p_b = 250 \text{ ppm (parts per million)} = 250 \times 10^{-6},$$
$$\lambda_b = 15,000 \text{ fr}/10^6 \text{ hr} = 15,000 \times 10^{-6} \text{ fr/hr},$$
$$\lambda_g = 0.5 \text{ fr}/10^6 \text{ hr} = 0.5 \times 10^{-6} \text{ fr/hr},$$

and

$$p_g = 1 - p_b = 999,750 \times 10^{-6}.$$

Determine the optimum burn-in time for a specified post-burn-in mission reliability goal of $R_G = 0.9949$ for a mission of $t = 10^4$ hr.

SOLUTION TO EXAMPLE 10–1

First, the feasible domain for Eq. (10.17) needs to be determined. According to Eq. (10.20),

$$\left(p_b e^{-\lambda_b t} + p_g e^{-\lambda_g t}\right) < R_G < e^{-\lambda_g t},$$

or

$$0.9948 < R_G < 0.9950$$

for a fixed mission time of $t = 10^4$ hr. Now, because the given reliability goal, $R_G = 0.9949$, is within this domain, an optimum burn-in time exists.

Substituting the given information into Eq. (10.17) yields

$$T_b^* = \frac{1}{(15,000 - 0.5) \times 10^{-6}} \cdot \log_e \left[\frac{250 \times 10^{-6} \left(0.9949 - e^{-15,000 \times 10^{-6} \times 10^4}\right)}{999,750 \times 10^{-6} \left(e^{-0.5 \times 10^{-6} \times 10^4} - 0.9949\right)} \right],$$

or

$$T_b^* = 52.92 \text{ hr, or } 53 \text{ hr.}$$

Figure 10.2 illustrates the behavior of T_b^* versus R_G in the range of $0.9948 < R_G < 0.9950$. This figure is based on Eq. (10.17), or

$$T_b^*(R_G) = \frac{1}{(15,000 - 0.5) \times 10^{-6}} \cdot \log_e \left[\frac{250 \times 10^{-6} \left(R_G - e^{-15,000 \times 10^{-6} \times 10^4}\right)}{999,750 \times 10^{-6} \left(e^{-0.5 \times 10^{-6} \times 10^4} - R_G\right)} \right],$$

or

$$T_b^*(R_G) = \left(\frac{2 \times 10^6}{29,999}\right) \log_e \left[\frac{R_G - e^{-150}}{3,999 \left(e^{-0.005} - R_G\right)} \right].$$

It may be seen from Fig. 10.2 that as the mission reliability goal increases, the required burn-in time increases accordingly.

10.5 THE OPTIMUM BURN-IN TIME FOR A SPECIFIED MEAN RESIDUAL LIFE (MRL) GOAL

The mean residual life (MRL) of a bimodal mixed-exponential life distribution after a T_b-hr burn-in is given by

$$MRL(T_b) = \int_0^\infty R(t|T_b) \, dt,$$

OPTIMUM BURN-IN TIME FOR A SPECIFIED MRL

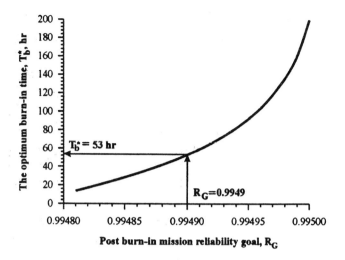

Fig. 10.2– The optimum burn-in time versus the post-burn-in mission reliability goal.

$$= \int_0^\infty \frac{R(T_b + t)}{R(T_b)} \, dt,$$

$$= \frac{1}{R(T_b)} \int_0^\infty R(T_b + t) \, dt,$$

or

$$MRL(T_b) = \frac{1}{R(T_b)} \int_{T_b}^\infty R(T) \, dT. \qquad (10.21)$$

Substituting Eq. (10.6) into Eq. (10.21) yields

$$MRL(T_b) = \frac{1}{p_b \, e^{-\lambda_b \, T_b} + p_g \, e^{-\lambda_g \, T_b}}$$
$$\cdot \int_{T_b}^\infty (p_b \, e^{-\lambda_b \, T} + p_g \, e^{-\lambda_g \, T}) \, dT,$$

or

$$MRL(T_b) = \frac{\frac{p_b}{\lambda_b} e^{-\lambda_b \, T_b} + \frac{p_g}{\lambda_g} e^{-\lambda_g \, T_b}}{p_b \, e^{-\lambda_b \, T_b} + p_g \, e^{-\lambda_g \, T_b}}. \qquad (10.22)$$

For a specified MRL goal, MRL_G, the optimum burn-in time can be determined by solving the following equation for T_b:

$$\frac{\frac{p_b}{\lambda_b} e^{-\lambda_b \, T_b} + \frac{p_g}{\lambda_g} e^{-\lambda_g \, T_b}}{p_b \, e^{-\lambda_b \, T_b} + p_g \, e^{-\lambda_g \, T_b}} = MRL_G. \qquad (10.23)$$

Multiplying both sides of Eq. (10.23) by $(p_b e^{-\lambda_b T_b} + p_g e^{-\lambda_g T_b})$ yields

$$\frac{p_b}{\lambda_b} e^{-\lambda_b T_b} + \frac{p_g}{\lambda_g} e^{-\lambda_g T_b} = MRL_G (p_b e^{-\lambda_b T_b} + p_g e^{-\lambda_g T_b}).$$

(10.24)

Dividing both sides of Eq. (10.24) by $e^{-\lambda_g T_b}$ yields

$$\frac{p_b}{\lambda_b} e^{-(\lambda_b - \lambda_g) T_b} + \frac{p_g}{\lambda_g} = MRL_G (p_b e^{-(\lambda_b - \lambda_g) T_b} + p_g).$$

Rearranging and factoring yields

$$p_g \left(\frac{1}{\lambda_g} - MRL_G \right) = p_b \left(MRL_G - \frac{1}{\lambda_b} \right) e^{-(\lambda_b - \lambda_g) T_b},$$

or

$$T_b^* = \left(\frac{1}{\lambda_b - \lambda_g} \right) \log_e \left[\frac{p_b \left(MRL_G - \frac{1}{\lambda_b} \right)}{p_g \left(\frac{1}{\lambda_g} - MRL_G \right)} \right],$$

(10.25)

which is the optimum burn-in time for the specified MRL goal, MRL_G. It may be seen from Eq. (10.25) that the feasible solution for T_b^* exists only for

$$\frac{1}{\lambda_b} < MRL_G < \frac{1}{\lambda_g},$$

(10.26)

and

$$\frac{p_b \left(MRL_G - \frac{1}{\lambda_b} \right)}{p_g \left(\frac{1}{\lambda_g} - MRL_G \right)} > 1,$$

or

$$MRL_G > p_b \left(\frac{1}{\lambda_b} \right) + p_g \left(\frac{1}{\lambda_g} \right).$$

(10.27)

Note that the right side of Eq. (10.27) is greater than $1/\lambda_b$ because

$$\left[p_b \left(\frac{1}{\lambda_b} \right) + p_g \left(\frac{1}{\lambda_g} \right) \right] - \frac{1}{\lambda_b}$$

$$= p_g \left(\frac{1}{\lambda_g} \right) - (1 - p_b) \left(\frac{1}{\lambda_b} \right),$$

$$= p_g \left(\frac{1}{\lambda_g} \right) - p_g \left(\frac{1}{\lambda_b} \right),$$

$$= p_g \left(\frac{1}{\lambda_g} - \frac{1}{\lambda_b} \right) > 0,$$

OPTIMUM BURN-IN TIME FOR A SPECIFIED MRL

due to $\lambda_g < \lambda_b$. Combining Eqs. (10.26) and (10.27) yields the feasible domain for Eq. (10.25); i.e.,

$$\left[p_b\left(\frac{1}{\lambda_b}\right) + p_g\left(\frac{1}{\lambda_g}\right)\right] < MRL_G < \frac{1}{\lambda_g}. \qquad (10.28)$$

EXAMPLE 10-2

Rework Example 10-1 for a specified MRL goal of $MRL_G = 1.9998 \times 10^6$ hr.

SOLUTION TO EXAMPLE 10-2

According to Example 10-1, the bimodal mixed-exponential life distribution parameters were

$$p_b = 250 \times 10^{-6}, \qquad p_g = 999,750 \times 10^{-6},$$
$$\lambda_b = 15,000 \times 10^{-6} \text{ fr/hr}, \quad \lambda_g = 0.5 \times 10^{-6} \text{ fr/hr}.$$

First the feasible domain for Eq. (10.25) needs to be determined. According to Eq. (10.28),

$$\left[p_b\left(\frac{1}{\lambda_b}\right) + p_g\left(\frac{1}{\lambda_g}\right)\right] < MRL_G < \frac{1}{\lambda_g},$$

or

$$1.9995 \times 10^6 \text{ hr} < MRL_G < 2.0000 \times 10^6 \text{ hr}.$$

Now the given MRL goal, $MRL_G = 1.9998 \times 10^6$ hr, falls in this domain. Then, the corresponding optimum burn-in time exists.

Substituting the given information into Eq. (10.25) yields

$$T_b^* = \frac{1}{(15,000 - 0.5) \times 10^{-6}} \cdot \log_e\left[\frac{250 \times 10^{-6}\left(1.9998 \times 10^6 - \frac{1}{15,000} \times 10^6\right)}{999,750 \times 10^{-6}\left(\frac{1}{0.5} \times 10^6 - 1.9998 \times 10^6\right)}\right],$$

or

$$T_b^* = 61.10 \text{ hr or } 61 \text{ hr}.$$

Figure 10.3 illustrates the behavior of T_b^* versus MRL_G in the domain of 1.9995×10^6 hr $< MRL_G < 2.0000 \times 10^6$ hr.

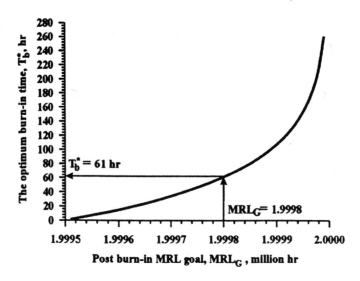

Fig. 10.3 – The optimum burn-in time versus the MRL goal.

Note that this figure is based on Eq. (10.25), or

$$T_b^*(MRL_G) = \frac{1}{(15{,}000 - 0.5) \times 10^{-6}} \cdot \log_e \left[\frac{250 \times 10^{-6} \left(MRL_G - \frac{1}{15{,}000} \times 10^6 \right)}{999{,}750 \times 10^{-6} \left(\frac{1}{0.5} \times 10^6 - MRL_G \right)} \right],$$

or

$$T_b^*(MRL_G) = \left(\frac{2 \times 10^6}{29{,}999} \right) \log_e \left[\frac{MRL_G - 200/3}{3{,}999 \, (2 \times 10^6 - MRL_G)} \right].$$

It may be seen from Fig. 10.3 that as the MRL goal increases, the required burn-in time increases accordingly.

10.6 THE OPTIMUM BURN-IN TIME FOR A SPECIFIED POST-BURN-IN FAILURE RATE GOAL

The reliability function for a mission t of a component with a bimodal mixed-exponential life distribution after T_b hours of burn-in is given by Eq. (10.14); i.e.,

$$R(t|T_b) = \frac{R(T_b + t)}{R(T_b)} = \frac{p_b \, e^{-\lambda_b (T_b + t)} + p_g \, e^{-\lambda_g (T_b + t)}}{p_b \, e^{-\lambda_b T_b} + p_g \, e^{-\lambda_g T_b}}. \quad (10.29)$$

The corresponding *pdf* is

$$f(t|T_b) = -\frac{dR(t|T_b)}{dt} = \frac{p_b \lambda_b e^{-\lambda_b (T_b+t)} + p_g \lambda_g e^{-\lambda_g (T_b+t)}}{p_b e^{-\lambda_b T_b} + p_g e^{-\lambda_g T_b}}.$$

(10.30)

Then, the corresponding failure rate function is

$$\lambda(t|T_b) = \frac{f(t|T_b)}{R(t|T_b)} = \frac{p_b \lambda_b e^{-\lambda_b (T_b+t)} + p_g \lambda_g e^{-\lambda_g (T_b+t)}}{p_b e^{-\lambda_b (T_b+t)} + p_g e^{-\lambda_g (T_b+t)}}.$$

(10.31)

For a specified post-burn-in failure rate goal at the end of a t–hr mission, λ_G, the optimum burn-in time can be determined by letting Eq. (10.31) equal to λ_G and solving for T_b; i.e.,

$$\frac{p_b \lambda_b e^{-\lambda_b (T_b+t)} + p_g \lambda_g e^{-\lambda_g (T_b+t)}}{p_b e^{-\lambda_b (T_b+t)} + p_g e^{-\lambda_g (T_b+t)}} = \lambda_G. \qquad (10.32)$$

Multiplying both sides of Eq. (10.32) by $[p_b e^{-\lambda_b (T_b+t)} + p_g e^{-\lambda_g (T_b+t)}]$ yields

$$p_b \lambda_b e^{-\lambda_b (T_b+t)} + p_g \lambda_g e^{-\lambda_g (T_b+t)} = \lambda_G [p_b e^{-\lambda_b (T_b+t)} + p_g e^{-\lambda_g (T_b+t)}].$$

(10.33)

Dividing both sides of Eq. (10.33) by $e^{-\lambda_g (T_b+t)}$ yields

$$p_b \lambda_b e^{-(\lambda_b-\lambda_g)(T_b+t)} + p_g \lambda_g = \lambda_G [p_b e^{-(\lambda_b-\lambda_g)(T_b+t)} + p_g].$$

Rearranging and factoring yields

$$p_b (\lambda_b - \lambda_G) e^{-(\lambda_b-\lambda_g)(T_b+t)} = p_g (\lambda_G - \lambda_g),$$

or

$$T_b^* = \left(\frac{1}{\lambda_b - \lambda_g}\right) \log_e \left[\frac{p_b (\lambda_b - \lambda_G)}{p_g (\lambda_G - \lambda_g)}\right] - t, \qquad (10.34)$$

which is the optimum burn-in time for a specified post-burn-in failure rate goal at the end of the t–hr mission, λ_G. It may be seen from Eq. (10.34) that the feasible solution of T_b^* exists only when the following three conditions are satisfied:

1. First condition,
$$\lambda_g < \lambda_G < \lambda_b. \qquad (10.35)$$

2. Second condition,
$$\frac{p_b (\lambda_b - \lambda_G)}{p_g (\lambda_G - \lambda_g)} > 1,$$

or
$$\lambda_G < p_b \lambda_b + p_g \lambda_g = (1 - p_g) \lambda_b + p_g \lambda_g,$$

or
$$\lambda_G < \lambda_b - p_g (\lambda_b - \lambda_g), \qquad (10.36)$$

which is less than λ_b due to $(\lambda_b - \lambda_g) > 0$.

3. Third condition, $T_b^* \geq 0$, or
$$\left(\frac{1}{\lambda_b - \lambda_g}\right) \log_e \left[\frac{p_b (\lambda_b - \lambda_G)}{p_g (\lambda_G - \lambda_g)}\right] \geq t,$$

or
$$\lambda_G \leq \lambda_b - \frac{p_g (\lambda_b - \lambda_g)}{p_g + p_b \, e^{-(\lambda_b - \lambda_g) t}}, \qquad (10.37)$$

which is also less than λ_b due to $(\lambda_b - \lambda_g) > 0$.

Combining Eqs. (10.35), (10.36) and (10.37) yields the feasible domain for Eq. (10.34) as

$$\lambda_g < \lambda_G < min\left\{\left[\lambda_b - \frac{p_g (\lambda_b - \lambda_g)}{p_g + p_b \, e^{-(\lambda_b - \lambda_g) t}}\right], [\lambda_b - p_g (\lambda_b - \lambda_g)]\right\}, \qquad (10.38)$$

or
$$\lambda_g < \lambda_G < \left[\lambda_b - \frac{p_g (\lambda_b - \lambda_g)}{p_g + p_b \, e^{-(\lambda_b - \lambda_g) t}}\right], \qquad (10.39)$$

OPTIMUM BURN-IN TIME FOR FAILURE RATE GOAL

due to

$$p_g + p_b \, e^{-(\lambda_b - \lambda_g) \, t} < p_g + p_b \times 1 = 1$$

in the denominator of the first term on the right side of Eq. (10.38).

EXAMPLE 10-3

Rework Example 10-1 for a specified post-burn-in failure rate goal of $\lambda_G = 10^{-6}$ fr/hr at the end of a 100-hr mission.

SOLUTION TO EXAMPLE 10-3

According to Example 10-1, the estimated parameters for the bimodal mixed-exponential life distribution were

$$p_b = 250 \times 10^{-6}, \qquad p_g = 999,750 \times 10^{-6},$$
$$\lambda_b = 15,000 \times 10^{-6} \text{ fr/hr}, \ \lambda_g = 0.5 \times 10^{-6} \text{ fr/hr}.$$

First the feasible domain for Eq. (10.34) needs to be determined. According to Eq. (10.39),

$$\lambda_g < \lambda_G < \left[\lambda_b - \frac{p_g \, (\lambda_b - \lambda_g)}{p_g + p_b \, e^{-(\lambda_b - \lambda_g) \, t}} \right],$$

or

$$0.5 \times 10^{-6} \text{ fr/hr} < \lambda_G < \lambda_{bb},$$

where

$$\lambda_{bb} = \left[15,000 \times 10^{-6} - \frac{999,750 \times 10^{-6}(15,000 - 0.5) \times 10^{-6}}{999,750 \times 10^{-6} + 250 \times 10^{-6} \, e^{-(15,000 - 0.5) \times 10^{-6} \times 10^4}} \right],$$

or

$$\lambda_{bb} = 1.3369 \times 10^{-6} \text{ fr/hr}.$$

Then,

$$5 \times 10^{-7} \text{ fr/hr} < \lambda_G < 1.3369 \times 10^{-6} \text{ fr/hr}.$$

Now the specified post-burn-in failure rate goal at the end of a 100-hr mission, $\lambda_G = 10^{-6}$ fr/hr, falls in this domain. Then, the corresponding optimum burn-in time exists.

Substituting the given information into Eq. (10.34) yields

$$T_b^* = \frac{1}{(15,000 - 0.5) \times 10^{-6}} \cdot \log_e \left[\frac{250 \times 10^{-6} \, (15,000 - 1) \times 10^{-6}}{999,750 \times 10^{-6} \, (1 - 0.5) \times 10^{-6}} \right] - 100,$$

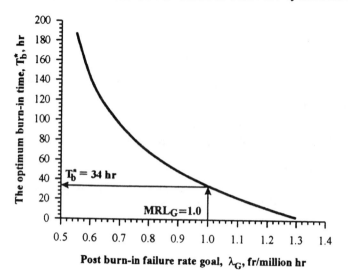

Fig. 10.4– The optimum burn-in time versus the post-burn-in failure rate goal.

or

$$T_b^* = 34.34 \text{ hr, or } 34 \text{ hr.}$$

Figure 10.4 illustrates the behavior of T_b^* versus λ_G in the domain of 5×10^{-7} fr/hr $< \lambda_G < 1.3369 \times 10^{-6}$ fr/hr. Note that this figure is based on Eq. (10.34), or

$$T_b^*(\lambda_G) = \frac{1}{(15,000 - 0.5) \times 10^{-6}} \cdot \log_e \left[\frac{250 \times 10^{-6} \ (15,000 \times 10^{-6} - \lambda_G)}{999,750 \times 10^{-6} \ (\lambda_G - 0.5 \times 10^{-6})} \right] - 100,$$

or

$$T_b^*(\lambda_G) = \left(\frac{2 \times 10^6}{29,999} \right) \log_e \left[\frac{0.015 - \lambda_G}{3,999 \ (\lambda_G - 5 \times 10^{-7})} \right] - 100.$$

It may be seen from Fig. 10.4 that as the failure rate goal restriction is relaxed, or the failure rate goal increases, the required burn-in time decreases accordingly.

10.7 THE OPTIMUM BURN-IN TIME FOR A SPECIFIED BURN-IN EFFICIENCY

Equation (10.13) lends itself to calculating the burn-in time, T_b, under various stress conditions. To perform this calculation, extract the time factor T from the equation [1]. First, put the equation into the form

$$\lambda(T_b) = \lambda_g (1 + \varepsilon), \tag{10.40}$$

where

$\lambda(T_b)$ = failure rate of the components at the end of burn-in,

and

$$\varepsilon = \frac{\lambda(T_b) - \lambda_g}{\lambda_g},$$

which is called the *Burn-in Residue*, and correspondingly $(1 - \varepsilon)$ is called the *Burn-in Efficiency*, and is a function of the burn-in time T_b. The reason for selecting this form of the failure rate equation is that ε can be made arbitrarily small, and this provides the possibility of controlling how close to get to λ_g in the burn-in process. Thus, in accordance with Eq. (10.40), preselect a desired residue value ε and obtain components of a failure rate $\lambda(T_b)$ corresponding to the preselected ε in a burn-in process of T_b hours duration.

Further Eq. (10.13) can be rewritten as

$$\lambda(T_b) = \lambda_g \left[1 + \left(\frac{1}{\lambda_g}\right) \frac{D\, p_b\, e^{-D\, T_b}}{p_g + p_b\, e^{-D\, T_b}}\right], \tag{10.41}$$

where

$$D = \lambda_b - \lambda_g \tag{10.42}$$

is the difference of the failure rates between the weak and the good components in the lot. From Eqs. (10.40) and (10.41) the *Burn-in Residue* is given by

$$\varepsilon = \left(\frac{1}{\lambda_g}\right) \frac{D\, p_b\, e^{-D\, T_b}}{p_g + p_b\, e^{-D\, T_b}}. \tag{10.43}$$

For the sake of simplicity, define the ratio of weak components to good components in the initial unburned-in lot as

$$\alpha = p_b/p_g, \tag{10.44}$$

and the ratio of the early failure rate to the good components' failure rate as

$$\beta = \lambda_b/\lambda_g, \tag{10.45}$$

and write Eq. (10.43) in the form

$$\varepsilon = \frac{\alpha\,(\beta-1)\,e^{-D\,T_b}}{\alpha\,e^{-D\,T_b}+1}. \tag{10.46}$$

Rearranging Eq. (10.46) yields

$$e^{-D\,T_b}\left[\frac{\alpha\,(\beta-1-\varepsilon)}{\varepsilon}\right] = 1, \tag{10.47}$$

and, finally, taking the logarithm of both sides yields the burn-in time T_b as

$$T_b^* = \frac{1}{D}\log_e\left[\frac{\alpha\,(\beta-1-\varepsilon)}{\varepsilon}\right], \tag{10.48}$$

which is a function of the preselected *Burn-in Residue*, ε, and of the difference D between the weak components and good components failure rates. Thus, a T_b-hr burn-in for a preselected value of ε will yield components with a failure rate given by Eq. (10.40); i.e.,

$$\lambda(T_b) = \lambda_g\,(1+\varepsilon).$$

Let

$$E = 1 - \varepsilon,$$

denote the burn-in efficiency. Then, the optimum burn-in time given by Eq. (10.48) can be expressed in terms of E as follows:

$$T_b^* = \frac{1}{D}\log_e\left[\frac{\alpha\,(\beta+E-2)}{1-E}\right]. \tag{10.49}$$

Under normal field operating conditions D will be a comparatively small number which makes T_b^* large. But under accelerated burn-in, T_b^* can be drastically reduced. In other words, self-elimination of early failures in the field, without a preceding burn-in, may be a very protracted process of low equipment reliability, while in an accelerated burn-in process the elimination of early failures can occur very fast.

EXAMPLE 10-4

Rework Example 10-1 for a specified burn-in efficiency of $E = 0.9$ or 90%.

SOLUTION TO EXAMPLE 10-4

According to Example 10-1, the estimated parameters for the bimodal mixed-exponential life distribution were

$$p_b = 250 \times 10^{-6}, \qquad p_g = 999,750 \times 10^{-6},$$
$$\lambda_b = 15,000 \times 10^{-6} \text{ fr/hr}, \ \lambda_g = 0.5 \times 10^{-6} \text{ fr/hr}.$$

Then, D, α and β in Eq. (10.49) can be evaluated using Eqs. (10.42), (10.44) and (10.45); i.e.,

$$D = \lambda_b - \lambda_g = (15,000 - 0.5) \times 10^{-6} = 2 \times 10^6/29,999,$$
$$\alpha = p_b/p_g = 250 \times 10^{-6}/(999,750 \times 10^{-6}) = 1/3,999,$$

and

$$\beta = \lambda_b/\lambda_g = 15,000 \times 10^{-6}/(0.5 \times 10^{-6}) = 30,000.$$

Therefore, the optimum burn-in time for a specified burn-in efficiency of 90% is given by Eq. (10.49), or

$$T_b^* = \frac{2 \times 10^6}{29,999} \log_e \left[\frac{(30,000 + 0.9 - 2)}{3,999 \ (1 - 0.9)} \right],$$

or

$$T_b^* = 287.8564 \text{ hr, or } 288 \text{ hr}.$$

Figure 10.5 illustrates the behavior of T_b^* versus E for $0 \leq E \leq 1$. Note that this figure is based on Eq. (10.49), or

$$T_b^*(E) = \frac{2 \times 10^6}{29,999} \log_e \left[\frac{(30,000 + E - 2)}{3,999 \ (1 - E)} \right],$$

or

$$T_b^*(E) = \frac{2 \times 10^6}{29,999} \log_e \left[\frac{(29,998 + E)}{3,999 \ (1 - E)} \right].$$

It may be seen from Fig. 10.5 that the higher the desired burn-in efficiency, the longer the required burn-in time.

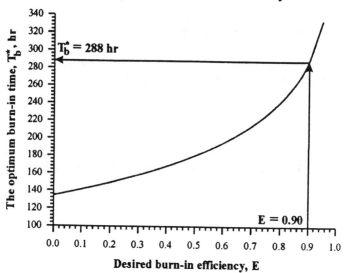

Fig. 10.5– The optimum burn-in time versus the desired burn-in efficiency.

10.8 THE OPTIMUM BURN-IN TIME FOR A SPECIFIED POWER FUNCTION

It may be seen from Eq. (10.40) that in a burn-in of duration T_b the failure rate of a population of components drops from an initial value of λ_i to a value of

$$\lambda(T_b) = \lambda_g(1+\varepsilon).$$

Thus, the difference

$$\lambda_i - \lambda_g(1+\varepsilon)$$

is a measure of how good the burn-in is. To formalize this as a *Power Function*, form the ratio of the actual burn-in result [1]

$$\lambda_i - \lambda_g(1+\varepsilon)$$

to the ideal case of

$$\lambda_i - \lambda_g,$$

in which all failures are eliminated. This ratio is called the *Power Function of the Burn-in*, PB; i.e.,

$$PB = \frac{\lambda_i - \lambda_g\,(1+\varepsilon)}{\lambda_i - \lambda_g}. \tag{10.50}$$

OPTIMUM BURN-IN TIME FOR POWER FUNCTION GOAL 397

Using the preceding results for λ_i and ε, the PB equation may be written in the explicit form of

$$PB = \frac{p_g \left(1 - e^{-D\, T_b}\right)}{p_g \left(1 - e^{-D\, T_b}\right) + e^{-D\, T_b}}. \tag{10.51}$$

The PB function supplies the answer to the question of "How good is a burn-in?" This function, as given by Eq. (10.50), becomes unity, or 100%, only if

$$\varepsilon = 0,$$

which is not achievable practically.

The optimum burn-in time for a specified burn-in power function, PB, can be obtained by solving Eq. (10.51) for T_b which yields

$$T_b^* = \frac{1}{D} \log_e \left[\frac{PB + p_g\,(1 - PB)}{p_g\,(1 - PB)} \right]. \tag{10.52}$$

EXAMPLE 10-5

Rework Example 10-1 for a specified *Power Function* of 96%.

SOLUTION TO EXAMPLE 10-5

According to Example 10-1, the estimated parameters for the bi-modal mixed-exponential life distribution were

$$p_b = 250 \times 10^{-6}, \qquad p_g = 999{,}750 \times 10^{-6},$$
$$\lambda_b = 15{,}000 \times 10^{-6} \text{ fr/hr}, \quad \lambda_g = 0.5 \times 10^{-6} \text{ fr/hr}.$$

Then, D in Eq. (10.52) can be evaluated using Eq. (10.42); i.e.,

$$D = \lambda_b - \lambda_g = (15{,}000 - 0.5) \times 10^{-6} = 29{,}999/(2 \times 10^6).$$

Therefore, the optimum burn-in time for a specified burn-in power function of 96% is given by Eq. (10.52), or

$$T_b^* = \left(\frac{2 \times 10^6}{29{,}999} \right) \log_e \left[\frac{0.96 + 999{,}750 \times 10^{-6}\,(1 - 0.96)}{999{,}750 \times 10^{-6}\,(1 - 0.96)} \right],$$

or

$$T_b^* = 214.6149 \text{ hr, or } 215 \text{ hr}.$$

Figure 10.6 illustrates the behavior of T_b^* versus PB for $0 \le PB < 1$. Note that this figure is based on Eq. (10.52), or

$$T_b^*(PB) = \left(\frac{2 \times 10^6}{29{,}999} \right) \log_e \left[\frac{PB + 999{,}750 \times 10^{-6}\,(1 - PB)}{999{,}750 \times 10^{-6}\,(1 - PB)} \right],$$

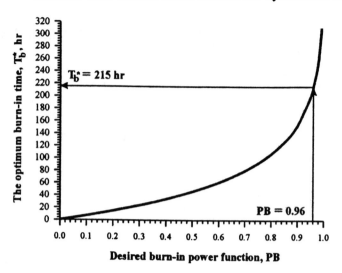

Fig. 10.6– The optimum burn-in time versus the desired burn-in power function.

or

$$T_b^*(PB) = \left(\frac{2 \times 10^6}{29,999}\right) \log_e \left[\frac{PB + 0.99975\,(1-PB)}{0.99975\,(1-PB)}\right].$$

It may be seen from Fig. 10.6 that the higher the desired burn-in power function, the longer the required burn-in time.

10.9 BURN-IN TIME INTERVAL FOR THE SPECIFIED BURN-IN RISKS

10.9.1 DEFINITION OF BURN-IN RISKS AND THE CORRESPONDING BURN-IN INTERVAL

To avoid unnecessary burn-in to the strong subpopulation and in the mean time to eliminate the desired percentage of the weak subpopulation, a criterion can be established for the determination of the optimum burn-in time [2].

Let $T_{1-\alpha}$ and T_β denote two burn-in times such that

$$P(T > T_{1-\alpha} | \text{weak subpopulation}) = \alpha, \tag{10.53}$$

and

$$P(T < T_\beta | \text{strong subpopulation}) = \beta, \tag{10.54}$$

where

α = risk of burn-in due to the incomplete elimination of the weak subpopulation,

and

β = risk of burn-in due to the undesired damage to the strong subpopulation.

The physical meaning of α and β risks can be explained as follows: Given probabilities α and β, burn-in can take place only if

$$T_{1-\alpha} \leq T_b \leq T_\beta. \tag{10.55}$$

This criterion implies that a T_b-hr burn-in will eliminate at least $100(1-\alpha)\%$ of the weak subpopulation but no more than $100\beta\%$ of the strong subpopulation. The interval $(T_{1-\alpha}, T_\beta)$ which satisfies Eq. (10.55) is defined as the *double-risk burn-in interval* which is valid only when

$$T_\beta \geq T_{1-\alpha}. \tag{10.56}$$

10.9.2 RESTRICTIONS ON THE FAILURE RATES FOR A VALID BURN-IN INTERVAL

For a bimodal mixed-exponential life distribution,

$$\alpha = P(T > T_{1-\alpha} | \text{weak subpopulation}),$$
$$= R(T_{1-\alpha} | \text{weak subpopulation}),$$

or

$$\alpha = e^{-\lambda_b T_{1-\alpha}}.$$

Then,

$$T_{1-\alpha} = -\frac{1}{\lambda_b} \log_e \alpha. \tag{10.57}$$

Similarly,

$$\beta = P(T < T_\beta | \text{strong subpopulation}),$$
$$= F(T_\beta | \text{strong subpopulation}),$$

or

$$T_\beta = 1 - e^{-\lambda_g T_\beta}.$$

Then,

$$T_\beta = -\frac{1}{\lambda_g} \log_e(1-\beta). \qquad (10.58)$$

To satisfy the condition of a feasible burn-in interval, the following must be satisfied according to Eq. (10.56):

$$\frac{T_\beta}{T_{1-\alpha}} = \frac{-\frac{1}{\lambda_g}\log_e(1-\beta)}{-\frac{1}{\lambda_b}\log_e \alpha} \geq 1,$$

or

$$\frac{\lambda_b}{\lambda_g} \geq \frac{\log_e \alpha}{\log_e(1-\beta)}. \qquad (10.59)$$

For example, for $\alpha = \beta = 0.1$, the following must be satisfied:

$$\frac{\lambda_b}{\lambda_g} \geq \frac{\log_e 0.1}{\log_e(1-0.1)} = 21.8543.$$

For $\alpha = \beta = 0.01$, the following must be satisfied:

$$\frac{\lambda_b}{\lambda_g} \geq \frac{\log_e 0.01}{\log_e(1-0.01)} = 458.2106. \qquad (10.60)$$

For $\alpha = \beta = 0.001$, the following must be satisfied:

$$\frac{\lambda_b}{\lambda_g} \geq \frac{\log_e 0.001}{\log_e(1-0.001)} = 6,904.3008, \qquad (10.61)$$

which requires a ratio of about 7,000 for the two failure rates. Hence, the issue of interference between the week and the strong subpopulations is essential in analyzing the burn-in process.

EXAMPLE 10-6

Rework Example 10-1 for the specified burn-in risks of $\alpha = \beta = 0.001$.

SOLUTION TO EXAMPLE 10-6

According to Example 10-1, the estimated parameters for the bimodal mixed-exponential life distribution were

$$p_b = 250 \times 10^{-6}, \qquad p_g = 999,750 \times 10^{-6},$$
$$\lambda_b = 15,000 \times 10^{-6} \text{ fr/hr}, \ \lambda_g = 0.5 \times 10^{-6} \text{ fr/hr}.$$

The failure rate ratio in this case is

$$\frac{\lambda_b}{\lambda_g} = \frac{15{,}000 \times 10^{-6}}{0.5 \times 10^{-6}} = 30{,}000 > 6{,}904.3008,$$

which satisfies the requirement of Eq. (10.61). Therefore, the valid burn-in interval exists.

Substituting the given information into Eqs. (10.57) and (10.58) yields

$$T_{1-\alpha} = -\frac{1}{0.015} \log_e 0.001 = 460.5170 \text{ hr, or } 461 \text{ hr,}$$

and

$$T_\beta = -\frac{1}{0.5 \times 10^{-6}} \log_e(1 - 0.001) = 2{,}001.0007 \text{ hr, or } 2{,}001 \text{ hr.}$$

Then the burn-in interval in this case is (461 hr; 2,001 hr).

Note that the obtained burn-in interval is a very wide range. This is because the two subpopulations in this case are well separated with two quite distinctive mean lives of

$$MTBF_b = \frac{1}{\lambda_b} = \frac{1}{0.015} \cong 66.6667 \text{ hr,}$$

and

$$MTBF_g = \frac{1}{\lambda_g} = \frac{1}{0.5 \times 10^{-6}} = 2 \times 10^6 \text{ hr.}$$

Any burn-in time between 461 hr and 2,001 hr will satisfy the specified burn-in risks of $\alpha = \beta = 0.001$.

EXAMPLE 10–7

Given are the estimated parameters for the bimodal mixed-exponential life distribution of

$p_b = 0.005, \quad p_g = 0.995,$
$\lambda_b = 0.02 \text{ fr/hr}, \quad \lambda_g = 4 \times 10^{-5} \text{ fr/hr.}$

Determine the burn-in interval for the specified burn-in risks of $\alpha = \beta = 0.01$.

SOLUTION TO EXAMPLE 10–7

The failure rate ratio in this case is given by

$$\frac{\lambda_b}{\lambda_g} = \frac{0.02}{4 \times 10^{-5}} = 500 > 458.2106,$$

which satisfies the requirement of Eq. (10.60). Therefore, the valid burn-in interval exists.

Substituting the given information into Eqs. (10.57) and (10.58) yields

$$T_{1-\alpha} = -\frac{1}{0.02} \log_e 0.01 = 230.2585 \text{ hr, or } 230 \text{ hr,}$$

and

$$T_\beta = -\frac{1}{4 \times 10^{-5}} \log_e(1 - 0.01) = 251.2584 \text{ hr, or } 251 \text{ hr.}$$

Then the burn-in interval in this case is (230 hr; 251 hr).

Note that the obtained burn-in interval has a very narrow range. This is because the two subpopulations in this case are not as well separated as those in Example 10-6. The mean lives of the two subpopulations are

$$MTBF_b = \frac{1}{\lambda_b} = \frac{1}{0.02} = 50 \text{ hr,}$$

and

$$MTBF_g = \frac{1}{\lambda_g} = \frac{1}{4 \times 10^{-5}} = 25,000 \text{ hr.}$$

Any burn-in time between 230 hr and 251 hr will satisfy the specified burn-in risks of $\alpha = \beta = 0.01$.

10.10 THE NUMBER AND COST OF FAILURES

While in a Poisson process, with an exponential underlying life distribution where the failure rate λ is constant, the expected number of failures in time T is given by

$$H(T) = \lambda T, \tag{10.62}$$

the case is different when the failure rate is not constant but is a function of time, such as that given by Eq. (10.13). To calculate the expected number of failures for such a case, Renewal Theory [3] should be applied. First the renewal rate, $h(T)$, of the *Renewal Process of the Burn-in*, needs to be evaluated. Then, the number of failures in T, $H(T)$, can be obtained by integration of $h(T)$ [1].

The *pdf* for the bimodal mixed-exponential life distribution is given by Eq. (10.7), or

$$f(T) = p_b \, \lambda_b \, e^{-\lambda_b T} + p_g \, \lambda_g \, e^{-\lambda_g T}.$$

The Laplace transform of this *pdf*, $f^*(s)$, is

$$f^*(s) = \frac{p_b \lambda_b}{s + \lambda_b} + \frac{p_g \lambda_g}{s + \lambda_g}. \qquad (10.63)$$

According to Renewal Theory [3], the Laplace transform of the renewal rate function, $h^*(s)$, can be obtained by

$$h^*(s) = \frac{f^*(s)}{1 - f^*(s)}. \qquad (10.64)$$

Substituting Eq. (10.63) into Eq. (10.64) and simplifying yields

$$h^*(s) = \frac{s\,(p_b \lambda_b + p_g \lambda_g) + \lambda_b \lambda_g}{s^2 + s\,(p_b \lambda_g + p_g \lambda_b)}, \qquad (10.65)$$

which can be rearranged as

$$h^*(s) = \frac{\lambda_b \lambda_g}{A}\left(\frac{1}{s} - \frac{1}{s+A}\right) + (p_b \lambda_b + p_g \lambda_g)\frac{1}{s+A},$$

or

$$h^*(s) = \frac{\lambda_b \lambda_g}{A}\frac{1}{s} + \left(p_b \lambda_b + p_g \lambda_g - \frac{\lambda_b \lambda_g}{A}\right)\frac{1}{s+A}, \qquad (10.66)$$

where

$$A = p_b \lambda_g + p_g \lambda_b. \qquad (10.67)$$

Taking the inverse Laplace transform of Eq. (10.66) yields

$$h(T) = \frac{\lambda_b \lambda_g}{A} + \left(p_g \lambda_g + p_b \lambda_b - \frac{\lambda_b \lambda_g}{A}\right) e^{-A\,T}. \qquad (10.68)$$

The initial value of $h(T)$, h_i, is

$$h_i = h(0) = p_g \lambda_g + p_b \lambda_b, \qquad (10.69)$$

while the limiting final value is

$$h_f = h(\infty) = \frac{\lambda_b \lambda_g}{A}, \qquad (10.70)$$

which occurs as T approaches infinity.

Using this notation we may rewrite Eq. (10.68) in the form

$$h(T) = h_f + (h_i - h_f)\,e^{-A\,T}, \qquad (10.71)$$

and the integral of the renewal rate function becomes the expected number of renewals which equals the expected number of failures in T; i.e.,

$$H(T) = h_f \, T + \frac{1}{A}(h_i - h_f)(1 - e^{-A\,T}). \tag{10.72}$$

At $T = 0$ the number of failures is zero, and grows continuously with T.

As to the cost of failures, $C(T)$, in time T, counted from $T = 0$, this is a linear function of the number of failures $H(T)$; i.e.,

$$C(T) = H(T) \cdot c, \tag{10.73}$$

where c is the average cost of a single failure.

10.11 THE OPTIMUM BURN-IN TIME FOR THE MINIMUM COST

As stated earlier, the objective of component burn-in is to eliminate the substandard components by subjecting them to stresses that induce failures in these components. As the burn-in duration increases, the probability of substandard components passing the burn-in decreases, thereby reducing the fraction of substandard components in the burned-in population. The reduction in the fraction of substandard components depends on the burn-in duration. The longer the burn-in duration the larger the reduction. However, as the burn-in duration increases beyond a certain point, the savings in field repair costs may not be commensurate with the increasing burn-in costs. The optimal burn-in duration can be obtained by minimizing the life-cycle cost [4; 5].

Let the burn-in duration be T_b. Then, the expected number of renewals for a non-burned-in component, is given by Eq. (10.72), or

$$H(T_b) = h_f \, T_b + \frac{1}{A}(h_i - h_f)(1 - e^{-A\,T_b}), \tag{10.74}$$

where A, h_i and h_f are given by Eqs. (10.67), (10.69) and (10.70), respectively, or

$$A = p_b \, \lambda_g + p_g \, \lambda_b,$$
$$h_i = h(0) = p_g \, \lambda_g + p_b \, \lambda_b,$$

and

$$h_f = h(\infty) = \frac{\lambda_b \, \lambda_g}{A}.$$

OPTIMUM BURN-IN TIME FOR THE MINIMUM COST

The probability of a component passing the burn-in is

$$R(T_b) = p_b \, R_b(T_b) + p_g \, R_g(T_b). \tag{10.75}$$

The probability that the component passing the burn-in is substandard is

$$p'_b = \frac{p_b \, R_b(T_b)}{p_b \, R_b(T_b) + p_g \, R_g(T_b)}. \tag{10.76}$$

The probability that the component passing the burn-in is standard is

$$p'_g = 1 - p'_b = \frac{p_g \, R_g(T_b)}{p_b \, R_b(T_b) + p_g \, R_g(T_b)}. \tag{10.77}$$

Therefore, the field reliability, during mission time t, of a component burned-in for T_b hr is

$$R'(t|T_b) = p'_b \frac{R'_b(T'_b + t)}{R_b(T_b)} + p'_g \frac{R'_g(T'_b + t)}{R_g(T_b)}, \tag{10.78}$$

where

R'_b = field reliability function of bad components which escaped the burn-in,

R'_g = field reliability function of good components which passed the burn-in,

and

T'_b = equivalent burn-in time under use stress level such that $R(T_b) = R'(T'_b)$.

Substituting Eqs. (10.76) and (10.77) into Eq. (10.78) yields

$$R'(t|T_b) = \frac{p_b \, R'_b(T'_b + t) + p_g \, R'_g(T'_b + t)}{p_b \, R_b(T_b) + p_g \, R_g(T_b)}. \tag{10.79}$$

The probability density function of the burned-in component in the field is given by

$$f'(t|T_b) = -\frac{d[R'(t|T_b)]}{dt},$$

or

$$f'(t|T_b) = \frac{p_b \, f'_b(T'_b + t) + p_g \, f'_g(T'_b + t)}{p_b \, R_b(T_b) + p_g \, R_g(T_b)}, \tag{10.80}$$

where

R' = reliability function of burned-in components under use stress level,
f' = pdf of burned-in components under use stress level,
f'_b = pdf of burned-in bad components under use stress level,

and

f'_g = pdf of burned-in good components under use stress level.

If the burn-in stress level is the same as the use stress level, then the components' life distribution parameters during burn-in and field operation are identical. Therefore, Eqs. (10.79) and (10.80) become

$$R'(t|T_b) = R(t|T_b) = \frac{p_b\, R_b(T_b+t) + p_g\, R_g(T_b+t)}{p_b\, R_b(T_b) + p_g\, R_g(T_b)}, \qquad (10.81)$$

and

$$f'(t|T_b) = f(t|T_b) = \frac{p_b\, f_b(T_b+t) + p_g\, f_g(T_b+t)}{p_b\, R_b(T_b) + p_g\, R_g(T_b)}. \qquad (10.82)$$

If a mixed-exponential distribution with parameters of λ_b and λ_g is used, then Eqs. (10.75), (10.76), (10.77), (10.81) and (10.82) become

$$R(T_b) = p_b\, e^{-\lambda_b T_b} + p_g\, e^{-\lambda_g T_b}, \qquad (10.83)$$

$$p'_b = \frac{p_b\, e^{-\lambda_b T_b}}{p_b\, e^{-\lambda_b T_b} + p_g\, e^{-\lambda_g T_b}}, \qquad (10.84)$$

$$p'_g = \frac{p_g\, e^{-\lambda_g T_b}}{p_b\, e^{-\lambda_b T_b} + p_g\, e^{-\lambda_g T_b}}, \qquad (10.85)$$

$$R'(t|T_b) = \frac{p_b\, e^{-\lambda_b (T_b+t)} + p_g\, e^{-\lambda_g (T_b+t)}}{p_b\, e^{-\lambda_b T_b} + p_g\, e^{-\lambda_g T_b}}, \qquad (10.86)$$

and

$$f'(t|T_b) = \frac{p_b\, \lambda_b\, e^{-\lambda_b (T_b+t)} + p_g\, \lambda_g\, e^{-\lambda_g (T_b+t)}}{p_b\, e^{-\lambda_b T_b} + p_g\, e^{-\lambda_g T_b}}, \qquad (10.87)$$

respectively.

OPTIMUM BURN-IN TIME FOR THE MINIMUM COST 407

Upon failure during field operation, components are replaced by new ones from the burned-in population. The number of renewals of a burned-in component during field operation can then be obtained using renewal theory as given by Eq. (10.72), or

$$H(t|T_b) = \int_0^t h'(\tau)\,d\tau = h'_f\,t + \frac{1}{A'}(h'_i - h'_f)(1 - e^{-A'\,t}), \quad (10.88)$$

where

$H(t|T_b)$ = number of renewals of a component in time t given that it has already been burned-in for T_b hr,

$h'(\tau)$ = renewal rate function for a burned-in component as given by Eq. (10.71),

$$h'(\tau) = h'_f + (h'_i - h'_f)\,e^{-A'\,\tau}, \quad (10.89)$$

h'_i = initial value of renewal rate $h'(T)$, $h'(0)$, as given by Eq. (10.69),

$$h'_i = p'_g\,\lambda_g + p'_b\,\lambda_b, \quad (10.90)$$

h'_f = limiting final value of renewal rate $h'(T)$, $h'(\infty)$, as given by Eq. (10.70),

$$h'_f = \frac{\lambda_b\,\lambda_g}{A'}, \quad (10.91)$$

and

A' = a constant as given by Eq. (10.67) for the burned-in components,

$$A' = p'_b\,\lambda_g + p'_g\,\lambda_b. \quad (10.92)$$

Equation (10.88) will be used for the number of field component failures in the life-cycle cost model.

As the burn-in duration tends to infinity, the renewal density tends to the failure rate of good components and the number of renewals becomes equal to the expected number of renewals for an exponential process; i.e., from Eq. (10.76)

$$\lim_{T_b \to \infty} p'_b = \lim_{T_b \to \infty} \frac{p_b\,e^{-\lambda_b\,T_b}}{p_b\,e^{-\lambda_b\,T_b} + p_g\,e^{-\lambda_g\,T_b}},$$

$$= \lim_{T_b \to \infty} \frac{p_b\,e^{-(\lambda_b - \lambda_g)\,T_b}}{p_b\,e^{-(\lambda_b - \lambda_g)\,T_b} + p_g},$$

$$= \frac{0}{0 + p_g},$$

or
$$\lim_{T_b \to \infty} p'_b = 0.$$

Correspondingly,
$$\lim_{T_b \to \infty} p'_g = \lim_{T_b \to \infty} (1 - p'_b) = 1,$$

$$\lim_{T_b \to \infty} h'_i = \lim_{T_b \to \infty} (p'_g \lambda_g + p'_b \lambda_b),$$
$$= 1 \times \lambda_g + 0 \times \lambda_b,$$

or
$$\lim_{T_b \to \infty} h'_i = \lambda_g.$$

$$\lim_{T_b \to \infty} A' = \lim_{T_b \to \infty} (p'_b \lambda_g + p'_g \lambda_b),$$
$$= 0 \times \lambda_g + 1 \times \lambda_b,$$

or
$$\lim_{T_b \to \infty} A' = \lambda_b,$$

and
$$\lim_{T_b \to \infty} h'_f = \lim_{T_b \to \infty} (\frac{\lambda_b \lambda_g}{A'}),$$
$$= \frac{\lambda_b \lambda_g}{\lambda_b},$$

or
$$\lim_{T_b \to \infty} h'_f = \lambda_g.$$

Therefore,
$$\lim_{T_b \to \infty} H(t|T_b) = \lim_{T_b \to \infty} [h'_f \, t + \frac{1}{A'}(h'_i - h'_f)(1 - e^{-A' \, t})],$$
$$= \lambda_g \, t + \frac{1}{\lambda_b}(\lambda_g - \lambda_g)(1 - e^{-\lambda_b \, t}),$$

or
$$\lim_{T_b \to \infty} H(t|T_b) = \lambda_g \, t. \tag{10.93}$$

In other words, as the burn-in duration increases, substandard components are eliminated and the population consists of good components only. The field savings asymptotically approach that of an exponential renewal process with the failure rate of the good components. However, after a certain duration, the decrease in field expenses may not be commensurate with the increasing burn-in costs. The optimal duration is obtained by considering the trade-offs between savings and burn-in costs.

Often burn-in is done at higher stress levels (accelerated stress levels). The failure rates of substandard and good components will be different during and after burn-in. In such a case, the fraction of substandard components after burn-in is computed with the accelerated failure rates, whereas the field failure rates are used for the computation of the number of failures in the field. In other words, Eq. (10.84) is computed with accelerated failure rates.

10.11.1 COST MODEL FORMULATION

The life-cycle cost model is composed of the following three cost categories:

1. The total cost of burn-in.

2. The total cost of the failed components during burn-in.

3. The warranty cost.

The expected life-cycle cost is given by

$$C(T_b) = N \left[C_0 + C_b T_b + C_{fb} H_b(T_b) + C_{fw} H_w(T_w|T_b) \right], \quad (10.94)$$

where

T_b = burn-in time,
T_w = warranty time,
N = number of components to be burned-in,
C_0 = fixed cost of burn-in per component,
C_b = burn-in cost per component per hour,
C_{fb} = cost of replacing a failed component during burn-in, which includes the cost of the failed component and its labor,
C_{fw} = cost of replacing a component that fails when operating in the field during the warranty period, which includes the cost of the failed

component, its labor, and production and reputation loss,

$H_b(T_b)$ = expected number of renewals of a component during burn-in as given by Eq. (10.74) ,

and

$H_w(T_w|T_b)$ = expected number of renewals of a burned-in component during the warranty period as given by Eq. (10.88) .

10.11.2 COST MODEL MINIMIZATION

The optimum burn-in time for the minimum life-cycle cost, T_b^*, may be obtained by taking the first partial derivative of Eq. (10.94) with respect to T_b, setting it equal to zero, and solving for T_b; i.e.,

$$\frac{\partial C(T_b)}{\partial T_b} = 0,$$

or

$$N \left[0 + C_b + C_{fb} \frac{\partial H_b(T_b)}{\partial T_b} + C_{fw} \frac{\partial H_w(T_w|T_b)}{\partial T_b} \right] = 0. \quad (10.95)$$

The two associated partial derivatives of Eq. (10.95) are obtained as follows:

(1) $\partial H_b(T_b)/\partial T_b$, using Eq. (10.74), is given by

$$\frac{\partial H_b(T_b)}{\partial T_b} = \frac{\partial \left[h_f T_b + \frac{1}{A}(h_i - h_f)(1 - e^{-A\,T_b}) \right]}{\partial T_b},$$

or

$$\frac{\partial H_b(T_b)}{\partial T_b} = h_f + (h_i - h_f)\, e^{-A\,T_b}.$$

(2) $\partial H_w(T_w|T_b)/\partial T_b$, using Eq. (10.88), is given by

$$\frac{\partial H_w(T_w|T_b)}{\partial T_b} = \frac{\partial \left[h'_f T_w + \frac{1}{A'}(h'_i - h'_f)(1 - e^{-A'\,T_w}) \right]}{\partial T_b},$$

$$= \left(\frac{\partial h'_f}{\partial T_b} \right) T_w - \left[\frac{1}{(A')^2} \right] \frac{\partial A'}{\partial T_b}(h'_i - h'_f)(1 - e^{-A'\,T_w})$$

$$+ \frac{1}{A'} \left(\frac{\partial h'_i}{\partial T_b} - \frac{\partial h'_f}{\partial T_b} \right) (1 - e^{-A'\,T_w})$$

$$+ \frac{1}{A'} (h'_i - h'_f)\, T_w\, e^{-A'\,T_w} \frac{\partial A'}{\partial T_b},$$

where, from Eq. (10.92),

$$\frac{\partial A'}{\partial T_b} = \frac{\partial (p'_b \lambda_g + p'_g \lambda_b)}{\partial T_b} = \lambda_g \frac{\partial p'_b}{\partial T_b} + \lambda_b \frac{\partial p'_g}{\partial T_b},$$

which, from Eqs. (10.84) and (10.85), is

$$\frac{\partial A'}{\partial T_b} = \lambda_g \frac{\partial}{\partial T_b}\left[\frac{p_b\, e^{-\lambda_b T_b}}{p_b\, e^{-\lambda_b T_b} + p_g\, e^{-\lambda_g T_b}}\right]$$
$$+ \lambda_b \frac{\partial}{\partial T_b}\left[\frac{p_g\, e^{-\lambda_g T_b}}{p_b\, e^{-\lambda_b T_b} + p_g\, e^{-\lambda_g T_b}}\right],$$

or

$$\frac{\partial A'}{\partial T_b} = \frac{p_b\, p_g\, (\lambda_b - \lambda_g)^2\, e^{-(\lambda_b + \lambda_g) T_b}}{(p_b\, e^{-\lambda_b T_b} + p_g\, e^{-\lambda_g T_b})^2}; \qquad (10.96)$$

from Eq. (10.91),

$$\frac{\partial h'_f}{\partial T_b} = \frac{\partial (\lambda_b \lambda_g / A')}{\partial T_b} = -\frac{\lambda_b \lambda_g}{(A')^2} \frac{\partial A'}{\partial T_b},$$

which, after substituting A' and $\partial A'/\partial T_b$, from Eqs. (10.92) and (10.96), respectively, and rearranging, becomes

$$\frac{\partial h'_f}{\partial T_b} = -\frac{(p_b\, \lambda_b)(p_g\, \lambda_g)(\lambda_b - \lambda_g)^2\, e^{-(\lambda_b + \lambda_g) T_b}}{(p_b\, \lambda_g\, e^{-\lambda_b T_b} + p_g\, \lambda_b\, e^{-\lambda_g T_b})^2}; \qquad (10.97)$$

and from Eq. (10.90),

$$\frac{\partial h'_i}{\partial T_b} = \frac{\partial (p'_g \lambda_g + p'_b \lambda_b)}{\partial T_b} = \left(\frac{\partial p'_g}{\partial T_b}\right) \lambda_g + \left(\frac{\partial p'_b}{\partial T_b}\right) \lambda_b,$$

or from Eqs. (10.84) and (10.85),

$$\frac{\partial h'_i}{\partial T_b} = -\frac{p_b\, p_g\, (\lambda_b - \lambda_g)^2\, e^{-(\lambda_b - \lambda_g) T_b}}{(p_b\, e^{-\lambda_b T_b} + p_g\, e^{-\lambda_g T_b})^2}, \qquad (10.98)$$

or

$$\frac{\partial h'_i}{\partial T_b} = -\frac{\partial A'}{\partial T_b}. \qquad (10.99)$$

It may be observed that T_b^* cannot be obtained analytically by simply substituting the preceding derivatives into Eq. (10.95) and solving for T_b. However, the optimum burn-in time, T_b^*, can be obtained by solving the following univariate minimization problem:

$$\underset{T_b > 0}{\text{Min}} \quad C(T_b). \tag{10.100}$$

This minimization can be achieved using any one-dimensional nonlinear search method, or by evaluating $C(T_b)$ directly. A computer program in FORTRAN language was developed, which is listed in Appendix 10A, to minimize Eq. (10.94) using the Davidon-Fletcher-Power Minimization Method [6]. This method takes advantage of the first derivative of the objective function, the total cost function in this case, and converges very fast.

Note that Eq. (10.94) is derived based on the assumption that the burn-in is conducted at the operating stress level. In practical situations, burn-ins are conducted at higher stress levels. Assume that the equivalent acceleration factor of an accelerated burn-in is A_{FB}, which may be obtained using the techniques presented in the following chapters. Then, a T_b-hr burn-in at the operating stress level is equivalent to a burn-in at the elevated stress level for a duration of

$$T_{ba} = T_b / A_{FB}. \tag{10.101}$$

Correspondingly, the optimum burn-in time at the operating stress level, T_b^*, is related to the optimum burn-in time at the elevated stress level, T_{ba}^*, via the following equation:

$$T_{ba}^* = T_b^* / A_{FB}. \tag{10.102}$$

In the mean time, increasing the burn-in stress level will introduce additional costs per hour of burn-in. In this case, the life-cycle cost model of Eq. (10.94) should be modified as follows:

$$C(T_{ba}) = N \left[C_0 + C_{ba} T_{ba} + C_{fb} H_{ba}(T_{ba}) + C_{fw} H_w(T_w | T_{ba}^e) \right],$$

$$\tag{10.103}$$

where

C_{ba} = actual burn-in cost per component per hour of burn-in at the elevated stress level,

$H_{ba}(T_{ba})$ = expected number of failures in T_{ba} hours of burn-in at the elevated stress level, which is equal to the expected number of failures in T_b hours of burn-in at the operating stress

level, according to Eq. (10.101), or
$$H_{ba}(T_{ba}) = H_b(T_b), \tag{10.104}$$
T_{ba}^e = burn-in time at the elevated stress level equivalent to that at the operating stress level,

or

$$T_{ba}^e = T_b. \tag{10.105}$$

Since the renewal functions presented in this chapter are derived from the life distribution at the operating stress level, then it is desirable to express the life-cycle cost function in terms of the burn-in time at the operating stress level, T_b. Substituting Eqs. (10.101), (10.104) and (10.105) into Eq. (10.103) yields

$$C(T_b) = N \left[C_0 + C_{ba} \left(\frac{T_b}{A_{FB}} \right) + C_{fb} H_b(T_b) + C_{fw} H_w(T_w|T_b) \right], \tag{10.106}$$

The optimum burn-in time for the minimum life-cycle cost, T_b^*, may be obtained by taking the first partial derivative of Eq. (10.106) with respect to T_b, setting it equal to zero, and solving for T_b; i.e.,

$$\frac{\partial C(T_b)}{\partial T_b} = 0,$$

or

$$N \left[0 + \frac{C_{ba}}{A_{FB}} + C_{fb} \frac{\partial H_b(T_b)}{\partial T_b} + C_{fw} \frac{\partial H_w(T_w|T_b)}{\partial T_b} \right] = 0. \tag{10.107}$$

The two associated partial derivatives of Eq. (10.107), $\frac{\partial H_b(T_b)}{\partial T_b}$ and $\frac{\partial H_w(T_w|T_b)}{\partial T_b}$, are the same as those in Eq. (10.95).

It may be observed that T_b^* cannot be obtained analytically by simply substituting the preceding derivatives into Eq. (10.107) and solving for T_b. However, the optimum burn-in time, T_b^*, can be obtained by solving the following univariate minimization problem:

$$\text{Min}_{T_b > 0} \quad C(T_b). \tag{10.108}$$

This minimization can be achieved using any one-dimensional nonlinear search method, or by evaluating $C(T_b)$ directly. The computer program given in the Appendix 10A covers this case also.

Therefore, minimizing Eq. (10.106) will yield the equivalent operating times for the optimum burn-in time at the elevated stress level, T_b^*. Then, the actual burn-in time at the elevated stress level, T_{ba}^*, can be obtained from Eq. (10.102). The example that follows illustrates the application of Eqs. (10.94) and (10.106).

MIXED-EXPONENTIAL BURN-IN QUANTIFICATION

EXAMPLE 10-8

A total of $N = 1,000$ components are subjected to burn-in. The mixed-exponential times-to-failure distribution parameters, under operational stress level burn-in, are determined to be the following:

$$p_b = 0.01; \ p_g = 0.99; \ \lambda_b = 10^{-2} \text{ fr/hr}; \ \lambda_g = 10^{-7} \text{ fr/hr}.$$

The associated costs are the following:

$$C_0 = \$0.20 \text{ per component}; \ C_b = \$0.04 \text{ per component per hour};$$
$$C_{fb} = \$7.00 \text{ per failure}; \quad C_{fw} = \$1,000 \text{ per failure}.$$

The warranty period is $T_w = 3,000$ hr. Find the optimum burn-in duration and the associated life-cycle cost for these components, T_b^* and $C(T_b^*)$, respectively, for the following cases:

Case 1: The burn-in stress level is the same as the operating stress.

Case 2: The stress level is elevated such that the equivalent acceleration factor is $A_{FB} = 3$. The actual burn-in cost due to the stress increase becomes $C_{ba} = \$0.05$.

Case 3: The stress level is elevated such that the equivalent acceleration factor is $A_{FB} = 5$. The actual burn-in cost due to the stress increase becomes $C_{ba} = \$0.06$.

SOLUTIONS TO EXAMPLE 10-8

Case 1: Since the burn-in stress level is the same as the operating stress level, then the cost function given by Eq. (10.94) should be used here. This cost function is minimized using the computer program given in Appendix 10A. The optimal solution for the given data is

$$T_b^* = 91.964 \text{ hr, or } 92 \text{ hr,}$$

and

$$C(T_b^*) = \$8,248.940, \text{ or } \$8,249.$$

Case 2: Since the equivalent acceleration factor is $A_{FB} = 3$, then, minimizing Eq. (10.106) yields the optimum burn-in time at the operating stress level, which is equivalent to the actual optimum burn-in time at the given elevated stress level and under the corresponding hourly burn-in cost. This cost function is minimized using the computer program given in Appendix 10A also. The optimal solution for the given data is

$$T_b^* = 179.487 \text{ hr, or } 179 \text{ hr,}$$

which is the optimum burn-in time at the operating stress level. Then, the optimum burn-in time at the given elevated stress level is

$$T_{ba}^* = T_b^*/A_{FB} = 179.487/3,$$

or

$$T_{ba}^* = 59.829 \text{ hr, or } 60 \text{ hr.}$$

The corresponding minimum cost is

$$C(T_b^*) = \$5,229.095, \text{ or } \$5,229.$$

Case 3: Since the equivalent acceleration factor is $A_{FB} = 5$, then, minimizing Eq. (10.106) yields the optimum burn-in time at the operating stress level, which is equivalent to the actual optimum burn-in time at the given elevated stress level and under the corresponding hourly burn-in cost. This cost function is minimized using the computer program given in Appendix 10A also. The optimal solution for the given data is

$$T_b^* = 212.343 \text{ hr, or } 212 \text{ hr,}$$

which is the optimum burn-in time at the operating stress level. Then, the optimum burn-in time at the given elevated stress level is

$$T_{ba}^* = T_b^*/A_{FB} = 212.343/5,$$

or

$$T_{ba}^* = 42.469 \text{ hr, or } 42 \text{ hr.}$$

The corresponding minimum cost is

$$C(T_b^*) = \$4,318.986, \text{ or } \$4,319.$$

It may be seen from these three cases that for the given hourly burn-in cost at both operating and elevated stress levels, increasing the burn-in stress levels will decrease the burn-in times and the corresponding life-cycle cost substantially; i.e.,

Case (2) versus **Case (1)**: 60 hr < 92 hr; \$5,229 < \$8,249.

Case (3) versus **Case (1)**: 42 hr < 92 hr; \$4,319 < \$8,249.

10.12 BAYESIAN APPROACH TO BURN-IN

10.12.1 WHY THE BAYESIAN APPROACH?

The failure rate function, $\lambda(T)$, of mixed-exponential distributions given by Eq. (10.9) yields a decreasing failure rate with increasing T which approaches λ_g as T tends to infinity. The classical approaches, such as the one presented in Section 10.7 was to specify some small value for *Burn-in Residue*, ε, and choose the burn-in duration, T_b, sufficiently large so that the difference between $\lambda(T)$ and λ_g is less than ε. To accomplish this, the failure rates λ_b and λ_g have to be specified in advance of testing. The difficulty with this approach is that it requires specifying exactly unknown quantities such as λ_b and λ_g, and ignores costs. This may lead to unreasonable burn-in duration times. To overcome these shortcomings of the classical approach, Barlow et al [7] proposed a Bayesian approach which assumes a joint prior distribution for the unknown parameters.

The Bayesian approach to decision analysis requires the specification of a loss, or utility, as a function of decision variables and logically possible outcomes. Since outcomes cannot be known in advance, a probability assessment must be made for possible outcomes. The expected value of the loss function is then computed relative to this probability assessment and that decision is made which minimizes the expected loss. Let

T_b^* = optimal duration for the burn-in under a given stress level l.

Now introduce the notation and discuss the costs involved in stressing and stopping the burn-in. Let

C_b = cost of having a "bad" component escape the burn-in,

and

C_g = cost of having a "good" component destroyed by the burn-in.

The costs C_b and C_g are "decision" costs in the sense that they describe the cost of wrong "decisions" regarding the component. Since the major concern in burn-in is to purify the population of components in the lot, the cost of C_b is usually much larger than C_g. In any event, it will only be necessary to specify the ratio C_b/C_g. We will not consider the cost of burn-in as a function of time T.

10.12.2 THE JOINT PRIOR DISTRIBUTION FOR PARAMETERS OF THE BIMODAL MIXED-EXPONENTIAL LIFE DISTRIBUTION

Since in the mixture *pdf* given by Eq. (10.7), p, λ_b and λ_g are in general unknown, we will need to specify a joint prior distribution for them. The design problem to be solved is the determination of $T_b = T_b^*$ in such a way that *the expected total cost with respect to the joint prior distribution for p, λ_b and λ_g is minimized*. Consider families of prior distributions which are convenient and large enough to accommodate different shades of opinion. The prior joint *pdf's* for λ_b and λ_g are conditional to the stress level l.

Assume that p is independent of λ_b and λ_g. The beta family of prior distributions with parameters of α and β can be used for p. Note that the uniform *pdf* on (0,1) is a special case of the beta *pdf*. An exponential *pdf* with parameter θ is used as the prior distribution for λ_g and another exponential *pdf* with parameter τ, shifted by λ_g, is used as the conditional distribution for λ_b given λ_g. Then, the joint prior distribution for p, λ_b and λ_g, given $l = 1$, is

$$f(p, \lambda_b, \lambda_g) = f(\theta) \, f(\lambda_b, \lambda_g),$$
$$= f(\theta) \, f(\lambda_g) \, f(\lambda_b | \lambda_g),$$
$$= \left[\frac{\Gamma(\alpha+\beta)}{\Gamma(\alpha)\,\Gamma(\beta)} \, p^{\alpha-1} \, (1-p)^{\beta-1} \right] \left(\theta \, e^{-\theta \, \lambda_g} \right) \left(\frac{\tau \, e^{-\tau \, \lambda_b}}{e^{-\tau \, \lambda_g}} \right),$$

or

$$f(p, \lambda_b, \lambda_g) = \theta \, \tau \, \frac{\Gamma(\alpha+\beta)}{\Gamma(\alpha)\Gamma(\beta)} p^{\alpha-1} \, (1-p)^{\beta-1} \, e^{-(\theta-\tau) \, \lambda_g} \, e^{-\tau \, \lambda_b},$$

(10.109)

for $0 < p < 1$ and $0 < \lambda_g < \lambda_b$, where

α, β = parameters of the beta *pdf* for p,
θ = parameter of the exponential prior *pdf* for λ_g such that $1/\theta$ is the prior expected value of λ_g,

and

τ = parameter of the exponential prior *pdf* for λ_b such that $1/\tau$ is the prior expected value of λ_b.

It follows that

$$E(\lambda_g) = 1/\theta, \qquad (10.110)$$

$$E(\lambda_b|\lambda_g) = \lambda_g + 1/\tau, \qquad (10.111)$$

and

$$E(\lambda_b) = \frac{\theta + \tau}{\theta\,\tau}. \qquad (10.112)$$

Equation (10.111) can be derived as follows:

$$E(\lambda_b|\lambda_g) = \int_{\lambda_g}^{\infty} \lambda_b\, f(\lambda_b|\lambda_g)\, d\lambda_b,$$

or

$$E(\lambda_b|\lambda_g) = \int_{\lambda_g}^{\infty} \lambda_b\, \tau\, e^{-\tau\,(\lambda_b - \lambda_g)}\, d\lambda_b.$$

Let

$$x = \lambda_b - \lambda_g,$$

then

$$E(\lambda_b|\lambda_g) = \int_0^{\infty} (x + \lambda_g)\, \tau\, e^{-\tau\, x}\, dx,$$

$$= \lambda_g \int_0^{\infty} \tau\, e^{-\tau\, x}\, dx + \int_0^{\infty} x\, \tau\, e^{-\tau\, x}\, dx,$$

or

$$E(\lambda_b|\lambda_g) = \lambda_g + 1/\tau.$$

Equation (10.112) can be derived as follows: The marginal distribution of λ_b is

$$f(\lambda_b) = \int_0^{\lambda_b} f(\lambda_b, \lambda_g)\, d\lambda_g,$$

$$= \tau\, e^{-\tau\, \lambda_b} \int_0^{\lambda_b} \theta\, e^{-\lambda_g\, (\theta - \tau)}\, d\lambda_g,$$

$$= \tau\, e^{-\tau\, \lambda_b} \frac{\theta}{\theta - \tau} \left[1 - e^{-\lambda_b\, (\theta - \tau)}\right],$$

or

$$f(\lambda_b) = \frac{\theta\, \tau}{\theta - \tau} \left(e^{-\tau\, \lambda_b} - e^{-\theta\, \lambda_b}\right). \tag{10.113}$$

Therefore,

$$E(\lambda_b) = \int_0^\infty \lambda_b\, f(\lambda_b)\, d\lambda_b,$$

$$= \int_0^\infty \lambda_b\, \frac{\theta\, \tau}{\theta - \tau} \left(e^{-\tau\, \lambda_b} - e^{-\theta\, \lambda_b}\right) d\lambda_b,$$

$$= \frac{\theta\, \tau}{\theta - \tau} \int_0^\infty \left[\frac{1}{\tau}\left(\lambda_b\, \tau\, e^{-\tau\, \lambda_b}\right) - \frac{1}{\theta}\left(\lambda_b\, \theta\, e^{-\theta\, \lambda_b}\right)\right] d\lambda_b,$$

$$= \frac{\theta\, \tau}{\theta - \tau} \left(\frac{1}{\tau^2} - \frac{1}{\theta^2}\right),$$

$$= \frac{\theta\, \tau}{\theta - \tau} \left(\frac{\theta^2 - \tau^2}{\theta^2 \tau^2}\right),$$

$$= \frac{\theta\, \tau}{\theta - \tau} \left[\frac{(\theta + \tau)(\theta - \tau)}{\theta^2 \tau^2}\right],$$

or

$$E(\lambda_b) = \frac{\theta + \tau}{\theta\, \tau}.$$

It may be seen that the marginal distribution of λ_b, $f(\lambda_b)$, given by Eq. (10.113), has a mode at

$$\lambda_b = [\log_e(\tau/\theta)]/(\tau - \theta)$$

for $\theta > \tau$.

In practical situations, where λ_g is much smaller than λ_b, the value of θ will be chosen much larger than the value of τ. The conditional exponential *pdf* of λ_b then becomes, if compared with the prior exponential *pdf* for λ_g, practically flat. The prior uncertainty for λ_g can be expressed through prior *pdf's* that make use of knowledge about production process standards, as contained in publications such as in Military Standards. On the other hand, the analyst is able to express his relatively much greater "ignorance" about λ_b, given λ_g, through an almost flat prior conditional *pdf* which is nevertheless proper. For values of θ much larger than the value of τ, the following relation shows how small the prior probability of having λ_b "close" to λ_g is – even if the mode of the marginal prior *pdf*, $f(\lambda_b, \lambda_g)$, is the origin; namely,

$$P(\lambda_b < M \, \lambda_g) = 1 - \frac{\theta}{\theta + \tau(M-1)}, \qquad (10.114)$$

for every $M > 1$. This relationship can be derived as follows:

$$P(\lambda_b < M \, \lambda_g) = \int_0^\infty \left[\int_{\lambda_g}^{M\lambda_g} f(\lambda_b, \lambda_g) \, d\lambda_b \right] d\lambda_g,$$

$$= \int_0^\infty \left[\int_{\lambda_g}^{M\lambda_g} \tau \, e^{-\tau \lambda_b} \, d\lambda_b \right] \theta \, e^{-\lambda_g (\theta - \tau)} \, d\lambda_g,$$

$$= \int_0^\infty \left(e^{-\tau \lambda_g} - e^{-M \tau \lambda_g} \right) \theta \, e^{-\lambda_g (\theta - \tau)} \, d\lambda_g,$$

$$= \int_0^\infty \left\{ \theta \, e^{-\theta \lambda_g} - \theta \, e^{-[\theta + (M-1)\tau] \lambda_g} \right\} d\lambda_g,$$

or

$$P(\lambda_b < M \lambda_g) = 1 - \frac{\theta}{\theta + \tau(M-1)},$$

for every $M > 1$.

10.12.3 THE BAYESIAN COST MODEL AND THE OPTIMAL DURATION OF BURN-IN

There is a proportion p of substandard components in the lot with N total components. But the inspection of any component does not reveal whether it is substandard or not. This fact makes all components look similar and entails a judgement of exchangeability of the components with respect to quality and behavior under burn-in. In particular, for any component in the lot, the probability that it is substandard is

$$E(p) = \alpha/(\alpha + \beta), \tag{10.115}$$

where E stands for expectation with respect to the beta prior for p.

The *conditional cost* per component of a burn-in of duration T_b at stress level $l = 1$ is, therefore, derived as follows:

$$\begin{aligned}C(T_b|p, \lambda_b, \lambda_g) &= [(\text{cost of having a "bad" component escape the} \\ &\quad \text{burn-in}) \times P(\text{any component drawn at random} \\ &\quad \text{from the lot is a "bad" component}) \\ &\quad \times P(\text{a "bad" component escapes the burn-in})] \\ &\quad + [(\text{cost of having a "good" component destroyed} \\ &\quad \text{by burn-in}) \times P(\text{any component drawn at} \\ &\quad \text{random from the lot is a "good" component }) \\ &\quad \times P(\text{a "good" component fails during burn-in})], \\ &= C_b\, p\, e^{-\lambda_b T_b} + C_g\, (1-p)\, [1 - e^{-\lambda_g T_b}],\end{aligned}$$

or

$$C(T_b|p, \lambda_b, \lambda_g) = p\, C_b\, [e^{-\lambda_b T_b}] + (1-p)\, C_g\, [1 - e^{-\lambda_g T_b}]. \tag{10.116}$$

The expected total cost with respect to the prior distribution for $(p, \lambda_b, \lambda_g)$ is

$$\begin{aligned}C(T_b) &= \int_{p=0}^{1} \int_{\lambda_b=0}^{\infty} \int_{\lambda_g=0}^{\lambda_b} C(T_b|p, \lambda_b, \lambda_g)\, f(p, \lambda_b, \lambda_g)\, dp\, d\lambda_b\, d\lambda_g, \\ &= \int_{p=0}^{1} \int_{\lambda_b=0}^{\infty} \int_{\lambda_g=0}^{\lambda_b} \left[p\, C_b\, e^{-\lambda_b T_b} + (1-p)\, C_g\left(1 - e^{-\lambda_g T_b}\right)\right] \\ &\quad \cdot f(p)\, f(\lambda_b, \lambda_g)\, dp\, d\lambda_b\, d\lambda_g,\end{aligned}$$

$$= C_g \int_{p=0}^{1} \int_{\lambda_b=0}^{\infty} \int_{\lambda_g=0}^{\lambda_b} (1-p) \, f(p) \, f(\lambda_b, \lambda_g) \, dp \, d\lambda_b \, d\lambda_g$$

$$+ C_b \int_{p=0}^{1} \int_{\lambda_b=0}^{\infty} \int_{\lambda_g=0}^{\lambda_b} p \, e^{-\lambda_b T_b} \, f(p) \, f(\lambda_b, \lambda_g) \, dp \, d\lambda_b \, d\lambda_g$$

$$- C_g \int_{p=0}^{1} \int_{\lambda_b=0}^{\infty} \int_{\lambda_g=0}^{\lambda_b} (1-p) \, e^{-\lambda_g T_b} \, f(p) \, f(\lambda_b, \lambda_g) \, dp \, d\lambda_b \, d\lambda_g,$$

or

$$C(T_b) = C_g \int_0^1 (1-p) \, f(p) \, dp \int_{\lambda_b=0}^{\infty} \int_{\lambda_g=0}^{\lambda_b} f(\lambda_b, \lambda_g) \, d\lambda_b \, d\lambda_g$$

$$+ C_b \int_0^1 p \, f(p) \, dp \int_{\lambda_b=0}^{\infty} \int_{\lambda_g=0}^{\lambda_b} e^{-\lambda_b T_b} \, f(\lambda_b, \lambda_g) \, d\lambda_b \, d\lambda_g$$

$$- C_g \int_0^1 (1-p) \, f(p) \, dp \int_{\lambda_b=0}^{\infty} \int_{\lambda_g=0}^{\lambda_b} e^{-\lambda_g T_b} \, f(\lambda_b, \lambda_g) \, d\lambda_b \, d\lambda_g.$$

(10.117)

But the first term in Eq. (10.117) is

$$C_g \int_0^1 (1-p) \, f(p) \, dp \int_{\lambda_b=0}^{\infty} \int_{\lambda_g=0}^{\lambda_b} f(\lambda_b, \lambda_g) \, d\lambda_b \, d\lambda_g$$

$$= C_g \left[\int_0^1 f(p) \, dp - \int_0^1 p \, f(p) \, dp \right] \times 1,$$

$$= C_g \left(1 - \frac{\alpha}{\alpha + \beta} \right),$$

$$= C_g \left(\frac{\beta}{\alpha + \beta} \right).$$

The second term in Eq. (10.117) is

$$C_b \int_0^1 p\, f(p)\, dp \int_{\lambda_b=0}^{\infty} \int_{\lambda_g=0}^{\lambda_b} e^{-\lambda_b T_b}\, f(\lambda_b, \lambda_g)\, d\lambda_b\, d\lambda_g$$

$$= C_b \left(\frac{\alpha}{\alpha+\beta}\right) \int_{\lambda_b=0}^{\infty} \int_{\lambda_g=0}^{\lambda_b} e^{-\lambda_b T_b}\, \theta\, \tau\, e^{-\lambda_g(\theta-\tau)} e^{-\lambda_b \tau}\, d\lambda_b\, d\lambda_g,$$

$$= C_b \left(\frac{\alpha}{\alpha+\beta}\right) \int_{\lambda_b=0}^{\infty} e^{-\lambda_b T_b}\, \frac{\theta\, \tau}{\theta - \tau} \left[e^{-\tau \lambda_b} - e^{-\theta \lambda_b}\right] d\lambda_b,$$

$$= C_b \left(\frac{\alpha}{\alpha+\beta}\right) \int_{\lambda_b=0}^{\infty} \frac{\theta\, \tau}{\theta - \tau} \left[e^{-(\tau+T_b)\lambda_b} - e^{-(\theta+T_b)\lambda_b}\right] d\lambda_b,$$

$$= C_b \left(\frac{\alpha}{\alpha+\beta}\right) \frac{\theta\, \tau}{\theta - \tau} \left(\frac{1}{\tau+T_b} - \frac{1}{\theta+T_b}\right),$$

or

$$= C_b \left(\frac{\alpha}{\alpha+\beta}\right) \frac{\theta\, \tau}{(\tau+T_b)(\theta+T_b)}.$$

The third term in Eq. (10.117) is

$$C_g \int_0^1 (1-p)\, f(p)\, dp \int_{\lambda_b=0}^{\infty} \int_{\lambda_g=0}^{\lambda_b} e^{-\lambda_g T_b} f(\lambda_b, \lambda_g)\, d\lambda_b\, d\lambda_g$$

$$= C_g \left(\frac{\beta}{\alpha+\beta}\right) \int_{\lambda_b=0}^{\infty} \int_{\lambda_g=0}^{\lambda_b} e^{-\lambda_g T_b}\, \theta\, \tau\, e^{-\lambda_g(\theta-\tau)} e^{-\lambda_b \tau}\, d\lambda_b\, d\lambda_g,$$

$$= C_g \left(\frac{\beta}{\alpha+\beta}\right) \int_{\lambda_b=0}^{\infty} \theta\, \tau\, e^{-\lambda_b \tau} \left[\int_{\lambda_g=0}^{\lambda_b} e^{-\lambda_g(T_b+\theta-\tau)}\, d\lambda_g\right] d\lambda_b,$$

$$= C_g \left(\frac{\beta}{\alpha+\beta}\right) \int_{\lambda_b=0}^{\infty} \left(\frac{\theta \tau}{T_b+\theta-\tau}\right) e^{-\lambda_b \tau} \left[1 - e^{-\lambda_b (T_b+\theta-\tau)}\right] d\lambda_b,$$

$$= C_g \left(\frac{\beta}{\alpha+\beta}\right) \left(\frac{\theta \tau}{T_b+\theta-\tau}\right) \int_{\lambda_b=0}^{\infty} \left[e^{-\lambda_b \tau} - e^{-\lambda_b (T_b+\theta)}\right] d\lambda_b,$$

$$= C_g \left(\frac{\beta}{\alpha+\beta}\right) \left(\frac{\theta \tau}{T_b+\theta-\tau}\right) \left(\frac{1}{\tau} - \frac{1}{T_b+\theta}\right),$$

or

$$= C_g \left(\frac{\beta}{\alpha+\beta}\right) \left(\frac{\theta}{T_b+\theta}\right).$$

Substituting the above into Eq. (10.117) yields

$$C(T_b) = C_g \left(\frac{\beta}{\alpha+\beta}\right) + C_b \left(\frac{\alpha}{\alpha+\beta}\right) \frac{\theta \tau}{(\tau+T_b)(\theta+T_b)}$$
$$- C_g \left(\frac{\beta}{\alpha+\beta}\right) \left(\frac{\theta}{T_b+\theta}\right). \tag{10.118}$$

To obtain the optimal T_b^* which minimizes the expected cost given by Eq. (10.118), differentiate both sides of Eq. (10.118) with respect to T_b, set it equal to zero, solve it for $T_b = T_b^*$ and get

$$\frac{dC(T_b)}{dT_b} = C_b \left(\frac{\alpha}{\alpha+\beta}\right) (\theta \tau) \left[\frac{-(\tau+T_b+\theta+T_b)}{(\tau+T_b)^2(\theta+T_b)^2}\right]$$
$$- C_g \left(\frac{\beta}{\alpha+\beta}\right) \theta \left[\frac{-1}{(T_b+\theta)^2}\right],$$
$$= 0.$$

Simplifying yields

$$C_b \alpha \tau (\theta + \tau + 2 T_b) = C_g \beta (T_b^2 + 2 \tau T_b + \tau^2),$$

or

$$\theta + \tau + 2 T_b = \left(\frac{C_g \beta}{C_b \alpha \tau}\right) (T_b^2 + 2 \tau T_b + \tau^2). \tag{10.119}$$

Let

$$K = \frac{1}{\tau} (b/a) (C_g/C_b). \tag{10.120}$$

Then, Eq. (10.119) becomes

$$\theta + \tau + 2\,T_b = K\,(T_b^2 + 2\,\tau\,T_b + \tau^2),$$

or

$$K\,T_b^2 + 2\,(\tau\,K - 1)\,T_b + (K\,\tau^2 - \theta - \tau) = 0.$$

Solving this equation yields

$$T_b = \frac{-(K\,\tau - 1) \pm \sqrt{K\,\theta - K\,\tau + 1}}{K},$$

or

$$T_b^* = \frac{1}{K}\,[1 - K\,\tau + \sqrt{1 + K\,(\theta - \tau)}]. \qquad (10.121)$$

If the value of T_b^* in Eq. (10.121) is negative, then the optimal decision is, of course, not to burn-in the lot. Note that T_b^* increases with θ for fixed τ and costs; i.e., the lower the failure rate for good components, λ_g, the longer the burn-in time would be.

Note that for the general case of $l \neq 1$, T_b^* would be written in the form $T_b^* = T_o^*/l$, with T_o^* being a constant independent of the acceleration of time; that is, T_o^* has the same value for all l. The minimum total expected cost of the burn-in process of optimal duration T_b^* for the whole lot is $C(T_b^*, l)$ multiplied by N.

10.12.4 OTHER BURN-IN MEASURES BASED ON BAYESIAN RESULTS

Now consider other measures of "goodness" relative to a burn-in design. *The expected probability that a substandard component will escape the burn-in of duration T_b is*

$$E(e^{-\lambda_b T_b}) = \int_{\lambda_b=0}^{\infty} \int_{\lambda_g=0}^{\lambda_b} e^{-\lambda_b T_b} f(\lambda_b, \lambda_g)\,d\lambda_b\,d\lambda_g,$$

$$= \int_{\lambda_b=0}^{\infty} \int_{\lambda_g=0}^{\lambda_b} e^{-\lambda_b T_b}\,\theta\,\tau\,e^{-\lambda_g(\theta-\tau)} e^{-\lambda_b \tau}\,d\lambda_b\,d\lambda_g,$$

$$= \int_{\lambda_b=0}^{\infty} e^{-\lambda_b T_b} \frac{\theta\,\tau}{\theta - \tau}\,[e^{-\tau\,\lambda_b} - e^{-\theta\,\lambda_b}]\,d\lambda_b,$$

$$= \int_{\lambda_b=0}^{\infty} \frac{\theta\tau}{\theta-\tau}\left[e^{-(\tau+T_b)\lambda_b} - e^{-(\theta+T_b)\lambda_b}\right]d\lambda_b,$$

$$= \frac{\theta\tau}{\theta-\tau}\left(\frac{1}{\tau+T_b} - \frac{1}{\theta+T_b}\right),$$

or

$$E(e^{-\lambda_b T_b}) = \frac{\theta\tau}{(\tau+T_b)(\theta+T_b)}, \qquad (10.122)$$

where E stands for expectation with respect to the prior joint *pdf* of λ_b and λ_g. Recall that $\lambda_b > \lambda_g$ so that Eq. (10.122) depends on both θ and τ. The expected number of substandard components that will escape from burn-in, also called the *Remaining Defect Density*, denoted by $D_R(T_b)$, can be obtained by substituting N for C_b in the second term of Eq. (10.118), as follows:

$$D_R(T_b) = N\left(\frac{\alpha}{\alpha+\beta}\right)\frac{\theta\tau}{(\tau+T_b)(\theta+T_b)}. \qquad (10.123)$$

The probability that a substandard component will not escape from the burn-in of duration T_b is called the *Burn-in Strength*, denoted by $BS(T_b)$, and can be obtained from Eq. (10.122) as follows:

$$BS(T_b) = 1 - E(e^{-\lambda_b T_b}),$$

or

$$BS(T_b) = 1 - \frac{\theta\tau}{(\tau+T_b)(\theta+T_b)}. \qquad (10.124)$$

On the other hand, the *probability that a good component will survive the burn-in* is

$$E(e^{-\lambda_g T_b}) = \int_{\lambda_b=0}^{\infty}\int_{\lambda_g=0}^{\lambda_b} e^{-\lambda_g T_b} f(\lambda_b, \lambda_g)\, d\lambda_b\, d\lambda_g,$$

$$= \int_{\lambda_b=0}^{\infty}\int_{\lambda_g=0}^{\lambda_b} e^{-\lambda_g T_b}\, \theta\tau\, e^{-\lambda_g(\theta-\tau)}\, e^{-\lambda_b \tau}\, d\lambda_b\, d\lambda_g,$$

$$= \int_{\lambda_b=0}^{\infty} \theta\tau\, e^{-\lambda_b \tau}\left[\int_{\lambda_g=0}^{\lambda_b} e^{-\lambda_g(T_b+\theta-\tau)}\, d\lambda_g\right] d\lambda_b,$$

$$= \int_{\lambda_b=0}^{\infty} \left(\frac{\theta \tau}{T_b + \theta - \tau}\right) e^{-\lambda_b \tau} \left[1 - e^{-\lambda_b (T_b + \theta - \tau)}\right] d\lambda_b,$$

$$= \left(\frac{\theta \tau}{T_b + \theta - \tau}\right) \int_{\lambda_b=0}^{\infty} \left[e^{-\lambda_b \tau} - e^{-\lambda_b (T_b + \theta)}\right] d\lambda_b,$$

$$= \left(\frac{\theta \tau}{T_b + \theta - \tau}\right) \left(\frac{1}{\tau} - \frac{1}{T_b + \theta}\right),$$

or

$$E(e^{-\lambda_g T_b}) = \frac{\theta}{T_b + \theta}. \tag{10.125}$$

The *expected number of good components remaining in the lot after the burn-in*, denoted by $G_R(T_b)$, can be obtained by substituting N for C_g in the third term of Eq. (10.118), as follows:

$$G_R(T_b) = N \left(\frac{\beta}{\alpha + \beta}\right) \left(\frac{\theta}{T_b + \theta}\right).$$

Another measure of interest in the Military Standards literature is the *Yield*, defined as the prior probability of having zero substandard components remaining in the lot after the burn-in. The *Yield*, denoted by $Y(T_b)$, is approximately

$$Y(T_b) \cong exp\left\{-\frac{\theta \tau [N\alpha/(\alpha + \beta)]}{[(\theta + T_b)(\tau + T_b)]}\right\} = e^{-D_R (T_b)}.$$

The parameter of the approximating Poisson distribution is the *Remaining Defect Density*, $D_R(T_b)$, given by Eq. (10.123). Note that

$$D_R(T_b) = [1 - BS(T_b)] D_{IN},$$

where

D_{IN} = expected number of substandard components before the burn-in,
  = *Incoming Defect Density*,

or

$$D_{IN} = N \left(\frac{\alpha}{\alpha + \beta}\right),$$

and

$$1 - BS(T_b) = \frac{\theta \tau}{(\theta + T_b)(\tau + T_b)},$$

from Eq. (10.124).

EXAMPLE 10-9

Assume $l = 1$ and
$$E(\lambda_b) = 1/\tau + 1/\theta \cong 10^{-2} \text{ fr/hr},$$
while
$$E(\lambda_g) = 1/\theta = 10^{-5} \text{ fr/hr}.$$

Find the optimal burn-in time, T_b^*, using the Bayesian approach.

SOLUTION TO EXAMPLE 10-9

From Eq. (10.121)
$$T_b^* = \frac{1}{K}\left[1 - K\tau + \sqrt{1 + K(\theta - \tau)}\right],$$
then,
$$(T_b^* + \tau)K - 1 = \sqrt{1 + K(\theta - \tau)}.$$

Squaring both sides yields
$$(T_b^* + \tau)^2 K^2 - 2(T_b^* + \tau)K + 1 = 1 + K(\theta - \tau),$$
or
$$(T_b^* + \tau)^2 K^2 = (2T_b^* + \theta + \tau)K.$$

Since $K \neq 0$, then
$$K = \frac{(2T_b^* + \theta + \tau)}{(T_b^* + \tau)^2}. \tag{10.126}$$

But from Eq. (10.120)
$$K = \frac{1}{\tau}(\beta/\alpha)(C_g/C_b),$$
or
$$K = \frac{1}{\tau}\left\{\frac{1}{[\alpha/(\alpha+\beta)]} - 1\right\}\frac{1}{C_b/C_g}. \tag{10.127}$$

Equating Eq. (10.127) to Eq. (10.126) yields
$$\frac{1}{\tau}\left\{\frac{1}{[\alpha/(\alpha+\beta)]} - 1\right\}\frac{1}{C_b/C_g} = \frac{(2T_b^* + \theta + \tau)}{(T_b^* + \tau)^2},$$

or

$$C_b/C_g = \frac{(T_b^* + \tau)^2}{\tau(2T_b^* + \theta + \tau)} \left\{ \frac{1}{[\alpha/(\alpha+\beta)]} - 1 \right\}. \qquad (10.128)$$

From the given information

$$\theta = 10^5 \text{ hr/fr},$$

and

$$\tau \cong 10^2 \text{ hr/fr}.$$

Substituting the values of θ and τ into Eq. (10.128) yields

$$C_b/C_g = \frac{(T_b^* + 100)^2}{100(2T_b^* + 100,000 + 100)} \left\{ \frac{1}{[\alpha/(\alpha+\beta)]} - 1 \right\},$$

or

$$C_b/C_g = \frac{(T_b^* + 100)^2}{200(T_b^* + 50,050)} \left\{ \frac{1}{[\alpha/(\alpha+\beta)]} - 1 \right\}. \qquad (10.129)$$

Plotting "C_b/C_g" versus "T_b^*" according to Eq. (10.129) with $\alpha/(\alpha+\beta)$ as the varying parameter yields Figure 10.7. The optimum burn-in time, T_b^*, for a given combination of C_b/C_g and $\alpha/(\alpha+\beta)$, can be obtained by drawing a horizontal line from the y-axis at the given value of C_b/C_g which intersects with the curve corresponding to the given value of $\alpha/(\alpha+\beta)$, and then drawing a vertical line downward which intersects with the x-axis at T_b^*. It is this final intersection point which gives the optimum burn-in time. In Table 10.1, Column 3 lists the optimum burn-in durations, T_b^*, computed from Eqs. (10.121) and (10.120), for various combinations of C_b/C_g and $\alpha/(\alpha+\beta)$.

The optimal duration for $l \neq 1$ is

$$T_b^{**} = T_b^*/l. \qquad (10.130)$$

Values of T_b^{**} under stress levels $l = 2$ and $l = 4$ for various combinations of C_b/C_g and $\alpha/(\alpha+\beta)$ are listed in Columns 4 and 5, respectively, of Table 10.1.

Optimal stress levels, l^*, can also be determined for fixed durations, T_b, since the contours are the same for any $l \, \theta/l \, \tau = 10^5/10^2 = 10^3$. Therefore,

$$l^* = T_b^*/T_b. \qquad (10.131)$$

For example, the optimum burn-in stress level for the specified burn-in duration $T_b = 100$ hr, when $C_b/C_g = 10$ and $\alpha/(\alpha+\beta) = 0.002$, is

$$l^* = 349.45/100 = 3.4945 \cong 3.5,$$

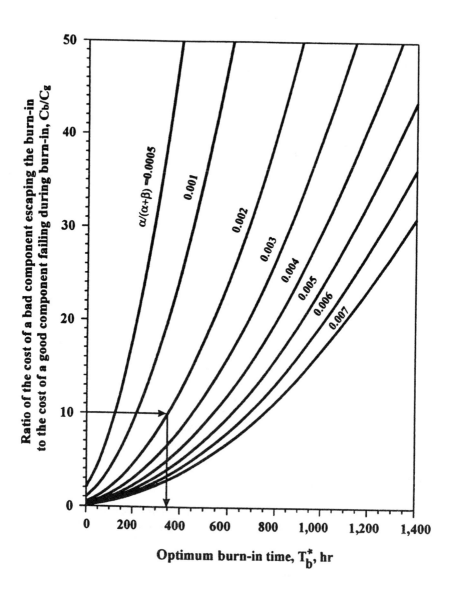

Fig. 10.7– Determination of the optimal burn-in duration, T_b^*, for various combinations of C_b/C_g and $\alpha/(\alpha+\beta)$, when $\theta = 10^5$ hr/fr and $\tau = 10^2$ hr/fr for Example 10-9.

TABLE 10.1– Optimum burn-in durations, T_b^*, for $l = 1$ and T_b^{**} for $l \neq 1$, in hours, for various combinations of C_b/C_g and $\alpha/(\alpha+\beta)$, for Example 10–9.

1	2	3	4	5
C_b/C_g	$\alpha/(\alpha+\beta)$	T_b^* ($l=1$)	T_b^{**} ($l=2$)	T_b^{**} ($l=4$)
5	0.0005	58.32	29.16	14.58
5	0.001	124.11	62.06	31.03
5	0.002	217.39	108.70	54.35
5	0.003	289.19	144.60	72.30
5	0.004	349.90	174.95	87.48
10	0.0005	124.05	62.03	31.01
10	0.001	217.23	108.62	54.31
10	0.002	349.45	174.73	87.36
10	0.003	451.29	225.65	112.82
10	0.004	537.44	268.72	134.36
20	0.0005	217.15	108.58	54.29
20	0.001	349.22	174.61	87.31
20	0.002	536.79	268.40	134.20
20	0.003	681.41	340.71	170.35
20	0.004	803.84	401.92	200.96

which is actually the optimum acceleration factor.

REFERENCES

1. Perlstein, H. J., Littlefield, J. W. and Bazovsky, I., "The Quantification of Environmental Stress Screening," *Proceedings of the Institute of Environmental Sciences*, San Jose, California, pp. 202-208, 1987.

2. Perlstein, D. and Welch, R. L. W., "A Bayesian Approach to the Analysis of Burn-in of Mixed Populations," *Proceedings of the Annual Reliability and Maintainability Symposium*, pp. 417-421, 1993.

3. Cox, D. R., *Renewal Theory*, Methuen Press, London, 142 pp., 1962.

4. Reddy, R. K. and Dietrich, D. L., "A Two-level *ESS* Model: A Mixed Distribution Approach," *IEEE Transactions on Reliability*, Vol. 43, No. 1, pp. 85-90, March 1994.

5. Kececioglu, Dimitri and Sun, Feng-Bin, *Environmental Stress Screening - Its Quantification, Optimization and Management*, Prentice Hall, Upper Saddle River, New Jersey 07458, 544 pp., 1995.

6. Press, W. H., Teukolsky, S. A., Vetterling, W. T., and Flannery, B. P., *Numerical Recipes in Fortran*, Second Edition, Cambridge University Press, 963 pp., 1992.

7. Barlow, R. E., Bazovsky, I. and Wechsler, S., "Classical and Bayes Approaches to Environmental Stress Screening (*ESS*): A Comparison," *Proceedings of the Annual Reliability and Maintainability Symposium*, Los Angeles, California, pp. 81-84, 1990.

ADDITIONAL REFERENCES

1. Canfield, R. V., "Cost Effective Burn-in and Replacement Times," *IEEE Transactions on Reliability*, Vol. R-24, No. 2, pp. 154-156, June 1975.

2. Kececioglu, Dimitri and Sun, Feng-Bin, "Mixed-Weibull Parameter Estimation for Burn-in Data Using the Bayesian Approach," *Microelectronics and Reliability*, Vol. 34, No. 10, pp. 1657-1679, 1994.

3. Kececioglu, Dimitri, *Reliability Engineering Handbook*, Prentice Hall, Upper Saddle River, NJ, Vol. 1, Fourth Printing, 726 pp., 1995.

4. Perlstein, H. J., Littlefield, J. W. and Bazovsky, I., "*ESS* Quantification for Complex Systems," *Proceedings of the Institute of Environmental Sciences*, King of Prussia, Pennsylvania, pp. 50-57, 1988.

5. Stewart, L. T. and Johnson, J. D., "Determining Optimum Burn-in and Replacement Times Using Bayesian Decision Theory," *IEEE Transactions on Reliability*, Vol. R-21, No. 3, pp. 170-175, August 1972.

APPENDIX 10A
COMPUTER PROGRAM FOR EXAMPLE 10-8

```
C$$$$$$$$$$$$$$$$$$$$$$$$$$$$$$$$$$$$$$$$$$$$$$$$$$$$$$$$$$$$$$$$$$
C     THIS IS THE MASTER PROGRAM FOR THE COST MINIMIZATION       $
C     USING THE DAVIDON-FLETCHER-POWELL MINIMIZATION METHOD      $
C     FOR EXAMPLE 10--8 (BIMODAL EXPONENTIAL BURN-IN) IN         $
C     THE BURN-IN BOOK                                            $
C                                                                 $
C     PB=PROPORTION OF THE WEAK SUBPOPULATION (p_b),              $
C     PG=PROPORTION OF THE STRONG SUBPOPULATION (p_g),            $
C     RB=FAILURE RATE OF THE WEAK SUBPOPULATION (lambda_b),       $
C     RG=FAILURE RATE OF THE STRONG SUBPOPULATION (lambda_g),     $
C     N=TOTAL NUMBER OF COMPONENTS TO BE PLACED ON BURN-IN,       $
C     TB=BURN-IN TIME (T_b),                                      $
C     TW=WARRANTY TIME (T_w),                                     $
C     CO=FIXED COST FOR BURN-IN PER COMPONNET (C_0),              $
C     CB=COST OF BURN-IN PER COMPONENT PER HOUR (C_b),            $
C     CFB=COST PER FAILURE DURING BURN-IN (C_fb),                 $
C     CFW=COST PER FAILURE DURING WARRANTY (C_fw),                $
C     NDIM=NUMBER OF UNKNOWNS TO BE DETERMINED (=1 HERE).         $
C                                                                 $
C     FILES NEEDED: EX10-8-1.FOR + DFPMIN.FOR + LNSRCH.FOR        $
C     EX10-8-1.FOR: THIS FILE.                                    $
C     DFPMIN.FOR: SEE PAGE 421 OF REFERENCE [6] OF THIS CHAPTER.  $
C     LNSRCH.FOR: SEE PAGE 378 OF REFERENCE [6] OF THIS CHAPTER.  $
C$$$$$$$$$$$$$$$$$$$$$$$$$$$$$$$$$$$$$$$$$$$$$$$$$$$$$$$$$$$$$$$$$$
C
      PARAMETER(NDIM=1,FTOL=1.0E-6)
      COMMON /PAR1/RB,RG
      COMMON /PAR2/PB,PG
      COMMON /PAR3/TBURN,TW,N
      COMMON /PAR4/CO,CB,CFB,CFW
      COMMON /PAR5/AFB
      COMMON /COST/COST
      COMMON /RENEW/HB,HW
      DIMENSION TB(NDIM)
      EXTERNAL FUNC, DFUNC
      OPEN(5, FILE='EX10-8.IN', STATUS='UNKNOWN')
      READ(5,*) N,PB,RB,PG,RG,TW,CO,CB,CFB,CFW,AFB
      WRITE(*,*)'                    THE INUT PARAMETERS'
      WRITE(*,*)('*',I=1,78)
      WRITE(*,*)'N=',N, '  PB=',PB, '  RB=',RB
```

```
WRITE(*,*)'PG=',PG, '   RG=',RG
WRITE(*,*)'TW=',TW, '   CO=',CO
WRITE(*,*)'CB=',CB, '   CFB=',CFB
WRITE(*,*)'CFW=',CFW, '   AFB=',AFB
WRITE(*,*)('*', I=1,78)
WRITE(*,*)'GIVE YOUR INITIAL GUESS FOR TB, PLEASE!'
READ(*,*) TB(1)
WRITE(*,*)'THE STARTING VALUE FOR TB IS'
WRITE(*,*)'TB(1)=',TB(1)
CALL DFPMIN(TB,NDIM,FTOL,ITER,FRET,FUNC,DFUNC)
TB(1)=TB(1)/AFB
WRITE(*,*)'                       RESULTS'
WRITE(*,*)('*', I=1,78)
WRITE(*,*)'THE OPTIMUM BURN-IN TIME IS'
WRITE(*,*)'          TBURN=',TB(1)
WRITE(*,*)'THE CONVERGENCE TOLERANCE ON COST FUNCTION IS'
WRITE(*,*)'             FTOL=',FTOL
WRITE(*,*)'THE NUMBER OF TOTAL ITERATIONS IS'
WRITE(*,*)'             ITER=',ITER
WRITE(*,*)'THE MINIMUM COST FUNCTION VALUE IS'
WRITE(*,*)'             CMIN=',FRET
WRITE(*,*)'COST=',COST
WRITE(*,*)'HB=',HB, '   HW=',HW
STOP
END

C       THIS SUBPROGRAM EVALUATES THE TOTAL COST FUNCTION
C       (THE OBJECTIVE FUNCTION) FOR EXAMPLE 10-8 IN THE
C       BURN-IN BOOK
C
FUNCTION FUNC(X)
COMMON /PAR1/RB,RG
COMMON /PAR2/PB,PG
COMMON /PAR3/TBURN,TW,N
COMMON /PAR4/CO,CB,CFB,CFW
COMMON /PAR5/AFB
COMMON /COST/COST
COMMON /RENEW/HB,HW
DIMENSION X(1)
WRITE(*,*)'X(1)=',X(1)
A=PB*RG + PG*RB
HF=RB*RG/A
HI=PB*RB + PG*RG
RELBP=PB*EXP(-RB*X(1))
```

APPENDIX 10A

```
RELGP=PG*EXP(-RG*X(1))
RELP=RELBP+RELGP
PB1=RELBP/RELP
PG1=RELGP/RELP
A1=PB1*RG+PG1*RB
HI1=PG1*RG+PB1*RB
HF1=RB*RG/A1
HB=HF*X(1) + (1.0/A)*(HI-HF)*(1.0-EXP(-A*X(1)))
HW=HF1*TW+(1.0/A1)*(HI1-HF1)*(1-EXP(-A1*TW))
COST=N*(CO+CB*(X(1)/AFB)+HB*CFB+HW*CFW)
FUNC=COST
WRITE(*,*)'TB=',X(1)
WRITE(*,*)'COST(TB)=',FUNC
RETURN
END

C       THIS SUBPROGRAM EVALUATES THE FIRST-ORDER DERIVATIVE
C       OF THE TOTAL COST FUNCTION, XI(1), WITH RESPECT
C       TO TB, OR X(1)
C
SUBROUTINE DFUNC(X,XI)
COMMON /PAR1/RB,RG
COMMON /PAR2/PB,PG
COMMON /PAR3/TBURN,TW,N
COMMON /PAR4/CO,CB,CFB,CFW
COMMON /PAR5/AFB
DIMENSION X(1),XI(1)
WRITE(*,*)'TB=',XI(1)
A=PB*RG + PG*RB
HF=RB*RG/A
HI=PB*RB + PG*RG
RELBP=PB*EXP(-RB*X(1))
RELGP=PG*EXP(-RG*X(1))
RELP=RELBP+RELGP
PB1=RELBP/RELP
PG1=RELGP/RELP
A1=PB1*RG+PG1*RB
HI1=PG1*RG+PB1*RB
HF1=RB*RG/A1
C       PARTIAL (A') OVER PARTIAL TB OR X(1)
A11=PB*PG*(RB-RG)**2.0*EXP(-(RB+RG)*X(1))/(RELP**2.0)
C       PARTIAL (Hi') OVER PARTIAL TB OR X(1)
HI11=-A11
C       PARTIAL (Hf') OVER PARTIAL TB OR X(1)
```

```
      RBG=PB*RG*EXP(-RB*X(1)) + PG*RB*EXP(-RG*X(1))
      HF11=-(PB*RB)*(PG*RG)*(RB-RG)**2.0*EXP(-(RB+RG)*X(1))
     1    /(RBG**2.0)
C         PARTIAL HB(TB) OVER PARTIAL TB OR X(1)
      HPB1=HF + (HI-HF)*EXP(-A*X(1))
C         PARTIAL HW(TW|TB) OVER PARTIAL TB OR X(1)
      HPW1=HF11*TW
     1    -(1.0/(A1**2.0))*A11*(HI1-HF1)*(1.0-EXP(-A1*TW))
     1    +(1.0/A1)*(HI11-HF11)*(1.0-EXP(-A1*TW))
     1    +(1.0/A1)*(HI11-HF11)*A11*TW*EXP(-A1*TW)
C         FIRST-ORDER DERIVATIVE OF TOTAL COST WITH RESPECT
C         TO TB OR X(1)
      XI(1)=N*(CB/AFB+HPB1*CFB+HPW1*CFW)
      RETURN
      END
```

Chapter 11

THE TOTAL-TIME-ON-TEST (TTT) TRANSFORM AND ITS APPLICATION TO BURN-IN TIME DETERMINATION

11.1 INTRODUCTION

The Total Time on Test (TTT) concept, which was introduced by Barlow, Bartholomew, Bremner and Brunk (BBBB) in Chapters 5 and 6 of [1] in 1972, has proved to be a very useful tool in many reliability applications, such as

1. model identification [2],

2. aging properties analysis [2],

3. age replacement policy [3],

4. dependent failure data analysis [3],

and so on. In [4], Bergman illustrated how the TTT transform, a certain function of the survival function, and an empirical counterpart, called the TTT plot, can be used in connection with optimization problems for some burn-in models. Though some of these models may be optimized in a fairly straightforward way, there are several advantages with the TTT technique. One is the possibility for sensitivity analysis. Another is the possibility to apply a similar technique when the exact *cdf* is not known but a sample of observed times to failure is available.

11.2 THE TTT TRANSFORMS, THE SCALED TTT TRANSFORMS AND THE SCALED TTT PLOTS

11.2.1 THE TTT STATISTICS AND THE SCALED TTT STATISTICS

Let

$$0 \leq T_{1:n} \leq T_{2:n} \leq \ldots \leq T_{n:n}$$

be an ordered sample of size n from a cumulative life distribution $F(T)$ with $F(0) = 0$, a survival function

$$R(T) = 1 - F(T)$$

and a finite mean

$$\mu = \int_0^\infty R(T) dT.$$

The TTT statistics, based on this sample, are then defined as

$$\begin{cases} T_i = \sum_{j=1}^{i} [n - (j-1)](T_{j:n} - T_{j-1:n}), \text{ for } i = 1, 2, \ldots, n, \\ T_0 = 0, \text{ for } i = 0, \end{cases} \quad (11.1)$$

which are equivalent to the cumulative unit-hours, or device-hours, (hence the name *total time on test*) up to the end of $T_{i:n}$ for $i = 0, 1, 2, \cdots, n$ where $T_{0:n} = 0$; i.e.,

$$\begin{cases} T_i = \sum_{j=1}^{i} T_{j:n} + (n-i) T_{i:n}, \text{ for } i = 1, 2, \ldots, n, \\ T_0 = 0, \text{ for } i = 0, \end{cases} \quad (11.2)$$

as shown in Fig. 11.1. The scaled TTT statistics are defined as

$$\varphi_i = T_i / T_n, \text{ for } i = 0, 1, 2, \ldots, n. \quad (11.3)$$

Note that φ_i is always less than or equal to one; i.e.,

$$0 \leq \varphi_i \leq 1.$$

TTT TRANSFORMS AND PLOTS

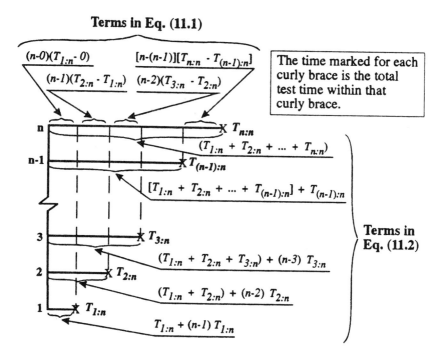

Fig. 11.1– Total time on test (TTT) calculation for an ordered sample from a life test.

11.2.2 THE TTT TRANSFORM AND THE SCALED TTT TRANSFORM

The TTT transform is defined as

$$\mu(T) = \int_0^T R(x)\,dx, \tag{11.4}$$

which is an increasing function of T with the minimum value of

$$\mu(0) = \int_0^0 R(x)\,dx = 0,$$

and the maximum value of

$$\mu(\infty) = \int_0^\infty R(x)\,dx = \mu,$$

the mean value of T. Therefore,

$$0 \leq \mu(T) \leq \mu. \tag{11.5}$$

The scaled TTT transform of $F(T)$ is defined as

$$\varphi(T) = \frac{\mu(T)}{\mu} = \frac{1}{\mu}\int_0^T R(x)\,dx, \text{ for } 0 \leq T < \infty, \tag{11.6}$$

or

$$\varphi(u) = \frac{1}{\mu}\int_0^{F^{-1}(u)} R(x)\,dx, \text{ for } 0 \leq u \leq 1, \tag{11.7}$$

where

$$\mu = \int_0^\infty R(T)\,dT, \tag{11.8}$$
$$u = F(T), \tag{11.9}$$
$$T = F^{-1}(u),$$

and

$$F^{-1}(u) = inf\{T : F(T) \geq u\}.$$

where $inf\{\ \}$ signifies the infimum, or the lowest value. Note that this transform is independent of scale, and $\varphi(T)$ is also a kind of a cumulative distribution function of T because it is a monotonically increasing function of T with the minimum value of

$$\varphi(T=0) = \frac{\mu(T=0)}{\mu} = \frac{0}{\mu} = 0,$$

and the maximum value of

$$\varphi(T = \infty) = \frac{\mu(T = \infty)}{\mu} = \frac{\mu}{\mu} = 1,$$

or

$$0 \leq \varphi(T) < 1, \text{ for } 0 \leq T < \infty. \tag{11.10}$$

Correspondingly, $\varphi(u)$ is also a kind of a cumulative distribution function of u because it is a monotonically increasing function of u with the minimum value of

$$\varphi(u = 0) = \frac{1}{\mu} \int_0^{F^{-1}(u=0)} R(x)\,dx = \frac{1}{\mu} \int_0^0 R(x)\,dx = 0,$$

and the maximum value of

$$\varphi(u = 1) = \frac{1}{\mu} \int_0^{F^{-1}(u=1)} R(x)\,dx = \frac{1}{\mu} \int_0^\infty R(x)\,dx = \frac{1}{\mu}\mu = 1,$$

or

$$0 \leq \varphi(u) \leq 1, \text{ for } 0 \leq u \leq 1. \tag{11.11}$$

Graphically, plotting $\varphi(u)$ versus u will yield a curve constrained in a square of width 1. Figure 11.2 illustrates the scaled TTT transforms for the following five different life distributions:

(1) Normal with $\overline{T} = 1$, $\sigma_T = 0.3$.

(2) Gamma distribution with shape parameter $\beta = 2.0$.

(3) Exponential distribution.

(4) Lognormal distribution with $\overline{T'} = 0$, $\sigma_{T'} = 1$.

(5) Pareto distribution with $f(T) = 2(1+T)^{-3}$ and $R(T) = (1+T)^{-2}$, $T \geq 0$.

The following observations may be made from Fig. 11.2:

1. If $F(T)$ has an increasing failure rate (IFR), such as distributions (1) and (2) do, then $\varphi(u)$ is concave.

2. If $F(T)$ has a constant failure rate (CFR), such as distribution (3) does, then $\varphi(u)$ is a straight line through points $(0,0)$ and $(1,1)$.

3. If $F(T)$ has a decreasing failure rate (DFR), such as distribution (5) does, then $\varphi(u)$ is convex.

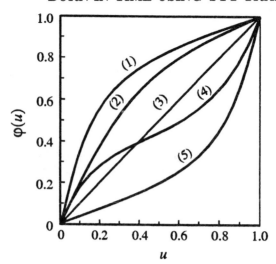

Fig. 11.2– Scaled TTT transforms from five different life distributions: (1) normal with $\overline{T} = 1$ and $\sigma_T = 0.3$; (2) gamma distribution with shape parameter $\beta = 2.0$; (3) exponential distribution; (4) lognormal distribution with $\overline{T'} = 0$ and $\sigma_{T'} = 1$; (5) Pareto distribution with $R(T) = (1+T)^{-2}$, $T \geq 0$.

4. If $F(T)$ has an increasing-then-decreasing failure rate ($IDFR$) or a decreasing-then-increasing failure rate ($DIFR$), such as distribution (4) does, then $\varphi(u)$ is "S"-shaped.

The theoretical explanations of these observations will be presented in the next section.

11.2.3   THE SCALED TTT PLOT FOR A GIVEN SAMPLE

For an ordered sample of size n,

$$0 \leq T_{1:n} \leq T_{2:n} \leq \cdots \leq T_{n:n},$$

the corresponding TTT statistics, T_i, $i = 0, 1, 2, \cdots, n$, can be evaluated using Eq. (11.1) or Eq. (11.2). Then, plotting T_i/T_n versus i/n, or equivalently plotting the points

$$(i/n, T_i/T_n), \text{ for } i = 0, 1, 2, ..., n,$$

and connecting them by line segments, yields a curve called the scaled TTT plot. It can be proved [2] that the scaled TTT plot, when n, the number of observations, goes to infinity, converges to the scaled TTT

GEOMETRICAL PROPERTIES OF TTT TRANSFORMS   443

transform of $F(T)$ defined by Eq. (11.7). Similar to the scaled TTT transform, $\varphi(u)$, the scaled TTT plot is a curve also constrained in a square of width 1 because

$$0 \leq i/n \leq 1,$$

and

$$0 \leq T_i/T_n \leq 1.$$

11.3 GEOMETRICAL PROPERTIES OF THE SCALED TTT TRANSFORMS

An important property of $\varphi(u)$ is

$$\frac{d[\varphi(u)]}{du} = \frac{1}{\mu \, \lambda(T)}. \tag{11.12}$$

This is so because from Eq. (11.7)

$$\frac{d[\varphi(u)]}{du} = \frac{d}{du}\left[\frac{1}{\mu}\int_0^{F^{-1}(u)} R(x)\,dx\right],$$

$$= \frac{1}{\mu}\left[\frac{d}{du}\int_0^{F^{-1}(u)} R(x)\,dx\right],$$

$$= \frac{1}{\mu}\left\{R[x=F^{-1}(u)]\frac{d}{du}\left[F^{-1}(u)\right] - R(x=0)\frac{d0}{du}\right\},$$

or

$$\frac{d[\varphi(u)]}{du} = \frac{1}{\mu}\left\{R[x=F^{-1}(u)]\frac{d}{du}\left[F^{-1}(u)\right]\right\}. \tag{11.13}$$

Note that in Eq. (11.13)

$$R[x=F^{-1}(u)] = 1 - F[x=F^{-1}(u)] = 1 - u = 1 - F(T) = R(T),$$

$$\tag{11.14}$$

and

$$\frac{d}{du}\left[F^{-1}(u)\right] = \frac{1}{[dF(T)/dT]} = \frac{1}{f(T)}. \tag{11.15}$$

Substituting Eqs. (11.14) and (11.15) into Eq. (11.13) yields

$$\frac{d[\varphi(u)]}{du} = \frac{1}{\mu}\left[R(T)\frac{1}{f(T)}\right],$$

or

$$\frac{d[\varphi(u)]}{du} = \frac{1}{\mu \lambda(T)}.$$

It may be seen from Eq. (11.12) that $\frac{d[\varphi(u)]}{du}$ is inversely proportional to the failure rate, $\lambda(T)$. It follows that

1. if $F(T)$ has an increasing failure rate (IFR) function, $\lambda(T)$, then $\frac{d[\varphi(u)]}{du}$ is a decreasing function of u, or $\varphi(u)$ is concave in $0 \leq u \leq 1$;

2. if $F(T)$ has a decreasing failure rate (DFR) function, $\lambda(T)$, then $\frac{d[\varphi(u)]}{du}$ is an increasing function of u, or $\varphi(u)$ is convex in $0 \leq u \leq 1$;

3. if $F(T)$ has a constant failure rate (CFR) function, or

$$\lambda(T) = \lambda = \frac{1}{\mu},$$

then T is exponentially distributed and

$$\frac{d[\varphi(u)]}{du} = \frac{1}{\mu \lambda} = 1,$$

or equivalently

$$\varphi(u) = u, \tag{11.16}$$

which is a linear function through points $(0,0)$ and $(1,1)$.

4. if $F(T)$ has a decreasing-then-increasing ($DIFR$), or increasing-then-decreasing ($IDFR$) failure rate function, $\lambda(T)$, then $\frac{d[\varphi(u)]}{du}$ will be increasing-then-decreasing or decreasing-then-increasing function of u, and consequently, $\varphi(u)$ will be concave-then-convex or convex-then-concave, or $\varphi(u)$ is "S"-shaped in $0 \leq u \leq 1$.

These conclusions are consistent with the observations made from Fig. 11.2 which are summarized in the preceding section.

11.4 THE SCALED TTT TRANSFORMS OF SOME FREQUENTLY USED LIFE DISTRIBUTIONS

In this section, the scaled TTT transforms of several frequently used life distributions will be derived from the original definition of $\varphi(u)$ given by Eq. (11.7). These distributions include the exponential, the Weibull, the gamma, the truncated normal, the lognormal and the mixed life distributions.

11.4.1 THE EXPONENTIAL LIFE DISTRIBUTION

The *pdf* and the mean value of the exponential life distribution are

$$f(T) = \lambda e^{-\lambda T}, \quad \lambda > 0,$$

and

$$\mu = \frac{1}{\lambda},$$

respectively. Its *cdf* and reliability function are

$$F(T) = 1 - e^{-\lambda T},$$

and

$$R(T) = 1 - F(T) = e^{-\lambda T},$$

respectively. Let

$$u = F(T) = 1 - e^{-\lambda T}. \tag{11.17}$$

Solving Eq. (11.17) for T yields

$$T = F^{-1}(u) = \frac{1}{\lambda} \log_e \left(\frac{1}{1-u}\right).$$

Then, the scaled TTT transform for the exponential life distribution is

$$\begin{aligned}
\varphi(u) &= \frac{1}{\mu} \int_0^{F^{-1}(u)} R(x)\, dx, \\
&= \lambda \int_0^{\frac{1}{\lambda} \log_e \left(\frac{1}{1-u}\right)} e^{-\lambda x}\, dx, \\
&= \left(-e^{-\lambda x}\right) \Big|_0^{\frac{1}{\lambda} \log_e \left(\frac{1}{1-u}\right)}, \\
&= 1 - e^{-\lambda \left[\frac{1}{\lambda} \log_e \left(\frac{1}{1-u}\right)\right]}, \\
&= 1 - (1-u),
\end{aligned}$$

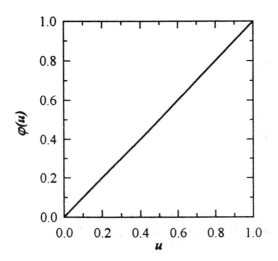

Fig. 11.3– The scaled TTT transform of the exponential life distribution.

or

$$\varphi(u) = u, \text{ for } 0 \leq u \leq 1, \tag{11.18}$$

which turns out to be a straight line through points (0, 0) and (1, 1) and is illustrated in Fig. 11.3. Note that Eq. (11.18) is consistent with Eq. (11.16) which was derived from Eq. (11.12).

11.4.2 THE WEIBULL LIFE DISTRIBUTION

The *pdf* and the mean value of Weibull life distribution are

$$f(T) = \frac{\beta}{\eta} \left(\frac{T}{\eta}\right)^{\beta-1} e^{-\left(\frac{T}{\eta}\right)^{\beta}}, \quad \beta > 0, \quad \eta > 0,$$

and

$$\mu = \eta \, \Gamma\left(\frac{1}{\beta} + 1\right),$$

respectively, where

$$\Gamma(n) = \int_0^\infty x^{n-1} e^{-x} \, dx$$

is the complete gamma function. Its *cdf* and reliability function are

$$F(T) = 1 - e^{-\left(\frac{T}{\eta}\right)^{\beta}},$$

and
$$R(T) = 1 - F(T) = e^{-\left(\frac{T}{\eta}\right)^\beta},$$
respectively. Let
$$u = F(T) = 1 - e^{-\left(\frac{T}{\eta}\right)^\beta}. \tag{11.19}$$
Solving Eq. (11.19) for T yields
$$T = F^{-1}(u) = \eta \left[\log_e \left(\frac{1}{1-u}\right)\right]^{\frac{1}{\beta}}.$$
Then, the scaled TTT transform for the Weibull life distribution is
$$\varphi(u) = \frac{1}{\mu} \int_0^{F^{-1}(u)} R(x)\, dx,$$
or
$$\varphi(u) = \frac{1}{\eta\, \Gamma\left(\frac{1}{\beta}+1\right)} \int_0^{\eta\, \left[\log_e\left(\frac{1}{1-u}\right)\right]^{\frac{1}{\beta}}} e^{-\left(\frac{x}{\eta}\right)^\beta}\, dx. \tag{11.20}$$

Let
$$y = \left(\frac{x}{\eta}\right)^\beta. \tag{11.21}$$

Then, in Eq. (11.20)
$$x = 0$$
corresponds to
$$y = \left(\frac{0}{\eta}\right)^\beta = 0$$
and
$$x = \eta \left[\log_e \left(\frac{1}{1-u}\right)\right]^{\frac{1}{\beta}}$$
corresponds to
$$y = \left\{\eta \left[\log_e \left(\frac{1}{1-u}\right)\right]^{\frac{1}{\beta}}/\eta\right\}^\beta = \log_e \left(\frac{1}{1-u}\right).$$

Also, from Eq. (11.21)
$$x = \eta y^{\frac{1}{\beta}}$$
and
$$dx = \frac{\eta}{\beta} y^{\frac{1}{\beta}-1} dy.$$

Substituting these into Eq. (11.20) yields

$$\varphi(u) = \frac{1}{\eta \, \Gamma\left(\frac{1}{\beta}+1\right)} \int_0^{\log_e\left(\frac{1}{1-u}\right)} e^{-y} \left(\frac{\eta}{\beta} y^{\frac{1}{\beta}-1} dy\right),$$

$$= \frac{1}{\beta \, \Gamma\left(\frac{1}{\beta}+1\right)} \int_0^{\log_e\left(\frac{1}{1-u}\right)} y^{\frac{1}{\beta}-1} e^{-y} dy,$$

$$= \frac{1}{\beta \left[\frac{1}{\beta}\Gamma\left(\frac{1}{\beta}\right)\right]} \Gamma\left[\frac{1}{\beta}; \log_e\left(\frac{1}{1-u}\right)\right],$$

or

$$\varphi(u) = \Gamma\left[\frac{1}{\beta}; \log_e\left(\frac{1}{1-u}\right)\right] / \Gamma\left(\frac{1}{\beta}\right), \qquad (11.22)$$

where

$$\Gamma(n; a) = \int_0^a y^{n-1} e^{-y} dy \qquad (11.23)$$

is the incomplete gamma function. Equation (11.22) is the final expression of the scaled TTT transform of the Weibull life distribution.

Note that the scaled TTT transform of the Weibull life distribution is only a function of probability, u, and the shape parameter, β. It is independent of the scale parameter, η. Figure 11.4 is an accurate graph of Eq. (11.22) for $\beta = 0.5, 1.0, 1.5, 2.0, 2.5, 3.0$. The following may be observed from Fig. 11.4:

1. For $\beta < 1$, or the Weibull life distribution has a DFR, then the scaled TTT transform is convex.

2. For $\beta > 1$, or the Weibull life distribution has an IFR, then the scaled TTT transform is concave.

3. For $\beta = 1$, or the Weibull life distribution has a CFR, then the scaled TTT transform is a straight line through points (0,0) and (1,1).

These observations are consistent with the conclusions drawn in the preceding section about the geometrical properties of the scaled TTT transforms in general.

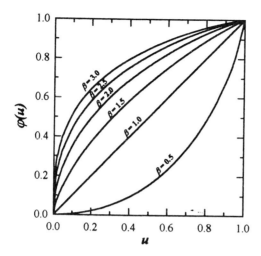

Fig. 11.4– The scaled TTT transform of the Weibull life distribution.

11.4.3 THE GAMMA LIFE DISTRIBUTION

The *pdf* and the mean value of the gamma life distribution are

$$f(T) = \frac{1}{\eta\,\Gamma(\beta)} \left(\frac{T}{\eta}\right)^{\beta-1} e^{-\frac{T}{\eta}}, \quad \beta > 0, \quad \eta > 0,$$

and

$$\mu = \beta\,\eta,$$

respectively. Its *cdf* and reliability function are

$$F(T) = \int_0^T f(x)\,dx = \Gamma\left(\beta;\frac{T}{\eta}\right) \Big/ \Gamma(\beta), \tag{11.24}$$

and

$$R(T) = 1 - F(T) = 1 - \Gamma\left(\beta;\frac{T}{\eta}\right) \Big/ \Gamma(\beta), \tag{11.25}$$

respectively, where $\Gamma\left(\beta;\frac{T}{\eta}\right)$ is the incomplete gamma function as defined by Eq. (11.23).

If β is a positive integer number, then Eqs. (11.24) and (11.25) simplify to

$$F(T) = 1 - \sum_{k=0}^{\beta-1} \frac{1}{k!} \left(\frac{T}{\eta}\right)^k e^{-\frac{T}{\eta}}, \tag{11.26}$$

and

$$R(T) = \sum_{k=0}^{\beta-1} \frac{1}{k!} \left(\frac{T}{\eta}\right)^k e^{-\frac{T}{\eta}}, \qquad (11.27)$$

respectively.

Then, the scaled TTT transform for the gamma life distribution is

$$\varphi(u) = \frac{1}{\mu} \int_0^{F^{-1}(u)} R(x)\, dx,$$

$$= \frac{1}{\beta\eta} \int_0^{F^{-1}(u)} \left[1 - \frac{\Gamma\left(\beta; \frac{x}{\eta}\right)}{\Gamma(\beta)}\right] dx,$$

or

$$\varphi(u) = \frac{1}{\beta\eta}\left[F^{-1}(u) - \frac{1}{\Gamma(\beta)} \int_0^{F^{-1}(u)} \Gamma\left(\beta; \frac{x}{\eta}\right) dx\right]. \qquad (11.28)$$

If β is a positive integer number, then

$$\varphi(u) = \frac{1}{\beta\eta} \int_0^{F^{-1}(u)} \left[\sum_{k=0}^{\beta-1} \frac{1}{k!} \left(\frac{x}{\eta}\right)^k e^{-\frac{x}{\eta}}\right] dx,$$

$$= \frac{1}{\beta} \sum_{k=0}^{\beta-1} \frac{1}{k!} \int_0^{F^{-1}(u)} \left(\frac{x}{\eta}\right)^k e^{-\frac{x}{\eta}} d\left(\frac{x}{\eta}\right),$$

or

$$\varphi(u) = \frac{1}{\beta} \sum_{k=0}^{\beta-1} \frac{1}{k!} \Gamma\left[k+1; \frac{F^{-1}(u)}{\eta}\right], \qquad (11.29)$$

where $\Gamma\left[k+1; \frac{F^{-1}(u)}{\eta}\right]$ is the incomplete gamma function as defined by Eq. (11.23). Equation (11.28) is the general expression of the scaled TTT transform of the gamma life distribution while Eq. (11.29) is a special case of Eq. (11.28) when β is a positive integer.

Figure 11.5 is an illustrative graph of Eq. (11.28) for $\beta = 1.0$, 2.0, 4.0, 6.0, 8.0 and 10.0, which behaves similar to Fig. 11.4 of the Weibull distribution and consistent with the conclusions drawn in the preceding section about the geometrical properties of the scaled TTT transforms in general.

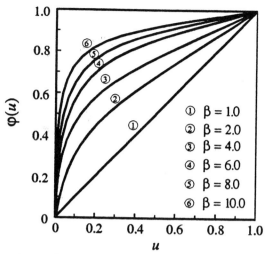

Fig. 11.5– The scaled TTT transform of the gamma life distribution.

11.4.4 THE TRUNCATED NORMAL LIFE DISTRIBUTION

The *pdf* and the mean value of the truncated normal life distribution are

$$f(T) = \frac{1}{\Phi\left(\frac{\overline{T}}{\sigma_T}\right)\sqrt{2\pi}\,\sigma_T} e^{-\frac{1}{2}\left(\frac{T-\overline{T}}{\sigma_T}\right)^2}, \quad \overline{T} > 0, \quad \sigma_T > 0,$$

and

$$\mu = \overline{T} + \sigma_T \frac{\phi\left(\frac{\overline{T}}{\sigma_T}\right)}{\Phi\left(\frac{\overline{T}}{\sigma_T}\right)},$$

respectively, where

$\phi(z)$ = *pdf* of the standardized normal distribution, $N(0,1)$,

$\phi(z) = \frac{1}{\sqrt{2\pi}} e^{-\frac{1}{2}z^2}$,

$\Phi(z)$ = *cdf* of the standardized normal distribution, $N(0,1)$,

and

$$\Phi(z) = \int_{-\infty}^{z} \phi(x)\,dx.$$

Its *cdf* and reliability function are

$$F(T) = \int_0^T f(x)\,dx = 1 - \Phi\left(\frac{\overline{T}-T}{\sigma_T}\right)\bigg/\Phi\left(\frac{\overline{T}}{\sigma_T}\right),$$

and

$$R(T) = \int_T^\infty f(x)\,dx = \Phi\left(\frac{\overline{T}-T}{\sigma_T}\right)\bigg/\Phi\left(\frac{\overline{T}}{\sigma_T}\right),$$

respectively. Let

$$u = F(T) = 1 - \Phi\left(\frac{\overline{T}-T}{\sigma_T}\right)\bigg/\Phi\left(\frac{\overline{T}}{\sigma_T}\right), \qquad (11.30)$$

Solving Eq. (11.30) for T yields

$$T = F^{-1}(u) = \overline{T} - \Phi^{-1}\left[(1-u)\,\Phi\left(\frac{\overline{T}}{\sigma_T}\right)\right]\sigma_T.$$

Then, the scaled TTT transform for the truncated normal life distribution is

$$\varphi(u) = \frac{1}{\mu}\int_0^{F^{-1}(u)} R(x)\,dx,$$

$$= \frac{1}{\overline{T} + \sigma_T\,\phi\left(\frac{\overline{T}}{\sigma_T}\right)\big/\Phi\left(\frac{\overline{T}}{\sigma_T}\right)} \int_0^{\overline{T}-\Phi^{-1}\left[(1-u)\Phi\left(\frac{\overline{T}}{\sigma_T}\right)\right]\sigma_T} \left[\frac{\Phi\left(\frac{\overline{T}-x}{\sigma_T}\right)}{\Phi\left(\frac{\overline{T}}{\sigma_T}\right)}\right] dx,$$

or

$$\varphi(u) = \frac{1}{\overline{T}\,\Phi\left(\frac{\overline{T}}{\sigma_T}\right) + \sigma_T\,\phi\left(\frac{\overline{T}}{\sigma_T}\right)} \int_0^{\overline{T}-\Phi^{-1}\left[(1-u)\Phi\left(\frac{\overline{T}}{\sigma_T}\right)\right]\sigma_T} \Phi\left(\frac{\overline{T}-x}{\sigma_T}\right) dx.$$

(11.31)

Integrating by parts yields

$$\varphi(u) = 1 - \frac{(1-u)\,\Phi\left(\frac{\overline{T}}{\sigma_T}\right)\,\Phi^{-1}\left[(1-u)\,\Phi\left(\frac{\overline{T}}{\sigma_T}\right)\right] + \phi\left\{\Phi^{-1}\left[(1-u)\,\Phi\left(\frac{\overline{T}}{\sigma_T}\right)\right]\right\}}{\left(\frac{\overline{T}}{\sigma_T}\right)\,\Phi\left(\frac{\overline{T}}{\sigma_T}\right) + \phi\left(\frac{\overline{T}}{\sigma_T}\right)},$$

or

$$\varphi(u) = 1 - \frac{(1-u)\,\Phi\left(\frac{1}{cov}\right)\,\Phi^{-1}\left[(1-u)\,\Phi\left(\frac{1}{cov}\right)\right] + \phi\left\{\Phi^{-1}\left[(1-u)\,\Phi\left(\frac{1}{cov}\right)\right]\right\}}{\left(\frac{1}{cov}\right)\,\Phi\left(\frac{1}{cov}\right) + \phi\left(\frac{1}{cov}\right)},$$

TTT TRANSFORMS OF SOME DISTRIBUTIONS

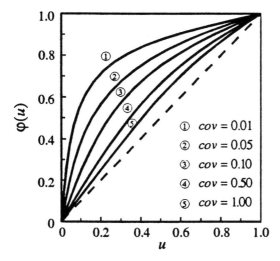

Fig. 11.6– The scaled TTT transform of the truncated normal life distribution.

(11.32)

where

$$cov = \frac{\sigma_T}{\overline{T}}$$

is the coefficient of variation of the truncated normal life distribution.

It may be seen from Eq. (11.32) that the scaled TTT transform of the truncated normal life distribution is a function of probability, u, and the coefficient of variation, cov. Figure 11.6 is an illustrative graph of Eq. (11.32) for $cov = 0.01, 0.05, 0.1, 0.5$ and 1.0. All of them are concave and above the 45° diagonal line indicating that the truncated normal life distribution has an IFR.

11.4.5 THE LOGNORMAL LIFE DISTRIBUTION

The *pdf* and the mean value of the lognormal life distribution are

$$f(T) = \frac{1}{\sqrt{2\pi}\, \sigma_{T'} T} e^{-\frac{1}{2}\left(\frac{\log_e T - \overline{T'}}{\sigma_{T'}}\right)^2}, \quad \overline{T'} > 0, \quad \sigma_{T'} > 0,$$

and

$$\mu = e^{\overline{T'} + \frac{1}{2}\sigma_{T'}^2},$$

respectively. Its *cdf* and reliability function are

$$F(T) = \int_0^T f(x)\, dx = \Phi\left(\frac{\log_e T - \overline{T'}}{\sigma_{T'}}\right),$$

and

$$R(T) = \int_T^\infty f(x)\, dx = \Phi\left(\frac{\overline{T'} - \log_e T}{\sigma_{T'}}\right),$$

respectively. Let

$$u = F(T) = \Phi\left(\frac{\log_e T - \overline{T'}}{\sigma_{T'}}\right). \tag{11.33}$$

Solving Eq. (11.33) for T yields

$$T = F^{-1}(u) = e^{\overline{T'} + \sigma_{T'}\, \Phi^{-1}(u)}.$$

Then, the scaled TTT transform for the lognormal life distribution is

$$\varphi(u) = \frac{1}{\mu} \int_0^{F^{-1}(u)} R(x)\, dx,$$

or

$$\varphi(u) = \frac{1}{e^{\overline{T'} + \frac{1}{2}\sigma_{T'}^2}} \int_0^{e^{\overline{T'} + \sigma_{T'}\, \Phi^{-1}(u)}} \Phi\left(\frac{\overline{T'} - \log_e x}{\sigma_{T'}}\right) dx. \tag{11.34}$$

Let

$$y = \log_e x.$$

Then,

$$x = e^y$$

and

$$dx = d(e^y).$$

Also, in Eq. (11.34)

$$x = 0$$

corresponds to

$$y = \log_e 0 = -\infty,$$

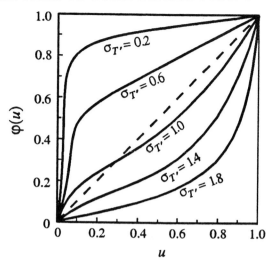

Fig. 11.7– The scaled TTT transform of the lognormal life distribution.

and

$$x = e^{\overline{T'} + \sigma_{T'} \, \Phi^{-1}(u)}$$

corresponds to

$$y = \log_e \left[e^{\overline{T'} + \sigma_{T'} \, \Phi^{-1}(u)} \right] = \overline{T'} + \sigma_{T'} \, \Phi^{-1}(u).$$

Substituting these into Eq. (11.34) yields

$$\varphi(u) = \frac{1}{e^{\overline{T'} + \frac{1}{2} \sigma_{T'}^2}} \int_{-\infty}^{\overline{T'} + \sigma_{T'} \, \Phi^{-1}(u)} \Phi\left(\frac{\overline{T'} - y}{\sigma_{T'}}\right) d(e^y). \qquad (11.35)$$

Integrating by parts yields

$$\varphi(u) = \Phi\left[\Phi^{-1}(u) - \sigma_{T'}\right] + (1 - u) \, e^{-\sigma_{T'} \left[\frac{1}{2} \sigma_{T'} - \Phi^{-1}(u)\right]}. \qquad (11.36)$$

It may be seen from Eq. (11.36) that the scaled TTT transform of the lognormal life distribution is only a function of the probability, u, and the log standard deviation, $\sigma_{T'}$. It is independent of the log mean, $\overline{T'}$. Figure 11.7 is an illustrative graph of Eq. (11.36) for $\sigma_{T'} = 0.2, 0.6, 1.0, 1.4$ and 1.8. For each $\sigma_{T'} > 0$, the graph crosses the 45° line once and from above. The abscissa values of the crossover points decrease as $\sigma_{T'}$ values increase.

11.4.6 THE MIXED LIFE DISTRIBUTION

The *pdf* and the mean value of a mixed life distribution are

$$f(T) = \sum_{i=1}^{n} p_i \, f_i(T),$$

and

$$\mu = \sum_{i=1}^{n} p_i \, \mu_i,$$

respectively, where

n = number of subpopulations,
$f_i(T)$ = *pdf* of subpopulation i,
μ_i = mean value of subpopulation i,
p_i = proportion of subpopulation i,

and

$$\sum_{i=1}^{n} p_i = 1.$$

Its *cdf* and reliability function are

$$F(T) = \int_0^T f(x) \, dx = \sum_{i=1}^{n} p_i \, F_i(T),$$

and

$$R(T) = \int_T^\infty f(x) \, dx = \sum_{i=1}^{n} p_i \, R_i(T),$$

respectively, where

$F_i(T)$ = *cdf* of subpopulation i,

and

$R_i(T)$ = reliability function of subpopulation i.

Then, the scaled TTT transform for the mixed life distribution is

$$\varphi(u) = \frac{1}{\mu} \int_0^{F^{-1}(u)} R(x)\, dx,$$

$$= \frac{1}{\mu} \int_0^{F^{-1}(u)} \left[\sum_{i=1}^{n} p_i\, R_i(x)\right] dx,$$

$$= \frac{1}{\sum_{i=1}^{n} p_i\, \mu_i} \sum_{i=1}^{n} p_i \int_0^{F^{-1}(u)} R_i(x)\, dx,$$

$$= \frac{1}{\sum_{i=1}^{n} p_i\, \mu_i} \sum_{i=1}^{n} p_i\, \mu_i \left[\frac{1}{\mu_i} \int_0^{F^{-1}(u)} R_i(x)\, dx\right],$$

or

$$\varphi(u) = \frac{\sum_{i=1}^{n} p_i\, \mu_i\, \varphi_i'(u)}{\sum_{i=1}^{n} p_i\, \mu_i}, \qquad (11.37)$$

where

$$\varphi_i'(u) = \frac{1}{\mu_i} \int_0^{F^{-1}(u)} R_i(x)\, dx.$$

Note that $\varphi_i'(u)$ is not the same as the scaled TTT transform of the ith subpopulation, $\varphi_i(u)$; i.e.,

$$\varphi_i'(u) \neq \varphi_i(u) = \frac{1}{\mu_i} \int_0^{F_i^{-1}(u)} R_i(x)\, dx,$$

because

$$F^{-1}(u) \neq F_i^{-1}(u).$$

Figure 11.8 is an illustrative graph of Eq. (11.37) for the scaled TTT transforms of a bimodal mixed-exponential and a bimodal mixed-Weibull life distributions, respectively.

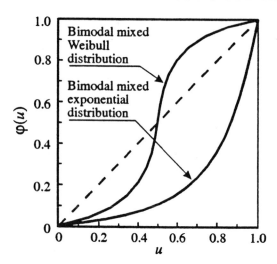

Fig. 11.8– The scaled TTT transforms of a bimodal mixed-exponential and a bimodal mixed-Weibull life distributions, respectively.

11.5 SCALED TTT PLOTS FOR COMPLETE AND INCOMPLETE LIFE DATA

11.5.1 COMPLETE, OR NONCENSORED, LIFE DATA

Case 1 – Ungrouped, or individual, data set

First, a data set without ties is considered. Let $0 \leq T_{1:n} < T_{2:n} < ... < T_{n:n}$ be an ordered and complete sample from a life test of n components. Then the TTT statistic, at each failure time, $T_{i:n}$, based on this sample, is given by Eq. (11.2), or

$$\begin{cases} T_i = \sum_{j=1}^{i} T_{j:n} + (n-i)\, T_{i:n}, \text{ for } i = 1, 2, ..., n, \\ T_0 = 0, \text{ for } i = 0. \end{cases} \quad (11.38)$$

The scaled TTT statistic at each failure time is defined by Eq. (11.3), or

$$\varphi_i = T_i/T_n, \text{ for } i = 0, 1, 2, ..., n, \quad (11.39)$$

where

$$T_n = \sum_{j=1}^{n} T_{j:n}.$$

The observed *cdf* at each failure time is defined as

$$u_i = \hat{F}(T_{i:n}) = \frac{i}{n}, \text{ for } i = 0, 1, 2, \cdots, n. \tag{11.40}$$

Therefore, plotting φ_i versus u_i, or equivalently plotting the points

$$(u_i, \varphi_i) = \left(\frac{i}{n}, \frac{T_i}{T_n}\right), \text{ for } i = 0, 1, 2, \cdots, n, \tag{11.41}$$

and connecting these points with line segments yields the scaled *TTT* plot for this particular sample which should converge to the scaled *TTT* transform of the underlying life distribution, $\varphi(u)$, as n goes to infinity.

Second, a data set with ties is considered. In this case, n components are life tested until all fail. A total of $m < n$ distinct failure times are observed at $W_1 < W_2 < \cdots < W_m$, respectively. Let r_i be the number of failures which occurred at W_i, $i = 1, 2, \cdots, m$. Therefore,

$$n = \sum_{i=1}^{m} r_i.$$

Then, the *TTT* statistic at each distinct failure time, W_i, based on this sample, is

$$\begin{cases} T_i = \sum_{j=1}^{i} r_j W_j + (n - \sum_{j=1}^{i} r_j) W_i, \text{ for } i = 1, 2, ..., m, \\ T_0 = 0, \text{ for } i = 0. \end{cases} \tag{11.42}$$

The scaled *TTT* statistic at each distinct failure time, W_i, is

$$\varphi_i = T_i/T_m, \text{ for } i = 0, 1, 2, ..., m, \tag{11.43}$$

where

$$T_m = \sum_{j=1}^{m} r_j W_j. \tag{11.44}$$

The observed *cdf* at each failure time is defined as

$$u_i = \hat{F}(W_i) = \begin{cases} \dfrac{\sum_{j=1}^{i} r_j}{n}, & \text{for } i = 1, 2, \cdots, m, \\ 0, & \text{for } i = 0. \end{cases} \tag{11.45}$$

Therefore, plotting φ_i versus u_i, or equivalently plotting the points

$$(u_i, \varphi_i) = \left(\frac{\sum_{j=1}^{i} r_j}{n}, \frac{T_i}{T_m} \right), \text{ for } i = 0, 1, 2, \cdots, m, \qquad (11.46)$$

and connecting these points with line segments yields the scaled TTT plot for this complete data set with ties which should converge to the scaled TTT transform of the underlying life distribution, $\varphi(u)$, as n goes to infinity.

Case 2 – Grouped data set

Often life data are recorded in terms of the number of failures within specified time intervals due to non-continuous monitoring. Assume a life test is started with n components. Let W_1, W_2, $\cdots$, and W_m be m time points at which the inspections are made and the number of failures recorded, where

$$0 < W_1 < W_2 < \cdots < W_m.$$

Let the r_i represent the number of failures in the ith time interval, (W_{i-1}, W_i), $i = 1, 2, \cdots, m$. Then, the n failures are recorded as follows:

$$[(W_0, W_1); r_1], [(W_1, W_2); r_2], \cdots, [(W_{m-1}, W_m); r_m],$$

where

$$W_0 = 0,$$

and

$$n = \sum_{i=1}^{m} r_i.$$

The TTT statistic, at the end of each time interval, W_i, based on this sample, is

$$\begin{cases} T_i = \sum_{j=1}^{i} r_j \left[\frac{1}{2} (W_{j-1} + W_j) \right] + (n - \sum_{j=1}^{i} r_j) W_i, \text{ for } i = 1, 2, ..., m, \\ T_0 = 0, \text{ for } i = 0, \end{cases}$$

$$(11.47)$$

TTT PLOTS FOR A GIVEN SAMPLE

which is based on the assumption that r_j failures occurred uniformly in the jth time interval, (W_{j-1}, W_j), or equivalently, r_j failures occurred at a common point, the midpoint of the interval, $\frac{1}{2}(W_{j-1} + W_j)$.

The scaled TTT statistic at the end of each time interval is defined by Eq. (11.3), or

$$\varphi_i = T_i/T_m, \text{ for } i = 0, 1, 2, ..., m, \tag{11.48}$$

where

$$T_m = \sum_{j=1}^{m} r_j \left[\frac{1}{2}(W_{j-1} + W_j)\right].$$

The observed *cdf* at the end of each time interval is defined as

$$u_i = \hat{F}(W_i) = \begin{cases} \dfrac{\sum_{j=1}^{i} r_j}{n}, & \text{for } i = 1, 2, \cdots, m, \\ 0, & \text{for } i = 0. \end{cases} \tag{11.49}$$

Therefore, plotting φ_i versus u_i, or equivalently plotting the points

$$(u_i, \varphi_i) = \left(\dfrac{\sum_{j=1}^{i} r_j}{n}, \dfrac{T_i}{T_m}\right), \text{ for } i = 0, 1, 2, \cdots, m, \tag{11.50}$$

and connecting these points with line segments yields the scaled TTT plot for this grouped data sample which should converge to the scaled TTT transform of the underlying life distribution, $\varphi(u)$, as n goes to infinity.

11.5.2 TIME-TERMINATED LIFE DATA

Case 1 – Ungrouped, or individual, data sets

First, a data set without ties is considered. Assume a life test is started with n components and terminated at time T_d. A total of $r < n$ failures are observed at $T_{1:r}, T_{2:r}, \cdots,$ and $T_{r:r}$ where

$$0 \leq T_{1:r} < T_{2:r} < ... < T_{r:r} \leq T_d.$$

Then, the TTT statistic, at each failure time, $T_{i:r}$, based on this sample, is

$$\begin{cases} T_i = \sum_{j=1}^{i} T_{j:r} + (n-i)\, T_{i:r}, & \text{for } i = 1, 2, ..., r, \\ T_0 = 0, & \text{for } i = 0. \end{cases} \tag{11.51}$$

The scaled TTT statistic at each failure time for this time-terminated data set is defined as

$$\varphi_i = T_i/T_r, \text{ for } i = 0, 1, 2, ..., r, \tag{11.52}$$

where

$$T_r = \sum_{j=1}^{r} T_{j:r} + (n-r) T_{r:r}.$$

The value of u at each failure time $T_{i:r}$, u_i, for this time-terminated case, is defined as

$$u_i = \frac{i}{r}, \text{ for } i = 0, 1, 2, \cdots, r, \tag{11.53}$$

which approaches the observed cdf at $T_{i:r}$, $\hat{F}(T_{i:r})$, as n goes to infinity. Therefore, plotting φ_i versus u_i, or equivalently plotting the points

$$(u_i, \varphi_i) = \left(\frac{i}{r}, \frac{T_i}{T_r}\right), \text{ for } i = 0, 1, 2, \cdots, r, \tag{11.54}$$

and connecting these points with line segments yields the scaled TTT plot for this time-terminated sample which should converge to the scaled TTT transform of the underlying life distribution, $\varphi(u)$, as n goes to infinity.

Second, a data set with ties is considered. In this case, n components are life tested and the test is terminated at T_d. A total of $r < n$ failures are observed at m distinct times,

$$W_1 < W_2 < \cdots < W_m,$$

respectively, where

$$m < r < n$$

and

$$W_m \leq T_d.$$

Also let r_i be the number of failures which occurred at W_i, $i = 1, 2, \cdots, m$. Therefore,

$$r = \sum_{i=1}^{m} r_i.$$

Then, the *TTT* statistic, at each distinct failure time, W_i, based on this sample, is

$$\begin{cases} T_i = \sum_{j=1}^{i} r_j W_j + (n - \sum_{j=1}^{i} r_j) W_i, & \text{for } i = 1, 2, ..., m, \\ T_0 = 0, & \text{for } i = 0. \end{cases} \quad (11.55)$$

The scaled *TTT* statistic at each distinct failure time, W_i, is defined as

$$\varphi_i = T_i/T_m, \text{ for } i = 0, 1, 2, ..., m, \quad (11.56)$$

where

$$T_m = \sum_{j=1}^{m} r_j W_j + (n-r) W_m. \quad (11.57)$$

The value of u at each distinct failure time W_i, u_i, for this time-terminated data set, is defined as

$$u_i = \begin{cases} \dfrac{\sum_{j=1}^{i} r_j}{r}, & \text{for } i = 1, 2, \cdots, m, \\ 0, & \text{for } i = 0, \end{cases} \quad (11.58)$$

which approaches the observed *cdf* at W_i, $\hat{F}(W_i)$, as n goes to infinity.

Therefore, plotting φ_i versus u_i, or equivalently plotting the points

$$(u_i, \varphi_i) = \left(\dfrac{\sum_{j=1}^{i} r_j}{r}, \dfrac{T_i}{T_m} \right), \text{ for } i = 0, 1, 2, \cdots, m, \quad (11.59)$$

and connecting these points with line segments yields the scaled *TTT* plot for this time-terminated data set with ties which should converge to the scaled *TTT* transform of the underlying life distribution, $\varphi(u)$, as n goes to infinity.

Case 2 – Grouped data set

In this case, n components are life tested and the test is terminated at T_d. A total of $r < n$ failures are observed and recorded in m time intervals as follows:

$$[(W_0, W_1); r_1], [(W_1, W_2); r_2], \cdots, [(W_{m-1}, W_m); r_m],$$

where

$W_0 = 0$,
$r_i =$ number of failures in interval (W_{i-1}, W_i),
$$r = \sum_{i=1}^{m} r_i,$$
$W_m = T_d$,

and

$$m < r < n.$$

Then, the *TTT* statistic, at the end of each time interval, W_i, based on this sample, is

$$\begin{cases} T_i = \sum_{j=1}^{i} r_j \left[\tfrac{1}{2}(W_{j-1} + W_j)\right] + (n - \sum_{j=1}^{i} r_j) W_i, & \text{for } i = 1, 2, ..., m, \\ T_0 = 0, & \text{for } i = 0, \end{cases}$$

(11.60)

which is based on the assumption that r_j failures occurred uniformly in the jth time interval, (W_{j-1}, W_j), or equivalently, r_j failures occurred at a common point, the midpoint of the interval, $\tfrac{1}{2}(W_{j-1} + W_j)$.

The scaled *TTT* statistic at the end of each time interval is defined by Eq. (11.3), or

$$\varphi_i = T_i/T_m, \quad \text{for } i = 0, 1, 2, ..., m, \tag{11.61}$$

where

$$T_m = \sum_{j=1}^{m} r_j \left[\tfrac{1}{2}(W_{j-1} + W_j)\right] + (n - r) W_m.$$

The value of u at the end of each time interval W_i, u_i, is defined as

$$u_i = \begin{cases} \dfrac{\sum_{j=1}^{i} r_j}{r}, & \text{for } i = 1, 2, \cdots, m, \\ 0, & \text{for } i = 0, \end{cases} \tag{11.62}$$

which approaches the observed *cdf* at W_i, $\hat{F}(W_i)$, as n goes to infinity.

TTT PLOTS FOR A GIVEN SAMPLE

Therefore, plotting φ_i versus u_i, or equivalently plotting the points

$$(u_i, \varphi_i) = \left(\frac{\sum_{j=1}^{i} r_j}{r}, \frac{T_i}{T_m} \right), \text{ for } i = 0, 1, 2, \cdots, m, \qquad (11.63)$$

and connecting these points with line segments yields the scaled TTT plot for this time-terminated and grouped data set which should converge to the scaled TTT transform of the underlying life distribution, $\varphi(u)$, as n goes to infinity.

11.5.3 FAILURE-TERMINATED LIFE DATA

Case 1 – Ungrouped, or individual, data sets

First, a data set without ties is considered. Assume a life test is started with n components and terminated at the rth failure. These r failures are observed at $T_{1:r}, T_{2:r}, \cdots$, and $T_{r:r}$ where

$$0 \leq T_{1:r} < T_{2:r} < \ldots < T_{r:r}.$$

Then, the TTT statistic at each failure time, $T_{i:r}$, based on this sample, is

$$\begin{cases} T_i = \sum_{j=1}^{i} T_{j:r} + (n-i)\, T_{i:r}, \text{ for } i = 1, 2, \ldots, r, \\ T_0 = 0, \text{ for } i = 0. \end{cases} \qquad (11.64)$$

The scaled TTT statistic at each failure time is

$$\varphi_i = T_i / T_r, \text{ for } i = 0, 1, 2, \ldots, r, \qquad (11.65)$$

where

$$T_r = \sum_{j=1}^{r} T_{j:r} + (n-r)\, T_{r:r}.$$

The value of u at each failure time $T_{i:r}$, u_i, for this failure-terminated data set, is defined as

$$u_i = \frac{i}{r}, \text{ for } i = 0, 1, 2, \cdots, r, \qquad (11.66)$$

which approaches the observed cdf at $T_{i:r}$, $\hat{F}(T_{i:r})$, as n goes to infinity.

Therefore, plotting φ_i versus u_i, or equivalently plotting the points

$$(u_i, \varphi_i) = \left(\frac{i}{r}, \frac{T_i}{T_r}\right), \text{ for } i = 0, 1, 2, \cdots, r, \tag{11.67}$$

and connecting these points with line segments yields the scaled TTT plot for this failure-terminated sample without ties which should converge to the scaled TTT transform of the underlying life distribution, $\varphi(u)$, as n goes to infinity.

Second, a data set with ties is considered. In this case, n components are life tested and the test is terminated at the rth failure. These r failures are observed at m distinct times,

$$W_1 < W_2 < \cdots < W_m,$$

respectively, where

$$m < r < n$$

and

W_m = time to the last, or the rth, failure.

Also let r_i be the number of failures which occurred at W_i, $i = 1, 2, \cdots, m$. Therefore,

$$r = \sum_{i=1}^{m} r_i.$$

Then, the TTT statistic at each distinct failure time, W_i, based on this sample, is

$$\begin{cases} T_i = \sum_{j=1}^{i} r_j W_j + \left(n - \sum_{j=1}^{i} r_j\right) W_i, \text{ for } i = 1, 2, ..., m, \\ T_0 = 0, \text{ for } i = 0. \end{cases} \tag{11.68}$$

The scaled TTT statistic at each distinct failure time, W_i, is

$$\varphi_i = T_i/T_m, \text{ for } i = 0, 1, 2, ..., m, \tag{11.69}$$

where

$$T_m = \sum_{j=1}^{m} r_j W_j + (n-r) W_m. \tag{11.70}$$

TTT PLOTS FOR A GIVEN SAMPLE

The value of u at each distinct failure time W_i, u_i, for this failure-terminated data set, is defined as

$$u_i = \begin{cases} \dfrac{\sum_{j=1}^{i} r_j}{r}, & \text{for } i = 1, 2, \cdots, m, \\ 0, & \text{for } i = 0, \end{cases} \qquad (11.71)$$

which approaches the observed *cdf* at W_i, $\hat{F}(W_i)$, as n goes to infinity.

Therefore, plotting φ_i versus u_i, or equivalently plotting the points

$$(u_i, \varphi_i) = \left(\dfrac{\sum_{j=1}^{i} r_j}{r}, \dfrac{T_i}{T_m} \right), \quad \text{for } i = 0, 1, 2, \cdots, m, \qquad (11.72)$$

and connecting these points with line segments yields the scaled *TTT* plot for this failure-terminated data set with ties which should converge to the scaled *TTT* transform of the underlying life distribution, $\varphi(u)$, as n goes to infinity.

Case 2 – Grouped data set

In this case, n components are life tested and the test is terminated at the rth failure. These r failures are observed and recorded in m time intervals as follows:

$$[(W_0, W_1); r_1], [(W_1, W_2); r_2], \cdots, [(W_{m-1}, W_m); r_m],$$

where

$W_0 = 0$,
$r_i = $ number of failures in interval (W_{i-1}, W_i),
$r = \sum_{i=1}^{m} r_i,$
$W_m = $ time to the last, or the rth, failure,

and

$$m < r < n.$$

Then, the *TTT* statistic at the end of each time interval, W_i, based on this sample, is

$$\begin{cases} T_i = \sum_{j=1}^{i} r_j \left[\tfrac{1}{2}(W_{j-1} + W_j) \right] + (n - \sum_{j=1}^{i} r_j) W_i, & \text{for } i = 1, 2, ..., m, \\ T_0 = 0, & \text{for } i = 0, \end{cases}$$

(11.73)

which is based on the assumption that r_j failures occurred uniformly in the jth time interval, (W_{j-1}, W_j), or equivalently, r_j failures occurred at a common point, the midpoint of the interval, $\frac{1}{2}(W_{j-1} + W_j)$.

The scaled TTT statistic at the end of each time interval is defined by Eq. (11.3), or

$$\varphi_i = T_i/T_m, \text{ for } i = 0, 1, 2, ..., m, \qquad (11.74)$$

where

$$T_m = \sum_{j=1}^{m} r_j \left[\frac{1}{2}(W_{j-1} + W_j) \right] + (n - r) W_m.$$

The value of u at the end of each time interval W_i, u_i, for this failure-terminated and grouped data set, is defined as

$$u_i = \begin{cases} \dfrac{\sum_{j=1}^{i} r_j}{r}, & \text{for } i = 1, 2, \cdots, m, \\ 0, & \text{for } i = 0, \end{cases} \qquad (11.75)$$

which approaches the observed cdf at W_i, $\hat{F}(W_i)$, as n goes to infinity.

Therefore, plotting φ_i versus u_i, or equivalently plotting the points

$$(u_i, \varphi_i) = \left(\frac{\sum_{j=1}^{i} r_j}{r}, \frac{T_i}{T_m} \right), \text{ for } i = 0, 1, 2, \cdots, m, \qquad (11.76)$$

and connecting these points with line segments yields the scaled TTT plot for this failure-terminated and grouped data set which should converge to the scaled TTT transform of the underlying life distribution, $\varphi(u)$, as n goes to infinity.

11.6 DESIRED STRUCTURE OF THE OBJECTIVE FUNCTIONS TO BE OPTIMIZED USING THE TTT TRANSFORM AND THE GRAPHICAL OPTIMIZATION PROCEDURE

Past studies [1, 2, 3] demonstrate that every problem which can be transformed into a problem of maximizing or minimizing an expression

OBJECTIVE FUNCTION STRUCTURE

of the form
$$C(T) = \frac{\alpha + \beta\,\mu(T)/\mu}{\theta + \delta\,F(T)}, \tag{11.77}$$
or
$$C(u) = \frac{\alpha + \beta\,\varphi(u)}{\theta + \delta\,u} \tag{11.78}$$
can be analyzed by the TTT transform technique.

Now let's first look at the maximum or minimum point of Eq. (11.78), u^*. Once u^* is determined, the optimum point of Eq. (11.77), T^*, can be obtained by solving
$$F(T^*) = u^*$$
for T^*. Note that the maximization or minimization of Eq. (11.78) is equivalent to solving the following differential equation for u:
$$C'(u) = \frac{dC(u)}{du} = 0,$$
which leads to
$$\frac{[\beta\,\varphi'(u^*)](\theta + \delta\,u^*) - [\alpha + \beta\,\varphi(u^*)]\,\delta}{(\theta + \delta\,u^*)^2} = 0,$$
or
$$[\beta\,\varphi'(u^*)](\theta + \delta\,u^*) - [\alpha + \beta\,\varphi(u^*)]\,\delta = 0. \tag{11.79}$$
Dividing both sides of Eq. (11.79) by $(\beta\,\delta)$ and rearranging yields
$$\varphi'(u^*) = \frac{\varphi(u^*) + \alpha/\beta}{u^* + \theta/\delta} = \frac{\varphi(u^*) - (-\alpha/\beta)}{u^* - (-\theta/\delta)}. \tag{11.80}$$
Note that the left side of Eq. (11.80), $\varphi'(u^*)$, is the slope of $\varphi(u)$ at Point $[u^*, \varphi(u^*)]$, and the right side of Eq. (11.80),
$$\frac{\varphi(u^*) - (-\alpha/\beta)}{u^* - (-\theta/\delta)},$$
is the slope of a straight line through Points $[u^*, \varphi(u^*)]$ and $(-\theta/\delta, -\alpha/\beta)$. Therefore, Eq. (11.80) implies that the maximum or minimum value of $C(u)$ occurs at a point C where the tangent line of $\varphi(u)$ goes through Point $(-\theta/\delta, -\alpha/\beta)$ as shown in Fig. 11.9. In Fig. 11.9, three cases are presented; i.e., (a) for a convex $\varphi(u)$, (b) for a concave $\varphi(u)$, and (c) for an 'S-shaped' $\varphi(u)$. Note that for the 'S-shaped' $\varphi(u)$, there are two optimum points in which one is a minimum point and the other is a maximum point.

Consequently, the optimization of Eq. (11.80) is equivalent to conducting the following two steps:

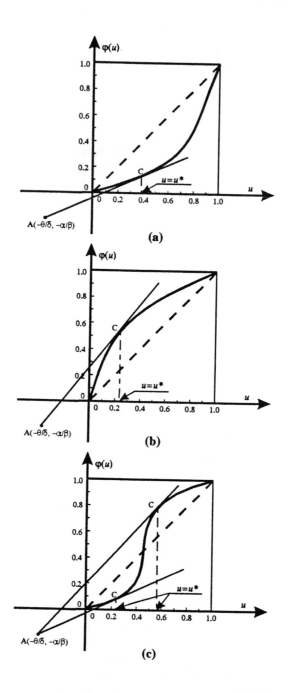

Fig. 11.9– Graphical optimization using the TTT transforms for three different $\varphi(u)'s$; i.e., (a) for a convex $\varphi(u)$, (b) for a concave $\varphi(u)$, and (c) for an 'S-shaped' $\varphi(u)$.

OBJECTIVE FUNCTION STRUCTURE

Step 1: Find u^* by drawing a line, on the "$\varphi(u)$ versus u" plane, through Point A, $(-\frac{\theta}{\delta}, -\frac{\alpha}{\beta})$, which touches the scaled TTT transform at Point C, $[u^*, \varphi(u^*)]$, and has the largest or the smallest slope.

Step 2: Determine T^* by solving

$$F(T^*) = u^*.$$

To determine if Point C is a maximum or a minimum point, the second derivative of $C(u)$, $C''(u)$, at Point C needs to be evaluated and its sign examined. Now

$$C''(u)\Big|_{u=u^*} = \frac{dC'(u)}{du}\Big|_{u=u^*},$$

$$= d\left[\frac{[\beta\,\varphi'(u^*)](\theta + \delta\,u^*) - [\alpha + \beta\,\varphi(u^*)]\,\delta}{(\theta + \delta\,u^*)^2}\right]\Big/du^*,$$

or

$$C''(u^*) = \{[\beta\,\varphi''(u^*)\,(\theta + \delta\,u^*) + \beta\,\varphi'(u^*)\,\delta - \beta\,\varphi'(u^*)\,\delta](\theta + \delta\,u^*)^2$$
$$-\{\beta\,\varphi'(u^*)\,(\theta + \delta\,u^*) - [\alpha + \beta\,\varphi(u^*)\,\delta]\}[2\,(\theta + \delta\,u^*)\,\delta\}$$
$$/(\theta + \delta\,u^*)^4. \quad (11.81)$$

Note that at Point C, Eq. (11.79) holds. Then, substituting Eq. (11.79) into Eq. (11.81) and rearranging yields

$$C''(u^*) = \frac{\beta\,\varphi''(u^*)\,(\theta + \delta\,u^*)^3 - 0}{(\theta + \delta\,u^*)^4},$$

or

$$C''(u^*) = \frac{\varphi''(u^*)}{\left(\frac{\theta}{\beta} + \frac{\delta}{\beta}u^*\right)}. \quad (11.82)$$

Recall that if $C''(u^*) > 0$, then Point C is a minimum point, and if $C''(u^*) < 0$, then Point C is a maximum point. Therefore, the nature of Point C can be determined as follows:

1. If $\varphi(u)$ is concave at Point C, or $\varphi''(u^*) < 0$, then it is a maximum point of $C(u)$ if $[\theta/\beta + (\delta/\beta)u^*] > 0$, because $C''(u^*) < 0$ in this case according to Eq. (11.82); and a minimum point of $C(u)$ if $[\theta/\beta + (\delta/\beta)u^*] < 0$, because $C''(u^*) > 0$ in this case according to Eq. (11.82).

2. If $\varphi(u)$ is convex at Point C, or $\varphi''(u^*) > 0$, then it is a minimum point of $C(u)$ if $[\theta/\beta + (\delta/\beta)u^*] > 0$, because $C''(u^*) > 0$ in this case according to Eq. (11.82); and a maximum point of $C(u)$ if $[\theta/\beta + (\delta/\beta)u^*] < 0$, because $C''(u^*) < 0$ in this case according to Eq. (11.82).

Some applications of the TTT transforms to the optimum burn-in time determination, for the minimum cost per unit time of useful operation, for the maximum profit, and for the maximum post-burn-in mean residual life (MRL), are presented next.

11.7 OPTIMUM BURN-IN TIME DETERMINATION USING THE TTT TRANSFORM FOR THE MINIMUM COST OR FOR THE MAXIMUM PROFIT

11.7.1 THE BERGMAN-KLEFSJO'S COST MODEL [4]

This model was established for non-repairable components. The expected long-run cost per unit time of useful operation, $C(T_b)$, is expressed as

$$C(T_b) = -C_b + \frac{C_{ff}}{\mu} \left(\frac{\xi - F(T_b)}{1 - (\int_0^{T_b} R(T)\,dT)/\mu} \right), \qquad (11.83)$$

which is a transformed version in the form of Eq. (11.77) of the original cost model where

C_b = cost per unit time of burn-in,
C_{ff} = cost of field failure after burn-in,
$$\mu = \int_0^\infty R(T)\,dT,$$
$$\xi = (1 + C_0 + C_b\mu + C_{ff})/C_{ff},$$

and

C_0 = fixed cost of burn-in per product.

It may be seen that $C(T_b)$ and

$$K(T_b) = \frac{\xi - F(T_b)}{1 - (\int_0^{T_b} R(T)\,dT)/\mu} \qquad (11.84)$$

are minimized for the same value of T_b because

$$C(T_b) = -C_b + \frac{C_{ff}}{\mu} K(T_b), \qquad (11.85)$$

and in this equation the first term, $-C_b$, and the multiplier of the second term, C_{ff}/μ, are both constants and C_{ff}/μ is positive. Substituting

$$u = F(T_b)$$

and

$$\varphi(u) = \left[\int_0^{T_b} R(T) \, dT \right] / \mu$$

into Eqs. (11.84) and (11.85) yields

$$K(u) = \frac{\xi - u}{1 - \varphi(u)}, \qquad (11.86)$$

and

$$C(u) = -C_b + \frac{C_{ff}}{\mu} K(u) = -C_b + \frac{C_{ff}}{\mu} \left[\frac{\xi - u}{1 - \varphi(u)} \right], \qquad (11.87)$$

respectively. Then, the minimization of $K(T_b)$ is equivalent to the minimization of $K(u)$ which is equivalent to the maximization of $1/K(u)$. Now

$$\frac{1}{K(u)} = \frac{1 - \varphi(u)}{\xi - u} \qquad (11.88)$$

which has the same structure as Eq. (11.78) with

$$\alpha = 1,$$
$$\beta = -1,$$
$$\theta = \xi,$$

and

$$\delta = -1.$$

Therefore, the optimum burn-in time, T_b^*, which minimizes $C(T_b)$ given by Eq. (11.83), can be obtained using the following two steps:

Step 1: Determine the value of u^*, for which

$$\frac{1}{K(u)} = \frac{1-\varphi(u)}{\xi - u} \tag{11.89}$$

is maximized. This can be done by drawing a line, on the "$\varphi(u)$ versus u" plane, from point

$$\left(-\frac{\theta}{\delta}, -\frac{\alpha}{\beta}\right) = \left(-\frac{\xi}{(-1)}, -\frac{1}{(-1)}\right) = (\xi, 1),$$

which touches the scaled TTT transform at

$$[u^*, \varphi(u^*)]$$

and has the largest slope as shown in Fig. 11.10.

For u^* to be a maximum point for $1/K(u)$, the following must be satisfied to assure $d^2[1/K(u)]/du^2 < 0$:

1. If $\varphi(u)$ is concave at $[u^*, \varphi(u^*)]$, or $\varphi''(u^*) < 0$, then

$$\frac{\theta}{\beta} + \frac{\delta}{\beta} u^* = \frac{\xi}{-1} + \frac{-1}{-1} u^* = u^* - \xi$$

has to be positive, or

$$u^* - \xi > 0.$$

2. If $\varphi(u)$ is convex at $[u^*, \varphi(u^*)]$, or $\varphi''(u^*) > 0$, then $(u^* - \xi)$ has to be negative, or

$$u^* - \xi < 0.$$

Step 2: Determine the optimal value of T_b, T_b^*, by solving the equation

$$F(T_b^*) = u^*, \tag{11.90}$$

where u^* was obtained from **Step 1**.

When T_b^*, which minimizes $C(T_b)$, is determined we have to compare the corresponding long-run cost, $C(T_b^*)$, with that without any burn-in at all; i.e., $u^* = T_b^* = 0$. If $C(T_b = 0) < C(T_b^*)$, or equivalently $C(u = 0) < C(u^*)$, then no burn-in needs to be conducted. A graphical comparison method is presented first here.

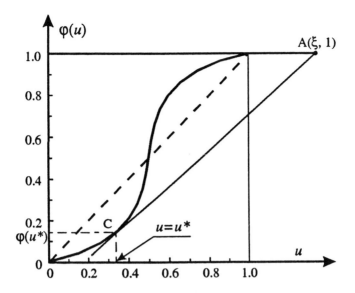

Fig. 11.10— Graphical determination of T_b^* which minimizes $C(T_b)$ using the Bergman-Klefsjo's Model.

Note that if no burn-in is conducted, then the cost C_0 does not appear and ξ becomes

$$\xi' = (1 + C_{ff} + C_b\mu)/C_{ff}. \tag{11.91}$$

Correspondingly, Eqs. (11.86) and (11.87) become

$$K(u=0) = \frac{\xi' - 0}{1 - \varphi(0)} = \xi',$$

and

$$C(u=0) = -C_b + \frac{C_{ff}}{\mu} K(u=0) = -C_b + \frac{C_{ff}}{\mu} \xi'. \tag{11.92}$$

respectively. Then, the difference between $C(u=0)$ and $C(u^*)$ is given by

$$C(u=0) - C(u^*) = \left[-C_b + \frac{C_{ff}}{\mu}\xi'\right] - \left\{-C_b + \frac{C_{ff}}{\mu}\left[\frac{\xi - u^*}{1 - \varphi(u^*)}\right]\right\},$$
$$= \frac{C_{ff}}{\mu}\left[\xi' - \frac{\xi - u^*}{1 - \varphi(u^*)}\right],$$

or

$$C(u=0) - C(u^*) = \frac{C_{ff}}{\mu}\left[\frac{\xi'-0}{1-0} - \frac{\xi-u^*}{1-\varphi(u^*)}\right].$$

If

$$C(u=0) - C(u^*) < 0,$$

or

$$\frac{\xi'-0}{1-0} < \frac{\xi-u^*}{1-\varphi(u^*)},$$

or

$$\frac{1-0}{\xi'-0} > \frac{1-\varphi(u^*)}{\xi-u^*}, \qquad (11.93)$$

then no burn-in needs to be conducted. Note that in Eq. (11.93)

$$\frac{1-0}{\xi'-0}$$

is the slope of the line through Points $(\xi', 1)$ and $(0, 0)$, and

$$\frac{1-\varphi(u^*)}{\xi-u^*}$$

is the slope of the line through Points $(\xi, 1)$ and $[u^*, \varphi(u^*)]$, respectively.

Therefore, the comparison can be performed in the following graphical way: Draw a line through the points $(0, 0)$ and $(\xi', 1) = [(1+C_{ff}+C_b\mu)/C_{ff}, 1]$. If this line has a slope which is larger than that of the optimum line through points $(\xi, 1)$ and $[u^*, \varphi(u^*)]$, then no burn-in needs to be performed. This is shown in Fig. 11.11.

In addition to the above graphical comparison method, a direct comparison can also be conducted. From Eq. (11.92),

$$C(u=0) = -C_b + \frac{C_{ff}}{\mu}\xi'.$$

Substituting Eq. (11.91) into this equation yields

$$C(u=0) = -C_b + \frac{C_{ff}}{\mu}\left(\frac{1+C_{ff}+C_b\mu}{C_{ff}}\right),$$
$$= -C_b + \frac{1+C_{ff}}{\mu} + C_b,$$

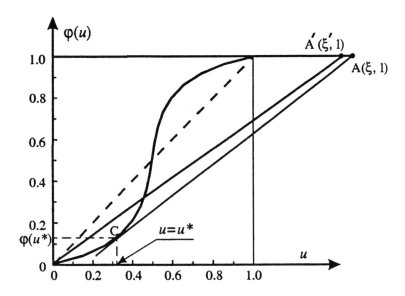

Fig. 11.11— Check if the optimum line through point A has a larger slope than the line through the points $[(1 + C_{ff} + C_b \mu)/C_{ff}, 1]$ and $(0, 0)$.

or

$$C(u=0) = \frac{1 + C_{ff}}{\mu}. \tag{11.94}$$

Therefore, if

$$C(u^*) = -C_b + \frac{C_{ff}}{\mu}\left[\frac{\xi - u^*}{1 - \varphi(u^*)}\right] > C(u=0) = \frac{1 + C_{ff}}{\mu}, \tag{11.95}$$

then no burn-in needs to be performed.

EXAMPLE 11-1

The life distribution of unrepairable components is Weibull, with

$$R(T) = e^{-\left(\frac{T}{147.3613}\right)^{0.6}}, \quad T \geq 0,$$

and with a mean life of $\mu = 222$ hr. The associated costs are $C_0 = \$1$, $C_{ff} = \$32.50$, and $C_b = 11/\mu = \$0.05/\mathrm{hr}$. Determine the optimum burn-in time T_b^* by the TTT transform technique and using Bergman-Klefsjo's cost model.

SOLUTION TO EXAMPLE 11-1

Step 1– Determine the scaled TTT transform of the given life distribution.

The scaled TTT transform corresponding to

$$R(T) = e^{-(\frac{T}{147.3613})^{0.6}}, \quad T \geq 0,$$

is given by Eq. (11.22), or

$$\varphi(u) = \Gamma\left[\frac{1}{\beta}; \log_e\left(\frac{1}{1-u}\right)\right] / \Gamma\left(\frac{1}{\beta}\right),$$

$$= \Gamma\left[\frac{1}{0.6}; \log_e\left(\frac{1}{1-u}\right)\right] / \Gamma\left(\frac{1}{0.6}\right),$$

or

$$\varphi(u) = 1.1074 \, \Gamma\left[\frac{1}{0.6}; \log_e\left(\frac{1}{1-u}\right)\right], \tag{11.96}$$

which is shown in Fig. 11.12.

Step 2– Determine the optimum value for u, u^*.

In this case

$$\xi = (1 + C_0 + C_b\mu + C_{ff})/C_{ff},$$
$$= \left(1 + 1 + \frac{11}{222} \, 222 + 32.50\right) / 32.50,$$

or

$$\xi = 1.4.$$

Drawing a line through point A, $(\xi, 1) = (1.4, 1)$, which touches the scaled TTT transform and has the largest possible slope, yields approximately

$$u^* \cong 0.5.$$

Note that at $u = u^* = 0.5$, $\varphi(u)$ is convex, or $\varphi''(u^*) > 0$, and

$$u^* - \xi = 0.5 - 1.4 = -0.9 < 0.$$

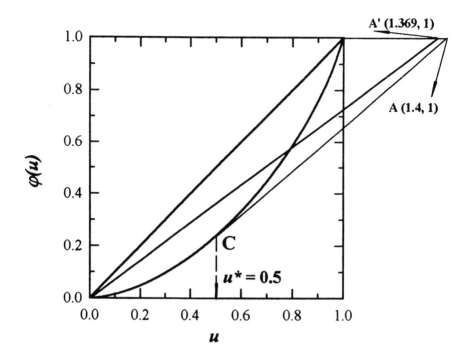

Fig. 11.12– The scaled TTT transform of $R(T) = e^{-(\frac{T}{147.3613})^{0.6}}$, $T \geq 0$, and the determination of T_b^*, for Example 11–1.

Then,

$$\left[\frac{1}{K(u)}\right]''\bigg|_{u=u^*} = \frac{\varphi''(u^*)}{u^* - \xi} < 0,$$

which implies that u^* is the maximum point of $\frac{1}{K(u)}$, or equivalently the minimum point of $K(u)$.

Step 3– Determine the optimum burn-in time, T_b^*.

The value of the optimum burn-in period, T_b^*, which minimizes $C(T_b)$, is the solution to Eq. (11.90); i.e.,

$$F(T_b^*) = 0.5$$

or

$$1 - e^{-\left(\frac{T_b^*}{147.3613}\right)^{0.6}} = 0.5,$$

which yields

$$T_b^* = 80.0 \text{ hr.}$$

Step 4– Cost comparison between the case with $T_b^* = 80.0$ hr and the case with no burn-in at all.

Draw a line $\overline{A'0}$ through point A', $(\xi', 1) = (1.369, 1)$, and point $(0,0)$ where

$$\xi' = (1 + C_{ff} + C_b \, \mu)/C_{ff},$$
$$= (1 + 32.50 + 11)/32.50,$$

or

$$\xi' = 1.369.$$

It may be seen from Fig. 11.12 that the slope of this line is less than that of the optimum line $\overline{AC}$. Therefore, the long-run cost per unit time of useful operation when $T_b^* = 80.0$ hr is less than that when no burn-in is conducted.

11.7.2 THE WEISS-DISHON'S MODEL 1 – THE COST MODEL [6]

This model was established for repairable components. After a burn-in of length T_b the unit is checked. If it has not failed it is used in operation at the condition it has after the burn-in. If the unit has failed it is repaired to an 'as-good-as-new' condition, or replaced by a new one, at cost C_{fb} before it is used. But the unit is not exposed to a new burn-in any more. The expected long-run cost per unit time of useful operation can be expressed as [6]

$$C(T_b) = \frac{(1 + C_0 + C_{ff}) + C_{fb} F(T_b) + C_b \int_0^{T_b} R(T)\, dT}{\mu F(T_b) + \int_{T_b}^{\infty} R(T)\, dT}, \quad (11.97)$$

where C_0, C_{ff}, C_b and μ are the same as defined before in Bergman-Klefsjo's Model, and

C_{fb} = cost of failure during burn-in.

It can be proved [6] that Eq. (11.97) is minimized for the value of T_b that minimizes the expression

$$K(T_b) = \frac{(\zeta - 1) + \frac{1}{\mu} \int_0^{T_b} R(T)\, dT}{\zeta + F(T_b)}, \quad (11.98)$$

or

$$K(u) = \frac{(\zeta - 1) + \varphi(u)}{\zeta + u}, \quad (11.99)$$

where

$\zeta = (1 + C_0 + C_{ff} + C_b \mu)/(C_b \mu + C_{fb})$,
$u = F(T_b)$,

and

$$\varphi(u) = \frac{1}{\mu} \int_0^{T_b} R(T)\, dT = \frac{1}{\mu} \int_0^{F^{-1}(u)} R(T)\, dT.$$

Note that Eq. (11.99) has the same structure as Eq. (11.78) with

$\alpha = \zeta - 1$,
$\beta = 1$,
$\theta = \zeta$,

and

$\delta = 1$.

Then, the optimum burn-in time, T_b^*, can be obtained in the same way as in Bergman-Klefsjo's cost model; i.e.,

Step 1: Determine the value u^* which minimizes Eq. (11.99) by drawing a straight line, on the "$\varphi(u)$ versus u" plane, through point B,

$$\left(-\frac{\theta}{\delta}, -\frac{\alpha}{\beta}\right) = \left[-\frac{\zeta}{1}, -\left(\frac{\zeta-1}{1}\right)\right] = [-\zeta, -(\zeta-1)],$$

which touches the scaled TTT transform at $[u^*, \varphi(u^*)]$ and has minimal slope.

For u^* to be a minimum point for $K(u)$, the following must be satisfied, according to Eq. (11.82), to assure $\frac{d^2[K(u)]}{du^2} > 0$:

1. If $\varphi(u)$ is concave at $[u^*, \varphi(u^*)]$, or $\varphi''(u^*) < 0$, then

$$\frac{\theta}{\beta} + \frac{\delta}{\beta}u^* = \frac{\zeta}{1} + \frac{1}{1}u^* = \zeta + u^*$$

has to be negative, or

$$\zeta + u^* < 0.$$

2. If $\varphi(u)$ is convex at $[u^*, \varphi(u^*)]$, or $\varphi''(u^*) > 0$, then $\zeta + u^*$ has to be positive, or

$$\zeta + u^* > 0.$$

Step 2: Determine T_b^* by solving the equation

$$F(T_b^*) = u^*,$$

where u^* is obtained from **Step 1**. This is illustrated graphically in Fig. 11.13. Note that $\zeta - 1 > 0$ for all values of costs involved.

After the determination of T_b^*, the total expected long-run cost per unit time of operation with burn-in, $C(T_b^*)$, which is given by Eq. (11.97), should be compared with the long-run cost without any burn-in, $C(T_b = 0)$, which is equivalent to comparing $K(u^*)$ with $K(u = 0)$ because $K(T_b)$ and $K(u)$ are minimized for the same value of T_b that minimizes $C(T_b)$. Note that if no burn-in is conducted, then the cost C_0 does not appear and ζ becomes

$$\zeta' = (1 + C_{ff} + C_b\,\mu)/(C_b\,\mu + C_{fb}).$$

Correspondingly, Eq. (11.99) becomes

$$K(u=0) = \frac{(\zeta'-1) + \varphi(0)}{\zeta' + 0} = \frac{(\zeta'-1) + 0}{\zeta' + 0}.$$

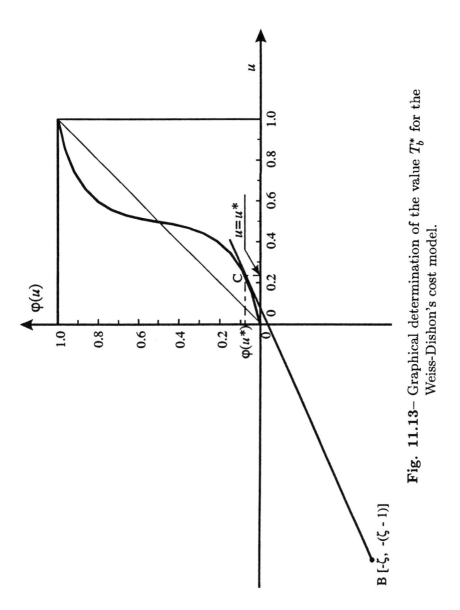

Fig. 11.13– Graphical determination of the value T_b^* for the Weiss-Dishon's cost model.

If
$$K(u=0) - K(u^*) < 0,$$
or
$$\frac{(\zeta' - 1) + 0}{\zeta' + 0} < \frac{(\zeta - 1) + \varphi(u^*)}{\zeta + u^*},$$
or
$$\frac{0 - [-(\zeta' - 1)]}{0 - (-\zeta')} < \frac{\varphi(u^*) - [-(\zeta - 1)]}{u^* - (-\zeta)}, \qquad (11.100)$$

then no burn-in needs to be conducted. Note that in Eq. (11.100)
$$\frac{0 - [-(\zeta' - 1)]}{0 - (-\zeta')}$$
is the slope of the line through Points $[-\zeta', -(\zeta' - 1)]$ and $(0,0)$, and
$$\frac{\varphi(u^*) - [-(\zeta - 1)]}{u^* - (-\zeta)}$$
is the slope of the line through Points $[u^*, \varphi(u^*)]$ and $[-\zeta, -(\zeta - 1)]$. Therefore, the comparison can be performed as follows: Draw a line through Points B', $[-\zeta', -(\zeta' - 1)]$, and $(0,0)$. If this line has a slope which is less than the optimum line through Points $[u^*, \varphi(u^*)]$ and $[-\zeta, -(\zeta - 1)]$, then no burn-in needs to be performed. This is shown in Fig. 11.14.

EXAMPLE 11–2

Rework Example 11–1 using the Weiss-Dishon's Model 1 given C_{fb} = \$5.00.

SOLUTION TO EXAMPLE 11–2

Step 1– Determine the scaled *TTT* transform of the given life distribution.

According to Example 11–1, the scaled *TTT* transform corresponding to
$$R(T) = e^{-\left(\frac{T}{147.3613}\right)^{0.6}}, \quad T \geq 0,$$
is given by Eq. (11.96), or
$$\varphi(u) = 1.1074\,\Gamma\left[\frac{1}{0.6}; \log_e\left(\frac{1}{1-u}\right)\right],$$
and is replotted in Fig. 11.15.

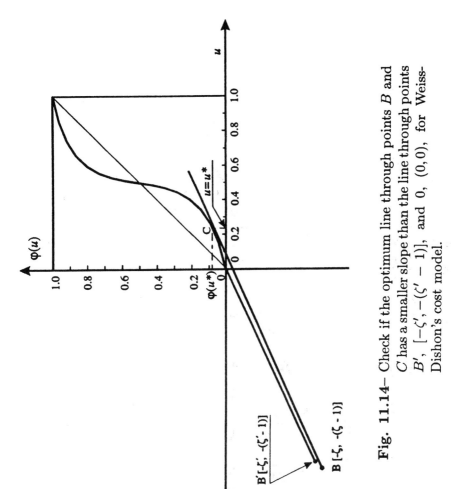

Fig. 11.14— Check if the optimum line through points B and C has a smaller slope than the line through points B', $[-\zeta', -(\zeta'-1)]$, and 0, $(0,0)$, for Weiss-Dishon's cost model.

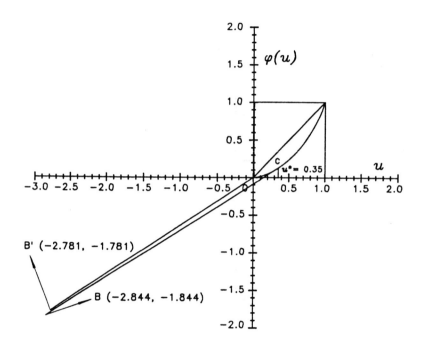

Fig. 11.15– The scaled TTT transform of $R(T) = e^{-(\frac{T}{147.3613})^{0.6}}$, $T \geq 0$, and the determination of T_b^*, for Example 11-2.

Step 2– Determine the optimum value for u, u^*.

In this case

$$\zeta = (1 + C_0 + C_{ff} + C_b \mu)/(C_b \mu + C_{fb}),$$
$$= (1 + 1 + 32.50 + 11)/(11 + 5),$$

or

$$\zeta = 2.844.$$

Drawing a line through point B, $[-\zeta, -(\zeta-1)] = (-2.844, -1.844)$, which touches the scaled TTT transform at Point C and has the minimum slope, yields approximately

$$u^* \cong 0.35.$$

Note that at $u = u^* = 0.35$, $\varphi(u)$ is convex, or $\varphi''(u^*) > 0$, and

$$\zeta + u^* = 2.844 + 0.35 = 3.194 > 0.$$

Then,

$$[K(u)]''\Big|_{u=u^*} = \frac{\varphi''(u^*)}{\zeta + u^*} > 0,$$

which implies that u^* is the minimum point of $K(u)$, or equivalently the minimum point of $C(T_b)$.

Step 3– Determine the optimum burn-in time, T_b^*.

The value of the optimum burn-in period, T_b^*, which minimizes $C(T_b)$, is the solution to the following equation:

$$F(T_b^*) = u^* = 0.35,$$

or

$$1 - e^{-\left(\frac{T_b^*}{147.3613}\right)^{0.6}} = 0.35,$$

which yields

$$T_b^* = 36.2088 \text{ hr, or } 36 \text{ hr.}$$

Step 4– Cost comparison between the case with $T_b^* = 36$ hr and the case with no burn-in at all.

Draw a line $\overline{B'0}$ through point B', $[\zeta', -(\zeta'-1)] = (-2.781, -1.781)$, and point $(0,0)$ where

$$\zeta' = (1 + C_{ff} + C_b \, \mu)/(C_b \, \mu + C_{fb}),$$
$$= (1 + 32.50 + 11)/(11 + 5),$$

or

$$\zeta' = 2.781.$$

It may be seen from Fig. 11.15 that the slope of this line is larger than that of the optimum line $\overline{BC}$. Therefore, the long-run cost per unit time of useful operation when $T_b^* = 36$ hr is less than that when no burn-in is conducted.

11.7.3 THE WEISS-DISHON'S MODEL 2 – THE PROFIT MODEL

This model was first discussed by Weiss and Dishon [6], in which the burn-in time is so determined that the expected profit per unit from burn-in is maximized. It was then proved by Bergman and Klefsjo [3] that the optimization can be performed by first determining u^* which maximizes

$$Q(u) = \left[\frac{K\mu - C_{fb}}{(K + C_b)\mu}\right] u - \varphi(u), \tag{11.101}$$

and then solving

$$F(T_b^*) = u^* \tag{11.102}$$

for T_b^* where C_{fb} and C_b are the same as defined in Bergman-Klefsjo's Model and Weiss-Dishon's Model 1, and

K = profit per unit time of useful operation.

Note that C_{fb} should be less than $K\,\mu$, otherwise we should not perform any burn-in, since $Q(u)$ will be a decreasing function of u if $C_{fb} \geq K\,\mu$.

The value of u^* which maximizes $Q(u)$ is the one that maximizes the distance between Points D, $[u^*, \frac{K\mu - C_{fb}}{(K+C_b)\mu} u^*]$, and C, $[u^*, \varphi(u^*)]$, and with point D lying above C, as shown in Fig. 11.16. This can be

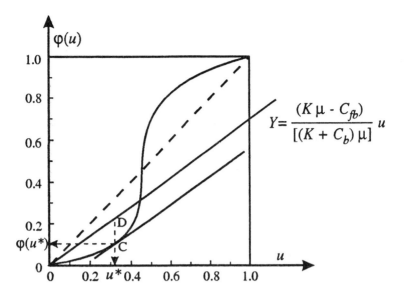

Fig. 11.16– Graphical determination of value T_b^* for Weiss-Dishon's profit model.

established by drawing a tangent line of $\varphi(u)$ which is paralell to and lying below the line

$$Y = \left[\frac{K\mu - C_{fb}}{(K + C_b)\mu}\right] u \tag{11.103}$$

and touches $\varphi(u)$ at Point C, $[u^*, \varphi(u^*)]$. Then, Point C is the optimum point. This can also be verified by taking the first derivative of Eq. (11.101) and letting it equal to zero; i.e.,

$$Q'(u) = \frac{K\mu - C_{fb}}{(K + C_b)\mu} - \varphi'(u) = 0,$$

which yields

$$\varphi'(u^*) = \frac{K\mu - C_{fb}}{(K + C_b)\mu}.$$

That is, the tangent line at the optimum point, u^*, has a slope of $\left[\frac{K\mu - C_{fb}}{(K+C_b)\mu}\right]$, which is the same as the slope of Eq. (11.103).

To determine if u^* is the real maximum point of $Q(u)$, taking the second derivative of Eq. (11.101) yields

$$Q''(u) = 0 - \varphi''(u) = -\varphi''(u).$$

Therefore, if $\varphi''(u)$ is concave at u^*, or $\varphi''(u^*) < 0$, then $Q''(u^*) > 0$ and u^* is the minimum point of $Q(u)$ instead of the maximum point. Only if $\varphi(u)$ is convex at u^*, or $\varphi''(u^*) > 0$, then $Q''(u^*) < 0$ and u^* is the maximum point of $Q(u)$.

EXAMPLE 11-3

Rework Example 11-1 using the Weiss-Dishon's Model 2 given $K = 88/\mu = \$0.40/\text{hr}$ and $C_{fb} = \$5.00$.

SOLUTION TO EXAMPLE 11-3

Step 1– Determine the scaled TTT transform of the given life distribution.

According to Example 11-1, the scaled TTT transform corresponding to

$$R(T) = e^{-\left(\frac{T}{147.3613}\right)^{0.6}}, \quad T \geq 0,$$

is given by Eq. (11.96), or

$$\varphi(u) = 1.1074\, \Gamma\left[\frac{1}{0.6}; \log_e\left(\frac{1}{1-u}\right)\right],$$

and is replotted in Fig. 11.17.

Step 2– Determine the optimum value for u, u^*.

First, draw the line

$$Y = \left[\frac{K\mu - C_{fb}}{(K + C_b)\mu}\right] u = \left[\frac{88 - 5}{88 + 11}\right] u = 0.8384\, u,$$

which can be obtained by connecting two points on the $\varphi(u)$ plane at $u = 0$ and $u = 1$; i.e., connecting point 0, (0,0), and point A, (1, 0.8384), as shown in Fig. 11.17.

Second, draw a tangent line to $\varphi(u)$ which is paralell to and lying below line $\overline{0A}$ and touches $\varphi(u)$ at point C, $[u^*, \varphi(u^*)]$. Reading from Fig. 11.17 yields approximately

$$u^* \cong 0.5,$$

which happens to be the same as that obtained in Example 11-1. Since $\varphi(u)$ is convex at $u^* = 0.5$, then $u^* = 0.5$ is the maximum point of $Q(u)$.

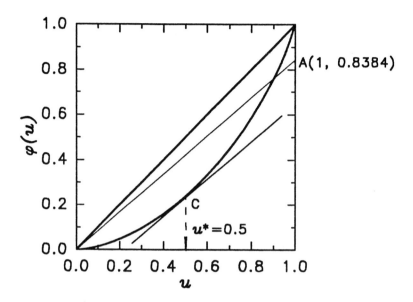

Fig. 11.17– The scaled TTT transform of $R(T) = e^{-\left(\frac{T}{147.3613}\right)^{0.6}}$, $T \geq 0$, and the determination of T_b^*, for Example 11-3.

Step 3– Determine the optimum burn-in time, T_b^*.

The value of the optimum burn-in period, T_b^*, which minimizes $C(T_b)$, is the solution to the following equation:

$$F(T_b^*) = u^* = 0.5,$$

or

$$1 - e^{-\left(\frac{T_b^*}{147.3613}\right)^{0.6}} = 0.5,$$

which yields

$$T_b^* = 80.0 \text{ hr.}$$

11.8 OPTIMUM BURN-IN TIME DETERMINATION USING THE TTT TRANSFORM FOR THE MAXIMUM POST-BURN-IN MEAN RESIDUAL LIFE (MRL)

According to Chapter 8, the mean residual life (MRL) of components with a reliability function of $R(T)$, after a T_b-hr burn-in, is

$$MRL(T_b) = \frac{\int_{T_b}^{\infty} R(T) \, dT}{R(T_b)}. \tag{11.104}$$

Let

$$u = F(T_b),$$
$$\mu = \int_0^{\infty} R(T) \, dT,$$

and

$$\varphi(u) = \frac{1}{\mu} \int_0^{F^{-1}(u)} R(T) \, dT = \frac{1}{\mu} \int_0^{T_b} R(T) \, dT.$$

Then, Eq. (11.104) can be expressed in terms of u as follows:

$$MRL(u) = \frac{\int_0^{\infty} R(T) \, dT - \int_0^{T_b} R(T) \, dT}{1 - F(T_b)},$$

$$= \frac{\mu - \mu \left[\frac{1}{\mu} \int_0^{T_b} R(T) \, dT\right]}{1 - u},$$

$$= \mu \left[\frac{1 - \frac{1}{\mu} \int_0^{F^{-1}(u)} R(T) \, dT}{1 - u}\right],$$

or

$$MRL(u) = \mu \left[\frac{1-\varphi(u)}{1-u}\right]. \qquad (11.105)$$

Therefore, maximizing Eq. (11.104) is equivalent to maximizing Eq. (11.105) which is maximized for the same value of u as for

$$M(u) = \frac{1-\varphi(u)}{1-u}, \qquad (11.106)$$

because the multiplier of the right side of Eq. (11.105), μ, is a positive constant.

Note that Eq. (11.106) has the same format as Eq. (11.7) with

$\alpha = 1,$
$\beta = -1,$
$\theta = 1,$

and

$\delta = -1.$

Therefore, the optimum burn-in time, T_b^*, which maximizes $MRL(T_b)$ given by Eq. (11.104), can be obtained using the following two steps:

Step 1: Determine the value of u^*, for which

$$M(u) = \frac{1-\varphi(u)}{1-u}$$

is maximized. This can be done by drawing a line, on the "$\varphi(u)$ versus u" plane, from point

$$\left(-\frac{\theta}{\delta}, -\frac{\alpha}{\beta}\right) = \left(-\frac{1}{(-1)}, -\frac{1}{(-1)}\right) = (1,1),$$

which touches the scaled TTT transform at

$$[u^*, \varphi(u^*)]$$

and has the largest slope as shown in Fig. 11.18.

For u^* to be a maximum point for $M(u)$, the following must be satisfied, according to Eq. (11.82), to assure $d^2[M(u)]/du^2 < 0$:

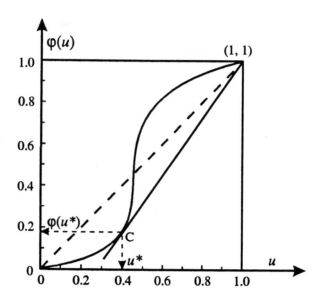

Fig. 11.18– Graphical determination of T_b^* which maximizes $MRL(T_b)$.

1. If $\varphi(u)$ is concave at $[u^*, \varphi(u^*)]$, or $\varphi''(u^*) < 0$, then

$$\frac{\theta}{\beta} + \frac{\delta}{\beta} u^* = \frac{1}{-1} + \frac{-1}{-1} u^* = u^* - 1$$

has to be positive, or

$$u^* > 1,$$

which is impossible because u^* is always less than or equal to 1, or

$$0 \leq u^* \leq 1.$$

Consequently, if $\varphi(u)$ is concave at point $[u^*, \varphi(u^*)]$, then u^* is not the maximum point of $M(u)$!

2. If $\varphi(u)$ is convex at $[u^*, \varphi(u^*)]$, or $\varphi''(u^*) > 0$, then $(u^* - 1)$ has to be negative, or

$$u^* < 1.$$

Consequently, if $\varphi(u)$ is convex at point $[u^*, \varphi(u^*)]$ and $u^* \neq 1$, then u^* is the maximum point of $M(u)$!

FROM A SAMPLE OF OBSERVED TIMES TO FAILURE   495

Step 2: Determine the optimal value of T_b, T_b^*, by solving the equation

$$F(T_b^*) = u^*,$$

where u^* was obtained from **Step 1**.

The above procedure will be illustrated by numerical examples in the next section where the optimum burn-in time for the maximum MRL is determined directly from the scaled TTT plot.

11.9 BURN-IN TIME DETERMINATION DIRECTLY FROM A SAMPLE OF OBSERVED TIMES TO FAILURE

11.9.1 GENERAL PROCEDURE

Suppose that T_b has to be estimated from an ordered, complete, sample $0 = T_{0:n} \leq T_{1:n} \leq T_{2:n} \leq ... \leq T_{n:n}$ of times to failure from an absolutely continuous life distribution, $F(T)$, and the objective function, such as the total cost per unit time of useful operation, or the post-burn-in MRL, has the structure of Eq. (11.7). Then, the TTT plot can be used, which does not require any life distribution assumption, therefore is *nonparametric*, and converges to the scaled TTT transform of $F(T)$ when the number of observations goes to infinity, to estimate T_b^* in the following two steps:

Step 1: Draw the TTT plot of the sample by connecting the adjacent points $(\frac{i-1}{n}, \frac{T_{i-1}}{T_n})$ and $(\frac{i}{n}, \frac{T_i}{T_n})$ for $i = 1, 2, ..., n$ by line segments.

Step 2: Draw the line through the point A,

$$[u_A, \varphi_A] = [u_A, \varphi(u_A)] = (-\theta/\delta, -\alpha/\beta),$$

which touches the TTT plot and has the largest slope. If the line touches the TTT plot at Point B, $(\frac{i^*}{n}, \frac{T_{i^*}}{T_n})$, where $1 \leq i^* \leq n$, then $T_{b:n}^* = T_{i^*:n}$ is the estimator of T_b^*.

This procedure is illustrated in Fig. 11.19 where the optimum burn-in time is determined for the minimum cost per unit time of useful operation using the Bergman-Klefsjo's cost model. For other types of data sets, such as complete (or censored) grouped data sets, censored individual data sets with or without ties, etc., the above procedure can

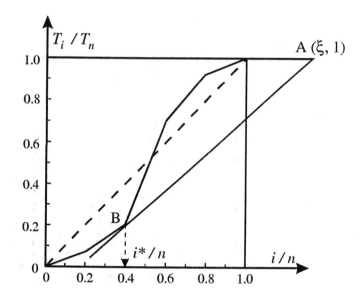

Fig. 11.19– Graphical estimation of T_b^* by TTT plot from a sample of observed times to failure for the minimum cost per unit time of useful operation using the Bergman-Klefsjo's cost model.

FROM A SAMPLE OF OBSERVED TIMES TO FAILURE

be applied similarly with the corresponding observed values for u and $\varphi(u)$.

It can be proved that this estimator, $T_{b:n}^*$, is a strongly consistent estimator of T_b^* if T_b^* is unique [6]. Correspondingly, $C_n(T_{b:n}^*)$ is a strongly consistent estimator of $C(T_b^*)$. For example, when Bergman-Klefsjö's cost model is used,

$$C_n(T_{b:n}^*) = -C_b + \frac{C_{ff}}{\hat{\mu}} \left(\frac{\xi - i^*/n}{1 - T_{i^*}/T_n} \right) \qquad (11.107)$$

is a strongly consistent estimator of $C(T_b^*)$ given by Eq. (11.83) where

$$\hat{\mu} = \frac{1}{n} \sum_{i=1}^{n} T_{i:n}.$$

If the line through points $(0,0)$ and $(\xi', 1) = [(1+C_{ff}+C_b\mu)/C_{ff}, 1]$ has a slope which is larger than that of the optimum line through points $(\xi, 1)$ and $(i^*/n, T_{i^*}/T_n)$, then no burn-in needs to be performed; or equivalently according to Eq. (11.95) if

$$C_n(T_{b:n}^*) > (1 + C_{ff})/\hat{\mu},$$

no burn-in needs to be performed.

It may be worth noting that the graphical procedure just described offers excellent possibilities for different sensitivity analyses; e.g., with respect to the costs (or times) involved. For example, the same estimator of T_b^* can be obtained from Fig. 11.19 for many values of ξ; i.e., for many different combinations of the involved costs.

11.9.2 REFINED PROCEDURE

The above determined $T_{b:n}^*$ from Point i^* can be regarded as a preliminary estimate of the optimum burn-in time for the minimum cost. To be more precise, a quadratic function can be fitted to the scaled TTT plot in the neighborhood of i^* using two points next to i^*, Points $i^* - 1$ and $i^* + 1$, as shown in Fig. 11.20, plus Point i^* itself; i.e.,

$$\varphi(u) = a\, u^2 + b\, u + c, \quad u_{i^*-1} \leq u \leq u_{i^*+1}, \qquad (11.108)$$

where a, b and c are three unknown constants to be determined from $[u_{i^*-1}, \varphi(u_{i^*-1})]$, $[u_{i^*}, \varphi(u_{i^*})]$, and $[u_{i^*+1}, \varphi(u_{i^*+1})]$. Note that the first constant, a, in Eq. (11.108) should always be a positive value, or

$$a > 0,$$

because the optimum burn-in time should be in the decreasing failure rate (DFR) region which corresponds to the convex region of the scaled TTT transform, or the convex region of the scaled TTT plot.

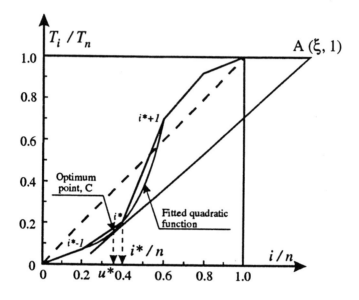

Fig. 11.20– A refined procedure for graphical estimation of T_b^* by the TTT plot from a sample of observed times to failure for the minimum cost per unit time of useful operation using the Bergman-Klefsjo's cost model.

Substituting $[u_{i^*-1}, \varphi(u_{i^*-1})]$, $[u_{i^*}, \varphi(u_{i^*})]$, and $[u_{i^*+1}, \varphi(u_{i^*+1})]$ into Eq. (11.108) yields

$$\begin{cases} \varphi(u_{i^*-1}) = a u_{i^*-1}^2 + b u_{i^*-1} + c, \\ \varphi(u_{i^*}) = a u_{i^*}^2 + b u_{i^*} + c, \\ \varphi(u_{i^*+1}) = a u_{i^*+1}^2 + b u_{i^*+1} + c. \end{cases}$$

Solving these three equations simultaneously yields

$$\begin{cases} a = D_a/D, \\ b = D_b/D, \\ c = D_c/D, \end{cases} \qquad (11.109)$$

where

$$\begin{aligned} D &= u_{i^*-1} u_{i^*} (u_{i^*-1} - u_{i^*}) + u_{i^*} u_{i^*+1} (u_{i^*} - u_{i^*+1}) \\ &\quad + u_{i^*+1} u_{i^*-1} (u_{i^*+1} - u_{i^*-1}), \\ D_a &= \varphi(u_{i^*-1})(u_{i^*} - u_{i^*+1}) + \varphi(u_{i^*})(u_{i^*+1} - u_{i^*-1}) \\ &\quad + \varphi(u_{i^*+1})(u_{i^*-1} - u_{i^*}), \\ D_b &= u_{i^*-1}^2 [\varphi(u_{i^*}) - \varphi(u_{i^*+1})] + u_{i^*}^2 [\varphi(u_{i^*+1}) - \varphi(u_{i^*-1})] \\ &\quad + u_{i^*+1}^2 [\varphi(u_{i^*-1}) - \varphi(u_{i^*})], \end{aligned}$$

and

$$\begin{aligned} D_c &= u_{i^*-1} u_{i^*} \varphi(u_{i^*+1}) (u_{i^*-1} - u_{i^*}) \\ &\quad + u_{i^*} u_{i^*+1} \varphi(u_{i^*-1}) (u_{i^*} - u_{i^*+1}) \\ &\quad + u_{i^*+1} u_{i^*-1} \varphi(u_{i^*}) (u_{i^*+1} - u_{i^*-1}). \end{aligned}$$

Then, the optimum value for u, u^*, can be determined by drawing a tangent line to the plot of Eq. (11.108) from Point A

$$[u_A, \varphi_A] = (-\theta/\delta, -\alpha/\beta)$$

which touches Eq. (11.108) at Point C, $[u^*, \varphi(u^*)]$. Since Point C belongs to both Eq. (11.108) and line $\overline{AC}$, then u^* can be analytically obtained by solving the following two equations simultaneously:

$$\begin{cases} \varphi'(u^*) = 2 a u^* + b = \frac{\varphi_A - \varphi(u^*)}{u_A - u^*}, \\ \varphi(u^*) = a (u^*)^2 + b u^* + c, \end{cases}$$

which yields

$$u^* = u_A \pm \sqrt{u_A^2 - \left[\frac{\varphi_A - b u_A - c}{a}\right]}. \qquad (11.110)$$

The choice of "+" or "-" sign in front of the square root term can be made so that the following relation is satisfied:

$$u_{i^*-1} \leq u^* \leq u_{i^*+1}.$$

To determine the optimum burn-in time, T_b^*, corresponding to $u = u^*$, the following equation needs to be evaluated:

$$T_b^* = F^{-1}(u^*). \qquad (11.111)$$

Now, the *cdf* for the components, $F(T)$, is unknown. But the observed *cdf* values at Points W_{i^*-1}, W_{i^*}, and W_{i^*+1}, are available; i.e.,

$$\hat{F}(W_{i^*-1}) = u_{i^*-1},$$
$$\hat{F}(W_{i^*}) = u_{i^*},$$

and

$$\hat{F}(W_{i^*+1}) = u_{i^*+1},$$

where W_{i^*-1}, W_{i^*}, and W_{i^*+1} are the end points of the (i^*-1)st, i^*th, and (i^*+1)st time intervals for the grouped data set; the (i^*-1)st, i^*th, and (i^*+1)st distinct failure times for the ungrouped data set. Also it is known that the optimum point is between Points W_{i^*-1} and W_{i^*+1}. Then, a quadratic function can also be fitted to $u = F(T)$ in (W_{i^*-1}, W_{i^*+1}) using these three points. Let

$$u = F(T) = A\,T^2 + B\,T + C, \quad W_{i^*-1} \leq T \leq W_{i^*+1}, \qquad (11.112)$$

where A, B and C are three unknown constants to be determined from $[W_{i^*-1}, u_{i^*-1}]$, $[W_{i^*}, u_{i^*}]$, and $[W_{i^*+1}, u_{i^*+1}]$. Substituting these three points into Eq. (11.112) yields

$$\begin{cases} u_{i^*-1} = A\,W_{i^*-1}^2 + B\,W_{i^*-1} + C, \\ u_{i^*} = A\,W_{i^*}^2 + B\,W_{i^*} + C, \\ u_{i^*+1} = A\,W_{i^*+1}^2 + B\,W_{i^*+1} + C. \end{cases}$$

Solving these three equations simultaneously yields

$$\begin{cases} A = K_A/K, \\ B = K_B/K, \\ C = K_C/K, \end{cases} \qquad (11.113)$$

where

$$K = W_{i^*-1} W_{i^*} (W_{i^*-1} - W_{i^*}) + W_{i^*} W_{i^*+1} (W_{i^*} - W_{i^*+1})$$
$$+ W_{i^*+1} W_{i^*-1} (W_{i^*+1} - W_{i^*-1}),$$
$$K_A = u_{i^*-1} (W_{i^*} - W_{i^*+1}) + u_{i^*} (W_{i^*+1} - W_{i^*-1})$$
$$+ u_{i^*+1} (W_{i^*-1} - W_{i^*}),$$
$$K_B = W_{i^*-1}^2 (u_{i^*} - u_{i^*+1}) + W_{i^*}^2 (u_{i^*+1} - u_{i^*-1})$$
$$+ W_{i^*+1}^2 (u_{i^*-1} - u_{i^*}),$$

and

$$K_C = W_{i^*-1} W_{i^*} u_{i^*+1} (W_{i^*-1} - W_{i^*})$$
$$+ W_{i^*} W_{i^*+1} u_{i^*-1} (W_{i^*} - W_{i^*+1})$$
$$+ W_{i^*+1} W_{i^*-1} u_{i^*} (W_{i^*+1} - W_{i^*-1}).$$

After Eq. (11.112) is determined, the optimum burn-in time, T_b^*, corresponding to $u = u^*$, can be obtained right away by solving

$$u^* = A (T_b^*)^2 + B T_b^* + C,$$

or

$$A (T_b^*)^2 + B T_b^* + (C - u^*) = 0,$$

which yields

$$T_b^* = \frac{-B \pm \sqrt{B^2 - 4 A (C - u^*)}}{2 A}, \qquad (11.114)$$

which is the more precise estimate of the optimum burn-in time. The choice of "+" or "-" sign in front of the square root term can be made so that the following relation is satisfied:

$$W_{i^*-1} \leq T_b^* \leq W_{i^*+1}.$$

A computer program is deveioped in FORTRAN language to assist the calculation in this refined procedure. This program is listed in Appendix 11A.

EXAMPLE 11-4

A life test on $n = 150$ electronic components yields the data in Table 11.1, where a total of $r = 19$ failures are observed at $m = 12$ distinctive failure times during $T_d = 4,500$ hr of operation. The associated costs are the same as in Example 11-1. Determine the optimum burn-in time using the scaled TTT plot approach

Table 11.1– **Raw failure data from a life test of $T_d = 4,500$ hr on $n = 150$ electronic components, for Example 11–4.**

Time order, i	Failure time, W_i, hr	Failure number, r_i
1	100	1
2	200	1
3	250	1
4	420	2
5	588	3
6	708	1
7	1,044	1
8	2,892	2
9	3,396	2
10	3,997	3
11	4,165	1
12	4,500	1

FROM A SAMPLE OF OBSERVED TIMES TO FAILURE   503

1. for the minimum cost per unit time of useful operation using the Bergman-Klefsjo's cost model, and

2. for the maximum post-burn-in MRL.

SOLUTION TO EXAMPLE 11–4

1.1 The general procedure is used first to get a preliminary estimate of the optimum burn-in time which minimizes the total cost per unit time of operation. Note that given in Table 11.1 is a time-terminated ungrouped data set with ties. Then, Eqs. (11.55), (11.56) and (11.58) should be used to determine the TTT points as follows:

The TTT statistic, at each distinct failure time, W_i, is given by Eq. (11.55), or

$$\begin{cases} T_i = \sum_{j=1}^{i} r_j\, W_j + (n - \sum_{j=1}^{i} r_j)\, W_i, \text{ for } i = 1, 2, ..., m, \\ T_0 = 0, \text{ for } i = 0, \end{cases}$$

or

$$\begin{cases} T_i = \sum_{j=1}^{i} r_j\, W_j + (150 - \sum_{j=1}^{i} r_j)\, W_i, \text{ for } i = 1, 2, ..., 12, \\ T_0 = 0, \text{ for } i = 0. \end{cases}$$

For example when $i = 2$,

$$\begin{aligned} T_2 &= \sum_{j=1}^{2} r_j\, W_j + (150 - \sum_{j=1}^{2} r_j)\, W_2, \\ &= (r_1\, W_1 + r_2\, W_2) + (150 - r_1 - r_2)\, W_2, \\ &= (1 \times 100 + 1 \times 200) + (150 - 1 - 1)\, 200, \end{aligned}$$

or

$$T_2 = 29,900 \text{ hr}.$$

The scaled TTT statistic at each distinct failure time, W_i, is given by Eq. (11.56), or

$$\varphi_i = T_i/T_m = T_i/T_{12}, \text{ for } i = 0, 1, 2, ..., 12,$$

where

$$T_m = \sum_{j=1}^{m} r_j W_j + (n-r)W_m,$$

or

$$T_{12} = \sum_{j=1}^{12} r_j W_j + (150 - 19)\, 4{,}500 = 627{,}638 \text{ hr}.$$

The observed *cdf* at each distinct failure time, W_i, is given by Eq. (11.58), or

$$u_i = \hat{F}(W_i) = \begin{cases} \dfrac{\sum_{j=1}^{i} r_j}{r}, & \text{for } i = 1, 2, \cdots, m, \\ 0 & \text{for } i = 0, \end{cases}$$

or

$$u_i = \hat{F}(W_i) = \begin{cases} \dfrac{\sum_{j=1}^{i} r_j}{19}, & \text{for } i = 1, 2, \cdots, 12, \\ 0 & \text{for } i = 0. \end{cases}$$

For example when $i = 2$,

$$u_2 = \dfrac{\sum_{j=1}^{2} r_j}{19} = \dfrac{r_1 + r_2}{19} = \dfrac{1+1}{19} = 0.1053.$$

The values for T_i, φ_i and u_i are calculated for $i = 0, 1, 2, \cdots, 12$ and listed in Table 11.2. Plotting points (u_i, φ_i) and connecting them with line segments yields the scaled TTT plot for this time-terminated data set with ties, given in Table 11.1, which is shown in Fig. 11.21.

Now, similar to Example 11-1 where the Bergman-Klefsjo's model is applied, draw the optimal line with the highest slope through Point A, $(u_A, \varphi_A) = (1.4, 1)$, which touches the plot at

$$[u^*, \varphi(u^*)] = (u_6, \varphi_6) = (0.4737, 0.1652).$$

Table 11.2– Values of the TTT statistic, T_i, scaled TTT statistic, φ_i, and observed cdf, u_i, evaluated for the data set given in Table 11.1 for Example 11–4.

Time order, i	Failure time, W_i, hr	T_i	φ_i	u_i
0	0	0	0	0
1	100	15,000	0.0239	0.0526
2	200	29,900	0.0476	0.1053
3	250	37,300	0.0594	0.1579
4	420	62,290	0.0992	0.2632
5	588	86,650	0.1381	0.4211
6	708	103,690	0.1652	0.4737
7	1,044	151,066	0.2407	0.5263
8	2,892	409,786	0.6529	0.6316
9	3,396	479,338	0.7637	0.7368
10	3,997	561,074	0.8939	0.8947
11	4,165	583,418	0.9295	0.9474
12	4,500	627,638	1.0000	1.0000

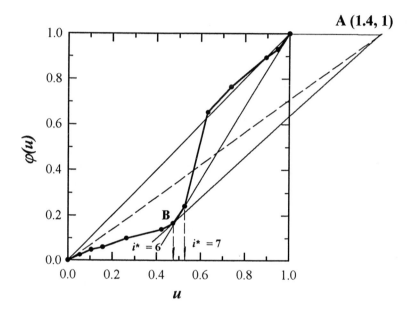

Fig. 11.21– The scaled TTT plot for the data set given in Table 11.1 and the graphical determination of the optimum burn-in time using the general procedure, for (1) the minimum cost per unit time of useful operation using the Bergman-Klefsjo's cost model, and (2) the maximum post-burn-in MRL, for Example 11–4.

Therefore,

$$i^* = 6,$$

and the estimated optimum burn-in time is

$$T_b^* = W_6 = 708 \text{ hr}.$$

Note that at $u^* = u_6 = 0.4737$, $\varphi(u)$ is convex, or $\varphi''(u^*) > 0$, and

$$u^* - \xi = 0.4737 - 1.4 = -0.9263 < 0.$$

Then,

$$\left[\frac{1}{K(u)}\right]''\bigg|_{u=u^*} = \frac{\varphi''(u^*)}{u^* - \xi} < 0,$$

which implies that $T_b^* = W_6 = 708$ hr is the optimum burn-in time which minimizes the total cost per unit time of useful operation.

1.2 The refined procedure is used next to get a more precise u^* value by fitting the quadratic function given in Eq. (11.108) in the neighborhood of $u_6 = 0.4737$ using the following three points given in Table 11.2: $[u_5, \varphi(u_5)] = (0.4211, 0.1381)$, $[u_6, \varphi(u_6)] = (0.4737, 0.1652)$ and $[u_7, \varphi(u_7)] = (0.5263, 0.2407)$. Substituting these three points into Eq. (11.108) yields

$$\begin{cases} 0.4211^2\, a + 0.4211\, b + c = 0.1381, \\ 0.4737^2\, a + 0.4737\, b + c = 0.1652, \\ 0.5263^2\, a + 0.5263\, b + c = 0.2407. \end{cases}$$

Solving these three equations simultaneously, or substituting $[u_5, \varphi(u_5)] = (0.4211, 0.1381)$, $[u_6, \varphi(u_6)] = (0.4737, 0.1652)$ and $[u_7, \varphi(u_7)] = (0.5263, 0.2407)$ into Eq. (11.109) directly, yields

$$\begin{cases} a = 8.7467, \\ b = -7.3113, \\ c = 1.6659. \end{cases}$$

Then, the fitted quadratic function for the scaled TTT plot in the neighborhood of $u_6 = 0.4737$ is

$$\varphi(u) = 8.7467\, u^2 - 7.3113\, u + 1.6659.$$

The optimum u value, u^*, is given by Eq. (11.110); i.e.,

$$u^* = u_A \pm \sqrt{u_A^2 - \left[\frac{\varphi_A - b\, u_A - c}{a}\right]},$$

$$= 1.4 \pm \sqrt{1.4^2 - \left[\frac{1 + 7.3113 \times 1.4 - 1.6659}{8.7467}\right]},$$

or

$$u^* = 0.4944, \text{ or } 2.3056.$$

Since u^* needs to be in the interval of $[u_5, u_7] = [0.4211, 0.5263]$, then only

$$u^* = 0.4944$$

is the real optimum value.

To determine the optimum burn-in time corresponding to $u^* = 0.4944$, fit the quadratic function given by Eq. (11.112) to $u = F(T)$ in the neighborhood of $T = W_6 = 708$ hr using the following three points: $(W_5, u_5) = (588 \text{ hr}, 0.4211)$, $(W_6, u_6) = (708 \text{ hr}, 0.4737)$ and $(W_7, u_7) = (1{,}044 \text{ hr}, 0.5263)$. Substituting these three points into Eq. (11.112) yields

$$\begin{cases} 588^2 A + 588 B + C & = 0.4211, \\ 708^2 A + 708 B + C & = 0.4737, \\ 1{,}044^2 A + 1{,}044 B + C & = 0.5263. \end{cases}$$

Solving these three equations simultaneously, or substituting $(W_5, u_5) = (588 \text{ hr}, 0.4211)$, $(W_6, u_6) = (708 \text{ hr}, 0.4737)$ and $(W_7, u_7) = (1{,}044 \text{ hr}, 0.5263)$ into Eq. (11.113) directly, yields

$$\begin{cases} A = -6.2 \times 10^{-7}, \\ B = 0.0012, \\ C = -0.0939. \end{cases}$$

Then, the fitted quadratic function for the observed *cdf* in the neighborhood of $W_6 = 708$ hr is

$$u(T) = -6.2 \times 10^{-7} T^2 + 0.0012 T - 0.0939.$$

Therefore, the optimum burn-in time, T_b^*, is given by Eq. (11.114); i.e.,

$$T_b^* = \frac{-B \pm \sqrt{B^2 - 4A(C - u^*)}}{2A},$$

$$= \frac{-0.0012 \pm \sqrt{0.0012^2 - 4(-6.2 \times 10^{-7})(-0.0939 - 0.4944)}}{2(-6.2 \times 10^{-7})},$$

or

$$T_b^* = 771.6695 \text{ hr, or } 1{,}233.6640 \text{ hr}.$$

Since T_b^* needs to be in the interval of $[W_5, W_7] = [588$ hr, $1{,}044$ hr$]$, then only

$$T_b^* = 771.6695 \text{ hr, or } 772 \text{ hr}$$

is the real optimum burn-in time.

2.1 The general procedure is used first to get a preliminary estimate of the optimum burn-in time which maximizes the post-burn-in *MRL*. Draw the optimal line with the highest slope through Point $(u_A, \varphi_A) = (1, 1)$, which touches the scaled *TTT* plot at

$$[u^*, \varphi(u^*)] = (u_7, \varphi_7) = (0.5263, 0.2407).$$

Therefore,

$$i^* = 7,$$

and the estimated optimum burn-in time is

$$T_b^* = W_7 = 1{,}044 \text{ hr}.$$

Note that at $u^* = u_7 = 0.5263$, $\varphi(u)$ is convex, or $\varphi''(u^*) > 0$, and

$$u^* - 1 = 0.5263 - 1 = -0.4737 < 0.$$

Then, $u^* = u_7 = 0.5263$ is the maximum point for $M(u)$. In turn, $T_b^* = W_7 = 1{,}044$ hr is the optimum burn-in time which maximizes the post-burn-in *MRL*. Note that this result is consistent with that in Example 8–8, where T_b^* was determined directly from the empirical *MRL* plot.

2.2 The refined procedure is used second to get a more precise u^* value by fitting the quadratic function given in Eq. (11.108) in the neighborhood of $u_7 = 0.5263$ using the following three points: $[u_6, \varphi(u_6)] = (0.4737, 0.1652)$, $[u_7, \varphi(u_7)] = (0.5263, 0.2407)$ and $[u_8, \varphi(u_8)] = (0.6316, 0.6529)$. Substituting these three points into Eq. (11.108) yields

$$\begin{cases} 0.4737^2\,a + 0.4737\,b + c = 0.1652, \\ 0.5263^2\,a + 0.5263\,b + c = 0.2407, \\ 0.6316^2\,a + 0.6316\,b + c = 0.6529. \end{cases}$$

Solving these three equations simultaneously, or substituting $[u_6, \varphi(u_6)] = (0.4737, 0.1652)$, $[u_7, \varphi(u_7)] = (0.5263, 0.2407)$ and $[u_8, \varphi(u_8)] = (0.6316, 0.6529)$ into Eq. (11.109) directly, yields

$$\begin{cases} a = 15.7009, \\ b = -14.2655, \\ c = 3.3996. \end{cases}$$

Then, the fitted quadratic function for the scaled TTT plot in the neighborhood of $u_7 = 0.5263$ is

$$\varphi(u) = 15.7009\,u^2 - 14.2655\,u + 3.3996.$$

The optimum u value, u^*, is given by Eq. (11.110); i.e.,

$$u^* = u_A \pm \sqrt{u_A^2 - \left[\frac{\varphi_A - b\,u_A - c}{a}\right]},$$

$$= 1 \pm \sqrt{1^2 - \left[\frac{1 + 14.2655 \times 1 - 3.3996}{15.7009}\right]},$$

or

$$u^* = 1.4942, \text{ or } 0.5058.$$

Since u^* needs to be in the interval of $[u_6, u_8] = [0.4737, 0.6316]$, then only

$$u^* = 0.5058$$

is the real optimum value.

To determine the optimum burn-in time corresponding to $u^* = 0.5058$, fit the quadratic function given by Eq. (11.112) to $u = F(T)$ in the neighborhood of $T = W_7 = 1,044$ hr using the following three points: $(W_6, u_6) = (708$ hr, $0.4737)$, $(W_7, u_7) = (1,044$ hr, $0.5263)$ and $(W_8, u_8) = (2,892$ hr, $0.6316)$. Substituting these three points into Eq. (11.112) yields

$$\begin{cases} 708^2\, A + 708\, B + C = 0.4737, \\ 1,044^2\, A + 1,044\, B + C = 0.5263, \\ 2,892^2\, A + 2,892\, B + C = 0.6316. \end{cases}$$

Solving these three equations simultaneously, or substituting $(W_6, u_6) = (708$ hr, $0.4737)$, $(W_7, u_7) = (1,044$ hr, $0.5263)$ and $(W_8, u_8) = (2,892$ hr, $0.6316)$ into Eq. (11.113) directly, yields

$$\begin{cases} A = -0.5 \times 10^{-7}, \\ B = 2.36 \times 10^{-4}, \\ C = 0.3292. \end{cases}$$

Then, the fitted quadratic function for the observed *cdf* in the neighborhood of $W_7 = 1,044$ hr is

$$u(T) = -0.5 \times 10^{-7}\, T^2 + 2.36 \times 10^{-4}\, T + 0.3292.$$

Therefore, the optimum burn-in time, T_b^*, is given by Eq. (11.114); i.e.,

$$\begin{aligned} T_b^* &= \frac{-B \pm \sqrt{B^2 - 4\, A\, (C - u^*)}}{2\, A}, \\ &= \frac{-2.36 \times 10^{-4} \pm \sqrt{(2.36 \times 10^{-4})^2 - 4\, (-0.5 \times 10^{-7})\, (0.3292 - 0.5058)}}{2\, (-0.5 \times 10^{-7})}, \end{aligned}$$

or

$$T_b^* = 904.9476 \text{ hr, or } 4{,}280.9168 \text{ hr.}$$

Since T_b^* needs to be in the interval of $[W_6, W_8] = [708$ hr, $2,892$ hr$]$, then only

$$T_b^* = 904.9476 \text{ hr, or } 905 \text{ hr}$$

is the real optimum burn-in time.

Table 11.3– Raw failure data from a life test of $T_d = 6{,}600$ hr on $n = 90$ electronic components, for Example 11-5.

Time interval number, i	Time interval, (W_i, W_{i+1}), hr	Failure number, r_i
1	(0, 600)	23
2	(600, 1,200)	9
3	(1,200, 1,800)	2
4	(1,800, 2,400)	3
5	(2,400, 3,000)	11
6	(3,000, 3,600)	6
7	(3,600, 4,200)	3
8	(4,200, 4,800)	6
9	(4,800, 5,400)	4
10	(5,400, 6,000)	3
11	(6,000, 6,600)	4

EXAMPLE 11-5

A life test on $n = 90$ electronic components yields the data in Table 11.3, where a total of $r = 74$ failures are observed in $m = 11$ time intervals during $T_d = 6{,}600$ hr of operation. The associated costs are the same as in Example 11-1. Determine the optimum burn-in time using the scaled TTT plot approach

1. for the minimum cost per unit time of useful operation using the Bergman-Kelfsjo's cost model, and

2. for the maximum post-burn-in MRL.

SOLUTIONS TO EXAMPLE 11-5

1.1 The general procedure is used first to get a preliminary estimate of the optimum burn-in time which minimizes the total cost per unit time of operation. Note that given in Table 11.3 is a time-terminated and grouped data set. Then, Eqs. (11.60), (11.61) and (11.62) should be used to determine the scaled TTT plot points as follows:

The TTT statistic, at the end of each time interval, W_i, is given by Eq. (11.60), or

$$\begin{cases} T_i = \sum_{j=1}^{i} r_j \left[\frac{1}{2}(W_{j-1} + W_j)\right] + (n - \sum_{j=1}^{i} r_j) W_i, \\ \quad \text{for } i = 1, 2, ..., m, \\ T_0 = 0, \text{ for } i = 0, \end{cases}$$

or

$$\begin{cases} T_i = \sum_{j=1}^{i} r_j \left[\frac{1}{2}(W_{j-1} + W_j)\right] + (90 - \sum_{j=1}^{i} r_j) W_i, \\ \quad \text{for } i = 1, 2, ..., 11, \\ T_0 = 0, \text{ for } i = 0. \end{cases}$$

For example, when $i = 2$,

$$T_2 = \sum_{j=1}^{2} r_j \left[\frac{1}{2}(W_{j-1} + W_j)\right] + (90 - \sum_{j=1}^{2} r_j) W_2,$$
$$= r_1 \left[\frac{1}{2}(W_0 + W_1)\right] + r_2 \left[\frac{1}{2}(W_1 + W_2)\right]$$
$$+ [90 - (r_1 + r_2)] W_2,$$
$$= 23 \left[\frac{1}{2}(0 + 600)\right] + 9 \left[\frac{1}{2}(600 + 1,200)\right]$$
$$+ [90 - (23 + 9)] 1,200,$$

or

$$T_2 = 84,600 \text{ hr}.$$

The scaled TTT statistic at the end of each time interval, W_i, is given by Eq. (11.61), or

$$\varphi_i = T_i/T_m, \text{ for } i = 0, 1, 2, ..., m,$$

or

$$\varphi_i = T_i/T_{11}, \text{ for } i = 0, 1, 2, ..., 11,$$

where

$$T_{11} = \sum_{j=1}^{11} r_j \left[\frac{1}{2}(W_{j-1} + W_j)\right] + (90 - 74)6{,}600 = 280{,}800 \text{ hr}.$$

For example, when $i = 2$,

$$\varphi_2 = T_2/T_{11} = 84{,}600/280{,}800 = 0.3013.$$

The observed *cdf* at the end of each time interval is given by Eq. (11.62), or

$$u_i = \hat{F}(W_i) = \begin{cases} \dfrac{\sum_{j=1}^{i} r_j}{r}, & \text{for } i = 1, 2, \cdots, m, \\ 0 & \text{for } i = 0, \end{cases}$$

or

$$u_i = \begin{cases} \dfrac{\sum_{j=1}^{i} r_j}{74}, & \text{for } i = 1, 2, \cdots, 11, \\ 0 & \text{for } i = 0. \end{cases}$$

For example, when $i = 2$,

$$u_2 = \frac{\sum_{j=1}^{2} r_j}{74} = \frac{r_1 + r_2}{74} = \frac{23 + 9}{74} = 0.4324.$$

The values for T_i, φ_i and u_i are calculated for $i = 0, 1, 2, \cdots, 11$ and listed in Table 11.4. Plotting points (u_i, φ_i) and connecting them with line segments yields the scaled TTT plot for this time-terminated and grouped data set given in Table 11.3 which is shown in Fig. 11.22.

Now similar to Example 11-1 where the Bergman-Klefsjo's model is applied, draw the optimal line with the highest slope through Point A, $(u_A, \varphi_A) = (1.4, 1)$, which touches the plot at

$$[u^*, \varphi(u^*)] = (u_1, \varphi_1) = (0.3108, 0.1677).$$

Therefore,

$$i^* = 1,$$

Table 11.4– Values of TTT statistic, T_i, scaled TTT statistic, φ_i, and observed cdf, u_i, evaluated for the data set given in Table 11.1 for Example 11–5.

Time order, i	Failure time, W_i, hr	T_i	φ_i	u_i
0	0	0	0	0
1	600	47,100	0.1677	0.3108
2	1,200	84,600	0.3013	0.4324
3	1,800	118,800	0.4231	0.4595
4	2,400	151,500	0.5395	0.5000
5	3,000	180,000	0.6410	0.6486
6	3,600	203,400	0.7244	0.7297
7	4,200	224,100	0.7981	0.7703
8	4,800	242,100	0.8622	0.8514
9	5,400	257,100	0.9156	0.9054
10	6,000	270,000	0.9615	0.9459
11	6,600	280,800	1.0000	1.0000

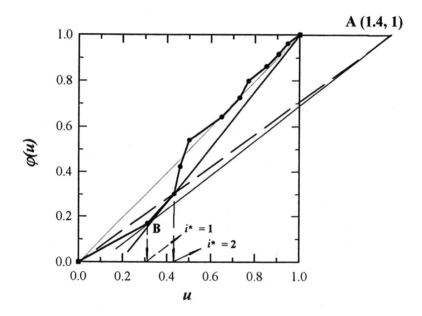

Fig. 11.22– The scaled TTT plot for the data set given in Table 11.3 and the graphical determination of the optimum burn-in time using the general procedure, for (1) the minimum cost per unit time of useful operation using the Bergman-Klefsjo's cost model, and (2) the maximum post-burn-in MRL, for Example 11–5.

and the estimated optimum burn-in time is

$$T_b^* = W_1 = 600 \text{ hr.}$$

Note that at $u^* = u_1 = 0.3108$, $\varphi(u)$ is convex, or $\varphi''(u^*) > 0$, and

$$u^* - \xi = 0.3108 - 1.4 = -1.0892 < 0.$$

Then,

$$\left[\frac{1}{K(u)}\right]''\bigg|_{u=u^*} = \frac{\varphi''(u^*)}{u^* - \xi} < 0,$$

which implies that $T_b^* = W_1 = 600$ hr is the optimum burn-in time which minimizes the total cost per unit time of operation.

1.2 The refined procedure is used second to get a more precise u^* value by fitting the quadratic function given in Eq. (11.108) to the scaled TTT plot in the neighborhood of $u_1 = 0.3108$ using the following three points: $[u_0, \varphi(u_0)] = (0, 0)$, $[u_1, \varphi(u_1)] = (0.3108, 0.1677)$ and $[u_2, \varphi(u_2)] = (0.4324, 0.3013)$. Substituting these three points into Eq. (11.108) yields

$$\begin{cases} 0^2 a + 0\, b + c & = 0, \\ 0.3108^2\, a + 0.3108\, b + c = 0.1677, \\ 0.4324^2\, a + 0.4324\, b + c = 0.3013. \end{cases}$$

Solving these three equations simultaneously, or substituting $[u_0, \varphi(u_0)] = (0, 0)$, $[u_1, \varphi(u_1)] = (0.3108, 0.1677)$ and $[u_2, \varphi(u_2)] = (0.4324, 0.3013)$ into Eq. (11.109) directly, yields

$$\begin{cases} a = 1.2930, \\ b = 0.1377, \\ c = 0. \end{cases}$$

Then, the fitted quadratic function for the scaled TTT plot in the neighborhood of $u_1 = 0.3108$ is

$$\varphi(u) = 1.2930\, u^2 + 0.1377\, u.$$

The optimum u value, u^*, is given by Eq. (11.110); i.e.,

$$u^* = u_A \pm \sqrt{u_A^2 - \left[\frac{\varphi_A - b\,u_A - c}{a}\right]},$$

$$= 1.4 \pm \sqrt{1.4^2 - \left[\frac{1 - 0.1377 \times 1.4 - 0}{1.2930}\right]},$$

or

$$u^* = 2.4131, \text{ or } 0.3869.$$

Since u^* needs to be in the interval of $[u_0, u_2] = [0, 0.4324]$, then only

$$u^* = 0.3869$$

is the real optimum value.

To determine the optimum burn-in time corresponding to $u^* = 0.3869$, fit the quadratic function given by Eq. (11.112) to $u = F(T)$ in the neighborhood of $T = W_1 = 600$ hr using the following three points: $(W_0, u_0) = (0\text{ hr}, 0)$, $(W_1, u_1) = (600\text{ hr}, 0.3108)$ and $(W_2, u_2) = (1{,}200\text{ hr}, 0.4324)$. Substituting these three points into Eq. (11.112) yields

$$\begin{cases} 0^2 A + 0\,B + C &= 0, \\ 600^2 A + 600\,B + C &= 0.3108, \\ 1{,}200^2 A + 1{,}200\,B + C &= 0.4324. \end{cases}$$

Solving these three equations simultaneously, or substituting $(W_0, u_0) = (0\text{ hr}, 0)$, $(W_1, u_1) = (600\text{ hr}, 0.3108)$ and $(W_2, u_2) = (1{,}200\text{ hr}, 0.4324)$ into Eq. (11.113) directly, yields

$$\begin{cases} A = -2.6 \times 10^{-7}, \\ B = 6.7567 \times 10^{-4}, \\ C = 0. \end{cases}$$

Then, the fitted quadratic function for the observed *cdf* in the neighborhood of $W_1 = 600$ hr is

$$u(T) = -2.6 \times 10^{-7}\,T^2 + 6.7567 \times 10^{-4}\,T.$$

Therefore, the optimum burn-in time, T_b^*, is given by Eq. (11.114); i.e.,

$$T_b^* = \frac{-B \pm \sqrt{B^2 - 4A(C - u^*)}}{2A},$$

$$= \frac{-6.7567 \times 10^{-4} \pm \sqrt{(6.7567 \times 10^{-4})^2 - 4(-2.6 \times 10^{-7})(0 - 0.3869)}}{2(-2.6 \times 10^{-7})},$$

or

$$T_b^* = 860.8014 \text{ hr, or } 1{,}710.4460 \text{ hr.}$$

Since T_b^* needs to be in the interval of $[W_0, W_2] = [0 \text{ hr}, 1{,}200 \text{ hr}]$, then only

$$T_b^* = 860.8014 \text{ hr, or } 861 \text{ hr}$$

is the real optimum burn-in time.

2.1 The general procedure is used first to get a preliminary estimate of the optimum burn-in time which maximizes the post-burn-in MRL. Draw the optimal line with the highest slope through Point $(u_A, \varphi_A) = (1, 1)$, which touches the scaled TTT plot at

$$[u^*, \varphi(u^*)] = (u_2, \varphi_2) = (0.4324, 0.3013).$$

Therefore,

$$i^* = 2,$$

and the estimated optimum burn-in time is

$$T_b^* = W_2 = 1{,}200 \text{ hr.}$$

Note that at $u^* = u_2 = 0.4324$, $\varphi(u)$ is convex, or $\varphi''(u^*) > 0$, and

$$u^* - 1 = 0.4324 - 1 = -0.5676 < 0.$$

Then, $u^* = u_2 = 0.4324$ is the maximum point for $M(u)$. In turn, $T_b^* = W_2 = 1{,}200$ hr is the optimum burn-in time which maximizes the post-burn-in MRL. Note that this result is consistent with that in Example 8-9, where T_b^* was determined directly from the empirical MRL plot.

2.2 The refined procedure is used second to get a more precise u^* value by fitting the quadratic function given in Eq. (11.108) to the scaled TTT plot in the neighborhood of $u_2 = 0.4324$ using the following three points: $[u_1, \varphi(u_1)] = (0.3108, 0.1677)$, $[u_2, \varphi(u_2)] = (0.4324, 0.3013)$ and $[u_3, \varphi(u_3)] = (0.4595, 0.4231)$. Substituting these three points into Eq. (11.108) yields

$$\begin{cases} 0.3108^2\, a + 0.3108\, b + c = 0.1677, \\ 0.4324^2\, a + 0.4324\, b + c = 0.3013, \\ 0.4595^2\, a + 0.4595\, b + c = 0.4231, \end{cases}$$

Solving these three equations simultaneously, or substituting $[u_1, \varphi(u_1)] = (0.3108, 0.1677)$, $[u_2, \varphi(u_2)] = (0.4324, 0.3013)$ and $[u_3, \varphi(u_3)] = (0.4595, 0.4231)$ into Eq. (11.109) directly, yields

$$\begin{cases} a = 22.8365, \\ b = -15.8734, \\ c = 2.8952. \end{cases}$$

Then, the fitted quadratic function for the scaled TTT plot in the neighborhood of $u_2 = 0.4324$ is

$$\varphi(u) = 22.8365\, u^2 - 15.8734\, u + 2.8952.$$

The optimum u value, u^*, is given by Eq. (11.110); i.e.,

$$u^* = u_A \pm \sqrt{u_A^2 - \left[\frac{\varphi_A - b\, u_A - c}{a}\right]},$$

$$= 1 \pm \sqrt{1^2 - \left[\frac{1 + 15.8734 \times 1 - 2.8952}{22.8365}\right]},$$

or

$$u^* = 1.6228, \text{ or } 0.3772.$$

Since u^* needs to be in the interval of $[u_1, u_3] = [0.3108, 0.4595]$, then only

$$u^* = 0.3772$$

is the real optimum value.

To determine the optimum burn-in time corresponding to $u^* = 0.3772$, fit the quadratic function given by Eq. (11.112) to $u = F(T)$ in the neighborhood of $T = W_2 = 1,200$ hr using the following three points: $(W_1, u_1) = (600 \text{ hr}, 0.3108)$, $(W_2, u_2) = (1,200 \text{ hr}, 0.4324)$ and $(W_3, u_3) = (1,800 \text{ hr}, 0.4595)$. Substituting these three points into Eq. (11.112) yields

$$\begin{cases} 600^2 A + 600 B + C = 0.3108, \\ 1,200^2 A + 1,200 B + C = 0.4324, \\ 1,800^2 A + 1,800 B + C = 0.4595. \end{cases}$$

Solving these three equations simultaneously, or substituting $(W_1, u_1) = (600 \text{ hr}, 0.3108)$, $(W_2, u_2) = (1,200 \text{ hr}, 0.4324)$ and $(W_3, u_3) = (1,800 \text{ hr}, 0.4595)$ into Eq. (11.113) directly, yields

$$\begin{cases} A = -1.3 \times 10^{-7}, \\ B = 4.3892 \times 10^{-4}, \\ C = 0.0947. \end{cases}$$

Then, the fitted quadratic function for the observed *cdf* in the neighborhood of $W_2 = 1,200$ is

$$u(T) = -1.3 \times 10^{-7} T^2 + 4.3892 \times 10^{-4} T + 0.0947.$$

Therefore, the optimum burn-in time, T_b^*, is given by Eq. (11.114); i.e.,

$$\begin{aligned} T_b^* &= \frac{-B \pm \sqrt{B^2 - 4A(C - u^*)}}{2A}, \\ &= \frac{-4.3892 \times 10^{-4} \pm \sqrt{(4.3892 \times 10^{-4})^2 - 4(-1.3 \times 10^{-7})(0.0947 - 0.3772)}}{2(-1.3 \times 10^{-7})}, \end{aligned}$$

or

$$T_b^* = 869.8477 \text{ hr, or } 2{,}474.2797 \text{ hr}.$$

Since T_b^* needs to be in the interval of $[W_1, W_3] = [600 \text{ hr}, 1{,}800 \text{ hr}]$, then only

$$T_b^* = 869.8477 \text{ hr, or } 870 \text{ hr}$$

is the real optimum burn-in time.

REFERENCES

1. Barlow, R. E., Bartholomew, D. J., Bremner, J. M. and Brunk, H. D., *Statistical Inference under Order Restrictions*, John Wiley & Sons, New York, 388 pp., 1972.

2. Barlow, R. E. and Campo, R., "Total Time on Test Processes and Applications to Failure Data Analysis," in Barlow, R. E., Fussell, J. and Singpurwalla, N. D., (Eds), *Reliability and Fault Tree Analysis*, SIAM, Philadelphia, pp. 451-481, 1975.

3. Bergman, B. and Klefsjo, B., "The Total Time on Test Concept and Its Use in Reliability Theory," *Operations Research*, Vol. 32, pp. 596-606, 1984.

4. Bergman, B. and Klefsjo, B., "Burn-in Models and TTT-transforms," *Quality and Reliability Engineering International*, Vol. 1, pp. 125-130, 1985.

5. Barlow, R. E., "Geometry of the Total Time on Test Transform," *Naval Research Logistics Quarterly*, Vol. 26, pp. 393-402, 1979.

6. Weiss, G. H. and Dishon, M., "Some Economic Problems Related to Burn-in Programs," *IEEE Transactions on Reliability*, Vol. R-20, No. 3, pp. 190-195, August 1971.

ADDITIONAL REFERENCES

1. Barlow, R. E. and Proschan, F., "Asymptotic Theory of Total Time on Test Processes with Applications to Life Testing," *Multivariate Analysis*, Vol. IV, pp. 227-237, 1977.

2. Barlow, R. E. and Proschan, F., "Analysis of Retrospective Failure Data Using Computer Graphics," *Proceedings of the Annual Reliability and Maintainability Symposium*, pp. 113-116, 1978.

3. Bergman, B., "Some Graphical Methods for Maintenance Planning," *Proceedings of the Annual Reliability and Maintainability Symposium*, pp. 467-471, 1977.

4. Bergman, B., "Crossings in the Total Time on Test Plot," *Scandinavian Journal of Statisticians*, Vol. 4, pp. 171-177, 1977.

5. Bergman, B., "On Age Replacement and the Total Time on Test Concept," *Scandinavian Journal of Statisticians*, Vol. 6, pp. 161-168, 1979.

6. Bergman, B., "On the Decision to Replace A Unit Early or Late – A Graphical Solution," *Microelectronics and Reliability*, Vol. 20, pp. 895-896, 1980.

7. Bergman, B. and Klefsjo, B., "A Graphical Method Applicable to Age-Replacement Problems," *IEEE Transactions on Reliability*, Vol. R-31, No. 3, pp. 478-481, 1982.

8. Bergman, B. and Klefsjo, B., "TTT-Transform and Age Replacements with Discounted Costs," *Naval Research Logistics Quarterly*, Vol. 30, pp. 631-639, 1983.

9. Klefsjo, B., "On Aging Properties and the Total Time on Test Transform," *Scandinavian Journal of Statisticians*, Vol. 9, pp. 37-41, 1982.

10. Klefsjo, B., "Some Tests Against Aging Based on the Total Time on Test Transform," *Communications of Statisticians*, Vol. 12, No. 8, pp. 907-927, 1983.

APPENDIX 11A
COMPUTER PROGRAM FOR THE REFINED PROCEDURE OF OPTIMUM BURN-IN TIME DETERMINATION USING THE SCALED TTT PLOT

```
C $$$$$$$$$$$$$$$$$$$$$$$$$$$$$$$$$$$$$$$$$$$$$$$$$$$$$$$
C $ OPTIMUM BURN-IN TIME DETERMINATION USING THE       $
C $ REFINED PROCEDURE OF THE SCALED TTT PLOT           $
C $ APPROACH FOR MINIMUM COST PER UNIT TIME            $
C $ OF USEFUL OPERATION OR FOR MAXIMUM MEAN            $
C $ RESIDUAL LIFE (MRL)                                $
C $      U0:=THETA/DELTA                               $
C $      PHI0:=ALPHA/BETA                              $
C $$$$$$$$$$$$$$$$$$$$$$$$$$$$$$$$$$$$$$$$$$$$$$$$$$$$$$$
C
      IMPLICIT DOUBLE PRECISION (A-H, O-Z)
      REAL  M1, M2, M3
      OPEN(5,FILE='ITTT.DAT',STATUS='UNKNOWN')
      OPEN(6,FILE='OTTT.DAT',STATUS='UNKNOWN')
C
C  (1) FIT A QUADRATIC FUNCTION TO THE SCALED TTT PLOT
C      IN THE NEIGHBORHOOD OF U2 USING THREE POINTS:
C      (U1, PHI1), (U2, PHI2), (U3, PHI3).
C  (2) DETERMINE THE OPTIMUM VALUE FOR U, U^*.
C
      WRITE(*,*)'ENTER U1, U2, U3 !'
      READ(5,*)T1,T2,T3
      WRITE(*,*)'U1=', T1, ' U2=', T2, ' U3=', T3
      WRITE(6,*)'U1=', T1, ' U2=', T2, ' U3=', T3
      WRITE(*,*)'ENTER PHI1, PHI2, PHI3 !'
      READ(5,*)M1,M2,M3
      WRITE(*,*)'PHI1=', M1, ' PHI2=', M2, ' PHI3=', M3
      WRITE(6,*)'PHI1=', M1, ' PHI2=', M2, ' PHI3=', M3
      WRITE(*,*)'ENTER U0, PHI0, PLEASE!'
      READ(5,*)U0, PHI0
      WRITE(*,*)'U0=', U0, ' PHI0=', PHI0
      WRITE(6,*)'U0=', U0, ' PHI0=', PHI0
      D=T1*T2*(T1-T2)+T2*T3*(T2-T3)+T3*T1*(T3-T1)
      DA=M1*(T2-T3)+M2*(T3-T1)+M3*(T1-T2)
      DB=T1*T1*(M2-M3)+T2*T2*(M3-M1)+T3*T3*(M1-M2)
      DC=T1*T2*M3*(T1-T2)+T2*T3*M1*(T2-T3)+T3*T1*M2*(T3-T1)
      A=DA/D
```

```
              B=DB/D
              C=DC/D
              WRITE(*,5) A, B, C
              WRITE(6,5) A, B, C
 5            FORMAT(2X, 'a=', F18.8, ' b=', F18.8, ' c=', F18.8)
              USTAR1=U0+SQRT(U0*U0-(U0-B*U0-C)/A)
              USTAR2=U0-SQRT(U0*U0-(U0-B*U0-C)/A)
              IF((USTAR1.LT.T1).OR.(USTAR1.GT.T3))GOTO 11
              USTAR=USTAR1
              GOTO 15
 11           USTAR=USTAR2
 15           WRITE(*,*)'U1^*=',USTAR1,'  U2^*=',USTAR2
              WRITE(6,*)'U1^*=',USTAR1,'  U2^*=',USTAR2
              WRITE(*,*) 'U^*=',USTAR
              WRITE(6,*) 'U^*=',USTAR
C
C     (1) FIT A QUADRATIC FUNCTION TO THE EMPIRICAL
C         CDF IN THE NEIGHBORHOOD OF W2 USING THREE
C         POINTS: (W1, U1), (W2, U2), (W3, U3).
C     (2) DETERMINE THE OPTIMUM BURN-IN TIME, TB^*.
C
              WRITE(*,*)'ENTER W1, W2, W3 !'
              READ(5,*) T1,T2,T3
              WRITE(*,*)'W1=', T1, ' W2=', T2, ' W3=', T3
              WRITE(6,*)'W1=', T1, ' W2=', T2, ' W3=', T3
              WRITE(*,*)'ENTER U1, U2, U3 !'
              READ(5,*) M1,M2,M3
              WRITE(*,*)'U1=', M1, ' U2=', M2, ' U3=', M3
              WRITE(6,*)'U1=', M1, ' U2=', M2, ' U3=', M3
              D=T1*T2*(T1-T2)+T2*T3*(T2-T3)+T3*T1*(T3-T1)
              DA=M1*(T2-T3)+M2*(T3-T1)+M3*(T1-T2)
              DB=T1*T1*(M2-M3)+T2*T2*(M3-M1)+T3*T3*(M1-M2)
              DC=T1*T2*M3*(T1-T2)+T2*T3*M1*(T2-T3)+T3*T1*M2*(T3-T1)
              A=DA/D
              B=DB/D
              C=DC/D
              TB1=(-B+SQRT(B*B-4.0*A*(C-USTAR)))/(2.0*A)
              TB2=(-B-SQRT(B*B-4.0*A*(C-USTAR)))/(2.0*A)
              IF((TB1.LT.T1).OR.(TB1.GT.T3))GOTO 21
              TB=TB1
              GOTO 25
 21           TB=TB2
 25           WRITE(*,85) A, B, C
              WRITE(6,85) A, B, C
```

```
         WRITE(*,87) TB1,TB2
         WRITE(6,87) TB1,TB2
         WRITE(*,90) TB
         WRITE(6,90) TB
85       FORMAT(2X, 'A=', F18.8, ' B=', F18.8, ' C=', F18.8)
87       FORMAT(2X, 'TB1=', F18.8, '   TB2=', F18.8)
90       FORMAT(2X, 'TB^*=', F18.8)
         STOP
         END
```

Chapter 12

ACCELERATED BURN-IN TESTING AND BURN-IN TIME REDUCTION ALGORITHMS

12.1 INTRODUCTION

Accelerating the test environment, or increasing the stresses, can yield a shorter burn-in test duration. If the test results can be extrapolated to the use stress levels, they yield estimates of the lives and reliabilities under use stresses. Such testing provides a saving in time and expense, since for many components and products, life under use conditions is so long that burning-in under those conditions is not time-wise and economically feasible.

If temperature is the only accelerating variable in burn-in, the Arrhenius and Eyring Accelerated Test Models may be used. If voltage is the only accelerating variable in burn-in, the Inverse Power Law Accelerated Test Model may be used. If both temperature and voltage are combined and accelerated in burn-in the Combination, Generalized Eyring and Bazovsky models may be used. In this Chapter only the Arrhenius, the Inverse Power Law and the Combination models are discussed. For more acceleration models, the readers are referred to [1] where various models for accelerated life tests and reliability determinations are presented, including the Arrhenius, Eyring, Inverse–Power law, Combination, Generalized Eyring, Bazovsky, Temperature–Humidity, Weibull Stress–Life, Log–log Stress-Life, Overload–Stress Reliability, Percent–Life, Deterioration Monitoring, Step–Stress, Distribution–Free Tolerance Limits, Non-parametric and Optimum Acceleration Models.

12.2 BURN-IN TIME REDUCTION USING AN ELEVATED TEMPERATURE – THE ARRHENIUS MODEL

12.2.1 THE ACCELERATION FACTOR WHEN THE LIFE DISTRIBUTION IS EXPONENTIAL

A widely used formula for modeling an electronic component's strength degradation with respect to increasing temperature, has for many years been the Arrhenius model. It estimates the characteristic life at use temperature from failure data obtained at high temperature. The Arrhenius model is stated as

$$V = Ae^{-\frac{E_A}{KT^*}}, \qquad (12.1)$$

where

V = reaction rate measured in moles/(volume×time),
A = proportionality constant, or the so called 'frequency factor',
E_A = activation energy measured in electron-volts, eV,
T^* = absolute temperature in °K,

and

K = Boltzmann's constant = 8.623×10^{-5} $eV/°K$.

Equation (12.1) expresses the time rate of device degradation as a function of operating temperature. The degradation is assumed to correspond to chemical or physical reactions among device constituents and contaminants. In those cases where the failure mechanism follows the Arrhenius relationship and the life distribution is exponential, the constant failure rate, $\lambda(T^*)$, is directly proportional to the reaction rate; thus, from Eq. (12.1)

$$\lambda(T^*) = D\, e^{-\frac{E_A}{KT^*}}, \qquad (12.2)$$

where D is a constant. Under the exponential life distribution assumption, the mean time to failure or mean life, $L(T^*)$, is given by

$$L(T^*) = \frac{1}{\lambda(T^*)} = C\, e^{\frac{E_A}{KT^*}}, \qquad (12.3)$$

where

$$C = \frac{1}{D} = \text{a constant}.$$

THE ARRHENIUS MODEL

Correspondingly, the failure rates and mean lives at operational temperature T_o^* and at accelerated temperature T_a^* are given by

$$\begin{cases} \lambda(T_o^*) = D\, e^{-\frac{E_A}{KT_o^*}}, \\ \lambda(T_a^*) = D\, e^{-\frac{E_A}{KT_a^*}}, \end{cases}$$

and

$$\begin{cases} L(T_o^*) = C\, e^{\frac{E_A}{KT_o^*}}, \\ L(T_a^*) = C\, e^{\frac{E_A}{KT_a^*}}, \end{cases}$$

respectively. The acceleration factor, $A_F(T_o^*, T_a^*)$, is defined as the ratio of the mean life at the operating temperature to that at the accelerated temperature, or

$$A_F(T_o^*, T_a^*) = \frac{L(T_o^*)}{L(T_a^*)} = e^{\frac{E_A}{K}\left(\frac{1}{T_o^*} - \frac{1}{T_a^*}\right)},$$

which is identical to the ratio of the failure rate at the accelerated temperature to that at the operating temperature; i.e.,

$$A_F(T_o^*, T_a^*) = \frac{L(T_o^*)}{L(T_a^*)} = \frac{\lambda(T_a^*)}{\lambda(T_o^*)} = e^{\frac{E_A}{K}\left(\frac{1}{T_o^*} - \frac{1}{T_a^*}\right)}. \tag{12.4}$$

Note that Eq. (12.4) is valid only when the activation energy, E_A, is independent of the applied temperature. This is not the case in reality as may be seen later on.

Given the failure rate or mean life at the accelerated temperature, $\lambda(T_a^*)$ or $L(T_a^*)$, the failure rate or mean life at the operating temperature is given by

$$\lambda(T_o^*) = \frac{\lambda(T_a^*)}{A_F(T_o^*, T_a^*)} = \lambda(T_a^*)\, e^{-\frac{E_A}{K}\left(\frac{1}{T_o^*} - \frac{1}{T_a^*}\right)},$$

or

$$L(T_o^*) = L(T_a^*)\, A_F(T_o^*, T_a^*) = L(T_a^*)\, e^{\frac{E_A}{K}\left(\frac{1}{T_o^*} - \frac{1}{T_a^*}\right)},$$

respectively.

In general, the acceleration factor between two different temperature levels, $T_1^* < T_2^*$, is given by

$$A_F(T_1^*, T_2^*) = \frac{L(T_1^*)}{L(T_2^*)} = \frac{\lambda(T_2^*)}{\lambda(T_1^*)} = e^{\frac{E_A}{K}\left(\frac{1}{T_1^*} - \frac{1}{T_2^*}\right)}, \tag{12.5}$$

from which the mean life or failure rate at one temperature level can be converted to that at another temperature level.

12.2.2 THE ACCELERATION FACTOR WHEN THE LIFE DISTRIBUTION IS NOT EXPONENTIAL

When the life distribution is not exponential; i.e., the failure rate is not a constant, say lognormal, Weibull, mixed-lognormal, mixed-Weibull, etc., then the Arrhenius model is often expressed in terms of the median or the mean lives, such as

$$L_{0.5}(T^*) = C\, e^{\frac{E_A}{K\,T^*}}, \tag{12.6}$$

or

$$\overline{L}(T^*) = C\, e^{\frac{E_A}{K\,T^*}}, \tag{12.7}$$

where

$$L_{0.5}(T^*) = \text{median life at temperature } T^*,$$

and

$$\overline{L}(T^*) = \text{mean life at temperature } T^*.$$

Correspondingly, the acceleration factor is defined as

$$A_F(T_o^*, T_a^*) = \frac{L_{0.5}(T_o^*)}{L_{0.5}(T_a^*)} = e^{\frac{E_A}{K}\left(\frac{1}{T_o^*} - \frac{1}{T_a^*}\right)},$$

or

$$A_F(T_o^*, T_a^*) = \frac{\overline{L}(T_o^*)}{\overline{L}(T_a^*)} = e^{\frac{E_A}{K}\left(\frac{1}{T_o^*} - \frac{1}{T_a^*}\right)}.$$

Similar to Eq. (12.5), the preceding two equations can be generalized to evaluate the acceleration factor between any two different temperature levels, $T_1^* < T_2^*$; i.e.,

$$A_F(T_1^*, T_2^*) = \frac{L_{0.5}(T_1^*)}{L_{0.5}(T_2^*)} = e^{\frac{E_A}{K}\left(\frac{1}{T_1^*} - \frac{1}{T_2^*}\right)},$$

or

$$A_F(T_1^*, T_2^*) = \frac{\overline{L}(T_1^*)}{\overline{L}(T_2^*)} = e^{\frac{E_A}{K}\left(\frac{1}{T_1^*} - \frac{1}{T_2^*}\right)},$$

from which the median life or the mean life at one temperature level can be converted to that at another temperature level.

THE ARRHENIUS MODEL

12.2.3 BURN-IN TIME REDUCTION USING AN ELEVATED TEMPERATURE

Given the required burn-in time at use temperature T_o^*, T_{bo}, the burn-in time at an elevated temperature T_a^*, T_{ba}, is given by

$$T_{ba} = T_{bo}/A_F(T_o^*, T_a^*) = T_{bo}\, e^{-\frac{E_A}{K}\left(\frac{1}{T_o^*} - \frac{1}{T_a^*}\right)}. \tag{12.8}$$

On the other hand, given the required burn-in time at an elevated temperature T_a^*, T_{ba}, the corresponding burn-in time at use temperature T_o^*, T_{bo}, is given by

$$T_{bo} = T_{ba}\, A_F(T_o^*, T_a^*) = T_{ba}\, e^{\frac{E_A}{K}\left(\frac{1}{T_o^*} - \frac{1}{T_a^*}\right)}. \tag{12.9}$$

In general, the relationship between the burn-in time at a lower temperature T_1^*, T_{b1}, and that at a higher temperature T_2^*, T_{b2}, is given by

$$T_{b2} = T_{b1}/A_F(T_1^*, T_2^*) = T_{b1}\, e^{\frac{E_A}{K}\left(\frac{1}{T_2^*} - \frac{1}{T_1^*}\right)},$$

or equivalently,

$$T_{b1} = T_{b2}\, A_F(T_1^*, T_2^*) = T_{b2}\, e^{\frac{E_A}{K}\left(\frac{1}{T_1^*} - \frac{1}{T_2^*}\right)}.$$

EXAMPLE 12-1

The required burn-in time of silicon photodiodes with a 32-elements configuration at the use temperature of 298°K (25°C) is determined to be 1,000 hr. Determine the corresponding burn-in times, at the elevated temperatures of 308°K (35°C), 318°K (45°C), $\cdots$, 398°K (125°C), using the Arrhenius model. Assume that the activation energy of this type of photodiodes remains constant at the value of 1.0 in the temperature range of 25°C – 125°C.

SOLUTION TO EXAMPLE 12-1

Substituting the given information into Eq. (12.8) yields Table 12.1. A sample calculation for $T_a^* = 328°$K (55°C) is given as follows:

$$\begin{aligned}
T_{ba} &= T_{bo}\, e^{-\frac{E_A}{K}\left(\frac{1}{T_o^*} - \frac{1}{T_a^*}\right)}, \\
&= (1,000)\, e^{-\frac{1.0}{8.623 \times 10^{-5}}\left(\frac{1}{298} - \frac{1}{328}\right)}, \\
&= (1,000)(2.8457 \times 10^{-2}),
\end{aligned}$$

TABLE 12.1– Equivalent burn-in times at various burn-in temperatures for Example 12–1.

T_a^*		T_{ba}, hr
°C	°K	
25	298	1,000.00
35	308	282.66
45	318	86.51
55	328	28.46
65	338	10.00
75	348	3.73
85	358	1.47
95	368	0.61
105	378	0.26
115	388	0.12
125	398	0.06

or

$T_{ba} = 28.46$ hr.

It may be seen from Table 12.1 that increasing the burn-in temperature will decrease the required burn-in time substantially.

12.2.4 A WORD OF CAUTION ON THE USE OF THE ARRHENIUS ACCELERATION MODEL

It should be noted that the values of E_A, D and C in the Arrhenius Model and in the derived acceleration factors can be assumed to be constant *provided that the temperature variation during the test is not significant* [2]. However, this last innocent sounding statement has unfortunately been overlooked by many reliability practitioners. The Arrhenius equation has been used to convert failure rates, or lifetimes, from one temperature to another using the same value for the activation energy, E_A. Admittedly, it is very difficult to state over *how large* a temperature variation the activation energy may be assumed constant. Now let us look at the real meaning of the activation energy and show the contradiction between the Arrhenius model and the age acceleration equation derived from the Arrhenius model, which is caused by the assumption that E_A is constant over any temperature range.

THE ARRHENIUS MODEL

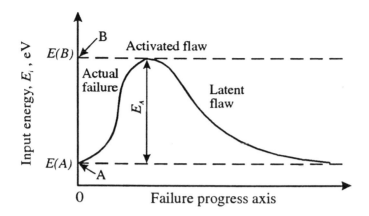

Fig. 12.1– Activation energy and the failure process [3].

12.2.5 THE PHYSICAL MEANING OF ACTIVATION ENERGY

The activation energy is what the term says it is; i.e., the energy input required to cause a molecule of a constituent to participate in the reaction. By the analogy with chemical reactions, Figure 12.1 illustrates its meaning [3].

In Fig. 12.1, the abscissa is the failure (reaction) progress axis and the ordinate is the energy axis. The values of energy constitute internal energy of defects at ambient temperature. These values may vary from defect to defect, depending on their nature, size, shape, location within the structure of the component, electric potentials across them, and the like, as well as ambient temperature. Values of the internal energy are always finite and greater than zero. In our case, the value of the activation energy, E_i, is

$$E(A) \leq E_i \leq E(B),$$

where

$E(B)$ = energy level at point B,

and

$E(A)$ = energy level at point A.

Let's take the case of $E_i = E(A)$, as an example. At this energy level, we received a product which was tested "good" but which we know has some flaws. We failed to detect them only because our means were limited. Since we cannot detect these flaws, we call them latent (inactive) flaws. To make a flaw detectable, we must raise its energy level to, at least, the level of $E_i = E(B)$, which means we would have to activate it by providing additional energy in some form, for example, heat. By maintaining the flaw at or above the level of activation, we intensify its effect to the point of failure. Thus, a flaw which was originally latent, according to our test, now becomes active and detectable.

The above considerations suggest that *activation energy, E_A, is an amount of energy necessary to surmount an energy barrier which separates latent flaws from actual failures.* This means that the condition $E_i \geq E(B)$ must be met to cause a failure. The values of the activation energy E_A can range from $E_A = 0$ to $E_A = \infty$.

12.2.6 EXPERIMENTAL DETERMINATION OF THE ACTIVATION ENERGY

In practice, E_A is determined experimentally by observing the times to failure of several batches of identical components at different temperature levels and evaluating the median (or mean) lives and the corresponding acceleration factors.

For example, let $L_{0.5}(T_1^*)$ and $L_{0.5}(T_2^*)$ represent the median lives at two different temperatures, $T_1^* < T_2^*$. According to Eq. (12.4),

$$\frac{L_{0.5}(T_1^*)}{L_{0.5}(T_2^*)} = e^{\frac{E_A}{K}\left(\frac{1}{T_1^*} - \frac{1}{T_2^*}\right)}.$$

Taking the logarithm on both sides of this equation and solving for E_A yields

$$\widehat{E}_A = K \left[\frac{\log_e L_{0.5}(T_1^*) - \log_e L_{0.5}(T_2^*)}{1/T_1^* - 1/T_2^*}\right], \quad (12.10)$$

which is a rough estimate of E_A based on the test results at only two temperature stress levels.

A more precise estimate of E_A may be obtained if test results at more temperature stress levels are available. For example, let $L_{0.5}(T_1^*)$, $L_{0.5}(T_2^*), \cdots, L_{0.5}(T_n^*)$ represent n median lives observed at n different temperature stress levels, $T_1^*, T_2^*, \cdots, T_n^*$, respectively. Then, the least-squares estimation method may be used to estimate E_A.

First, linearize Eq. (12.6) by taking the logarithm on both sides; i.e.,

$$\log_e L_{0.5}(T^*) = \log_e C + \frac{E_A}{K}\left(\frac{1}{T^*}\right). \quad (12.11)$$

Second, let
$$X = 1/T^*,$$
and
$$Y = \log_e L_{0.5}(T^*).$$
Then, Eq. (12.11) can be rewritten as
$$Y = \log_e C + \frac{E_A}{K} X. \tag{12.12}$$

Based on this linear equation, the least-squares estimate of E_A is given by
$$\widehat{E}_A = K \left(\frac{S_{XY}}{S_{XX}}\right) = 8.623 \times 10^{-5} \times \left(\frac{S_{XY}}{S_{XX}}\right), \tag{12.13}$$
where
$$S_{XY} = \sum_{i=1}^{n}(X_i - \overline{X})(Y_i - \overline{Y}),$$
$$S_{XX} = \sum_{i=1}^{n}(X_i - \overline{X})^2,$$
n = number of temperature stress levels applied,
$X_i = 1/T_i^*, \quad i = 1, 2, \cdots, n,$
$$\overline{X} = \frac{1}{n}\sum_{i=1}^{n} X_i,$$
$Y_i = \log_e L_{0.5}(T_i^*), \quad i = 1, 2, \cdots, n,$
$$\overline{Y} = \frac{1}{n}\sum_{i=1}^{n} Y_i,$$
T_i^* = ith applied temperature,

and
$$L_{0.5}(T_i^*) = \text{median life at } T_i^*.$$

Note that Eqs. (12.10) and (12.13) are valid only if E_A remains a constant in the test temperature range. This can be verified by plotting $\log_e L_{0.5}(T^*)$ versus $1/T^*$. If this plot yields a satisfactory straight line, or equivalently a constant slope, then E_A can be treated as a constant in the temperature range of test. Otherwise, E_A is not a constant in the applied temperature range.

In many cases, such as for semiconductor devices, $E_A=1$ eV has been popular as a general value to use in the useful life period when more precise data are not available. In the early life period, the component's freak failures tend to have lower values of activation energy, values of $E_A=0.2$—0.4 eV being quoted frequently.

For the non-screened components, a bimodal mixed life distribution is frequently observed during burn-in at each applied temperature stress level, as shown in Fig. 12.2, which is a plot of $\log_L(T^*)$ versus $1/T^*$, or from Eq. (12.3),

$$\log_e L(T^*) = \log_e C + \frac{E_A}{K}\left(\frac{1}{T^*}\right). \tag{12.14}$$

It may be seen that two subpopulations correspond to two different activation energies, the activation energy for the weak (bad) subpopulation, E_{Ab}, and the activation energy for the main (good) subpopulation, E_{Ag}, where $E_{Ag} > E_{Ab}$. These two activation energies can be evaluated by fitting a straight line through the median life points for each subpopulation at various temperature values, measure its slope, E_A/K, and multiply the measured slope by $K = 8.623 \times 10^{-5}$ eV/°K.

Table 12.2 [4, p. 80] gives values of E_A for some components, extracted from manufacturers' reliability reports. There are three things which should be noted here [4, p. 79]:

1. The E_A values reported in Table 12.2 are often obtained from laboratory tests at high temperatures, maybe 150° to 300°C, and while the value of E_A may be valid for the failure mechanism encountered in this temperature range, there is no guarantee that the same mechanism will be dominant under conditions of normal use.

2. Different manufacturers of the same component type may experience different failure mechanisms; thus different E_A values would apply.

3. The E_A values sometimes may be modified by stress other than temperature, such as electrical or mechanical stresses, which are encountered under use conditions. Consequently, it is advisable to determine E_A by test with the specific stresses involved.

EXAMPLE 12-2

A life test on CMOS 4007 microcircuits from a certain manufacturer was conducted at various temperature levels. The post-test failure analysis on the fallouts identified two subpopulations; i.e., weak

THE ARRHENIUS MODEL

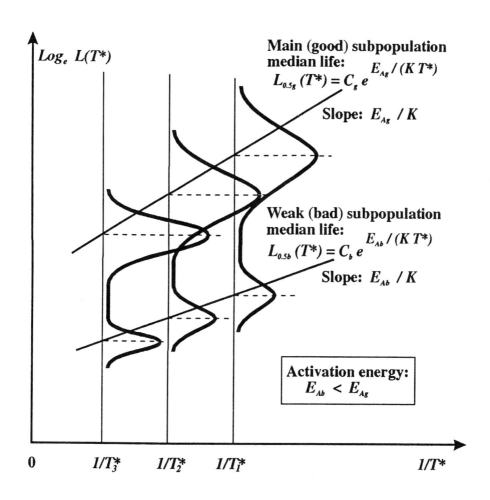

Fig. 12.2– Bimodal life distributions at various applied temperature stress levels and the activation energies corresponding to the weak and the main subpopulations. Note that in this figure, $T_1^* < T_2^* < T_3^*$ and therefore $1/T_3^* < 1/T_2^* < T_1^*$.

TABLE 12.2– Activation energies for various components and failure mechanisms [5]. Reproduced by permission of John Wiley & Sons.

Component and mechanism	Reported E_A, eV	
	Main subpopulation	Weak subpopulation
Silicon semiconductor devices		
Silicon oxide and Si/SiO$_2$ interface		
surface charge accumulation, bipolar	1.0–1.05	
surface charge accumulation, MOS	1.2–1.35	
Slow trapping charge injection	1.3–1.4	
Metalization		
Electromigration	0.5–1.2	
Corrosion (chemical, galvanic,		
electrolytic)	0.3–0.6	
Bonds		
Intermetallic growth Al/Au	1.0–1.05	
N-channel Si gate dynamic RAM		
Slow trapping	1.0	
Contamination	1.4	1.4
Surface charge	0.5–1.0	
Polarization	1.0	
Electro-migration	1.0	
Oxide defects	0.3	0.3
FAMOS transistors		
Charge loss	0.8	
Long-life bipolar transistors for		
submarine cable repeaters	1.4	
Integrated circuit MOSFETs		
Threshold voltage shift	1.2	
TTL ICs in general, beam lead		
transistors and diodes	0.8–2.0	0.3–1.0
Plastic encapsulated transistors	0.5	
MOS devices in general, including		
PMOS, CMOS	1.1–1.3	0.3–0.9
GaAs microwave devices		
Contact metal migration	2.3	
Opto-electronic devices		
Opto-coupler LED	0.4	
Opto-coupler photo-transistor	1.3	
Opto-coupler composite	0.6	
Discrete LEDs	0.8	
Carbon composition resistors	0.6	
Flexible printed circuits		
Surface insulation resistance loss		
below 75°C	0.4	
above 75°C	1.4	
Linear operation amplifier	1.6–1.8	0.7–1.1

THE ARRHENIUS MODEL

TABLE 12.3– Median lives at various life test temperatures for the CMOS 4007 microcircuits for Example 12–2.

Test No., i	1	2	3	4
Temperature, T_i^*, °C	125	200	225	250
Temperature, T_i^*, °K	398	473	498	523
Median life of the weak subpopulation, $L_{0.5b}(T_i^*)$, hr	110	30	0.8	0.5
Median life of the main subpopulation, $L_{0.5g}(T_i^*)$, hr	–	300	50	15

and main subpopulations. The statistical analysis on each one of these two subpopulations separately yields the median life for each subpopulation at the applied test temperatures which are listed in Table 12.3. Determine the activation energy for each subpopulation using the least-squares estimation method.

SOLUTION TO EXAMPLE 12–2

Another table, Table 12.4, is prepared to accommodate the least-squares estimation.

Applying Eq. (12.13) to the weak subpopulation yields

$$\widehat{E_{Ab}} = K \left(\frac{S_{XY_b}}{S_{XX}}\right) = 8.623 \times 10^{-5} \left(\frac{S_{XY_b}}{S_{XX}}\right),$$

where

$$S_{XX} = \sum_{i=1}^{4}(X_i - \overline{X})^2 = \sum_{i=1}^{4}(X_i - 2.1367 \times 10^{-3})^2,$$

or

$$S_{XX} = 2.0886 \times 10^{-7},$$

TABLE 12.4– Variable transformation for the least-squares estimation of the activation energies for Example 12–2.

i	1	2	3	4	Mean
$X_i = 1/T_i^*$, in 10^{-3}	2.5126	2.1142	2.0080	1.9120	$\overline{X}$=2.1367
$Y_{bi} = \log_e L_{0.5b}(T_i^*)$	4.7005	3.4012	-0.2231	-0.6931	$\overline{Y_b}$=1.7064
$X_i = 1/T_i^*$, in 10^{-3}	–	2.1142	2.0080	1.9120	$\overline{X}$=2.0460
$Y_{gi} = \log_e L_{0.5g}(T_i^*)$	–	5.7038	3.9120	2.7081	$\overline{Y_g}$=4.1080

and

$$S_{XY_b} = \sum_{i=1}^{4}(X_i - \overline{X})(Y_{bi} - \overline{Y_b}),$$

$$= \sum_{i=1}^{4}(X_i - 2.1367 \times 10^{-3})(Y_{bi} - 1.7064),$$

or

$$S_{XY_b} = 1.8748 \times 10^{-3}.$$

Then,

$$\widehat{E_{Ab}} = 8.623 \times 10^{-5} \left(\frac{1.8748 \times 10^{-3}}{2.0886 \times 10^{-7}} \right) = 0.7740 \text{ eV}.$$

Applying Eq. (12.13) to the main subpopulation yields

$$\widehat{E_{Ag}} = K \left(\frac{S_{XY_g}}{S_{XX}} \right) = 8.623 \times 10^{-5} \left(\frac{S_{XY_g}}{S_{XX}} \right),$$

where

$$\overline{X} = \frac{1}{3} \sum_{i=2}^{4} X_i = 2.0114 \times 10^{-3},$$

$$S_{XX} = \sum_{i=2}^{4}(X_i - \overline{X})^2 = \sum_{i=2}^{4}(X_i - 2.0114 \times 10^{-3})^2,$$

or

$$S_{XX} = 2.0460 \times 10^{-8},$$

and

$$S_{XY_g} = \sum_{i=2}^{4}(X_i - \overline{X})(Y_{gi} - \overline{Y_g}),$$

$$= \sum_{i=2}^{4}(X_i - 2.0114 \times 10^{-3})(Y_{bi} - 4.1080),$$

or

$$S_{XY_g} = 3.0386 \times 10^{-4}.$$

Then,

$$\widehat{E_{Ag}} = 8.623 \times 10^{-5} \left(\frac{3.0386 \times 10^{-4}}{2.0460 \times 10^{-8}}\right) = 1.2806 \text{ eV}.$$

It may be seen that the activation energy for the main subpopulation is larger than that of the weak subpopulation; i.e.,

$$E_{Ag} = 1.2806 \text{ eV} > E_{Ab} = 0.7740 \text{ eV}.$$

12.2.7 THE PITFALLS OF THE ARRHENIUS MODEL AND THE TEMPERATURE DEPENDENCE OF THE ACTIVATION ENERGIES

Now let us fix the temperature, T^*, at some arbitrary value and vary the value of E_A in Eq. (12.2) to compare the outcomes of both the Arrhenius model and the acceleration equation derived from it. It will be shown that the activation energy is not only failure mechanism dependent but also temperature dependent.

According to the Arrhenius model given by Eq. (12.1), if $E_A \longrightarrow 0$, or the energy to activate a particular flaw equals zero, then all the components with these flaws are certain to fail spontaneously without any action on our part. As the required activation energy increases, the rate of failure precipitation decreases. Particularly, if $E_A \longrightarrow \infty$, the failure rate drops to zero; thus, no component is going to fail no matter how much action (temperature) we apply. This brief analysis leads to the following conclusion:

Given the temperature level, the lower the value of the activation energy, the higher the probability of failure, and, consequently, the higher the failure rate. On the other hand, the higher the activation energy, the lower the probability of failure; therefore, the lower the failure rate.

Now, let us look at the aging acceleration obtained by the use of elevated temperature, as compared to that at the nominal operating temperature. If E_A is constant over any temperature range, then the aging acceleration factor is given by Eq. (12.4), or

$$A_F(T_o^*, T_a^*) = \frac{L(T_o^*)}{L(T_a^*)} = \frac{\lambda(T_a^*)}{\lambda(T_o^*)} = e^{\frac{E_A}{K}\left(\frac{1}{T_o^*} - \frac{1}{T_a^*}\right)}. \qquad (12.15)$$

It then follows from Eq. (12.15) that

$$L(T_a^*) = \frac{L(T_o^*)}{A_F(T_o^*, T_a^*)}. \qquad (12.16)$$

Given T_o^* and T_a^*, if $E_A \longrightarrow 0$, then from Eq. (12.15) $A_F(T_o^*, T_a^*) \longrightarrow 1$, which means no acceleration. From Eq. (12.16) the time of burn-in testing approaches the actual life time of the product in the field, regardless of the applied temperature, provided it is less than infinity. As the activation energy increases, the acceleration factor increases correspondingly. Then, the life at elevated temperature decreases; i.e., the probability of failure increases. Particularly, if $E_A \longrightarrow \infty$, then $A_F(T_o^*, T_a^*) \longrightarrow \infty$. From Eq. (12.16) the life at elevated temperature will go to zero, which means that the desired level of burn-in shall be accomplished instantly without any action on our part, regardless of the temperature, provided it is greater than zero. This analysis leads to the following conclusion:

Given the nominal and accelerated temperature levels, the lower the value of the activation energy, the lower the failure rate. On the other hand, the higher the activation energy, the higher the failure rate. This conclusion is obviously contradictory to that drawn from the basic Arrhenius law as stated earlier.

Therefore, the present use of the acceleration formula of Eq. (12.15) is inconsistent with the Arrhenius Law. According to the Arrhenius model, increasing the activation energy implies an increase both in time and in action to precipitate a predetermined fraction of defects. In contrast, according to the acceleration equation, increasing the activation energy implies a decrease both in time and in action to burn-in a predetermined fraction of defects. To avoid this inconsistency, the activation energy should be considered as the energy provided by the test condition. In other words, the *provided activation energy* is strictly temperature dependent, while the *required activation energy* is a characteristic constant for a given failure mechanism which is independent of the burn-in temperature.

Figure 12.3 [2] illustrates the relationship between the activation energy and the junction temperature for bipolar circuits, TTL (Transistor-Transistor Logic) and LSTTL (Low-power Schottky Transistor-Tran-sistor Logic), and for MOS (Metal Oxide Semiconductor) circuits,

THE ARRHENIUS MODEL

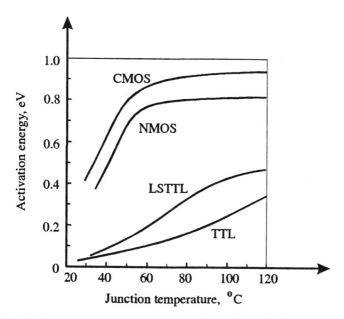

Fig. 12.3– Activation energy as a function of the junction temperature for bipolar circuits, TTL and LSTTL, and for MOS circuits, CMOS and NMOS [2].

CMOS (Complementary Metal Oxide Semiconductor) and NMOS (N-channel Metal Oxide Semiconductor). This figure was derived from the failure rate-junction temperature equations in MIL-HDBK-217D.

It may be seen that the activation energy is a nondecreasing function for these four types of circuits and the MOS circuits possess higher activation energies than the bipolar circuits. For CMOS and NMOS circuits, the activation energy increases significantly as the junction temperature increases in the lower temperature range, say $20°C \leq T^* \leq 70°C$, but keeps stable in the higher temperature range, say $T^* > 70°C$. For TTL and LSTTL circuits, the activation energy increases steadily as the junction temperature increases.

12.2.8 MODIFICATIONS AND PARAMETER ESTIMATIONS OF THE ACCELERATION FACTOR EQUATION

12.2.8.1 FIRST MODIFIED ARRHENIUS MODEL AND ITS PARAMETERS' ESTIMATION

Equation (12.15) can be modified by substituting E_A by $E_A(T^*)$ for $T^* = T_o^*$ and $T^* = T_a^*$, respectively, yielding

$$A_F(T_o^*, T_a^*) = e^{\frac{1}{K}\left[\frac{E_A(T_o^*)}{T_o^*} - \frac{E_A(T_a^*)}{T_a^*}\right]}. \tag{12.17}$$

This modification of the Arrhenius Model assumes that the activation energy, $E_A(T^*)$, is a function of the applied temperature, T^*. Then, the failure rate function given by Eq. (12.2) can be modified to

$$\lambda(T^*) = D\, e^{-\frac{E_A(T^*)}{KT^*}}. \tag{12.18}$$

Assume that the activation energy at any applied temperature T^*, $E_A(T^*)$, can be expressed by

$$E_A(T^*) = a\,(T^*)^b, \quad a > 0, b > 0, \tag{12.19}$$

which will be temperature independent only if $b = 0$. Substituting Eq. (12.19) into Eq. (12.18) yields

$$\lambda(T^*) = D\, e^{-\frac{a}{K}(T^*)^{b-1}}. \tag{12.20}$$

Note that the higher the applied temperature, the higher the failure rate; i.e., $\lambda(T^*)$ should be increasing as T^* increases. Now consider the following two cases of $\lambda(T^*)$ when b falls into two different regions:

Case 1 If the activation energy is a concave increasing function of T^*, such as the one shown in Fig. 12.4 (1), then $\frac{d[E_A(T^*)]}{dT^*} = a\,b\,(T^*)^{b-1}$ is a decreasing function of T^*, or equivalently $0 < b < 1$. Consequently, $\lambda(T^*)$ given by Eq. (12.20) is an increasing function of T^*.

Case 2 If the activation energy is a convex increasing function of T^*, such as the one shown in Fig. 12.4 (2), then $\frac{d[E_A(T^*)]}{dT^*} = a\,b\,(T^*)^{b-1}$ is an increasing function of T^*, or equivalently $b > 1$. Consequently, $\lambda(T^*)$ given by Eq. (12.20) is not an increasing function, but a decreasing function, of T^*.

THE ARRHENIUS MODEL

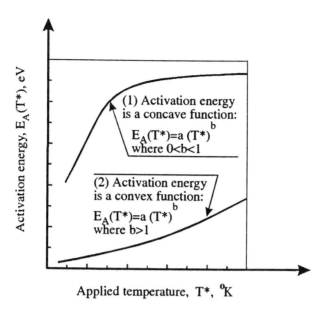

Fig. 12.4– The value of parameter b and its effect on the shape of an activation energy function.

Therefore, to maintain $\lambda(T^*)$ as an increasing function of T^* while keeping $a > 0$ and $b > 0$ at the same time, $\lambda(T^*)$ should be defined as

$$\lambda(T^*) = \begin{cases} D e^{-\frac{a}{K}(T^*)^{b-1}}, & \text{if } 0 < b < 1, \\ D e^{\frac{a}{K}(T^*)^{b-1}}, & \text{if } b \geq 1. \end{cases} \quad (12.21)$$

Based on this definition, the corresponding acceleration function equation, from Eqs. (12.17) and (12.19), becomes

$$A_F(T_o^*, T_a^*) = \frac{\lambda(T_a^*)}{\lambda(T_o^*)} = \begin{cases} e^{\frac{a}{K}[(T_o^*)^{b-1} - (T_a^*)^{b-1}]}, & \text{if } 0 < b < 1, \\ e^{\frac{a}{K}[(T_a^*)^{b-1} - (T_o^*)^{b-1}]}, & \text{if } b \geq 1. \end{cases}$$

$$(12.22)$$

However, before conducting the activation energy parameter estimation, it is not known a priori whether b is greater or less than 1. One solution to this is to start with the equation of A_F assuming $0 < b < 1$ (or $b > 1$) and determine the least-squares estimates for the parameters based on life test results. If the estimated b value, $\hat{b}$, matches the original assumption, that is $0 < \hat{b} < 1$ (or $\hat{b} > 1$), then the estimated a, $\hat{a}$, will be positive. However, if the estimated b value, $\hat{b}$, does not match the original assumption, that is $\hat{b} > 1$ (or $0 < \hat{b} < 1$), then the estimated a, $\hat{a}$, will be negative which implies that the other equation of $A_F(T_o^*, T_a^*)$ for $b > 1$ (or $0 < b < 1$) should have been used.

Substituting the estimated values of a and b, $\hat{a}$ and $\hat{b}$, directly into the selected equation, no matter whether $\hat{a}$ is positive or negative, will automatically yield the correct acceleration factor equation because the sign of parameter a will automatically justify the correctness of the equation ("+" sign validates the original assumption and keeps the assumed equation, while "-" sign will reverse the order of the two terms within the parentheses and bring the equation back to the right one.) From this point of view, Eq. (12.22) can be simplified to

$$A_F(T_o^*, T_a^*) = e^{\frac{a}{K}[(T_o^*)^{b-1} - (T_a^*)^{b-1}]}, \tag{12.23}$$
$$\text{for } (a > 0, 0 < b < 1) \text{ or } (a < 0, b > 1).$$

However, the absolute values of parameters a and b should be substituted into Eq. (12.19) to get the correct activation energy equation; i.e.,

$$E_A(T^*) = |\hat{a}|\,(T^*)^{\hat{b}}. \tag{12.24}$$

The next task is to estimate the parameters, a and b, using Eq. (12.23) and the life test results. Note that Eq. (12.23) can be rewritten as

$$A_F(T_o^*, T_a^*) = e^{[\frac{a}{K}(T_o^*)^{b-1} - \frac{a}{K}(T_a^*)^{b-1}]}, \tag{12.25}$$
$$\text{for } (a > 0, 0 < b < 1) \text{ or } (a < 0, b > 1).$$

Taking the natural logarithm of both sides of Eq. (12.25) yields

$$\log_e[A_F(T_o^*, T_a^*)] = \frac{a}{K}(T_o^*)^{b-1} - \frac{a}{K}(T_a^*)^{b-1}, \tag{12.26}$$
$$\text{for } (a > 0, 0 < b < 1) \text{ or } (a < 0, b > 1).$$

Let

$$G = \frac{a}{K}(T_o^*)^{b-1}. \tag{12.27}$$

Then,

$$\frac{a}{K} = G\left(\frac{1}{T_o^*}\right)^{b-1}.$$

Substituting these into Eq. (12.26) yields

$$\log_e[A_F(T_o^*, T_a^*)] = G - G\left(\frac{T_a^*}{T_o^*}\right)^{b-1}. \tag{12.28}$$

THE ARRHENIUS MODEL

Let

$$H = b - 1, \tag{12.29}$$
$$X = \frac{T_a^*}{T_o^*}, \tag{12.30}$$

and

$$Y = \log_e[A_F(T_o^*, T_a^*)]. \tag{12.31}$$

Then, Eq. (12.28) becomes

$$Y = G - G X^H, \text{ for } (G > 0, H < 0) \text{ or } (G < 0, H > 0). \tag{12.32}$$

For a given random sample $\{(T_{a1}^*, A_{F1}), (T_{a2}^*, A_{F2}), \cdots, (T_{an}^*, A_{Fn})\}$, where $A_{Fi} = A_F(T_o^*, T_{ai}^*)$ for $i = 1, 2, \cdots, n$, the least-squares estimates for parameters G and H, $\hat{G}$ and $\hat{H}$, can be obtained by minimizing the following sum of squares:

$$\text{MIN} \sum_{i=1}^{n}[Y_i - (G - G X_i^H)]^2, \text{ for } (G > 0, H < 0) \text{ or } (G < 0, H > 0), \tag{12.33}$$

where

$$X_i = \frac{T_{ai}^*}{T_o^*}, \tag{12.34}$$

and

$$Y_i = \log_e[A_F(T_o^*, T_{ai}^*)]. \tag{12.35}$$

This can be accomplished by any non-linear minimization method, such as the Quasi-Newton Method, the Simplex Method, the Hooke-Jeeves Pattern Moves Method, the Rosenbrock Pattern Search Method, etc [4].

After obtaining the least-squares estimates of G and H, $\hat{G}$ and $\hat{H}$, from Eq. (12.33), the corresponding least-squares estimates of the original parameters a and b, $\hat{a}$ and $\hat{b}$, can be solved from Eqs. (12.27) and (12.29), respectively; i.e.,

$$\begin{cases} \hat{b} = \hat{H} + 1, \\ \hat{a} = K \hat{G} (T_o^*)^{1-\hat{b}}, \end{cases} \tag{12.36}$$

where

$$K = 8.623 \times 10^{-5},$$

and

$T_o^* = $ operating temperature, $°K$.

Therefore, the modified acceleration factor Eq. (12.23), using the least-squares parameter estimates, becomes

$$A_F(T_o^*, T_a^*) = e^{\frac{\hat{a}}{K}[(T_o^*)^{\hat{b}-1} - (T_a^*)^{\hat{b}-1}]}. \tag{12.37}$$

Equation (12.37) can be generalized to evaluate the acceleration factor between any two different temperature levels, $T_1^* < T_2^*$. The acceleration factors for these two temperature levels, with respect to the operating temperature T_o^*, are given by

$$A_F(T_o^*, T_1^*) = \frac{L(T_o^*)}{L(T_1^*)} = e^{\frac{\hat{a}}{K}[(T_o^*)^{\hat{b}-1} - (T_1^*)^{\hat{b}-1}]},$$

and

$$A_F(T_o^*, T_2^*) = \frac{L(T_o^*)}{L(T_2^*)} = e^{\frac{\hat{a}}{K}[(T_o^*)^{\hat{b}-1} - (T_2^*)^{\hat{b}-1}]},$$

respectively. Then, the acceleration factor at T_2^* with respect to T_1^* is given by

$$A_F(T_1^*, T_2^*) = \frac{L(T_1^*)}{L(T_2^*)} = \frac{L(T_o^*)/L(T_2^*)}{L(T_o^*)/L(T_1^*)} = \frac{A_F(T_o^*, T_2^*)}{A_F(T_o^*, T_1^*)},$$

or

$$A_F(T_1^*, T_2^*) = \frac{e^{\frac{\hat{a}}{K}[(T_o^*)^{\hat{b}-1} - (T_2^*)^{\hat{b}-1}]}}{e^{\frac{\hat{a}}{K}[(T_o^*)^{\hat{b}-1} - (T_1^*)^{\hat{b}-1}]}},$$

or

$$A_F(T_1^*, T_2^*) = e^{\frac{\hat{a}}{K}[(T_1^*)^{\hat{b}-1} - (T_2^*)^{\hat{b}-1}]}. \tag{12.38}$$

Following is an example illustrating the parameter estimation of this first modified acceleration factor equation using the STATISTICA computer software [4].

EXAMPLE 12-3

To obtain the equivalent acceleration factor due to thermal cycling in burn-in, it is often necessary to quantify the aging acceleration due to

THE ARRHENIUS MODEL

TABLE 12.5– Results of two accelerated life tests on two different electronic components, for Example 12–3.

i	Component 1			Component 2		
	T^*_{ai}, °K	L_{ai}, hr	A_{Fi}†	T^*_{ai}, °K	L_{ai}, hr	A_{Fi}†
1	358	3,631.42	4.13×10^3	358	3,189.54	6.27×10^3
2	378	200.46	7.48×10^4	378	157.22	1.27×10^5
3	398	10.48	1.43×10^6	398	7.43	2.69×10^6
4	418	0.52	2.88×10^7	418	0.34	5.93×10^7
5	438	0.02	6.09×10^8	438	0.01	1.36×10^9

† $A_{Fi} = L_o/L_{ai}$.
$L_o = 1.5 \times 10^7$ hr for Component 1.
$L_o = 2.0 \times 10^7$ hr for Component 2.

the reaction-rate stress; i.e., the elevated temperature. To accomplish this, accelerated life tests should be so conducted that the acceleration factor equation's parameters can be estimated. Given in Table 12.5 are two random samples from the accelerated life tests on two different electronic components. The nominal operating temperature is $25°C$, or $298°K$. The nominal lives of these two electronic components under this temperature are $L_{o1} = 1.5 \times 10^7$ hr and $L_{o2} = 2.0 \times 10^7$ hr, respectively. Determine the acceleration factor equation1 by the least-squares method for each component using Eq. (12.33).

SOLUTION TO EXAMPLE 12–3

Transforming the given T_{ai} and A_{Fi} into X_i and Y_i using Eqs. (12.34) and (12.35) yields another table: Table 12.6.

Entering the X_i and Y_i into the NONLINEAR ESTIMATION spreadsheet of the STATISTICA software [4] and conducting non-linear least-squares estimation yields the following results:

1. **Component 1:**

$$\begin{cases} \widehat{G}_1 = -29.4368, \\ \widehat{H}_1 = 1.3582, \end{cases}$$

with a correlation coefficient of

$$R_{XY} \cong 1.0.$$

TABLE 12.6– X_i and Y_i values transformed from Table 12.5 using Eqs. (12.34) and (12.35), for Example 12-3.

i	Component 1		Component 2	
	$X_i = T^*_{ai}/T^*_o$	$Y_i = \log_e A_{Fi}$†	$X_i = T^*_{ai}/T^*_o$	$Y_i = \log_e A_{Fi}$†
1	1.2013	8.3260	1.2013	8.7435
2	1.2685	11.2226	1.2685	11.7519
3	1.3356	14.1732	1.3356	14.8051
4	1.4027	17.1759	1.4027	17.8981
5	1.4698	20.2273	1.4698	21.0308

† $A_{Fi} = A_F(T^*_o, A^*_{ai})$.

According to Eq. (12.36),

$$\hat{b}_1 = \hat{H}_1 + 1 = 2.3582 > 1,$$

and

$$\hat{a}_1 = K\,\hat{G}_1\,(T^*_o)^{1-\hat{b}_1},$$
$$= (8.623 \times 10^{-5})(-29.4368)(298)^{1-2.3582},$$

or

$$\hat{a}_1 = -1.1069 \times 10^{-6} < 0.$$

Note that $\hat{b}_1 > 1$ in this case which leads to $\hat{a}_1 < 0$. Substituting these parameters into Eq. (12.23) yields the corresponding modified acceleration factor equation for Component 1; i.e.,

$$A_{F1} = e^{\frac{\hat{a}_1}{K}[(T^*_o)^{\hat{b}_1-1} - (T^*_a)^{\hat{b}_1-1}]},$$
$$= e^{\frac{-1.1069 \times 10^{-6}}{8.623 \times 10^{-5}}[298^{2.3582-1} - (T^*_a)^{2.3582-1}]},$$

or

$$A_{F1} = e^{1.2837 \times 10^{-2}\,[(T^*_a)^{1.3582} - 2,293.4038]}. \tag{12.39}$$

THE ARRHENIUS MODEL

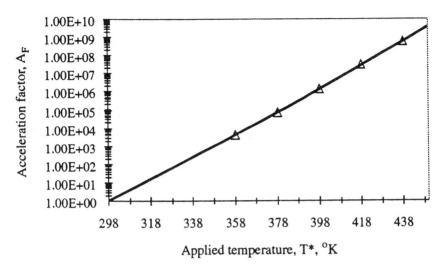

Fig. 12.5– The observed acceleration factor values and the fitted acceleration factor equation versus the applied temperatures for Component 1.

Figure 12.5 plots this fitted acceleration factor equation as well as the observed acceleration factor values versus the applied temperatures. It may be seen that this equation fits the original data extremely well. The activation energy function is given by Eq. (12.24), or

$$E_{A1}(T^*) = |\hat{a}_1|(T^*)^{\hat{b}_1} = 1.1069 \times 10^{-6} (T^*)^{2.3582}.$$

It may be seen that since

$$\hat{b}_1 = 2.3582 \neq 0,$$

the activation energy in this case is temperature dependent.

2. **Component 2:**

$$\begin{cases} \widehat{G}_2 = -33.2864, \\ \widehat{H}_2 = 1.2715, \end{cases}$$

with a correlation coefficient of

$$R_{XY} \cong 1.0.$$

According to Eq. (12.36),

$$\hat{b}_2 = \hat{H}_2 + 1 = 2.2715 > 1,$$

and

$$\hat{a}_2 = K\,\hat{G}_2\,(T_o^*)^{1-\hat{b}_2},$$
$$= (8.623 \times 10^{-5})(-33.2864)(298)^{1-2.2715},$$

or

$$\hat{a}_2 = -2.0504 \times 10^{-6} < 0.$$

Note that $\hat{b}_2 > 1$ in this case which leads to $\hat{a}_2 < 0$. Substituting these parameters into Eq. (12.23) yields the corresponding modified acceleration factor equation for Component 2; i.e.,

$$A_{F2} = e^{\frac{\hat{a}_2}{K}[(T_o^*)^{\hat{b}_2-1}-(T_a^*)^{\hat{b}_2-1}]},$$
$$= e^{\frac{-2.0504 \times 10^{-6}}{8.623 \times 10^{-5}}[298^{2.2715-1}-(T_a^*)^{2.2715-1}]},$$

or

$$A_{F2} = e^{2.3778 \times 10^{-2}\,[(T_a^*)^{1.2715}-1{,}399.4777]}. \qquad (12.40)$$

Figure 12.6 plots this fitted acceleration factor equation as well as the observed acceleration factor values versus the applied temperatures. It may be seen that this equation fits the original data extremely well. The activation energy function is given by Eq. (12.24), or

$$E_{A2}(T^*) = |\hat{a}_2|\,(T^*)^{\hat{b}_2} = 2.0504 \times 10^{-6}\,(T^*)^{2.2715}.$$

It may be seen that since

$$\hat{b}_2 = 2.2715 \neq 0,$$

the activation energy in this case is temperature dependent.

THE ARRHENIUS MODEL

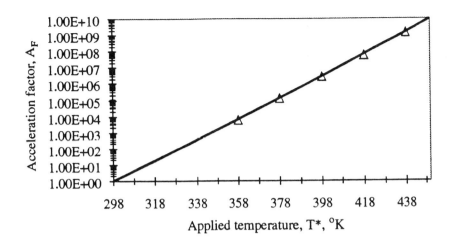

Fig. 12.6– The observed acceleration factor values and the fitted acceleration factor equation versus the applied temperatures for Component 2.

12.2.8.2 SECOND MODIFIED ARRHENIUS MODEL AND ITS PARAMETERS' ESTIMATION

Another modification on Eq. (12.15) was made by Pugacz-Muraszkiewicz [3] as follows:

$$A_F(T_o^*, T_a^*) = e^{\frac{[E_P(T_a^*)]^2}{E_R \, K}\left(\frac{1}{T_o^*}-\frac{1}{T_a^*}\right)}, \qquad (12.41)$$

where

$E_P(T_a^*)$ = energy provided by the test in eV, which is a function of the applied temperature, T_a^*,

and

E_R = energy required by the failure mechanism in eV, which is independent of the applied temperature, T_a^*.

It may be seen that, when $E_P(T^*) = E_R$, Eq. (12.41) reduces to Eq. (12.15).

Assume the provided activation energy, $E_P(T_a^*)$, can be expressed by

$$E_P(T_a^*) = a' \, (T_a^*)^{b'}, \qquad (12.42)$$

which will be temperature independent only if $b' = 0$. Then, Eq. (12.41) can be rewritten as

$$A_F(T_o^*, T_a^*) = e^{\frac{[E_P(T_a^*)]^2}{E_R K}\left(\frac{1}{T_o^*} - \frac{1}{T_a^*}\right)},$$

$$= e^{\frac{[a'(T_a^*)^{b'}]^2}{E_R K}\left(\frac{1}{T_o^*} - \frac{1}{T_a^*}\right)},$$

$$= e^{\left[\frac{(a')^2}{E_R K}\right](T_a^*)^{2b'}\left(\frac{1}{T_o^*} - \frac{1}{T_a^*}\right)},$$

or

$$A_F(T_o^*, T_a^*) = e^{C\,(T_a^*)^B\left(\frac{1}{T_o^*} - \frac{1}{T_a^*}\right)}, \tag{12.43}$$

where

C = a constant independent of the applied temperature,

or

$$C = \frac{(a')^2}{E_R K}, \tag{12.44}$$

and

$$B = 2\,(b'). \tag{12.45}$$

Note that only if $B = b' = 0$, then $E_P(T_a^*)$ will be temperature independent. Taking the logarithm twice of both sides of Eq. (12.43) yields

$$\log_e \log_e[A_F(T_o^*, T_a^*)] = \log_e\left[C\,(T_a^*)^B\left(\frac{1}{T_o^*} - \frac{1}{T_a^*}\right)\right],$$

$$= \log_e C + B\,\log_e T_a^* + \log_e\left(\frac{1}{T_o^*} - \frac{1}{T_a^*}\right),$$

or

$$\log_e \log_e[A_F(T_o^*, T_a^*)] - \log_e\left(\frac{1}{T_o^*} - \frac{1}{T_a^*}\right) = \log_e C + B\,\log_e T_a^*. \tag{12.46}$$

Let

$$Y = \log_e \log_e[A_F(T_o^*, T_a^*)] - \log_e\left(\frac{1}{T_o^*} - \frac{1}{T_a^*}\right), \tag{12.47}$$

$$A = \log_e C, \tag{12.48}$$

THE ARRHENIUS MODEL

and

$$X = \log_e T_a^*. \tag{12.49}$$

Then, Eq. (12.46) becomes

$$Y = A + BX. \tag{12.50}$$

For a given random sample $\{(T_{a1}^*, A_{F1}), (T_{a2}^*, A_{F2}), \cdots, (T_{an}^*, A_{Fn})\}$, where $A_{Fi} = A_F(T_o^*, T_{ai}^*)$ for $i = 1, 2, \cdots, n$, the least-squares estimates for $C = \frac{(a')^2}{E_R K}$ and B, or $\widehat{C}$ and $\widehat{B}$, are

$$\widehat{C} = e^{\widehat{A}}, \tag{12.51}$$

and

$$\widehat{B} = \frac{L_{XY}}{L_{XX}}, \tag{12.52}$$

with the correlation coefficient of

$$R_{XY} = \frac{L_{XY}}{\sqrt{L_{XX} L_{YY}}}, \tag{12.53}$$

where

$$\widehat{A} = \overline{Y} - \widehat{B}\overline{X}, \tag{12.54}$$

$$L_{XY} = \sum_{i=1}^{n} (X_i - \overline{X})(Y_i - \overline{Y}), \tag{12.55}$$

$$L_{XX} = \sum_{i=1}^{n} (X_i - \overline{X})^2, \tag{12.56}$$

$$L_{YY} = \sum_{i=1}^{n} (Y_i - \overline{Y})^2, \tag{12.57}$$

$$\overline{X} = \frac{1}{n}\sum_{i=1}^{n} X_i, \tag{12.58}$$

$$\overline{Y} = \frac{1}{n}\sum_{i=1}^{n} Y_i, \tag{12.59}$$

$$X_i = \log_e T_{ai}^*, \tag{12.60}$$

and

$$Y_i = \log_e \log_e A_{Fi} - \log_e \left(\frac{1}{T_o^*} - \frac{1}{T_{ai}^*}\right). \tag{12.61}$$

Therefore, the modified acceleration factor Eq. (12.41), using the least-squares parameter estimates, becomes

$$A_F(T_o^*, T_a^*) = e^{\widehat{C} \, (T_a^*)^{\widehat{B}} \left(\frac{1}{T_o^*} - \frac{1}{T_a^*}\right)}, \qquad (12.62)$$

Equation (12.62) can be generalized to evaluate the acceleration factor between any two different temperature levels, $T_1^* < T_2^*$. The acceleration factors for these two temperature levels, with respect to the operating temperature T_o^*, are given by

$$A_F(T_o^*, T_1^*) = \frac{L(T_o^*)}{L(T_1^*)} = e^{\widehat{C} \, (T_1^*)^{\widehat{B}} \left(\frac{1}{T_o^*} - \frac{1}{T_1^*}\right)},$$

and

$$A_F(T_o^*, T_2^*) = \frac{L(T_o^*)}{L(T_2^*)} = e^{\widehat{C} \, (T_2^*)^{\widehat{B}} \left(\frac{1}{T_o^*} - \frac{1}{T_2^*}\right)},$$

respectively. Then, the acceleration factor at T_2^* with respect to T_1^* is given by

$$A_F(T_1^*, T_2^*) = \frac{L(T_1^*)}{L(T_2^*)} = \frac{L(T_o^*)/L(T_2^*)}{L(T_o^*)/L(T_1^*)} = \frac{A_F(T_o^*, T_2^*)}{A_F(T_o^*, T_1^*)},$$

and, using Eq. (12.62),

$$A_F(T_1^*, T_2^*) = \frac{e^{\widehat{C} \, (T_2^*)^{\widehat{B}} \left(\frac{1}{T_o^*} - \frac{1}{T_2^*}\right)}}{e^{\widehat{C} \, (T_1^*)^{\widehat{B}} \left(\frac{1}{T_o^*} - \frac{1}{T_1^*}\right)}},$$

or

$$A_F(T_1^*, T_2^*) = e^{\widehat{C} \left[(T_2^*)^{\widehat{B}} \left(\frac{1}{T_o^*} - \frac{1}{T_2^*}\right) - (T_1^*)^{\widehat{B}} \left(\frac{1}{T_o^*} - \frac{1}{T_1^*}\right) \right]}. \qquad (12.63)$$

Following is an example illustrating the application of this least-squares estimation method.

EXAMPLE 12-4

Given are the same as in Example 12-3. Determine the acceleration factor equation by the least-squares method for each component using Eq. (12.41).

SOLUTION TO EXAMPLE 12-4

For the sake of convenience, Table 12.5 is reproduced here and identified as Table 12.7. Transforming the given T_{ai} and A_{Fi} into X_i and Y_i, using Eqs. (12.60) and (12.61), yields Table 12.8.

Substituting the calculated X_i and Y_i into Eqs. (12.54), (12.51), (12.52) and (12.53) yields the following results:

THE ARRHENIUS MODEL

TABLE 12.7 – Results of two accelerated life tests on two different electronic components, for Example 12–4.

	Component 1			Component 2		
i	T_{ai}^*, °K	L_{ai}, hr	A_{Fi}†	T_{ai}^*, °K	L_{ai}, hr	A_{Fi}†
1	358	3,631.42	4.13×10^3	358	3,189.54	6.27×10^3
2	378	200.46	7.48×10^4	378	157.22	1.27×10^5
3	398	10.48	1.43×10^6	398	7.43	2.69×10^6
4	418	0.52	2.88×10^7	418	0.34	5.93×10^7
5	438	0.02	6.09×10^8	438	0.01	1.36×10^9

† $A_{Fi} = L_o/L_{ai}$.
$L_o = 1.5 \times 10^7$ hr for Component 1.
$L_o = 2.0 \times 10^7$ hr for Component 2.

TABLE 12.8 – X_i and Y_i values transformed from Table 12.7 using Eqs. (12.60) and (12.61), for Example 12–4.

	Component 1		Component 2	
i	$X_i = \log_e T_{ai}^*$	Y_i†	$X_i = \log_e T_{ai}^*$	Y_i†
1	5.8805	8.3260	5.8805	9.6516
2	5.9349	11.2226	5.9349	9.7140
3	5.9865	14.1732	5.9865	9.7733
4	6.0355	17.1759	6.0355	9.8298
5	6.0822	20.2273	6.0822	9.8837

† $Y_i = \log_e \log_e A_{Fi} - \log_e \left(\frac{1}{T_o^*} - \frac{1}{T_{ai}^*} \right)$.

1. **Component 1:**

$$\hat{C}_1 = 12.7535,$$
$$\hat{B}_1 = 1.2000,$$

and

$$R_{XY} \cong 1.0.$$

Substituting these parameters into Eq. (12.43) yields the modified acceleration factor equation for Component 1; i.e.,

$$A_{F1} = e^{\hat{C}_1 \, (T_a^*)^{\hat{B}_1} \left(\frac{1}{T_o^*} - \frac{1}{T_a^*}\right)},$$

or

$$A_{F1} = e^{12.7535 \, (T_a^*)^{1.2000} \left(\frac{1}{298} - \frac{1}{T_a^*}\right)}. \qquad (12.64)$$

Figure 12.7 plots this fitted acceleration factor equation as well as the observed acceleration factor values versus the applied temperatures. It may be seen that this equation fits the original data extremely well.

Note that since

$$\hat{B}_1 = 1.2000 \neq 0,$$

the activation energy in this case is temperature dependent.

2. **Component 2:**

$$\hat{C}_2 = 17.8997,$$
$$\hat{B}_2 = 1.1507,$$

and

$$R_{XY} \cong 1.0.$$

Substituting these parameters into Eq. (12.43) yields the modified acceleration factor equation for Component 2; i.e.,

$$A_{F2} = e^{\hat{C}_2 \, (T_a^*)^{\hat{B}_2} \left(\frac{1}{T_o^*} - \frac{1}{T_a^*}\right)},$$

THE ARRHENIUS MODEL

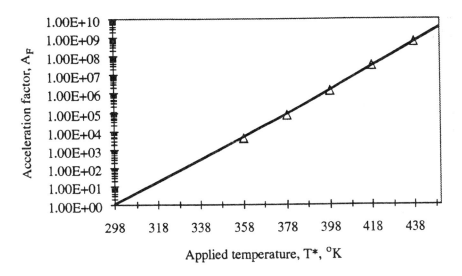

Fig. 12.7– The observed acceleration factor values and the fitted acceleration factor equation versus the applied temperatures for Component 1.

or

$$A_{F2} = e^{17.8997\ (T_a^*)^{1.1507}\left(\frac{1}{298}-\frac{1}{T_a^*}\right)}. \qquad (12.65)$$

Figure 12.8 plots this fitted acceleration factor equation as well as the observed acceleration factor values versus the applied temperatures. It may be seen that this equation fits the original data extremely well.

Note that since

$$\widehat{B}_2 = 1.1507 \neq 0,$$

the activation energy in this case is temperature dependent.

12.2.9 BURN-IN TIME REDUCTION USING AN ELEVATED TEMPERATURE AND THE MODIFIED ARRHENIUS MODEL

Given a required burn-in time at the use temperature T_o^*, T_{bo}, the corresponding burn-in time at an elevated temperature T_a^*, T_{ba}, is

$$T_{ba} = T_{bo}/A_F(T_o^*, T_a^*) = T_{bo}\ e^{-\frac{\hat{a}}{K}\ [(T_o^*)^{\hat{b}-1}-(T_a^*)^{\hat{b}-1}]}, \qquad (12.66)$$

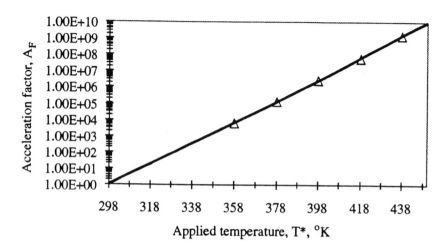

Fig. 12.8– The observed acceleration factor values and the fitted acceleration factor equation versus the applied temperatures for Component 2.

if the first modified Arrhenius Model, or Eq. (12.23), is used, and

$$T_{ba} = T_{bo}/A_F(T_o^*, T_a^*) = T_{bo} \, e^{-\widehat{C} \, (T_a^*)^{\widehat{B}} \left(\frac{1}{T_o^*} - \frac{1}{T_a^*}\right)}, \qquad (12.67)$$

if the second modified Arrhenius Model, or Eq. (12.43), is used.

In general, given a required burn-in time at a lower temperature T_1^*, T_{b1}, the corresponding burn-in time at a higher temperature T_2^*, T_{b2}, is given by

$$T_{b2} = T_{b1}/A_F(T_1^*, T_2^*) = T_{b1} \, e^{-\frac{\hat{a}}{\hat{K}} \, [(T_1^*)^{\hat{b}-1} - (T_2^*)^{\hat{b}-1}]}, \qquad (12.68)$$

if the first modified Arrhenius Model is used, and

$$T_{b2} = T_{b1}/A_F(T_1^*, T_2^*) = T_{b1} \, e^{-\widehat{C} \, \left[(T_2^*)^{\widehat{B}} \left(\frac{1}{T_o^*} - \frac{1}{T_2^*}\right) - (T_1^*)^{\widehat{B}} \left(\frac{1}{T_o^*} - \frac{1}{T_1^*}\right)\right]},$$

(12.69)

if the second modified Arrhenius Model is used.

EXAMPLE 12-5

If a 1,000-hr burn-in at the use temperature of 298°K (25°C) is required for each one of the two types of components given in Examples

THE ARRHENIUS MODEL

12-3 and 12-4, how long will it take to burn them in at the elevated temperature of 328°K (55°C)? Note that the activation energy for each one of these two components can not be assumed to be a constant here. Use both versions of the modified Arrhenius Model and compare the results.

SOLUTION TO EXAMPLE 12-5

In this case, the lower temperature is the use temperature, or

$$T_o^* = 298°K,$$

and the higher temperature is

$$T_a^* = 328°K.$$

Case 1– Use the first modified Arrhenius Model.

If the first modified Arrhenius Model is used, then the acceleration factor for Component 1, using Eq. (12.39) obtained in Example 12-3, is

$$A_{F1}(T_o^*, T_a^*) = e^{1.2837 \times 10^{-2}[(328)^{1.3582} - 2,293.4038]} = 60.1302,$$

and for Component 2, using Eq. (12.40), is

$$A_{F2}(T_o^*, T_a^*) = e^{2.3778 \times 10^{-2}[(328)^{1.2715} - 1,399.4777]} = 74.9174.$$

Therefore, a burn-in on Component 1 at 25°C for 1,000 hr is equivalent to a burn-in at 55°C for

$$T_{ba1} = \frac{T_{bo1}}{A_{F1}(T_o^*, T_a^*)},$$
$$= \frac{1,000}{60.1302},$$

or

$$T_{ba1} = 16.63 \text{ hr, or approximately 17 hr.}$$

Similarly, a burn-in on Component 2 at 25°C for 1,000 hr is equivalent to a burn-in at 55°C for

$$T_{ba2} = \frac{T_{bo2}}{A_{F2}(T_o^*, T_a^*)},$$
$$= \frac{1,000}{74.9174},$$

or

$$T_{ba2} = 13.35 \text{ hr, or approximately 13 hr.}$$

Case 2– Use the second modified Arrhenius Model.

If the second Modified Arrhenius Model is used, then the acceleration factor for Component 1, using Eq. (12.64) obtained in Example 12–4, is

$$A_{F1}(T_o^*, T_a^*) = e^{12.7535 \ (328)^{1.2000} \left(\frac{1}{298} - \frac{1}{328}\right)} = 59.7319,$$

and for Component 2, using Eq. (12.65), is

$$A_{F2}(T_o^*, T_a^*) = e^{17.8997 \ (328)^{1.1507} \left(\frac{1}{298} - \frac{1}{328}\right)} = 74.7483.$$

Therefore, a burn-in on Component 1 at 25°C for 1,000 hr is equivalent to a burn-in at 55°C for

$$T_{ba1} = \frac{T_{bo1}}{A_{F1}(T_o^*, T_a^*)},$$
$$= \frac{1,000}{59.7319},$$

or

$$T_{ba1} = 16.74 \text{ hr, or approximately 17 hr.}$$

Similarly, a burn-in on Component 2 at 25°C for 1,000 hr is equivalent to a burn-in at 55°C for

$$T_{ba2} = \frac{T_{bo2}}{A_{F2}(T_o^*, T_a^*)},$$
$$= \frac{1,000}{74.7483},$$

or

$$T_{ba2} = 13.37 \text{ hr, or approximately 13 hr.}$$

Comparing the required burn-in times at 55° using the two modified Arrhenius Models, it may be seen that both yield very close results.

12.3 BURN-IN TIME REDUCTION USING AN ELEVATED VOLTAGE BIAS STRESS – THE INVERSE POWER LAW MODEL

Standard semiconductor voltage bias stress level and associated failure mechanisms/rates are related by the Inverse Power Law. The characteristic life is given by

$$L(V) = \frac{1}{K\, V^n}, \qquad (12.70)$$

where V is the voltage bias stress, or a nonthermal accelerating stress, and K and n are positive parameters characteristic of the device and the test method.

The characteristic life, L_a, at accelerated stress, V_a, is given by

$$L_a = \frac{1}{K\, V_a^n}, \qquad (12.71)$$

and the characteristic life at use stress, V_o, is given by

$$L_o = \frac{1}{K\, V_o^n}. \qquad (12.72)$$

Dividing Eq. (12.72) by Eq. (12.71) yields the acceleration factor equation, and solving for L_o yields the Inverse Power Law for quantifying the use stress characteristic life; i.e.,

$$A_F(V_o, V_a) = \frac{L_o}{L_a} = \left(\frac{V_a}{V_o}\right)^n, \qquad (12.73)$$

and

$$L_o = L_a \left(\frac{V_a}{V_o}\right)^n. \qquad (12.74)$$

Note that the validity of Eqs. (12.73) and (12.74) requires that n remains unchanged when V changes from V_o to V_a.

In general, the acceleration factor at V_2 with respect to V_1 where $V_2 < V_1$ is given by

$$A_F(V_1, V_2) = \frac{L_1}{L_2} = \left(\frac{V_2}{V_1}\right)^n, \qquad (12.75)$$

from which the characteristic life at one stress level can be converted to that at another stress level.

The relationship between burn-in times at V_1 and V_2 is

$$T_{b1} = T_{b2} \left(\frac{V_2}{V_1}\right)^n,$$

or

$$T_{b2} = T_{b1} \left(\frac{V_1}{V_2}\right)^n.$$

EXAMPLE 12-6

According to the previous accelerated life tests, the mean time to first failure of the silicon photodiode with a 32-element configuration, as discussed in Example 12-1, is 240 hr at 85°C and -1 Vdc, and 10 hr at 85°C and -10 Vdc.

A 1,300-hr burn-in at 85°C and -2 Vdc was required for such silicon photodiodes. How long is it going to take if the burn-in is conducted at 85°C but -10 Vdc?

SOLUTION TO EXAMPLE 12-6

Using the Inverse Power Law, Eq. (12.73) yields

$$A_F(-1 \text{ Vdc}, -10 \text{ Vdc}) = \frac{L_{-1} \text{ Vdc}}{L_{-10} \text{ Vdc}} = \left(\frac{V = -10 \text{ Vdc}}{V = -1 \text{ Vdc}}\right)^n,$$

or

$$\frac{240}{10} = \left(\frac{-10}{-1}\right)^n.$$

Solving for n yields

$$n = 1.3802.$$

The burn-in time under the higher bias of -10 Vdc corresponding to the burn-in time of 1,300 hr under the lower bias of -2 Vdc will be

$$T_{b;-10 \text{ Vdc}} = T_{b;-2 \text{ Vdc}} \left(\frac{V = -2 \text{ Vdc}}{V = -10 \text{ Vdc}}\right)^n,$$

or

$$T_{b;-10 \text{ Vdc}} = (1,300) \left(\frac{-2}{-10}\right)^{1.3802},$$

or

$$T_{b;-10 \text{ Vdc}} = 141 \text{ hr}.$$

It may be seen that increasing the bias level 5 times yields almost a 9-fold decrease in the burn-in time.

12.4 BURN-IN TIME REDUCTION USING AN ELEVATED TEMPERATURE PLUS AN ELEVATED VOLTAGE BIAS – THE COMBINATION MODEL

12.4.1 WHEN THE TEMPERATURE DEPENDENCY OF THE ACTIVATION ENERGY IS NOT CONSIDERED – THE CONVENTIONAL COMBINATION MODEL

When both temperature and voltage, or a nonthermal stress, are accelerated in the burn-in test, the Arrhenius and the Inverse Power Law models can be combined to form a new model. This is known as the Combination Model. The relationship between the characteristic lives of both the operating stresses and the accelerated stresses is

$$L_o = L_a \left(\frac{V_a}{V_o}\right)^n e^{\frac{E_A}{K}\left(\frac{1}{T_o^*} - \frac{1}{T_a^*}\right)}, \tag{12.76}$$

or

$$L_a = L_o \left(\frac{V_o}{V_a}\right)^n e^{\frac{E_A}{K}\left(\frac{1}{T_a^*} - \frac{1}{T_o^*}\right)}. \tag{12.77}$$

Note that the validity of Eqs. (12.76) and (12.77) requires that both n and E_A remain unchanged at the operating stresses, (V_o, T_o^*), and at the accelerated stresses, (V_a, T_a^*).

In general, the relationship between the characteristic lives of any two different stress levels, (T_1^*, V_1) and (T_2^*, V_2), is

$$L_1 = L_2 \left(\frac{V_2}{V_1}\right)^n e^{\frac{E_A}{K}\left(\frac{1}{T_1^*} - \frac{1}{T_2^*}\right)}, \tag{12.78}$$

or

$$L_2 = L_1 \left(\frac{V_1}{V_2}\right)^n e^{\frac{E_A}{K}\left(\frac{1}{T_2^*} - \frac{1}{T_1^*}\right)}. \tag{12.79}$$

Similarly, the relationship between the burn-in times of any two different stress levels, (T_1^*, V_1) and (T_2^*, V_2), is

$$T_{b1} = T_{b2} \left(\frac{V_2}{V_1}\right)^n e^{\frac{E_A}{K}\left(\frac{1}{T_1^*} - \frac{1}{T_2^*}\right)}, \tag{12.80}$$

or

$$T_{b2} = T_{b1} \left(\frac{V_1}{V_2}\right)^n e^{\frac{E_A}{K}\left(\frac{1}{T_2^*} - \frac{1}{T_1^*}\right)}. \tag{12.81}$$

EXAMPLE 12-7

In Example 12-6, the burn-ins were conducted at two different bias levels, -2 Vdc and -10 Vdc, but at the same temperature level of 85°C. If the burn-in is going to be conducted at an elevated temperature of 100°C and an elevated bias of -10 Vdc, determine the corresponding burn-in time using the Combination Model by considering both temperature and voltage stresses at the same time.

SOLUTION TO EXAMPLE 12-7

Substituting the information given and generated in Examples 12-1 and 12-6 into Eq. (12.81) yields the burn-in time at the higher stress level of $T_2^* = 100°C$ (373°K) and $V_2 = -10\ V$, corresponding to a 1,300-hr burn-in at the lower stress level of $T_1^* = 85°C$ (358°K) and $V_1 = -2\ V$, as

$$T_{b2} = T_{b1} \left(\frac{V_1}{V_2}\right)^n e^{\frac{E_A}{K}(\frac{1}{T_2^*} - \frac{1}{T_1^*})},$$

$$= (1,300) \left(\frac{-2}{-10}\right)^{1.3802} e^{(\frac{1.0}{8.623 \times 10^{-5}})(\frac{1}{358} - \frac{1}{373})},$$

or

$$T_{b2} = 38\ \text{hr}.$$

12.4.2 WHEN THE TEMPERATURE DEPENDENCY OF THE ACTIVATION ENERGY IS CONSIDERED – THE MODIFIED COMBINATION MODEL

When the temperature dependency of the activation energy is to be considered, then the modified Arrhenius acceleration factor given by Eqs. (12.23) and (12.43) should be incorporated into the Combination Model.

Case 1– Use the first Modified Arrhenius Model given by Eq. (12.23); i.e.,

$$L_o = L_a \left(\frac{V_a}{V_o}\right)^n e^{\frac{a}{k}[(T_o^*)^{b-1} - (T_a^*)^{b-1}]}, \qquad (12.82)$$

or

$$L_a = L_o \left(\frac{V_o}{V_a}\right)^n e^{\frac{a}{k}[(T_a^*)^{b-1} - (T_o^*)^{b-1}]}. \qquad (12.83)$$

Case 2– Use the second modified Arrhenius Model given by Eq. (12.43); i.e.,

$$L_o = L_a \left(\frac{V_a}{V_o}\right)^n e^{C\,(T_a^*)^B \left(\frac{1}{T_o^*} - \frac{1}{T_a^*}\right)}, \qquad (12.84)$$

or

$$L_a = L_o \left(\frac{V_o}{V_a}\right)^n e^{C\,(T_a^*)^B \left(\frac{1}{T_a^*} - \frac{1}{T_o^*}\right)}. \qquad (12.85)$$

Equations (12.82) or (12.83), and (12.84) or (12.85) are called the Modified Combination Models. Based on these models, the relationship between the burn-in time at use stress level (T_o^*, V_o), T_{bo}, and that at an elevated stress level (T_a^*, V_a), T_{ba}, can be determined as follows:

Case 1– If the first Modified Arrhenius Model is used, then

$$T_{bo} = T_{ba} \left(\frac{V_a}{V_o}\right)^n e^{\frac{a}{K}[(T_o^*)^{b-1} - (T_a^*)^{b-1}]},$$

or

$$T_{ba} = T_{bo} \left(\frac{V_o}{V_a}\right)^n e^{\frac{a}{K}[(T_a^*)^{b-1} - (T_o^*)^{b-1}]}.$$

In general, the relationship between the characteristic lives of any two different stress levels, (T_1^*, V_1) and (T_2^*, V_2), is

$$L_1 = L_2 \left(\frac{V_2}{V_1}\right)^n e^{\frac{a}{K}[(T_1^*)^{b-1} - (T_2^*)^{b-1}]},$$

or

$$L_2 = L_1 \left(\frac{V_1}{V_2}\right)^n e^{\frac{a}{K}[(T_2^*)^{b-1} - (T_1^*)^{b-1}]}.$$

Similarly, the relationship between the burn-in times at any two different stress levels, (T_1^*, V_1) and (T_2^*, V_2), is given by

$$T_{b1} = T_{b2} \left(\frac{V_2}{V_1}\right)^n e^{\frac{a}{K}[(T_1^*)^{b-1} - (T_2^*)^{b-1}]}, \qquad (12.86)$$

or

$$T_{b2} = T_{b1} \left(\frac{V_1}{V_2}\right)^n e^{\frac{a}{K}[(T_2^*)^{b-1} - (T_1^*)^{b-1}]}. \qquad (12.87)$$

Case 2 – If the second Modified Arrhenius Model is used, then

$$T_{bo} = T_{ba} \left(\frac{V_a}{V_o}\right)^n e^{C\,(T_a^*)^B \left(\frac{1}{T_o^*} - \frac{1}{T_a^*}\right)},$$

or

$$T_{ba} = T_{bo} \left(\frac{V_o}{V_a}\right)^n e^{C\,(T_a^*)^B \left(\frac{1}{T_a^*} - \frac{1}{T_o^*}\right)}.$$

In general, the relationship between the characteristic lives of any two different stress levels, (T_1^*, V_1) and (T_2^*, V_2), is

$$L_1 = L_2 \left(\frac{V_2}{V_1}\right)^n e^{C\left[(T_2^*)^B \left(\frac{1}{T_o^*} - \frac{1}{T_2^*}\right) - (T_1^*)^B \left(\frac{1}{T_o^*} - \frac{1}{T_1^*}\right)\right]},$$

or

$$L_2 = L_1 \left(\frac{V_1}{V_2}\right)^n e^{C\left[(T_1^*)^B \left(\frac{1}{T_o^*} - \frac{1}{T_1^*}\right) - (T_2^*)^B \left(\frac{1}{T_o^*} - \frac{1}{T_2^*}\right)\right]}.$$

Similarly, the relationship between the burn-in times at any two different stress levels, (T_1^*, V_1) and (T_2^*, V_2), is given by

$$T_{b1} = T_{b2} \left(\frac{V_2}{V_1}\right)^n e^{C\left[(T_2^*)^B \left(\frac{1}{T_o^*} - \frac{1}{T_2^*}\right) - (T_1^*)^B \left(\frac{1}{T_o^*} - \frac{1}{T_1^*}\right)\right]}, \qquad (12.88)$$

or

$$T_{b2} = T_{b1} \left(\frac{V_1}{V_2}\right)^n e^{C\left[(T_1^*)^B \left(\frac{1}{T_o^*} - \frac{1}{T_1^*}\right) - (T_2^*)^B \left(\frac{1}{T_o^*} - \frac{1}{T_2^*}\right)\right]}. \qquad (12.89)$$

EXAMPLE 12-8

Rework Example 12-7 using the Modified Combination Models assuming that the parameters of the modified Arrhenius Models are the same as those for Component 2 obtained in Examples 12-3 and 12-4; i.e., $a = -2.0504 \times 10^{-6}$ and $b = 2.2715$ for the first Modified Arrhenius Model, and $C = 17.8997$ and $B = 1.1507$ for the second Modified Arrhenius Model. Compare the results obtained from these two models.

SOLUTION TO EXAMPLE 12-8

Case 1 – Use the first Modified Arrhenius Model

THE COMBINATION MODEL

Substituting $a = -2.0504 \times 10^{-6}$ and $b = 2.2715$ into Eq. (12.87) yields

$$T_{b2} = T_{b1} \left(\frac{V_1}{V_2}\right)^n e^{\frac{a}{K}[(T_2^*)^{b-1} - (T_1^*)^{b-1}]},$$

$$= (1,300) \left(\frac{-2}{-10}\right)^{1.3802} e^{-\frac{2.0504 \times 10^{-6}}{8.623 \times 10^{-5}}[(373)^{2.2715-1} - (358)^{2.2715-1}]},$$

or

$T_{b2} = 14.84$ hr, or approximately 15 hr.

Case 2– Use the second Modified Arrhenius Model

Substituting $C = 17.8997$ and $B = 1.1507$ into Eq. (12.89) yields

$$T_{b2} = T_{b1} \left(\frac{V_1}{V_2}\right)^n e^{C\left[(T_1^*)^B \left(\frac{1}{T_o^*} - \frac{1}{T_1^*}\right) - (T_2^*)^B \left(\frac{1}{T_o^*} - \frac{1}{T_2^*}\right)\right]},$$

$$= (1,300) \left(\frac{-2}{-10}\right)^{1.3802} e^{17.8997 \left[(358)^{1.1507}\left(\frac{1}{298} - \frac{1}{358}\right) - (373)^{1.1507}\left(\frac{1}{298} - \frac{1}{373}\right)\right]},$$

or

$T_{b2} = 14.81$ hr, or approximately 15 hr.

Comparing the required burn-in times using the two Modified Arrhenius Models, it may be seen that both yield very close results.

12.5 OPTIMUM BURN-IN TIME DETERMINATION BASED ON THE TEST RESULTS AT A HIGHER STRESS LEVEL

Since burn-in at use stress level is often too lengthy a process, it is generally conducted at an elevated stress level. If the life test results are available at a higher level stress, say at a higher temperature and/or a higher voltage bias, then the optimum burn-in time at this higher level stress for a specified, or for the maximum, mean residual life (MRL), post-burn-in mission reliability, and for a specified or for the minimum post-burn-in failure rate, and life-cycle cost, can be determined using the methods and techniques presented in the previous chapters. This burn-in time which is derived from the test results at the high level stress can be converted to the corresponding burn-in time at any other stress level using an appropriate acceleration factor as presented in this chapter; i.e.,

$$T_{bl} = T_{bh} \; A_F(S_l, S_h),$$

where

T_{bh} = burn-in time at a higher stress level, S_h,
T_{bl} = burn-in time at a lower stress level, S_l,

and

$A_F(S_l, S_h)$ = acceleration factor which is a function of the stress levels, S_l and S_h.

This relationship is graphically illustrated in Fig. 12.9. Note that the stress here, S, can be a single stress, such as temperature or voltage bias, and can be a stress combination, such as temperature plus voltage bias.

REFERENCES

1. Kececioglu, Dimitri, *Reliability & Life Testing Handbook*, Vol. 2, Prentice Hall, Upper Saddle River, NJ 07458, 894 pp., 1993.

2. Jensen, F., "Activation Energies and the Arrhenius Equation," *Quality and Reliability International*, Vol. 1, pp. 13-17, 1985.

3. Pugacz-Muraszkiewicz, I., "Arrhenius Law in Its Application to the Stress Screening of Electronic Hardware – A Model With a Pitfall," *Proceedings of the Institute of Environmental Sciences*, pp. 779-783, 1990.

4. StatSoft, *CSS: STATISTICA*, StatSoft Inc., 2325 East 13th Street, Tulsa, OK 74104, 925 pp., 1994.

5. Jensen, F. and Peterson, N. E., *Burn-in, An Engineering Approach to the Design and Analysis of Burn-in Procedures*, John Wiley & Sons, New York, 167 pp., 1982.

ADDITIONAL REFERENCES

1. Crook, D. L., "Method of Determining Reliability Screens for the Time Dependent Dielectric Breakdown," *Proceedings of the International Reliability Physics Symposium*, pp. 1-7, 1979.

2. Dodson, G. A., "Analysis of Accelerated Temperature Cycle Test Data Containing Different Failure Modes," *Proceedings of the International Reliability Physics Symposium*, pp. 238-246, 1979.

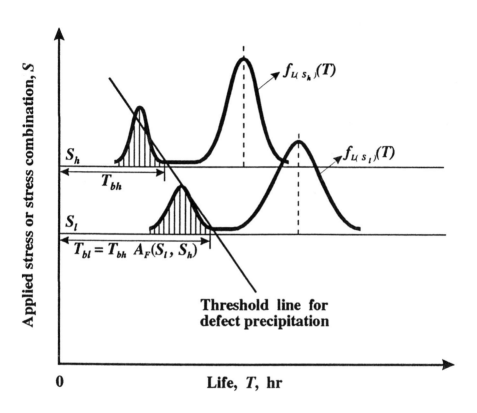

Fig. 12.9– Converting burn-in time at a higher stress level to that at a lower stress level.

3. Hart, L. J. and Bruning, K. L., "Power-Temperature Cycle Accelerated Testing," *Quality and Reliability Engineering International*, Vol. 11, pp. 41-48, 1995.

4. Hu, J. M., Pecht, M. and Dasgupta, A., "A Probabilistic Approach for Predicting Thermal Fatigue Life of Wire Bonding in Microelectronics," *Journal of Electronic Packaging*, Vol. 113, pp. 275-285, 1991.

5. Hu, J. M., Barker, D., Dasgupta, A. and Arora, A., "Role of Failure-Mechanism Identification in Accelerated Testing," *Journal of the IES*, pp. 39-45, July/August 1993.

6. Ingalls, M. W., "A New Reliability Test for Multilayer Ceramic Capacitors," *IEEE Transactions on Components, Packaging, and Manufacturing Technology*, Vol. 17, No. 3, pp. 344-347, September 1994.

7. Koh, J. S., Kim, C. H., Park, J. I. and Yum, B. J., "Determination of Effective Burn-in Time for Printed Board Assembly," *Microelectronics and Reliability*, Vol. 35, No. 6, pp. 893-902, 1995.

8. Kooi, C. F., "Physical Analysis of Stress Testing for Failure of Electronic Components," *IEEE Transactions on Reliability*, Vol. R-17, No.2, 1968.

9. Norris, R. H., "Run-in or Burn-in of Electronic Parts – A Comprehensive, Quantitative Basis for Choice of Temperature, Stresses and Duration," *Proceedings Ninth National Symposium on Reliability and Quality Control*, pp. 335-357, January 1963.

10. Pecht, M. and Lall, P., "A Physics of Failure Approach to IC Burn-in," *Advances in Electronic Packaging, ASME*, pp. 917-923, 1992.

11. Peck, D. S. and Zierdt C. H., "The Reliability of Semiconductor Devices in the Bell System," *Proceedings of the IEEE*, Vol. 62, No. 2, pp. 185-211, February 1974.

12. Stitch, M., Johnson, G. M., Kirk, B. P. and Brauer, J. B., "Microcircuit Accelerated Testing Using High Temperature Operating Tests," *IEEE Transactions on Reliability*, Vol. R-24, No. 4, pp. 238-250, 1975.

13. Suyko, A. and Sy, S. "Development of a Burn-in Time Reduction Algorithm Using the Principles of Accelerated Factors," *Proceedings of the International Reliability Physics Symposium*, pp. 264-270, 1991.

14. Trindade, D. C., "Can Burn-in Screen Wearout Mechanisms?: Reliability Modeling of Defective Subpopulations - A Case Study," *Proceedings of the International Reliability Physics Symposium*, pp. 260-263, 1991.

Chapter 13

ACCELERATED BURN-IN USING TEMPERATURE CYCLING

13.1 WHY TEMPERATURE CYCLING?

Temperature is generally considered to be a key parameter in the design of electronic components and equipment, and cautions concerning temperature and its relationship to reliability are widely documented. While some studies suggest that temperature is the most critical stress that influences microelectronic device failures, the actual failure mechanisms have generally not been quantified in terms of whether a steady-state temperature, temperature range, rate of temperature change, or a spatial temperature gradient induces failures.

Currently, temperature-dependent models such as the Arrhenius and the Eyring models, originally proposed to model the effect of temperature on chemical reaction rates, are widely applied to microelectronic devices. These models predict the effect of temperature on electronic component failure rates under the assumption that the dominant component failure mechanisms depend on steady-state temperature. However, the correlation between steady-state temperature and failure mechanisms needs to be confirmed using failure analysis. In fact, most of the microelectronic device failure mechanisms are not highly steady-state temperature dependent within the equipment operating temperature range [1; 2, pp. 61-152]. Besides, detailed assessment of the role of steady-state temperature, after failure analysis of field failure returns, often does not provide clear evidence that a lowered temperature would have prevented that failure, or that a higher temperature would have caused an earlier failure. Therefore, the use of burn-in without attention to the dominant failure mechanisms, and the nature of their temperature dependencies, such as on steady-state temperature, temperature range, temperature gradient, and temperature change rate, is a misapplication of the reliability concepts which may result in failure avoidance efforts without yielding the anticipated overall results.

Failures in microelectronic devices are often classified into the following three categories:

1. **Electrical failures**, which occur predominantly at the device (i.e., die) level, and include slow trapping, hot electrons, electrical overstress, electrostatic discharge, dielectric failure, oxide breakdown, and electromigration. Some of these electrical failures are temperature dependent.

2. **Mechanical failures**, which are predominately package related and arise from differential thermal expansion between bonded materials, large time-dependent temperature changes, and large spatial temperature gradients, all of which can cause tensile, compressive, bending, fatigue and fracture failures.

3. **Corrosion induced failures**, which include electrolyte formation, and galvanic and ionic corrosion. Corrosion failures are complex functions of contamination, temperature, humidity and bias.

Table 13.1 lists various failure sites, their failure mechanisms and the corresponding nature of temperature dependence. It may be seen from this table that most of the microelectronic device failure mechanisms are not highly steady-state temperature dependent in the equipment operating temperature range, but dependent on the temperature range and the rate of temperature change which can be provided by temperature cycling. Therefore, a properly designed temperature cycling profile will provide a very effective stress environment to precipitate component defects.

In the following section, a general model for computing the age acceleration due to thermal cycling is presented. This model accounts for Arrhenius-type reaction effects during heating and cooling, as well as during the elevated temperature soak, and includes a further term to reflect ramp rate effects. The model for ramp rate is general and permits any monotonic heating or cooling profile. In addition, the model allows for non-constant activation energies.

TABLE 13.1– Temperature dependence of integrated circuit failure sites assuming a normal operating temperature range of $-55°C$ to $-125°C$ [1; 2, pp. 126-133].

Failure site	Failure mechanism	Dominant temperature dependence†	Nature of temperature dependence under normal operating temperature
Device	Ionic contamination	$(T^*)^{-1}$	Steady-state temperature dependent above 200°C. Failed devices, if baked at a temperature in the neighborhood of 250°C, are found to recover their characteristics.
	Forward second breakdown	T^*	Independent of steady-state temperature below 160°C.
	Reverse second breakdown	$(T^*)^{-1}$	Insignificant dependence on steady-state temperature. The breakdown voltage increases from 650 V to 680 V when the temperature increases from 25°C to 150°C.
	Surface charge spreading	T^*	Steady-state temperature dependent above 150°C.
Device oxide	Slow trapping	T^*	Steady-state temperature dependent above 175°C.
	Electrostatic discharge	T^*	ESD voltage (i.e., resistance to ESD) reduces with temperature increase (from 25°C to 125°C). Not a dominant mechanism in properly protected devices.
	Time dependent dielectric breakdown (TDDB)	$(T^*)^{-1}$	Steady-state temperature dependence is very weak; TDDB is a dominant function of voltage.
Device substrate/ oxide interface	Hot electrons	$(T^*)^{-1}$	Steady-state temperature dependence decreases above $-55°C$. Temperature independent in the range of 20°C to 100°C.
Die	Fracture	$\Delta T^*, \nabla T^*$	Primarily dependent on temperature cycling.
	Electrical over-stress	T^*	Independent of steady-state temperature below 160°C (i.e., the temperature at which the thermal coefficient of resistance changes sign).
Die adhesive	Fatigue	ΔT^*	Independent of steady-state temperature.

† T^* = steady-state temperature; ΔT^* = temperature range;
∇T^* = spatial temperature gradient; dT^*/dt = temperature change rate.

TABLE 13.1– Continued.

Failure site	Failure mechanism	Dominant temperature dependence†	Nature of temperature dependence under normal operating temperature
Die metalization	Corrosion	dT^*/dt	Only occurs above dew point temperature. Mildly steady-state temperature dependent under normal operation.
	Electromigration	∇T^*	Steady-state temperature dependent above 150°C. Dominant temperature gradient dependence.
	Hillock formation	T^*	Hillocks in die metalization can form as a result of: electromigration, or due to extended periods under temperature cycling conditions (thermal aging). Extended periods in the neighborhood of 400°C produce hillocks.
	Metalization migration	T^*	Independent of steady-state temperature below 500°C.
	Contact spiking	T^*	Independent of steady-state temperature below 400°C.
	Constraint cavitation	T^*	Steady-state temperature dependent above 25°C.
Encapsulant	Reversion	T^*	Independent of steady-state temperature below 300°C (glass transition for typical epoxy molding resin for plastic packages).
	Cracking	ΔT^*	Independent of steady-state temperature below the glass transition temperature of the encapsulant. It is a ΔT^*, ∇T^* driven mechanism.
Package	Stress corrosion	dT^*/dt	Mildly steady-state temperature dependent.
Wire	Flexure fatigue	ΔT^*	Independent of steady-state temperature.
	Shear fatigue	ΔT^*	Independent of steady-state temperature.
Wire bond	Kirkendall voiding	T^*	Independent of steady-state temperature below 150°C; Independent of steady-state temperature above lower temperatures ($T^* < 150°C$) in the presence of halogenated compounds.

† T^* = steady-state temperature; ΔT^* = temperature range; ∇T^* = spatial temperature gradient; dT^*/dt = temperature change rate.

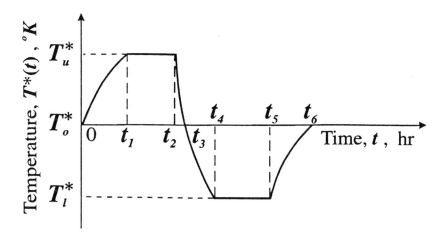

Fig. 13.1– A typical temperature cycle in burn-in.

13.2 TEMPERATURE PROFILE AND THE MODEL FOR THE ACCELERATION FACTOR

Consider a typical thermal cycle, as shown in Fig. 13.1. On the time scale, heating occurs in the intervals $(0, t_1)$ and (t_5, t_6) and cooling occurs during (t_2, t_4). The interval (t_1, t_2) is the elevated temperature dwell (or soak) time. The interval (t_4, t_5) is the lower temperature dwell time. T_o^* represents the nominal operating temperature, and T_u^* and T_l^* are the upper and the lower temperature extremes, respectively. If the ramps displayed in Fig. 13.1 are linear, then the heating and cooling processes have uniform rates as shown in Fig. 13.2. In reality, temperature ramps are rarely uniform.

In general, ramps are monotonic but may have various forms. To represent a general temperature profile over time, let a typical temperature cycle be represented by the following:

$$T^*(t) = \begin{cases} T_o^* + (T_u^* - T_o^*)(t/t_1)^{\beta_h}, & 0 \leq t < t_1, \\ T_u^*, & t_1 \leq t < t_2, \\ T_u^* - (T_u^* - T_l^*)[(t-t_2)/(t_4-t_2)]^{\beta_c}, & t_2 \leq t < t_4, \\ T_l^*, & t_4 \leq t < t_5, \\ T_l^* + (T_o^* - T_l^*)[(t-t_5)/(t_6-t_5)]^{\beta_h}, & t_5 \leq t \leq t_6, \end{cases} \quad (13.1)$$

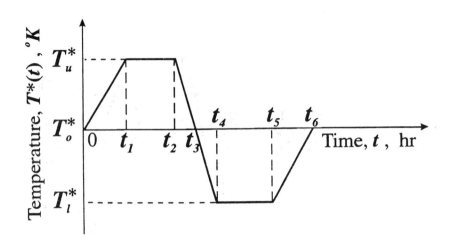

Fig. 13.2– A temperature cycle in burn-in with uniform heating and cooling.

where

T_o^* = nominal operating temperature,
T_u^* = upper temperature extreme,
T_l^* = lower temperature extreme,
β_h = shape parameter of the heating profile,

and

β_c = shape parameter of the cooling profile.

Equation (13.1) represents many monotonic heating and cooling profiles. Different choices of the shape parameters, β_h and β_c, yield different forms for the heating and cooling ramps. It is suggested that the shape provided by $\beta_h = 2$ with various choices of β_c occur frequently [3]. Note that $\beta_h = \beta_c = 1$ yields uniform heating and cooling.

It is plausible that different rates of expansion and contraction in the constituent materials of electronic devices result in mechanical degradation of the devices during heating and cooling. This may be responsible for the frequently observed failures during temperature changes. To represent this behavior, heating and cooling rates are viewed as additional stresses. They are treated as being independent of the reaction rate stress, as suggested by Nachlas, Binney and Gruber [4]. Therefore, the aging acceleration of thermal cycling is the combined effect of two independent stresses; i.e., reaction rate stress governed by the Arrhenius Law, and the temperature change rate stress during heating and cooling. *The equivalent acceleration factor in one typical temperature cycle is the average value of the products of the acceleration factors from these two independent stresses*; i.e.,

$$A_F^* = \frac{1}{t_6} \int_0^{t_6} A_{F1}[T^*(t)] \cdot A_{F2}\left[\frac{dT^*(t)}{dt}\right] dt, \tag{13.2}$$

where

A_F^* = equivalent acceleration factor in one typical temperature cycle,

$A_{F1}[T^*(t)]$ = acceleration factor due to reaction rate stress, which is a function of $T^*(t)$,

and

$A_{F2}\left[\dfrac{dT^*(t)}{dt}\right]$ = acceleration factor due to temperature change rate stress, which is a function of $\dfrac{dT^*(t)}{dt}$.

The acceleration factor due to reaction rate stress, $A_{F1}[T^*(t)]$, can be obtained readily by replacing T_a^* by $T^*(t)$ in Eq. (12.17), the first modified Arrhenius Model; i.e.,

$$A_{F1}[T^*(t)] = e^{\frac{1}{k}\left[\frac{E_A(T_o^*)}{T_o^*} - \frac{E_A[T^*(t)]}{T^*(t)}\right]}, \tag{13.3}$$

or in Eq. (12.41), the second modified Arrhenius Model; i.e.,

$$A_{F1}[T^*(t)] = e^{\frac{\{E_P[T^*(t)]\}^2}{E_R K}\left(\frac{1}{T_o^*} - \frac{1}{T^*(t)}\right)}. \tag{13.4}$$

The acceleration factor due to temperature change rate stress, A_{F2}, can be evaluated from [4]

$$A_{F2}\left[\frac{dT^*(t)}{dt}\right] = e^{\eta\left|\frac{dT^*(t)}{dt}\right|}, \tag{13.5}$$

where

η = a constant.

Substituting Eqs. (13.3) and (13.5) into Eq. (13.2) yields the equivalent acceleration factor using Eqs. (13.3), the *first modified form for the acceleration factor* due to reaction rate stress only; i.e.,

$$A_F^* = \frac{1}{t_6} \int_0^{t_6} e^{\frac{1}{K}\left\{\frac{E_A(T_o^*)}{T_o^*} - \frac{E_A[T^*(t)]}{T^*(t)}\right\}} e^{\eta \left|\frac{dT^*(t)}{dt}\right|} dt,$$

or

$$A_F^* = \frac{1}{t_6} e^{\frac{E_A(T_o^*)}{KT_o^*}} \int_0^{t_6} e^{-\left\{\frac{E_A[T^*(t)]}{KT^*(t)} - \eta \left|\frac{dT^*(t)}{dt}\right|\right\}} dt. \qquad (13.6)$$

Substituting Eqs. (13.4) and (13.5) into Eq. (13.2) yields the equivalent acceleration factor using Eqs. (13.4), the *second modified form for the acceleration factor* due to reaction rate stress only; i.e.,

$$A_F^* = \frac{1}{t_6} \int_0^{t_6} e^{\frac{\{E_P[T^*(t)]\}^2}{E_R K}\left[\frac{1}{T_o^*} - \frac{1}{T^*(t)}\right]} e^{\eta \left|\frac{dT^*(t)}{dt}\right|} dt,$$

or

$$A_F^* = \frac{1}{t_6} \int_0^{t_6} e^{\frac{\{E_P[T^*(t)]\}^2}{E_R K}\left[\frac{1}{T_o^*} - \frac{1}{T^*(t)}\right] + \eta \left|\frac{dT^*(t)}{dt}\right|} dt. \qquad (13.7)$$

Note that since $dT^*(t)/dt = 0$ in the intervals (t_1, t_2) and (t_4, t_5), the integrals implied by Eqs. (13.6) and (13.7) in these two intervals correspond to the modified Arrhenius models given by Eqs. (12.17) and (12.41) evaluated at $T_a^* = T_u^*$ and $T_a^* = T_l^*$, respectively.

13.3 EQUIVALENT ACCELERATION FACTOR EVALUATION USING EQ. (13.6)

If Eq. (13.6) is used and the activation energy is assumed to be in the form of Eq. (12.19), or

$$E_A(T^*) = a(T^*)^b; \qquad (13.8)$$

then, Eq. (13.6) becomes

$$A_F^* = \frac{1}{t_6} e^{\frac{E_A(T_o^*)}{KT_o^*}} \int_0^{t_6} e^{-\left\{\frac{E_A[T^*(t)]}{KT^*(t)} - \eta \left|\frac{dT^*(t)}{dt}\right|\right\}} dt,$$

or

$$A_F^* = \frac{1}{t_6} e^{\frac{a(T_o^*)^{b-1}}{K}} \int_0^{t_6} e^{-\left\{\frac{a[T^*(t)]^{b-1}}{K} - \eta \left|\frac{dT^*(t)}{dt}\right|\right\}} dt. \quad (13.9)$$

Now, using Eqs. (13.1) yields

$$\left|\frac{dT^*(t)}{dt}\right| = \begin{cases} \beta_h(T_u^* - T_o^*)(t/t_1)^{\beta_h-1}\left(\frac{1}{t_1}\right), & 0 \le t < t_1, \\ 0, & t_1 \le t < t_2, \\ \beta_c(T_u^* - T_l^*)[(t-t_2)/(t_4-t_2)]^{\beta_c-1}\left(\frac{1}{t_4-t_2}\right), & t_2 \le t < t_4, \\ 0, & t_4 \le t < t_5, \\ \beta_h(T_o^* - T_l^*)[(t-t_5)/(t_6-t_5)]^{\beta_h-1}\left(\frac{1}{t_6-t_5}\right), & t_5 \le t \le t_6. \end{cases}$$

(13.10)

Therefore, substituting Eqs. (13.1) and (13.10) into Eq. (13.9) yields

$$A_F^* = \frac{1}{t_6} e^{\frac{a(T_o^*)^{b-1}}{K}} \Bigg\{ \int_0^{t_1} exp\left\{-\left\{\frac{a}{K}\left[T_o^* + (T_u^* - T_o^*)\left(\frac{t}{t_1}\right)^{\beta_h}\right]^{b-1}\right.\right.$$

$$\left.\left. - \frac{\eta \beta_h (T_u^* - T_o^*)}{t_1}\left(\frac{t}{t_1}\right)^{\beta_h-1}\right\}\right\} dt$$

$$+ \int_{t_1}^{t_2} e^{-\frac{a}{K}(T_u^*)^{b-1}} dt$$

$$+ \int_{t_2}^{t_4} exp\left\{-\left\{\frac{a}{K}\left[T_u^* - (T_u^* - T_l^*)\left(\frac{t-t_2}{t_4-t_2}\right)^{\beta_c}\right]^{b-1}\right.\right.$$

$$\left.\left. - \frac{\eta \beta_c (T_u^* - T_l^*)}{t_4 - t_2}\left(\frac{t-t_2}{t_4-t_2}\right)^{\beta_c-1}\right\}\right\} dt$$

$$+ \int_{t_4}^{t_5} e^{-\frac{a}{K}(T_l^*)^{b-1}} dt$$

$$+ \int_{t_5}^{t_6} exp\left\{-\left\{\frac{a}{K}\left[T_l^* + (T_o^* - T_l^*)\left(\frac{t-t_5}{t_6-t_5}\right)^{\beta_h}\right]^{b-1}\right.\right.$$

$$\left.\left. - \frac{\eta \beta_h (T_o^* - T_l^*)}{t_6 - t_5}\left(\frac{t-t_5}{t_6-t_5}\right)^{\beta_h-1}\right\}\right\} dt \Bigg\},$$

or

$$
\begin{aligned}
A_F^* = \frac{1}{t_6} e^{\frac{a(T_o^*)^{b-1}}{K}} &\left\{ \int_0^{t_1} exp\left\{ -\left\{ \frac{a}{K}\left[T_o^* + (T_u^* - T_o^*)\left(\frac{t}{t_1}\right)^{\beta_h}\right]^{b-1} \right.\right.\right. \\
&\left.\left. -\frac{\eta\beta_h(T_u^* - T_o^*)}{t_1}\left(\frac{t}{t_1}\right)^{\beta_h-1} \right\} dt \\
&+ (t_2 - t_1)e^{-\frac{a}{K}(T_u^*)^{b-1}} \\
&+ \int_{t_2}^{t_4} exp\left\{ -\left\{ \frac{a}{K}\left[T_u^* - (T_u^* - T_l^*)\left(\frac{t-t_2}{t_4-t_2}\right)^{\beta_c}\right]^{b-1} \right.\right. \\
&\left.\left. -\frac{\eta\beta_c(T_u^* - T_l^*)}{t_4 - t_2}\left(\frac{t-t_2}{t_4-t_2}\right)^{\beta_c-1} \right\} dt \\
&+ (t_5 - t_4)e^{-\frac{a}{K}(T_l^*)^{b-1}} \\
&+ \int_{t_5}^{t_6} exp\left\{ -\left\{ \frac{a}{K}\left[T_l^* + (T_o^* - T_l^*)\left(\frac{t-t_5}{t_6-t_5}\right)^{\beta_h}\right]^{b-1} \right.\right. \\
&\left.\left.\left. -\frac{\eta\beta_h(T_o^* - T_l^*)}{t_6-t_5}\left(\frac{t-t_5}{t_6-t_5}\right)^{\beta_h-1} \right\} dt \right\}. \quad (13.11)
\end{aligned}
$$

If $\beta_h = \beta_c = 1$, or the heating and cooling processes are uniform, then Eq. (13.11) becomes

$$
\begin{aligned}
A_F^* = \frac{1}{t_6} e^{\frac{a(T_o^*)^{b-1}}{K}} &\left\{ \int_0^{t_1} e^{-\left\{\frac{a}{K}\left[T_o^* + (T_u^* - T_o^*)\left(\frac{t}{t_1}\right)\right]^{b-1} - \frac{\eta(T_u^* - T_o^*)}{t_1}\right\}} dt \right. \\
&+ (t_2 - t_1)e^{-\frac{a}{K}(T_u^*)^{b-1}} \\
&+ \int_{t_2}^{t_4} e^{-\left\{\frac{a}{K}\left[T_u^* - (T_u^* - T_l^*)\left(\frac{t-t_2}{t_4-t_2}\right)\right]^{b-1} - \frac{\eta(T_u^* - T_l^*)}{t_4-t_2}\right\}} dt \\
&+ (t_5 - t_4)e^{-\frac{a}{K}(T_l^*)^{b-1}} \\
&\left. + \int_{t_5}^{t_6} e^{-\left\{\frac{a}{K}\left[T_l^* + (T_o^* - T_l^*)\left(\frac{t-t_5}{t_6-t_5}\right)\right]^{b-1} - \frac{\eta(T_o^* - T_l^*)}{t_6-t_5}\right\}} dt \right\},
\end{aligned}
$$

or

$$A_F^* = \frac{1}{t_6} e^{\frac{a(T_o^*)^{b-1}}{K}} \Biggl\{ e^{\frac{\eta(T_u^*-T_o^*)}{t_1}} \int_0^{t_1} e^{-\frac{a}{K}\left[T_o^* + \left(\frac{T_u^*-T_o^*}{t_1}\right)t\right]^{b-1}} dt$$

$$+ (t_2 - t_1) e^{-\frac{a}{K}(T_u^*)^{b-1}}$$

$$+ e^{\frac{\eta(T_u^*-T_l^*)}{t_4-t_2}} \int_{t_2}^{t_4} e^{-\frac{a}{K}\left[T_u^* - \left(\frac{T_u^*-T_l^*}{t_4-t_2}\right)(t-t_2)\right]^{b-1}} dt$$

$$+ (t_5 - t_4) e^{-\frac{a}{K}(T_l^*)^{b-1}}$$

$$+ e^{\frac{\eta(T_o^*-T_l^*)}{t_6-t_5}} \int_{t_5}^{t_6} e^{-\frac{a}{K}\left[T_l^* + (T_o^*-T_l^*)\left(\frac{t-t_5}{t_6-t_5}\right)\right]^{b-1}} dt \Biggr\}. \quad (13.12)$$

Admittedly, both Eqs. (13.11) and (13.12) are two complicated expressions. In fact, there is no closed form solution for the general model. However, Eqs. (13.11) and (13.12) are well behaved and can be integrated numerically with little effort.

As a special case, if $b = 1$ in Eq. (13.12), or the activation energy is a linear function of the applied temperature, then Eq. (13.8) becomes

$$E_A(T^*) = a\, T^*,$$

and Eq. (13.3) becomes

$$A_{F1}[T^*(t)] = e^{\frac{a}{K}\left[\frac{T_o^*}{T_o^*} - \frac{T^*(t)}{T^*(t)}\right]} \equiv 1,$$

or the reaction rate stress has no effect on the aging acceleration no matter how high the applied temperature is. In this case, a closed form solution for Eq. (13.12) can be obtained as follows:

$$A_F^* = \frac{1}{t_6} e^{\frac{a}{K}} \Biggl\{ e^{\frac{\eta(T_u^*-T_o^*)}{t_1}} \int_0^{t_1} e^{-\frac{a}{K}} dt$$

$$+ (t_2 - t_1) e^{-\frac{a}{K}}$$

$$+ e^{\frac{\eta(T_u^*-T_l^*)}{t_4-t_2}} \int_{t_2}^{t_4} e^{-\frac{a}{K}} dt$$

$$+ (t_5 - t_4) e^{-\frac{a}{K}}$$

$$+ e^{\frac{\eta(T_o^*-T_l^*)}{t_6-t_5}} \int_{t_5}^{t_6} e^{-\frac{a}{K}} dt \Biggr\},$$

$$= \frac{1}{t_6} e^{\frac{a}{K}} \left\{ e^{\frac{\eta(T_u^* - T_o^*)}{t_1}} \left[t_1 e^{-\frac{a}{K}} \right] \right.$$
$$+ (t_2 - t_1) e^{-\frac{a}{K}}$$
$$+ e^{\frac{\eta(T_u^* - T_l^*)}{t_4 - t_2}} \left[(t_4 - t_2) e^{-\frac{a}{K}} \right]$$
$$+ (t_5 - t_4) e^{-\frac{a}{K}}$$
$$\left. + e^{\frac{\eta(T_o^* - T_l^*)}{t_6 - t_5}} \left[(t_6 - t_5) e^{-\frac{a}{K}} \right] \right\},$$

or

$$A_F^* = \frac{1}{t_6} \left[t_1 e^{\frac{\eta(T_u^* - T_o^*)}{t_1}} + (t_2 - t_1) \right.$$
$$+ (t_4 - t_2) e^{\frac{\eta(T_u^* - T_l^*)}{t_4 - t_2}} + (t_5 - t_4)$$
$$\left. + (t_6 - t_5) e^{\frac{\eta(T_o^* - T_l^*)}{t_6 - t_5}} \right]. \tag{13.13}$$

In summary, Eqs. (13.11), (13.12) and (13.13) are equivalent acceleration factor equations using the *first modified form for the reaction rate stress acceleration factor,* or Eq. (13.6). Equation (13.11) is the general model. Equation (13.12) is a special case of Eq. (13.11) when $\beta_h = \beta_c = 1$, or the heating and cooling processes are uniform. Equation (13.13) is another special case of Eq. (13.11) when $\beta_h = \beta_c = 1$ and $b = 1$, or the heating and cooling processes are uniform and the activation energy is a linear function of the applied temperature.

13.4 EQUIVALENT ACCELERATION FACTOR EVALUATION USING EQ. (13.7)

If Eq. (13.7) is used and the provided activation energy is assumed to be in the form of Eq. (12.42), or

$$E_P(T^*) = a' (T^*)^{b'}, \tag{13.14}$$

then the acceleration factor term, due to reaction rate stress only, in Eq. (13.7), should be substituted by Eq. (12.43) or

$$A_{F1}[T^*(t)] = e^{C \, [T^*(t)]^B \left[\frac{1}{T_o^*} - \frac{1}{T^*(t)} \right]},$$

and Eq. (13.7) becomes

$$A_F^* = \frac{1}{t_6} \int_0^{t_6} e^{\frac{\{E_P[T^*(t)]\}^2}{E_R K} \left[\frac{1}{T_o^*} - \frac{1}{T^*(t)}\right] + \eta \left|\frac{dT^*(t)}{dt}\right|} dt,$$

$$= \frac{1}{t_6} \int_0^{t_6} e^{C[T^*(t)]^B \left[\frac{1}{T_o^*} - \frac{1}{T^*(t)}\right] + \eta \left|\frac{dT^*(t)}{dt}\right|} dt,$$

$$= \frac{1}{t_6} \left\{ \int_0^{t_1} exp \left\{ C \left\{ \left[T_o^* + (T_u^* - T_o^*)\left(\frac{t}{t_1}\right)^{\beta_h}\right]^B \left(\frac{1}{T_o^*}\right) \right. \right. \right.$$

$$\left. - \left[T_o^* + (T_u^* - T_o^*)\left(\frac{t}{t_1}\right)^{\beta_h}\right]^{B-1} \right\}$$

$$+ \frac{\eta \beta_h (T_u^* - T_o^*)}{t_1} \left(\frac{t}{t_1}\right)^{\beta_h - 1} \right\} dt$$

$$+ \int_{t_1}^{t_2} e^{C(T_u^*)^B \left(\frac{1}{T_o^*} - \frac{1}{T_u^*}\right)} dt$$

$$+ \int_{t_2}^{t_4} exp \left\{ C \left\{ \left[T_u^* - (T_u^* - T_l^*)\left(\frac{t - t_2}{t_4 - t_2}\right)^{\beta_c}\right]^B \left(\frac{1}{T_o^*}\right) \right. \right.$$

$$\left. - \left[T_u^* - (T_u^* - T_l^*)\left(\frac{t - t_2}{t_4 - t_2}\right)^{\beta_c}\right]^{B-1} \right\}$$

$$+ \frac{\eta \beta_c (T_u^* - T_l^*)}{t_4 - t_2} \left(\frac{t - t_2}{t_4 - t_2}\right)^{\beta_c - 1} \right\} dt$$

$$+ \int_{t_4}^{t_5} e^{C(T_l^*)^B \left(\frac{1}{T_o^*} - \frac{1}{T_l^*}\right)} dt$$

$$+ \int_{t_5}^{t_6} exp \left\{ C \left\{ \left[T_l^* + (T_o^* - T_l^*)\left(\frac{t - t_5}{t_6 - t_5}\right)^{\beta_h}\right]^B \left(\frac{1}{T_o^*}\right) \right. \right.$$

$$\left. - \left[T_l^* + (T_o^* - T_l^*)\left(\frac{t - t_5}{t_6 - t_5}\right)^{\beta_h}\right]^{B-1} \right\}$$

$$+ \frac{\eta \beta_h (T_o^* - T_l^*)}{t_6 - t_5} \left(\frac{t - t_5}{t_6 - t_5}\right)^{\beta_h - 1} \right\} dt \right\},$$

or

$$A_F^* = \frac{1}{t_6} \left\{ \int_0^{t_1} exp \left\{ C \left\{ \left[T_o^* + (T_u^* - T_o^*) \left(\frac{t}{t_1}\right)^{\beta_h} \right]^B \left(\frac{1}{T_o^*}\right) \right. \right.\right.$$

$$\left. - \left[T_o^* + (T_u^* - T_o^*) \left(\frac{t}{t_1}\right)^{\beta_h} \right]^{B-1} \right\}$$

$$\left. + \frac{\eta \beta_h (T_u^* - T_o^*)}{t_1} \left(\frac{t}{t_1}\right)^{\beta_h - 1} \right\} dt$$

$$+ (t_2 - t_1) e^{C(T_u^*)^B \left(\frac{1}{T_o^*} - \frac{1}{T_u^*}\right)}$$

$$+ \int_{t_2}^{t_4} exp \left\{ C \left\{ \left[T_u^* - (T_u^* - T_l^*) \left(\frac{t - t_2}{t_4 - t_2}\right)^{\beta_c} \right]^B \left(\frac{1}{T_o^*}\right) \right.\right.$$

$$\left. - \left[T_u^* - (T_u^* - T_l^*) \left(\frac{t - t_2}{t_4 - t_2}\right)^{\beta_c} \right]^{B-1} \right\}$$

$$\left. + \frac{\eta \beta_c (T_u^* - T_l^*)}{t_4 - t_2} \left(\frac{t - t_2}{t_4 - t_2}\right)^{\beta_c - 1} \right\} dt$$

$$+ (t_5 - t_4) e^{-C(T_l^*)^B \left(\frac{1}{T_l^*} - \frac{1}{T_o^*}\right)}$$

$$+ \int_{t_5}^{t_6} exp \left\{ C \left\{ \left[T_l^* + (T_o^* - T_l^*) \left(\frac{t - t_5}{t_6 - t_5}\right)^{\beta_h} \right]^B \left(\frac{1}{T_o^*}\right) \right.\right.$$

$$\left. - \left[T_l^* + (T_o^* - T_l^*) \left(\frac{t - t_5}{t_6 - t_5}\right)^{\beta_h} \right]^{B-1} \right\}$$

$$\left.\left. + \frac{\eta \beta_h (T_o^* - T_l^*)}{t_6 - t_5} \left(\frac{t - t_5}{t_6 - t_5}\right)^{\beta_h - 1} \right\} dt \right\}.$$

(13.15)

If $\beta_h = \beta_c = 1$, or the heating and cooling processes are uniform, then Eq. (13.15) becomes

$$A_F^* = \frac{1}{t_6} \left\{ \int_0^{t_1} exp \left\{ C \left\{ \left[T_o^* + (T_u^* - T_o^*) \left(\frac{t}{t_1}\right) \right]^B \left(\frac{1}{T_o^*}\right) \right.\right.\right.$$

$$-\left[T_o^* + (T_u^* - T_o^*)\left(\frac{t}{t_1}\right)\right]^{B-1}\bigg\} + \frac{\eta(T_u^* - T_o^*)}{t_1}\bigg\}dt$$

$$+(t_2 - t_1)e^{C(T_u^*)^B\left(\frac{1}{T_o^*} - \frac{1}{T_u^*}\right)}$$

$$+\int_{t_2}^{t_4} exp\left\{C\left\{\left[T_u^* - (T_u^* - T_l^*)\left(\frac{t-t_2}{t_4-t_2}\right)\right]^B \left(\frac{1}{T_o^*}\right)\right.\right.$$

$$-\left[T_u^* - (T_u^* - T_l^*)\left(\frac{t-t_2}{t_4-t_2}\right)\right]^{B-1}\bigg\} + \frac{\eta(T_u^* - T_l^*)}{t_4 - t_2}\bigg\}dt$$

$$+(t_5 - t_4)e^{-C(T_l^*)^B\left(\frac{1}{T_l^*} - \frac{1}{T_o^*}\right)}$$

$$+\int_{t_5}^{t_6} exp\left\{C\left\{\left[T_l^* + (T_o^* - T_l^*)\left(\frac{t-t_5}{t_6-t_5}\right)\right]^B \left(\frac{1}{T_o^*}\right)\right.\right.$$

$$-\left[T_l^* + (T_o^* - T_l^*)\left(\frac{t-t_5}{t_6-t_5}\right)\right]^{B-1}\bigg\} + \frac{\eta(T_o^* - T_l^*)}{t_6 - t_5}\bigg\}dt\bigg\},$$

or

$$A_F^* = \frac{1}{t_6}\left\{e^{\frac{\eta(T_u^* - T_o^*)}{t_1}}\int_0^{t_1} exp\left\{C\left\{\left[T_o^* + (T_u^* - T_o^*)\left(\frac{t}{t_1}\right)\right]^B \left(\frac{1}{T_o^*}\right)\right.\right.\right.$$

$$-\left[T_o^* + (T_u^* - T_o^*)\left(\frac{t}{t_1}\right)\right]^{B-1}\bigg\}\bigg\}dt$$

$$+(t_2 - t_1)e^{C(T_u^*)^B\left(\frac{1}{T_o^*} - \frac{1}{T_u^*}\right)}$$

$$+e^{\frac{\eta(T_u^* - T_l^*)}{t_4 - t_2}}\int_{t_2}^{t_4} exp\left\{C\left\{\left[T_u^* - (T_u^* - T_l^*)\left(\frac{t-t_2}{t_4-t_2}\right)\right]^B \left(\frac{1}{T_o^*}\right)\right.\right.$$

$$-\left[T_u^* - (T_u^* - T_l^*)\left(\frac{t-t_2}{t_4-t_2}\right)\right]^{B-1}\bigg\}\bigg\}dt$$

$$+(t_5 - t_4)e^{-C(T_l^*)^B\left(\frac{1}{T_l^*} - \frac{1}{T_o^*}\right)}$$

$$+e^{\frac{\eta(T_o^* - T_l^*)}{t_6 - t_5}}\int_{t_5}^{t_6} exp\left\{C\left\{\left[T_l^* + (T_o^* - T_l^*)\left(\frac{t-t_5}{t_6-t_5}\right)\right]^B \left(\frac{1}{T_o^*}\right)\right.\right.$$

$$-\left[T_l^* + (T_o^* - T_l^*)\left(\frac{t-t_5}{t_6-t_5}\right)\right]^{B-1}\bigg\}\bigg\}dt\bigg\}.$$

(13.16)

Admittedly, both Eqs. (13.15) and (13.16) are also two complicated expressions. In fact, there is no closed form solution for the general model. However, Eqs. (13.15) and (13.16) are also well behaved and can be integrated numerically with little effort.

As a special case, if $b = 0.5$ or $B = 2b = 1$ in Eq. (13.16), then

$$A_{F1}[T^*(t)] = e^{C\,[T^*(t)]^1\left[\frac{1}{T_o^*} - \frac{1}{T^*(t)}\right]},$$

or

$$A_{F1}[T^*(t)] = e^{C\left[\frac{T^*(t)}{T_o^*} - 1\right]}.$$

In this case, a closed form solution can be obtained as follows:

$$A_F^* = \frac{1}{t_6}\left\{ e^{\frac{\eta(T_u^* - T_o^*)}{t_1}} \int_0^{t_1} e^{C\left\{\left[T_o^* + (T_u^* - T_o^*)\left(\frac{t}{t_1}\right)\right]\left(\frac{1}{T_o^*}\right) - 1\right\}} dt \right.$$

$$+ (t_2 - t_1)e^{CT_u^*\left(\frac{1}{T_o^*} - \frac{1}{T_u^*}\right)}$$

$$+ e^{\frac{\eta(T_u^* - T_l^*)}{t_4 - t_2}} \int_{t_2}^{t_4} e^{C\left\{\left[T_u^* - (T_u^* - T_l^*)\left(\frac{t - t_2}{t_4 - t_2}\right)\right]\left(\frac{1}{T_o^*}\right) - 1\right\}} dt$$

$$+ (t_5 - t_4)e^{-CT_l^*\left(\frac{1}{T_l^*} - \frac{1}{T_o^*}\right)}$$

$$\left. + e^{\frac{\eta(T_o^* - T_l^*)}{t_6 - t_5}} \int_{t_5}^{t_6} e^{C\left\{\left[T_l^* + (T_o^* - T_l^*)\left(\frac{t - t_5}{t_6 - t_5}\right)\right]\left(\frac{1}{T_o^*}\right) - 1\right\}} dt \right\},$$

$$= \frac{1}{t_6}\left\{ e^{\left[\frac{\eta(T_u^* - T_o^*)}{t_1} - C\right]} \int_0^{t_1} e^{C\left[1 + \frac{T_u^* - T_o^*}{T_o^* t_1} t\right]} dt \right.$$

$$+ (t_2 - t_1)e^{CT_u^*\left(\frac{1}{T_o^*} - \frac{1}{T_u^*}\right)}$$

$$+ e^{\left[\frac{\eta(T_u^* - T_l^*)}{t_4 - t_2} - C\right]} \int_{t_2}^{t_4} e^{C\left[\frac{T_u^*}{T_o^*} - \frac{(T_u^* - T_l^*)}{T_o^*(t_4 - t_2)}(t - t_2)\right]} dt$$

$$+ (t_5 - t_4)e^{-CT_l^*\left(\frac{1}{T_l^*} - \frac{1}{T_o^*}\right)}$$

$$\left. + e^{\left[\frac{\eta(T_o^* - T_l^*)}{t_6 - t_5} - C\right]} \int_{t_5}^{t_6} e^{C\left[\frac{T_l^*}{T_o^*} + \frac{(T_o^* - T_l^*)}{T_o^*(t_6 - t_5)}(t - t_5)\right]} dt \right\},$$

$$= \frac{1}{t_6} \left\{ e^{\frac{\eta(T_u^* - T_o^*)}{t_1}} \int_0^{t_1} e^{\frac{C(T_u^* - T_o^*)}{T_o^* t_1} t} dt \right.$$

$$+ (t_2 - t_1) e^{CT_u^* \left(\frac{1}{T_o^*} - \frac{1}{T_u^*}\right)}$$

$$+ e^{\left[\frac{\eta(T_u^* - T_l^*)}{t_4 - t_2} - C + \frac{CT_u^*}{T_o^*}\right]} \int_{t_2}^{t_4} e^{-\frac{C(T_u^* - T_l^*)}{T_o^*(t_4 - t_2)}(t - t_2)} dt$$

$$+ (t_5 - t_4) e^{-CT_l^* \left(\frac{1}{T_l^*} - \frac{1}{T_o^*}\right)}$$

$$+ e^{\left[\frac{\eta(T_o^* - T_l^*)}{t_6 - t_5} - C + \frac{CT_l^*}{T_o^*}\right]} \int_{t_5}^{t_6} e^{\frac{C(T_o^* - T_l^*)}{T_o^*(t_6 - t_5)}(t - t_5)} dt \right\},$$

or

$$A_F^* = \frac{1}{t_6} \left\{ \frac{T_o^* t_1}{C(T_u^* - T_o^*)} e^{\frac{\eta(T_u^* - T_o^*)}{t_1}} \left[e^{\frac{C(T_u^* - T_o^*)}{T_o^*}} - 1 \right] \right.$$

$$+ (t_2 - t_1) e^{CT_u^* \left(\frac{1}{T_o^*} - \frac{1}{T_u^*}\right)}$$

$$+ \frac{T_o^*(t_4 - t_2)}{C(T_u^* - T_l^*)} e^{\left[\frac{\eta(T_u^* - T_l^*)}{t_4 - t_2} - C\right]} \left[e^{\frac{CT_u^*}{T_o^*}} - e^{\frac{CT_l^*}{T_o^*}} \right]$$

$$+ (t_5 - t_4) e^{-CT_l^* \left(\frac{1}{T_l^*} - \frac{1}{T_o^*}\right)}$$

$$+ \frac{T_o^*(t_6 - t_5)}{C(T_o^* - T_l^*)} e^{\left[\frac{\eta(T_o^* - T_l^*)}{t_6 - t_5} - C\right]} \left[e^C - e^{\frac{CT_l^*}{T_o^*}} \right] \right\}.$$

(13.17)

In summary, Eqs. (13.15), (13.16) and (13.17) are equivalent acceleration factor equations using the *second modified form for the reaction rate stress acceleration factor*, or Eq. (13.7). Equation (13.15) is the general model. Equation (13.16) is a special case of Eq. (13.11) when $\beta_h = \beta_c = 1$, or when the heating and cooling processes are uniform. Equation (13.17) is another special case of Eq. (13.15) when $\beta_h = \beta_c = 1$ and $b = 0.5$.

13.5 APPLICATION OF THE AGING ACCELERATION MODELS

The general model for age acceleration under thermal cycling has several possible uses. The first of these is to compute the acceleration attained in each cycle for any specific device. Given a particular component and the corresponding values of the model parameters, the acceleration may be computed. In addition, the temperature dependence of the activation energy, and/or the ramp rate effects may be assumed not to exist for a particular device and the model may be solved for the resulting simplified case to again compute the age acceleration attained in each cycle. The following is an example illustrating this application.

EXAMPLE 13-1

Given are the following parameters of a temperature cycle profile:

$T_o^* = 298°K$, $T_u^* = 338°K$, $T_l^* = 233°K$, $t_1 = \frac{1}{3}$ hr,
$t_2 = \frac{4}{3}$ hr, $t_4 = 2$ hr, $t_5 = \frac{7}{3}$ hr, $t_6 = \frac{8}{3}$ hr,
$\beta_h = 2$, and $\beta_c = 2$.

The constant, η, for the acceleration factor, due to the temperature change stress, is given by

$\eta = 0.01.$

The parameters for the acceleration factor due to the reaction rate stress, have been estimated in Examples 12-3 and 12-4 using the accelerated life test data of the second component; i.e.,

$$\begin{cases} a = -2.0504 \times 10^{-6}, \\ b = 2.2715, \end{cases}$$

for the first modified Arrhenius Model, and

$$\begin{cases} C = 17.8997, \\ B = 1.1507, \end{cases}$$

for the second modified Arrhenius Model. Do the following:

1. Evaluate the equivalent acceleration factor in one temperature cycle using Eq. (13.11).

2. Evaluate the equivalent acceleration factor in one temperature cycle using Eq. (13.15).

APPLICATION OF AGING ACCELERATION MODELS   591

3. Compare the results obtained in Cases 1 and 2.

4. Redo Cases 1 and 2, assuming $b = 0$ (constant activation energy).

5. Redo Cases 1 and 2, assuming $\eta = 0$ (ignore the ramp effect).

SOLUTION TO EXAMPLE 13-1

1. The computer program, given in Appendix 13A, has been developed for the evaluation of the equivalent acceleration factors in one temperature cycle for Eqs. (13.11) and (13.15). Entering the given input information into this computer program yields the equivalent acceleration factor using Eq. (13.11) which is

$$A_{F1}^* = 186.4579.$$

2. Applying the same computer program also yields the equivalent acceleration factor using Eq. (13.15), which is

$$A_{F2}^* = 186.1559.$$

3. By comparing the results in Cases 1 and 2, it may be seen that the equivalent acceleration factors using Eqs. (13.11) and (13.15) are very close.

   Taking the average value of these two acceleration factors yields

$$A_F^* = \frac{A_{F1}^* + A_{F2}^*}{2},$$
$$= \frac{186.4579 + 186.1559}{2},$$

or

$$A_F^* = 186.3069.$$

4. If the activation energy is assumed constant, or

$$b = 0,$$

and

$$B = 0.$$

Running the same computer program with $b = B = 0$, and the other parameters unchanged, yields

$$A^*_{F1} = 4.3454,$$

and

$$A^*_{F2} = 4.3299,$$

which are much lower than those when the activation energy is not constant but dependent on the applied temperature profile.

Taking the average value of these two acceleration factors yields

$$A^*_F = \frac{A^*_{F1} + A^*_{F2}}{2},$$
$$= \frac{4.3454 + 4.3299}{2},$$

or

$$A^*_F = 4.3377.$$

5. If the ramp effect is ignored, or

$$\eta = 0;$$

then, all of the acceleration is contributed solely by the reaction rate stress which is governed by the modified Arrhenius Law with temperature dependent activation energy.

Running the same computer program with $\eta = 0$, with the other parameters unchanged, yields

$$A^*_{F1} = 143.5862,$$

and

$$A^*_{F2} = 143.3663,$$

which are lower than those of Cases 1 and 2 when the ramp effect is also considered.

Taking the average value of these two acceleration factors yields

$$A_F^* = \frac{A_{F1}^* + A_{F2}^*}{2},$$
$$= \frac{143.5862 + 143.3663}{2},$$

or

$$A_F^* = 143.4763.$$

Other uses of the model can be defined in terms of the model parameters. These parameters may be grouped into information sets. For example, given the values of the other parameters, T_u^* and t_6 may be treated as decision variables. The model may then be used to select the elevated temperature and the cycle length required to achieve a specific acceleration. In this manner, the model will support a test design effort.

On the other hand, if the value of η is not known, or if the values of a and b (or B) are in question, the model can be applied to experimental data to provide estimates for these parameters.

Perhaps one of the most valuable uses of the model is in equipment selection. Assuming that the available thermal chambers can be characterized in terms of their achievable temperatures, T_u^*, and their heating and cooling specifications, β_h and β_c, the model may be used to determine which equipment will provide a desired acceleration. In addition, given appropriate cost information, the model can be used to support equipment purchase decisions.

In summary, taking the model parameters in information sets such as $\{T_u^*, t_6\}$, $\{\eta\}$, $\{a, b\}$ and $\{T_u^*, \beta_h, \beta_c\}$ allows the use of the model to analyze questions affected by the values of the parameters in any set. In each case, the same general approach is used and the analysis can be based upon any assumed type of equipment behavior.

13.6 OPTIMUM NUMBER OF THERMAL CYCLES FOR A SPECIFIED FIELD $MTBF$ GOAL

In the preceding sections, the mathematical models of equivalent acceleration factors for a thermal cycling profile using different activation energy forms have been established. The next task is to determine the optimum number of thermal cycles, N_{therm}, for a given temperature profile and for a specified field $MTBF$ goal.

13.6.1 USING THE MIXED-EXPONENTIAL LIFE DISTRIBUTION

Assume that the times to failure of certain electronic components under a given thermal cycling profile follow a bimodal mixed life distribution with the following *pdf* and reliability functions:

$$f(T) = p_b\, f_b(T) + p_g\, f_g(T),$$

and

$$R(T) = p_b\, R_b(T) + p_g\, R_g(T).$$

Let T_b be the burn-in time, then the post-burn-in reliability function for a mission time of t is given by

$$R(T_b, t) = \frac{R(T_b + t)}{R(T_b)} = \frac{p_b\, R_b(T_b + t) + p_g\, R_g(T_b + t)}{p_b\, R_b(T_b) + p_g\, R_g(T_b)}.$$

Then, the post-burn-in *MTBF* under the given thermal cycling profile environment, $MTBF_b$, is given by

$$MTBF_b = \int_0^\infty R(T_b, t)\, dt,$$

$$= \frac{p_b\, [\int_0^\infty R_b(T_b + t)\, dt] + p_g\, [\int_0^\infty R_g(T_b + t)\, dt]}{p_b\, R_b(T_b) + p_g\, R_g(T_b)}.$$

Let $\tau = T_b + t$, then $d\tau = dt$ and

$$MTBF_b = \frac{p_b\, \left[\int_{T_b}^\infty R_b(\tau)\, d\tau\right] + p_g\, \left[\int_{T_b}^\infty R_g(\tau)\, d\tau\right]}{p_b\, R_b(T_b) + p_g\, R_g(T_b)}.$$

Therefore, the equivalent *MTBF* in the field, $MTBF_{field}$, will be given by

$$MTBF_{field} = A_F\, MTBF_b,$$

or

$$MTBF_{field} = A_F\, \left\{ \frac{p_b\, \left[\int_{T_b}^\infty R_b(\tau)\, d\tau\right] + p_g\, \left[\int_{T_b}^\infty R_g(\tau)\, d\tau\right]}{p_b\, R_b(T_b) + p_g\, R_g(T_b)} \right\}. \quad (13.18)$$

If the bimodal mixed-exponential life distribution governs, then

$$R_b(T) = e^{-\lambda_b\, T},$$

OPTIMUM NUMBER OF THERMAL CYCLES 595

and
$$R_g(T) = e^{-\lambda_g T}.$$

Correspondingly,
$$\int_{T_b}^{\infty} R_b(\tau)\, d\tau = \int_{T_b}^{\infty} e^{-\lambda_b \tau}\, d\tau = \frac{e^{-\lambda_b T_b}}{\lambda_b},$$

and
$$\int_{T_b}^{\infty} R_g(\tau)\, d\tau = \int_{T_b}^{\infty} e^{-\lambda_g \tau}\, d\tau = \frac{e^{-\lambda_g T_b}}{\lambda_g}.$$

Consequently,
$$MTBF_{field} = A_F \left[\frac{p_b \left(\frac{e^{-\lambda_b T_b}}{\lambda_b} \right) + p_g \left(\frac{e^{-\lambda_g T_b}}{\lambda_g} \right)}{p_b\, e^{-\lambda_b T_b} + p_g\, e^{-\lambda_g T_b}} \right],$$

or
$$MTBF_{field} = A_F \left[\frac{\left(\frac{p_b\, e^{-\lambda_b T_b}}{\lambda_b} \right) + \left(\frac{p_g\, e^{-\lambda_g T_b}}{\lambda_g} \right)}{p_b\, e^{-\lambda_b T_b} + p_g\, e^{-\lambda_g T_b}} \right].$$

The optimum burn-in time, T_b^*, for a specified field $MTBF$ goal, $MTBF_{field}^*$, can be obtained by solving the following equation for T_b^*:

$$MTBF_{field}^* = A_F \left[\frac{\left(\frac{p_b\, e^{-\lambda_b T_b^*}}{\lambda_b} \right) + \left(\frac{p_g\, e^{-\lambda_g T_b^*}}{\lambda_g} \right)}{p_b\, e^{-\lambda_b T_b^*} + p_g\, e^{-\lambda_g T_b^*}} \right]. \quad (13.19)$$

Multiplying both sides of Eq. (13.19) by $p_b\, e^{-\lambda_b T_b^*} + p_g\, e^{-\lambda_g T_b^*}$ yields

$$(p_b\, e^{-\lambda_b T_b^*} + p_g\, e^{-\lambda_g T_b^*})\, MTBF_{field}^* = \frac{A_F}{\lambda_b}(p_b\, e^{-\lambda_b T_b^*}) + \frac{A_F}{\lambda_g}(p_g\, e^{-\lambda_g T_b^*}).$$

Rearranging this equation yields

$$\left(MTBF_{field}^* - \frac{A_F}{\lambda_b} \right) p_b\, e^{-\lambda_b T_b^*} = \left(\frac{A_F}{\lambda_g} - MTBF_{field}^* \right) p_g\, e^{-\lambda_g T_b^*},$$

or
$$e^{(\lambda_b - \lambda_g) T_b^*} = \left(\frac{p_b}{p_g} \right) \left(\frac{MTBF_{field}^* - \frac{A_F}{\lambda_b}}{\frac{A_F}{\lambda_g} - MTBF_{field}^*} \right).$$

Taking the natural logarithm on both sides of this equation yields

$$(\lambda_b - \lambda_g) T_b^* = \log_e \left[\left(\frac{p_b}{p_g} \right) \left(\frac{MTBF_{field}^* - \frac{A_F}{\lambda_b}}{\frac{A_F}{\lambda_g} - MTBF_{field}^*} \right) \right],$$

or

$$T_b^* = \frac{1}{\lambda_b - \lambda_g} \log_e \left[\left(\frac{p_b}{p_g} \right) \left(\frac{MTBF_{field}^* - \frac{A_F}{\lambda_b}}{\frac{A_F}{\lambda_g} - MTBF_{field}^*} \right) \right]. \qquad (13.20)$$

Correspondingly, the optimum number of thermal cycles, N_{therm}^*, is

$$N_{therm}^* = \frac{T_b^*}{t_6}. \qquad (13.21)$$

The application of Eqs. (13.20) and (13.21) is illustrated next by an example.

EXAMPLE 13-2

Given is a temperature cycle profile with the same parameters as given in Example 13-1; i.e.,

$T_o^* = 298°K$, $T_u^* = 338°K$, $T_l^* = 233°K$, $t_1 = \frac{1}{3}$ hr,
$t_2 = \frac{4}{3}$ hr, $t_4 = 2$ hr, $t_5 = \frac{7}{3}$ hr, $t_6 = \frac{8}{3}$ hr,
$\beta_h = 2$, and $\beta_c = 2$.

The constant, η, for the acceleration factor, due to the temperature change stress, is given by

$$\eta = 0.01.$$

The parameters for the acceleration factor due to the reaction rate stress, have been estimated in Examples 12-3 and 12-4 using the accelerated life test data of the second component; i.e.,

$$\begin{cases} a = -2.0504 \times 10^{-6}, \\ b = 2.2715, \end{cases}$$

for the first modified Arrhenius Model, and

$$\begin{cases} C = 17.8997, \\ B = 1.1507, \end{cases}$$

for the second modified Arrhenius Model.

The data analysis of the historical times to failure, which were observed during thermal cycling using the same temperature profile as given above, yields the mixed-exponential life distribution with the following parameters:

$$p_b = 0.1, \quad \lambda_b = 0.150 \text{ fr/hr},$$
$$p_g = 0.9, \quad \lambda_g = 0.0005 \text{ fr/hr}.$$

Determine the optimum thermal cycling time, T_b^*, or equivalently, the optimum number of thermal cycles, N_{therm}^*, for a specified field $MTBF$ of 370,000 hr.

SOLUTION TO EXAMPLE 13–2

According to Example 13–1, the equivalent acceleration factor produced by a typical temperature cycle of length $t_6 = 8/3$ hr is

$$A_F = 186.3069.$$

Substituting the given data into Eq. (13.20) yields the optimum thermal cycling time; i.e.,

$$T_b^* = \frac{1}{0.150 - 0.0005} \log_e \left[\left(\frac{0.1}{0.9}\right) \left(\frac{370,000 - \frac{186.3069}{0.150}}{\frac{186.3069}{0.0005} - 370,000} \right) \right],$$

or

$$T_b^* = 18.41 \text{ hr}.$$

Equivalently, the corresponding optimum number of thermal cycles is given by Eq. (13.21), or

$$N_{therm}^* = \frac{T_b^*}{t_6} = 18.41 \, (3/8) = 6.90, \text{ or approximately 7 cycles}.$$

13.6.2 USING THE WEIBULL LIFE DISTRIBUTION

Assume that the times to failure of certain electronic components under a given thermal cycling profile follow a 2-parameter Weibull life distribution with the following *pdf*, reliability and failure rate functions:

$$f(T) = \frac{\beta}{\eta} \left(\frac{T}{\eta}\right)^{\beta-1} e^{-\left(\frac{T}{\eta}\right)^\beta}, \quad 0 < \beta < 1, \eta > 0,$$

$$R(T) = e^{-\left(\frac{T}{\eta}\right)^\beta}, \quad 0 < \beta < 1, \eta > 0,$$

and

$$\lambda(T) = \frac{f(T)}{R(T)} = \frac{\beta}{\eta}\left(\frac{T}{\eta}\right)^{\beta-1}, \quad 0 < \beta < 1, \eta > 0,$$

where $0 < \beta < 1$ is required because a decreasing failure rate function is desired to justify a burn-in.

Let T_b be the burn-in time, then the post-burn-in reliability function for a mission time of t is given by

$$R(T_b, t) = \frac{R(T_b + t)}{R(T_b)} = \frac{e^{-\left(\frac{T_b+t}{\eta}\right)^\beta}}{e^{-\left(\frac{T_b}{\eta}\right)^\beta}},$$

or

$$R(T_b, t) = e^{\left(\frac{T_b}{\eta}\right)^\beta - \left(\frac{T_b+t}{\eta}\right)^\beta}.$$

Consequently, the post-burn-in $MTBF$ under the given thermal cycling profile environment, $MTBF_b$, is given by

$$MTBF_b = \int_0^\infty R(T_b, t)\, dt,$$

$$= \int_0^\infty e^{\left(\frac{T_b}{\eta}\right)^\beta - \left(\frac{T_b+t}{\eta}\right)^\beta}\, dt,$$

or

$$MTBF_b = e^{\left(\frac{T_b}{\eta}\right)^\beta} \int_0^\infty e^{-\left(\frac{T_b+t}{\eta}\right)^\beta}\, dt. \qquad (13.22)$$

Let

$$X = \left(\frac{T_b + t}{\eta}\right)^\beta.$$

Then,

$$t = \eta\, X^{\frac{1}{\beta}} - T_b,$$

and

$$dt = \frac{\eta}{\beta} X^{\frac{1}{\beta}-1}\, dX.$$

OPTIMUM NUMBER OF THERMAL CYCLES 599

Substituting these into Eq. (13.22) yields

$$MTBF_b = e^{\left(\frac{T_b}{\eta}\right)^\beta} \int_{\left(\frac{T_b}{\eta}\right)^\beta}^{\infty} e^{-X} \left(\frac{\eta}{\beta} X^{\frac{1}{\beta}-1}\right) dX,$$

$$= e^{\left(\frac{T_b}{\eta}\right)^\beta} \left(\frac{\eta}{\beta}\right) \int_{\left(\frac{T_b}{\eta}\right)^\beta}^{\infty} X^{\frac{1}{\beta}-1} e^{-X} dX,$$

or

$$MTBF_b = \frac{\eta}{\beta} \Gamma\left[\frac{1}{\beta}; \left(\frac{T_b}{\eta}\right)^\beta\right] e^{\left(\frac{T_b}{\eta}\right)^\beta}, \qquad (13.23)$$

where

$$\Gamma\left[\frac{1}{\beta}; \left(\frac{T_b}{\eta}\right)^\beta\right] = \int_{\left(\frac{T_b}{\eta}\right)^\beta}^{\infty} X^{\frac{1}{\beta}-1} e^{-X} dX,$$

$$= \text{incomplete gamma function}.$$

Therefore, the equivalent $MTBF$ in the field, $MTBF_{field}$, will be given by

$$MTBF_{field} = A_F \, MTBF_b,$$

or

$$MTBF_{field} = \frac{A_F \, \eta}{\beta} \Gamma\left[\frac{1}{\beta}; \left(\frac{T_b}{\eta}\right)^\beta\right] e^{\left(\frac{T_b}{\eta}\right)^\beta}.$$

The optimum burn-in time, T_b^*, for a specified field $MTBF$ goal, $MTBF_{field}^*$, can be obtained by solving the following equation for T_b^*:

$$MTBF_{field}^* = \frac{A_F \, \eta}{\beta} \Gamma\left[\frac{1}{\beta}; \left(\frac{T_b^*}{\eta}\right)^\beta\right] e^{\left(\frac{T_b^*}{\eta}\right)^\beta}. \qquad (13.24)$$

Solving this equation for T_b^* yields the following iteration relation:

$$T_b^* = \eta \left\{ \log_e \left\{ \frac{MTBF_{field}^* \, \beta}{A_F \, \eta \, \Gamma\left[\frac{1}{\beta}; \left(\frac{T_b^*}{\eta}\right)^\beta\right]} \right\} \right\}^{\frac{1}{\beta}}. \qquad (13.25)$$

A computer program in FORTRAN language has been developed for solving this equation and is listed in Appendix 13B. Once the optimum burn-in time is obtained, the corresponding optimum number of thermal cycles, N^*_{therm}, is given by

$$N^*_{therm} = \frac{T^*_b}{t_6}. \qquad (13.26)$$

The application of Eqs. (13.25) and (13.26) is illustrated next by an example.

EXAMPLE 13-3

Given is a temperature cycle profile with the same parameters as given in Example 13-1; i.e.,

$T^*_o = 298°K$, $T^*_u = 338°K$, $T^*_l = 233°K$, $t_1 = \frac{1}{3}$ hr,
$t_2 = \frac{4}{3}$ hr, $t_4 = 2$ hr, $t_5 = \frac{7}{3}$ hr, $t_6 = \frac{8}{3}$ hr,
$\beta_h = 2$, and $\beta_c = 2$.

The constant, η, for the acceleration factor, due to the temperature change stress, is given by

$$\eta = 0.01.$$

The parameters for the acceleration factor due to the reaction rate stress, have been estimated in Examples 12-3 and 12-4 using the accelerated life test data of the second component; i.e.,

$$\begin{cases} a = -2.0504 \times 10^{-6}, \\ b = 2.2715, \end{cases}$$

for the first modified Arrhenius Model, and

$$\begin{cases} C = 17.8997, \\ B = 1.1507, \end{cases}$$

for the second modified Arrhenius Model.

The data analysis of the historical times to failure, which were observed during thermal cycling using the same temperature profile as given above, yields the Weibull life distribution with the following parameters:

$$\beta = 0.8,$$

and

$$\eta = 1,710 \text{ hr.}$$

Determine the optimum thermal cycling time, T_b^*, or equivalently, the optimum number of thermal cycles, N_{therm}^*, for a specified field $MTBF$ of 367,240 hr.

SOLUTION TO EXAMPLE 13-3

According to Example 13-1, the equivalent acceleration factor produced by a typical temperature cycle of length $t_6 = 8/3$ hr is

$A_F = 186.3069.$

Entering the given data into the computer program in Appendix 13B yields the optimum thermal cycling time; i.e.,

$T_b^* = 18.43$ hr.

Equivalently, the corresponding optimum number of thermal cycles is given by Eq. (13.26), or

$$N_{therm}^* = \frac{T_b^*}{t_6} = 18.43 \, (3/8) = 6.91, \text{ or approximately 7 cycles.}$$

13.7 A SUMMARY OF SOME USEFUL THERMAL FATIGUE LIFE PREDICTION MODELS FOR ELECTRONIC EQUIPMENT

It is well known that the standard Eyring and Arrhenius models can be used to describe and quantify the effect of temperature on chemical reaction rates provided that the dominant component failure mechanisms depend on the steady-state temperature. However, it is often difficult to confirm the correlation between steady-state temperature and the failure mechanisms [2, p. 61]. Usually an electronic equipment contains various components of different thermal masses and different thermal characteristics, and various interconnections, which may result in all kinds of failures even under the same stress environment. Multiple failure mechanisms of an electronic equipment under a dynamic temperature environment may require specific physical models to describe the failure behavior and predict the corresponding time to failure for each failure mechanism.

During thermal cycling, mechanical failures occur from differential thermal expansion between bonded materials, rate and amplitude of time dependent temperature changes, and large spatial temperature gradients, all of which can cause tensile, compressive, bending, fatigue and fracture failures. The time to failure for each failure mechanism may be governed by multiple factors, such as geometry, physical properties, environmental stress levels, etc. Various physical models have been derived in the literature for the associated failure mechanisms to predict the mean time to failure, based on stress-strain analysis, cumulative damage evaluation and/or fracture mechanics analysis. The following is a summary of some useful life prediction models for mechanical failures under thermal cycling environment.

13.7.1 WIRE FATIGUE LIFE PREDICTION

Wires in an electronic equipment are usually used to connect die bond pads to leads, or die bond pads to die bond pads for hybrid packages. The cyclic temperature changes during temperature and power cycling cause repeated flexing of the wire due to the differential coefficient of thermal expansion between the wire and the package, which may lead to wire failures. It has been confirmed that the main failure position is at the heel of the wire [2; 5; 6; 7].

The average number of cycles to failure, N_f, for a wire undergoing temperature cycling is dependent on the temperature change magnitude, the geometry of the wire and the wire material properties, and is given by [2; 8]

$$N_f = A\, (\epsilon_f)^n, \qquad (13.27)$$

where

$$\epsilon_f = \frac{r}{\rho_0} \left[\frac{\cos^{-1}\{\cos \lambda_0 [1 - (\alpha_w - \alpha_s)\, \Delta T^*]\}}{\theta_0} - 1 \right],$$

r = wire radius, m (meter),

θ_0 = angle of the wire with the substrate, radians,

α_w = coefficient of thermal expansion of the wire, (m/m)/°C,

α_s = coefficient of thermal expansion of the substrate, (m/m)/°C,

ΔT^* = average magnitude of temperature change in °C during temperature cycling; i.e., the absolute difference between the maximum and the minimum temperature extremes,

ρ_0 = initial radius of curvature of the wire, m (meter),

A = material constant,

and

n = material constant.

The typical values for the average number of cycles to failure are the following [9]:

THERMAL FATIGUE LIFE MODELS 603

1. 18,000 cycles for 0.008 inch diameter 99.99% pure aluminum bonds on a 2N4863 power transistor subjected to temperatures from 25°C to 125 °C.

2. Over 200,000 power cycles for 0.002 inch diameter aluminum, 1% silicon, ultrasonic bonds with loop heights greater than 25% of the bond-to-bond spacing, for temperatures ranging from 38°C to 170°C.

13.7.2 WIRE BOND FATIGUE LIFE PREDICTION

In an electronic equipment, the wire bonds connect the bond pad and its associated die metalization to a wire which is connected, in turn, to either a lead, or to other bond pads in the case of a hybrid package. When subjected to thermal cycling a wire bond will experience cyclic shear stresses between the wire bond itself, and the wire and the substrate due to differential thermal expansion resulting in cumulative fatigue damage until a failure occurs [10; 11]. The fatigue damage induced by thermal cycling results from the initiation and growth of cracks to fracture. Cracks may be initiated from defects or microcracks inherent in the wire or bond pad. Some cracks in the bond pad or substrate may be introduced during the bonding process, or from ultrasonic vibrations or thermal energy applied at the wire-pad interface during manufacture and assembly. Usually these cracks go undetected during inspection and will propagate under the thermal cycling environment. It has been found that a significant number of modules fail due to wire bond failure during the thermal cycling process [11].

1. **Fatigue Life of the Bond Pad Due to Cyclic Shear Between the Bond Pad and the Substrate and Between the Bond Pad and the Wire**

    The average number of cycles to failure of the bond pad, $N_{f,p}$, due to cyclic shear between the bond pad and the substrate, and between the bond pad and the wire during thermal cycling, is given by [9]

    $$N_{f,p} = C_p \left(\tau_{p,max}\right)^{-m_p}, \qquad (13.28)$$

    where

    C_p, m_p = shear fatigue properties for the bond pad materials,
    $\tau_{p,max}$ = maximum shear stress in the bond pad due to shear between the bond pad and the substrate, and between the bond pad and the wire,

$$\tau_{p,max} = Q\,\Delta T^*,$$

ΔT^* = average magnitude of temperature changes during thermal cycling, °C,

$$Q = \left(\frac{G_p}{b_p Z}\right)\left\{(\alpha_w - \alpha_s) - \frac{(\alpha_s - \alpha_p)}{\left[1 + \frac{E_s A_s}{E_p A_p (1-\nu_s)}\right]}\right\},$$

G_p = shear modulus of the bond pad materials, N/m²,
b_p = bond pad thickness, m,
b_s = substrate thickness, m,
α_s = coefficient of thermal expansion for the substrate, (m/m)/°C,
α_p = coefficient of thermal expansion for the bond pad, (m/m)/°C,
α_w = coefficient of thermal expansion for the wire, (m/m)/°C,
E_p = modulus of elasticity of the pad material, N/m²,
E_s = modulus of elasticity of the substrate material, N/m²,
ν_s = Poisson's ratio for the substrate material, which is dimensionless,
A_p = cross-sectional area of the pad, m²,
A_s = effective cross-sectional area of the substrate, m²,
$$A_s = \frac{b_s\,(W_p + W_s)}{2},$$
W_p = width of the bond pad, m,

and

W_s = width of the substrate, m.

2. **Fatigue Life of the Wire Due to Cyclic Shear Between the Wire and the Bond Pad**

The average number of cycles to failure of the wire, $N_{f,w}$, due to cyclic shear between the wire and the bond pad during thermal cycling is given by [11]

$$N_{f,w} = C_w\,(\tau_{w,max})^{-m_w}, \qquad (13.29)$$

where

C_w, m_w = shear fatigue properties of the wire material,
$\tau_{w,max}$ = maximum shear stress between wire and bond pad,

$$\tau_{w,max} = \left\{ \frac{r^2}{4 Z^2 A_w^2} \left[\frac{\cosh(Z\, x_w)}{\cosh(Z\, l_w)} - 1 \right]^2 \right.$$

$$\left. + \frac{\sinh^2(Z\, x_w)}{\cosh^2(Z\, l_w)\, Q^2} \right\}^{\frac{1}{2}} \Delta T^*,$$

$$Z = \left\{ \left(\frac{G_p}{b_p} \right) \left[\frac{r}{E_w A_w} + \frac{(1-\nu_s) W_p}{E_s A_s} \right] \right\}^{\frac{1}{2}},$$

r = wire radius, m,
A_w = cross-sectional area of the wire,
E_w = modulus of elasticity of the wire,
x_w = location of the maximum shear stress in wire,

$$x_w = \pm \operatorname{arctanh}\left(\frac{A_w}{r} \right),$$

and

l_w = bonded length of the wire, m.

The rest of the nomenclature is as defined in the case of bond pad fatigue life prediction.

3. **Fatigue Life of the Substrate Due to Cyclic Shear Between the Substrate and the Bond Pad**

The average number of cycles to failure of the substrate, $N_{f,s}$, due to cyclic shear between the substrate and the bond pad during thermal cycling is given by [11]

$$N_{f,s} = C_s \left(\tau_{s,max} \right)^{-m_s}, \tag{13.30}$$

where

C_s, m_s = shear fatigue properties of the substrate material,
$\tau_{s,max}$ = maximum shear stress between substrate and bond pad,

$$\tau_{s,max} = \left\{ \left[\frac{W_p\, Q}{2\, Z\, A_s} \left(1 - \frac{\cosh(Z\, x_s)}{\cosh(Z\, l_s)} \right) \right. \right.$$

$$+\frac{(\alpha_s - \alpha_p)}{(1-\nu_s)/E_s + A_s/(E_p\, A_p)}\Bigg]^2$$

$$+Q^2\,\frac{\sinh^2(Z\,x_s)}{\cosh^2(Z\,l_s)}\Bigg\}^{\frac{1}{2}}\;\Delta T^*,$$

x_s = location of the maximum shear stress in the substrate

$$x_s = \mp \operatorname{arctanh}\left(\frac{A_s}{W_p}\right),$$

and

l_s = bonding length of the substrate, m.

The rest of the nomenclature is as defined in the previous two cases of fatigue life prediction for bond pad and wire, respectively.

13.7.3 DIE FRACTURE AND FATIGUE LIFE PREDICTION

Generally, die and substrate or lead frame have different thermal expansion coefficients because they are made of different materials. For example, dies are usually made of silicon, gallium arsenide or indium phosphide; but the substrate is typically alumina, berylia or copper. Tensile stress in the central portion of the die and shear stress at the edges of the die are developed during thermal and power cycling, which may lead to vertical and horizontal cracking of the die, respectively. As the surface cracks at the center or at the edge of the die propagate and finally reach their critical sizes, a sudden ultimate fracture of the brittle die may occur without any plastic deformation.

The average number of cycles to fracture failure, N_f, of the die is given by [2]

$$N_f = \frac{2}{(2-n)\,A\,(\Delta\sigma_{app})^n\,\pi^{n/2}}\left(a_c^{1-\frac{n}{2}} - a_i^{1-\frac{n}{2}}\right), \qquad (13.31)$$

where

A, n = die material dependent coefficient and exponent, respectively, which govern the die crack propagation behavior as represented by Paris' Power Law [11]; i.e.,

$$\frac{da}{dN} = A\,(\Delta K)^n$$

N = number of temperature cycles applied,

a = crack size, m,
a_i = initial crack size, m,
a_c = critical crack size, m,
K = stress intensity factor at or near the die crack tip,
$\Delta K = K_{max} - K_{min}$,
   $= \sigma_{app} (\pi a)^{\frac{1}{2}}$, for an infinite solid,
σ_{app} = applied stress amplitude, N/m^2,

$$\sigma_{app} = 10^{-6} k |\alpha_s - \alpha_d| \Delta T^* \sqrt{\frac{E_s E_a L}{X}},$$

k = dimensionless geometric constant,
α_s = coefficient of thermal expansion of substrate, (m/m)/°C,
α_d = coefficient of thermal expansion of die, (m/m)/°C,
E_s = tensile modulus of substrate, N/m^2,
E_a = tensile modulus of adhesive, N/m^2,
L = diagonal length of die, m,
ΔT^* = temperature change magnitude during thermal cycling, °C,

and

X = adhesive bond thickness, m.

13.7.4 DIE AND SUBSTRATE ADHESION FATIGUE LIFE PREDICTION

During a temperature or power cycling, temperature differences and temperature gradient differences will occur between the die, the die attach and the die substrate. The bond between the die and the substrate will experience fatigue failure due to the differential thermal expansion.

Voids are the most common die attach defects. These voids in the die attach, in particular when they are relatively large in size, will induce high longitudinal stresses during thermal or power cycling since the heat must flow around the void creating a large temperature gradient in the silicon. They act as microcracks which propagate and lead to debonding of the die from the substrate or the substrate from the case.

1. **Fatigue Life for the Die Attach**

   The average number of cycles to failure of the die attach, N_f, as a result of plastic strain cycling is given by the Coffin-Manson relation [2, p. 80] as follows:

   $$N_f = 0.5 \left(\frac{\gamma_a}{\gamma_f}\right)^{\frac{1}{c}}, \qquad (13.32)$$

   where

   $c =$ Coffin-Manson coefficient,
   $\gamma_f =$ fatigue ductility coefficient of the die attach which is defined as the shear strain required to cause failure in one load reversal,
   $\gamma_a =$ actual plastic strain amplitude,
   $$\gamma_a = \frac{L\,|\alpha_s - \alpha_d|\,\Delta T^*}{X},$$
   $L =$ diagonal die length, m,
   $\alpha_s =$ coefficient of thermal expansion of the substrate, (m/m)/°C,
   $\alpha_d =$ coefficient of thermal expansion of the die, (m/m)/°C,
   $\Delta T^* =$ temperature change amplitude during thermal cycling, °C,

   and

   $X =$ height of die attach, m.

2. **Fatigue Life for the Substrate Attach**

   The average number of cycles to fatigue failure of the substrate attach is also given by the Coffin-Manson relation [2, p. 81; 12; 13] as follows:

   $$N_f = 0.5 \left(\frac{\gamma_a}{\gamma_f}\right)^{\frac{1}{c}}, \qquad (13.33)$$

where

c = Coffin-Manson coefficient,
γ_f = fatigue ductility coefficient of the substrate which is defined as the shear strain required to cause failure in one load reversal,
γ_a = actual plastic strain amplitude,
$$\gamma_a = \frac{L_s \, |\alpha_c - \alpha_s| \, \Delta T^*}{h_{sa}},$$
L_s = diagonal length of the substrate, m,
α_s = coefficient of thermal expansion of the substrate, (m/m)/°C,
α_c = coefficient of thermal expansion of the case, (m/m)/°C,
ΔT = temperature change amplitude during thermal cycling, °C,

and

h_{sa} = thickness of the substrate attach, m.

13.7.5 SOLDER JOINT FATIGUE LIFE PREDICTION

It was indicated in [14] that most of the electronic components used for surface mounted technology (SMT) applications rely upon the structural integrity of the solder to support these components in various thermal, vibration, and shock environments. The SMT solder joint mechanically attaches the components to the printed wiring board (PWB) and provides electrical and thermal continuity. The major cause of failure is the low-cycle thermal fatigue of the solder joints [15], though vibration-induced high cycle fatigue is believed to make a secondary contribution to fatigue damage [15; 16; 17]. In low-cycle thermal fatigue, the thermal strains are caused by different coefficients of thermal expansion for materials used in PWB assembly and also by the temperature differentials due to internal heat generation or the external temperature variations caused by component, system load fluctuations, and/or on-off cycles, and by environmental temperature changes. Experience has shown [2, p. 263] that solder can creep at a rate of about 0.002 to 0.003 inches per year when subjected to compression loads.

The most often used solder has been the eutectic solder which is composed of 63% tin and 37% lead, the so-called Sn-Pb solder. The melting temperature of this Sn-Pb solder is only about 184°C, which is low enough to protect the sensitive electronic components from thermal damage when they are attached to the printed circuit board (PCB). When the stress levels are above approximately 800 psi at room temperature conditions, this solder tends to creep extensively. These creep effects can cause premature failures in the solder unless the stress levels are kept under about 400 psi in thermal cycling and vibration environments and they are temperature sensitive. When the temperature is above about 85°C, the creep increases very significantly. However, when the temperature is below about $-20°C$, the creep reduces sharply.

A widely accepted thermal fatigue life prediction model for near eutectic Sn-Pb solders under thermal cycling conditions was proposed by Engelmaier [14], which includes an empirical description of the effect of cyclic frequency and mean temperature value. This model can be expressed as

$$N_f = \frac{1}{2}\left(\frac{\Delta\gamma}{2\,\varepsilon_f}\right)^{\frac{1}{c}}, \qquad (13.34)$$

where

N_f = mean number of thermal cycles to failure,
$\Delta\gamma$ = total shear strain range in one cycle of loading,
ε_f = fatigue ductility coefficient = 0.325,
c = fatigue ductility exponent ,
$c = -0.442 - 6 \times 10^{-4}\,T^*_{mean} + 1.74 \times 10^{-2}\,\log_e(1+f)$,
T^*_{mean} = mean cyclic temperature in the solder joint, °C,
$T^*_{mean} = \frac{1}{2}(T^*_{max} + T^*_{min})$,
T^*_{max} = maximum cyclic temperature, °C,
T^*_{min} = minimum cyclic temperature, °C,

and

f = cyclic frequency, $1 < f < 1,000$ cycles/day.

13.7.6 COMMENTS

The physical models for the thermal fatigue life prediction are summarized for various failure mechanisms which may occur during thermal cycling. Each of them gives the average number of cycles to failure for a particular failure mechanism in terms of the associated physical characteristic parameters, geometries, applied temperature change amplitude during thermal cycling and temperature-induced stress. However, the effect of the rate of temperature change on the thermal fatigue life is not reflected in these models. Furthermore, the randomness in the physical characteristics, geometries and applied temperature profiles will cause randomness both in the temperature-induced stress and the fatigue life; i.e., the thermal fatigue life should be statistically distributed. Multiple failure mechanisms, each being governed by a life distribution in an electronic equipment during the thermal cycling, may lead to a mixed life distribution for failures of the whole equipment. More extensive investigation is needed in the future in these areas.

13.8 CONCLUSIONS

Conventional practice in evaluating the effect of elevated temperature in accelerated life testing and in burn-in testing is based upon the standard Arrhenius equation and thereby ignores several aspects of the stress. A general model is provided here to include these additional aspects of the stress. This model combines the basic reaction rate stress during the cooling and heating intervals and the temperature change stress (ramp effect), which yields an equivalent acceleration factor in one temperature cycle. The classical Arrhenius model is modified by incorporating a temperature-dependent and therefore cycling-time-dependent activation energy in place of the traditionally assumed temperature independent constant activation energy. The resultant models with two modified versions of the Arrhenius model are rather complicated expressions which must be evaluated numerically, but provide a more realistic portrayal of the effect of thermal cycling upon electronic devices. This expanded but more general and realistic model presented in this chapter provides greater visibility to the parameters of a thermal cycle.

A model is derived to determine the optimum number of thermal cycles for a given temperature profile and a specified field $MTBF$ goal. This model is based on the result of equivalent acceleration factor of a typical temperature cycle and assumes the knowledge of the equipment's life distribution at the burn-in stress level.

Finally, several useful physical models for thermal fatigue life prediction are presented. These models serve the purpose of predicting the average time to failure for various specific failure mechanisms in an electronic equipment during thermal cycling. The application of these models requires the knowledge of the associated physical properties and geometries.

REFERENCES

1. Pecht, M. and Lall, P., "A Physics of Failure Approach to IC Burn-in," *Advances in Electronic Packaging*, ASME, pp. 917-923, 1992.

2. Bar-Cohen, A. and Kraus, A. D., *Advances in Thermal Modeling of Electronic Components and Systems*, Vol. 3, ASME Press, New York, IEEE Press, New York, 402 pp., 1993.

3. Nachlas, J. A., "A General Model for Age Acceleration During Thermal Cycling," *Quality and Reliability International*, Vol. 2, pp. 3-6, 1986.

4. Nachlas, J. A., Binney, B. A. and Gruber, S. S., "Aging Acceleration Under Multiple Stresses," *Proceedings of the Annual Reliability and Maintainability Symposium*, Philadelphia, PA, pp. 438-440, 1985.

5. Gaffeny, J., "Internal Lead Fatigue Through Thermal Expansion in Semiconductor Devices," *IEEE Transactions on Electronic Devices*, ED-15, p. 617, 1968.

6. Ravi, K. V. and Philosky, E. M., "Reliability Improvement of Wire Bonds Subjected to Fatigue Stresses," *Proceedings of the 10th IEEE Annual Reliability Physics Symposium*, Las Vegas, NV, pp. 143-149, 1972.

7. Phillips, W. E., "Microelectronic Ultrasonic Bonding," edited by Harman, G. G., National Bureau of Standards (US), Special Publication 400-2, pp. 80-86, 1974.

8. Pecht, M., Lall, P. and Dasgupta, A., "A Failure Prediction Model for Wire Bonds," *Proceedings of 1989 International Symposium on Hybrid Microelectronics*, pp. 607-613, 1989.

9. Harman, G. G., "Metallurgical Failure Modes of Wire Bonds," *Proceedings of 12th Annual International Reliability Physics Symposium*, pp. 131-141, 1974.

10. Hu, J., Pecht, M. and Dasgupta, A., "A Probabilistic Approach for Predicting Thermal Fatigue Life of Wirebonding in Microelectronics," *Journal of Electronic Packaging*, Vol. 113, pp. 275-285, September 1991.

11. Paris, P. C., Gomez, M. P. and Anderson, W. E., "A Rational Analytical Theory of Fatigue," *The Trend in Engineering*, Vol. 13, University of Washington, Seattle, WA, pp. 9-24, 1961.

12. Bolger, J. C., "Polyimide Adhesive to Reduce Thermal Stress in LSI Ceramic Packages," *Proceedings of 14th National SAMPE Technical Conference*, pp. 12-14, October, 1982.

13. Bolger, J. C. and Mooney, C. T., "Die Attach in Hi-rel DIPS: Polyimides or Low Chloride Epoxies," *IEEE Transactions on Composite Hybrid Manufacturing Techniques*, CHMT-7 (4), December, 1984.

14. Engelmaier, W., "Fatigue Life of Leadless Chip Carrier Solder Joints During Power Cycling," *IEEE Transactions on Components, Hybrids, and Manufacturing Technology*, Vol. CHMT-6, No. 3, pp. 232-237, 1985.

15. Barker, D. B., Sharif, I., Dasgupta, A. and Pecht, M. G., "Effect of SMC Lead Dimensional Variabilities on Lead Compliance and Solder Joint Fatigue Life," *Journal of Electronic Packaging*, Vol. 114, pp. 177-184, 1992.

16. Blanks, H. S., "Accelerated Vibration Fatigue Life Testing of Leads and Solder Joint," *Microelectronics and Reliability*, Vol. 15, pp. 213-219, 1976.

17. Steinberg, D. S., *Vibration Analysis for Electronic Equipment*, John Wiley & Sons, New York, 443 pp., 1988.

APPENDIX 13A
COMPUTER PROGRAM FOR EXAMPLE 13-1

```
C
C $$$$$$$$$$$$$$$$$$$$$$$$$$$$$$$$$$$$$$$$$$$$$$$$$$$$$$$$$$$$$$
C $ THIS PROGRAM IS DEVELOPED FOR EVALUATING THE EQUIVALENT $
C $ ACCELERATION FACTORS WHEN THE REACTION RATE STRESS AND   $
C $ THE TEMPERATURE CHANGE ARE BOTH CONSIDERED.              $
C $                                                          $
C $ TWO MODIFICATION FORMS FOR THE ARRHENIUS EQUATION ARE    $
C $ USED WHICH SHOULD YIELD VERY CLOSE RESULTS.              $
C $$$$$$$$$$$$$$$$$$$$$$$$$$$$$$$$$$$$$$$$$$$$$$$$$$$$$$$$$$$$$$
C
      IMPLICIT DOUBLE PRECISION (A-H,O-Z)
      EXTERNAL FUNC1,FUNC2,FUNC3,FUNC4,FUNC5,FUNC6
C
C       'AS' == SMALL A, 'a', in the 1st modified model
C       'BS' == SMALL B, 'b', in the 1st modified model
C       'CB' == BIG C, 'C', in the 2nd modified model
C       'BB' == BIG B, 'B', in the 2nd modified model
C
      COMMON /PAR1/AS,BS,CB,BB,ETA
      COMMON /PAR2/T1,T2,T4,T5,T6,T0,TL,TU,BETAH,BETAC
      OPEN(5,FILE='BURN-DIX-13A.IN',STATUS='UNKNOWN')
      READ(5,*)AS,BS,CB,BB,ETA,T1,T2,T4,T5,T6,T0,TL,
     1 TU,BETAH,BETAC
      WRITE(*,*)'                 THE INPUT PARAMETERS'
      WRITE(*,*)('*',I=1,78)
      WRITE(*,*)'AS=',AS,'BS=',BS,'CB=',CB,'BB=',BB
      WRITE(*,*)'ETA=',ETA,'T1=',T1,'T2=',T2,'T4=',T4
      WRITE(*,*)'T5=',T5,'T6=',T6,'T0=',T0,'TL=',TL
      WRITE(*,*)'TU=',TU,'BETAH=',BETAH,'BETAC=',BETAC
      WRITE(*,*)('*',I=1,78)
      CALL ACC1(AF1)
      CALL ACC2(AF2)
      WRITE(*,*)
      WRITE(*,*)'                          RESULTS'
      WRITE(*,*)('*',I=1,78)
      WRITE(*,*)'EQUIVALENT ACCELERATION FACTOR USING
     1 FIRST MODIFICATION:'
      WRITE(*,*)'             AF1==',AF1
      WRITE(*,*)'EQUIVALENT ACCELERATION FACTOR USING
     1 SECOND MODIFICATION:'
```

```
          WRITE(*,*)'              AF2==',AF2
          STOP
          END
C
C     SUBPROGRAM EVALUATING THE EQUIVALENT ACCELERATION
C        FACTOR USING THE FIRST VERSION OF MODIFIED
C                  ARRHENIUS EQUATION
C
          SUBROUTINE ACC1(AF1)
          IMPLICIT DOUBLE PRECISION (A-H,O-Z)
          EXTERNAL FUNC1,FUNC2,FUNC3,FUNC4,FUNC5,FUNC6
          REAL K
          COMMON /PAR1/AS,BS,CB,BB,ETA
          COMMON /PAR2/T1,T2,T4,T5,T6,T0,TL,TU,BETAH,BETAC
          K=8.623E-5
          CALL QGAUS(FUNC1,0.0,T1,S1)
          E1=(T2-T1)*EXP(-AS/K*TU**(BS-1.0))
          CALL QGAUS(FUNC2,T2,T4,S2)
          E2=(T5-T4)*EXP(-AS/K*TL**(BS-1.0))
          CALL QGAUS(FUNC3,T5,T6,S3)
          E0=1.0/T6*EXP(AS/K*T0**(BS-1.0))
          AF1=E0*(S1+E1+S2+E2+S3)
          RETURN
          END
C
C     SUBPROGRAM EVALUATING THE EQUIVALENT ACCELERATION
C        FACTOR USING THE SECOND VERSION OF MODIFIED
C                  ARRHENIUS EQUATION
C
          SUBROUTINE ACC2(AF2)
          IMPLICIT DOUBLE PRECISION (A-H,O-Z)
          EXTERNAL FUNC1,FUNC2,FUNC3,FUNC4,FUNC5,FUNC6
          REAL K
          COMMON /PAR1/AS,BS,CB,BB,ETA
          COMMON /PAR2/T1,T2,T4,T5,T6,T0,TL,TU,BETAH,BETAC
          K=8.623E-5
          CALL QGAUS(FUNC4,0.0,T1,S1)
          E1=(T2-T1)*EXP(CB*TU**BB*(1.0/T0-1.0/TU))
          CALL QGAUS(FUNC5,T2,T4,S2)
          E2=(T5-T4)*EXP(-CB*TL**BB*(1.0/TL-1.0/T0))
          CALL QGAUS(FUNC6,T5,T6,S3)
          E0=1.0/T6
          AF2=E0*(S1+E1+S2+E2+S3)
          RETURN
```

```
            END
C
C       SUBPROGRAM EVALUATING THE INTEGRATION
C                USING GAUSS METHOD
C
        SUBROUTINE QGAUS(FUNC,A,B,SS)
        IMPLICIT DOUBLE PRECISION (A-H,O-Z)
        DIMENSION X(5),W(5)
        DATA X/0.1488743389,0.4333953941,0.6794095682,
     1       0.8650633666,0.9739065285/
        DATA W/0.2955242247,0.2692667193,0.2190863625,
     1       0.1494513491,0.0666713443/
        XM=0.5*(B+A)
        XR=0.5*(B-A)
        SS=0.0
        DO 11 J=1,5
        DX=XR*X(J)
        SS=SS+W(J)*(FUNC(XM+DX)+FUNC(XM-DX))
11         CONTINUE
        SS=XR*SS
        RETURN
        END
C
C       FUNC1 THROUGH FUNC3 ARE THE INTEGRANDS FOR
C                 SUBPROGRAM ACC1(AF1)
C
        FUNCTION FUNC1(Z)
        IMPLICIT DOUBLE PRECISION (A-H,O-Z)
        REAL K
        COMMON /PAR1/AS,BS,CB,BB,ETA
        COMMON /PAR2/T1,T2,T4,T5,T6,T0,TL,TU,BETAH,BETAC
        K=8.623E-5
        P1=AS/K*(T0+(TU-T0)*(Z/T1)**BETAH)**(BS-1.0)
        P2=ETA*BETAH*(TU-T0)/T1*(Z/T1)**(BETAH-1.0)
        FUNC1=EXP(-(P1-P2))
        RETURN
        END
C
        FUNCTION FUNC2(Z)
        IMPLICIT DOUBLE PRECISION (A-H,O-Z)
        REAL K
        COMMON /PAR1/AS,BS,CB,BB,ETA
        COMMON /PAR2/T1,T2,T4,T5,T6,T0,TL,TU,BETAH,BETAC
        K=8.623E-5
```

APPENDIX 13A

```
              P1=AS/K*(TU-(TU-TL)*((Z-T2)/(T4-T2))**BETAC)**(BS-1.0)
              P2=ETA*BETAC*(TU-TL)/(T4-T2)*((Z-T2)/(T4-T2))**(BETAC-1.0)
              FUNC2=EXP(-(P1-P2))
              RETURN
              END
C
              FUNCTION FUNC3(Z)
              IMPLICIT DOUBLE PRECISION (A-H,O-Z)
              REAL K
              COMMON /PAR1/AS,BS,CB,BB,ETA
              COMMON /PAR2/T1,T2,T4,T5,T6,T0,TL,TU,BETAH,BETAC
              K=8.623E-5
              P1=AS/K*(TL+(T0-TL)*((Z-T5)/(T6-T5))**BETAH)**(BS-1.0)
              P2=ETA*BETAH*(T0-TL)/(T6-T5)*((Z-T5)/(T6-T5))**(BETAH-1.0)
              FUNC3=EXP(-(P1-P2))
              RETURN
              END
C
C       FUNC4 THROUGH FUNC6 ARE THE INTEGRANDS
C              FOR SUBPROGRAM ACC2(AF2)
C
              FUNCTION FUNC4(Z)
              IMPLICIT DOUBLE PRECISION (A-H,O-Z)
              REAL K
              COMMON /PAR1/AS,BS,CB,BB,ETA
              COMMON /PAR2/T1,T2,T4,T5,T6,T0,TL,TU,BETAH,BETAC
              K=8.623E-5
              P1=(1.0/T0)*(T0+(TU-T0)*(Z/T1)**BETAH)**BB
              P2=(T0+(TU-T0)*(Z/T1)**BETAH)**(BB-1.0)
              P3=ETA*BETAH*(TU-T0)/T1*(Z/T1)**(BETAH-1.0)
              FUNC4=EXP(CB*(P1-P2)+P3)
              RETURN
              END
C
              FUNCTION FUNC5(Z)
              IMPLICIT DOUBLE PRECISION (A-H,O-Z)
              REAL K
              COMMON /PAR1/AS,BS,CB,BB,ETA
              COMMON /PAR2/T1,T2,T4,T5,T6,T0,TL,TU,BETAH,BETAC
              K=8.623E-5
              P1=(1.0/T0)*(TU-(TU-TL)*((Z-T2)/(T4-T2))**BETAC)**BB
              P2=(TU-(TU-TL)*((Z-T2)/(T4-T2))**BETAC)**(BB-1.0)
              P3=ETA*BETAC*(TU-TL)/(T4-T2)*((Z-T2)/(T4-T2))**(BETAC-1.0)
              FUNC5=EXP(CB*(P1-P2)+P3)
```

```
      RETURN
      END
C
      FUNCTION FUNC6(Z)
      IMPLICIT DOUBLE PRECISION (A-H,O-Z)
      REAL K
      COMMON /PAR1/AS,BS,CB,BB,ETA
      COMMON /PAR2/T1,T2,T4,T5,T6,TO,TL,TU,BETAH,BETAC
      K=8.623E-5
      P1=(1.0/TO)*(TL+(TO-TL)*((Z-T5)/(T6-T5))**BETAH)**BB
      P2=(TL+(TO-TL)*((Z-T5)/(T6-T5))**BETAH)**(BB-1.0)
      P3=ETA*BETAH*(TO-TL)/(T6-T5)*((Z-T5)/(T6-T5))**(BETAH-1.0)
      FUNC6=EXP(CB*(P1-P2)+P3)
      RETURN
      END
```

APPENDIX 13B
COMPUTER PROGRAM FOR EXAMPLE 13-3

```
$$$$$$$$$$$$$$$$$$$$$$$$$$$$$$$$$$$$$$$$$$$$$$$$$$$$
$   OPTIMUM TEMPERATURE CYCLING TIME FOR A SPECIFIED  $
$    FIELD MTBF GOAL USING THE 2-PARAMETER WEIBULL    $
$         LIFE DISTRIBUTION DURING BURN-IN            $
$$$$$$$$$$$$$$$$$$$$$$$$$$$$$$$$$$$$$$$$$$$$$$$$$$$$
      IMPLICIT DOUBLE PRECISION (A-H, O-Z)
      REAL INT, MG
      EXTERNAL FUNC, DQDAGI
      COMMON /PAR/BETA,ETA
C
C  BURN-IN TIME CORRESPONDING TO A SPECIFIED FIELD MTBF
C  GOAL, MG, USING THE 2-PARAMETER MIXED-WEIBULL MODEL
C
      WRITE(*,*) 'GIVE PARAMETERS OF WEIBULL
     1 DISTRIBUTION: BETA, ETA !!'
      READ(*,*) BETA, ETA
      WRITE(*,*) 'BETA=', BETA, 'ETA=', ETA
      WRITE(*,*) 'GIVE ACCELERATION FACTOR FOR A
     1 TYPICAL TEMPERATURE CYCLE, AF !'
      READ(*,*) AF
      WRITE(*,*) 'AF=',AF
      WRITE(*,*) 'GIVE THE FIELD MTBF GOAL, MG, PLEASE!'
      READ(*,*) MG
      WRITE(*,*)'FIELD MTBF GOAL ==',MG
      WRITE(*,*) 'ENTER DESIRED ACCURACY FOR Tb !'
      READ(*,*) ERROR
      WRITE(*,*) 'ENTER YOUR INITIAL GUESS FOR Tb !'
      READ(*,*) TB1
 250  TT=(TB1/ETA)**BETA
      CALL DQDAGI(FUNC,TT,1,0.0,0.00001,INT,ERREST)
      TB2=ETA*(DLOG((MG*BETA)/(AF*ETA*INT)))**(1.0/BETA)
      ERR=ABS((TB2-TB1)/TB1)
      IF(ERR.LT.ERROR)GOTO 260
      TB1=TB2
      GOTO 250
 260  WRITE(*,*)'THE CORRESPONDING BURN-IN TIME IS',TB2
      STOP
      END
C
```

```
C     THE INTEGRAND FOR THE INCOMPLETE GAMMA FUNCTION
C
      FUNCTION FUNC(Z)
      IMPLICIT DOUBLE PRECISION (A-H, O-Z)
      COMMON /PAR/BETA,ETA
      FUNC=Z**((1.0/BETA)-1)*EXP(-Z)
      RETURN
      END
```

Chapter 14

GUIDELINES FOR BURN-IN QUANTIFICATION AND OPTIMUM BURN-IN TIME DETERMINATION

14.1 INTRODUCTION

Burn-in data modeling and optimum burn-in time determination have been investigated extensively in this book. To facilitate the better understanding and appropriate application of the methodologies and mathematical models presented in this book, summarizing guidelines will be very helpful [1]. This chapter is developed to serve this purpose.

14.2 GENERAL PROCEDURE

Typically, a general procedure for the quantification of a burn-in test and the optimum burn-in time determination consists of the following steps:

Step 1- Times-to-failure data collection for a component under study.

Step 2- Initial data analysis.

Step 3- Formal data analysis parametrically or non-parametrically.

Step 4- Burn-in optimization objective function development.

Step 5- Optimum burn-in time determination parametrically or non-parametrically.

Step 6- Burn-in time justification and adjustment.

The above 6-step procedure is illustrated in Fig. 14.1.

If a burn-in is to be conducted at a stress level that is different from that at which the times-to-failure data were collected, say a higher stress level, then the acceleration factor needs to be determined for converting a burn-in time at one stress level to that at another stress level. Steps 7 through 10 in Fig. 14.1 are provided for the accelerated burn-in time determination:

Step 7- Burn-in acceleration: stress selection.

Step 8- Acceleration factor determination.

Step 9- Reduced burn-in time determination.

Step 10- Final burn-in time justification and adjustment.

The more detailed guidelines for implementing these steps are presented next.

14.3 TIMES-TO-FAILURE DATA COLLECTION

Times-to-failure data are the key to both burn-in quantification and optimum burn-in time determination. Typically, these data can be collected from

1. previous burn-in tests,

2. previous life tests, or

3. field operation records, such as failure analysis reports.

These burn-in tests or life tests may be conducted under normal stress levels or under accelerated stress levels. It should be noted that the optimum burn-in time determined directly from these data is only valid for the same stress level under which these tests were conducted and times-to-failure data collected. For a burn-in to be conducted under a different stress level, the optimum burn-in time should be the one converted from the previous stress level using an appropriate acceleration factor.

Depending on how the tests were planned and conducted, or how the field failures were monitored and recorded, the times-to-failure data can be

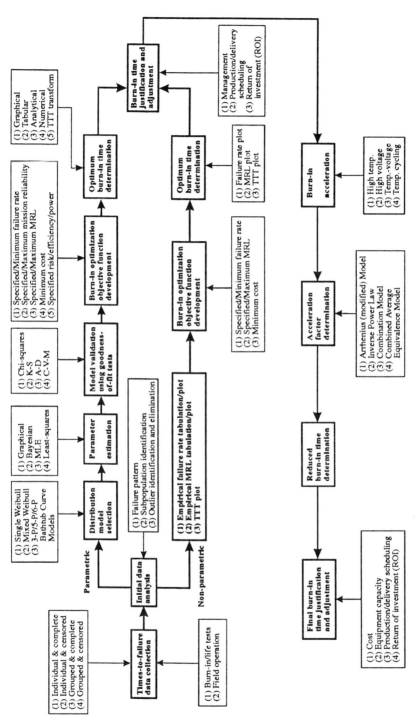

Fig. 14.1 – A flow diagram of the general procedure for burn-in quantification and optimization.

624 BURN-IN QUANTIFICATION GUIDELINE

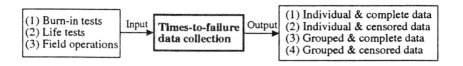

Fig. 14.2– Times-to-failure data collection.

1. individual (due to continuous monitoring) or grouped (due to periodic monitoring), and

2. censored (time terminated, failure terminated, or suspended) or non-censored (complete, or all tested to failure).

Figure 14.2 shows where we may collect data and what these collected data may look like. Note that the analysis methods and procedures may vary from one data format to another.

Data presented in Chapters 5 through 11 are from un-accelerated burn-in or un-accelerated life tests, while Chapters 12 and 13 present data under accelerated stress conditions. Various data formats, individual or grouped, complete or censored, appear in the examples throughout the whole book and are listed in Table 14.1.

14.4 INITIAL DATA ANALYSIS

Initial data analysis is especially important to avoid "garbage-in-garbage-out" in later analyses. Usually, graphical plotting is a very economic approach which gives a very straight forward indication about the failure pattern, such as decreasing failure rate (DFR), increasing failure rate (IFR), constant failure rate (CFR), mixed failure rate (MFR); and subpopulation behavior and existence of outliers. For example, the empirical failure rate plots, mean residual life (MRL) plots, total time on test (TTT) plots, and Weibull probability paper plots serve this purpose very well. The first three plotting methods do not require any distribution assumption and are non-parametric while the fourth method automatically assumes a Weibull or a mixed-Weibull life distribution for the plotted data. But keep in mind that the Weibull

INITIAL DATA ANALYSIS

TABLE 14.1– Data formats and the corresponding examples in which they appear.

No.	Data format	Example number
1	Individual & complete	None
2	Individual & censored	5-1; 5-2; 11-4
3	Grouped complete	8-10
4	Grouped & censored	5-3; 5-4; 5-5; 5-6; Chapter 6; 7-1; 7-2; 7-3; 7-4; 7-5; 7-6; 7-7; 7-8; 7-9; 8-3; 8-6; 8-7; 8-8; 8-9; 11-5

TABLE 14.2– Chapters containing various plotting techniques which may be used for initial burn-in data analysis.

No.	Plotting technique	Chapter/section/ subsection
1	Empirical failure rate plot	Chapter 5
2	MRL plot	8.4; 8.8
3	TTT plot	11.5; 11.9
4	Weibull probability paper plot	5.4.1; 7.1

distribution is a very flexible distribution and can treat other distributions as its special cases which makes it very universal. Chapters which contain these plotting procedures are listed in Table 14.2. Figure 14.3 is a flow diagram for the initial data analysis.

If any points appear to be outliers, the test procedure and data records need to be checked. Any confirmed outliers should be removed from the original data set before further analyses. Initial data analysis is also very helpful later in choosing the burn-in models.

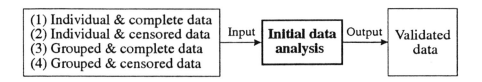

Fig. 14.3 – Initial data analysis.

14.5 PARAMETRIC BURN-IN DATA ANALYSIS AND OPTIMUM BURN-IN TIME DETERMINATION

14.5.1 PARAMETRIC BURN-IN DATA ANALYSIS

Parametric burn-in data analysis basically consists of three steps; i.e., model selection, parameter estimation, and goodness-of-fit testing. This is shown in Fig. 14.4.

1. Model Selection

The initial data analysis in Section 14.4 provides an indication about the failure pattern, and subpopulation behavior, which facilitates the selection of a proper model for describing the failure process during burn-in. The commonly used math models describing the failure process during burn-in can be classified into two major categories; i.e., the mixed (Weibull) distribution family (which includes the single Weibull distribution and the mixed-exponential distribution as its special cases) and the bathtub curve family. Chapters that cover these models are listed in Table 14.3.

2. Parameter Estimation

Once a math model is selected to describe the burn-in data, the next task is to estimate the model parameters. Basically, the parameter estimation methods for the burn-in models can be divided into the following three categories:

TABLE 14.3– Models and their selection criteria for burn-in data analysis.

No.	Model	Chapter /section	Selection criteria
1	Single Weibull model ($\beta < 1$)	13.6.2	(1) Failure rate is decreasing, or (2) MRL plot is increasing, or (3) TTT plot is convex.
2	Mixed-exponential model	10.2; 10.3	(1) Failure rate is decreasing and/or then becomes stable, or (2) MRL plot is increasing and/or then becomes stable, or (3) TTT plot is convex.
3	Mixed-Weibull model	5.1; 5.4	(1) Multiple failure modes or subpopulations are observed, or (2) MRL plot is concave, or (3) TTT plot is "S" shaped or "multi-S" shaped.
4	Bathtub curve models	5.2	(1) Multiple failure modes or subpopulations are observed and the failure rate plot indicates a bathtub curve pattern, or (2) MRL plot indicates an upside-down bathtub curve pattern, or (3) TTT plot is "S" shaped or "multi-S" shaped.

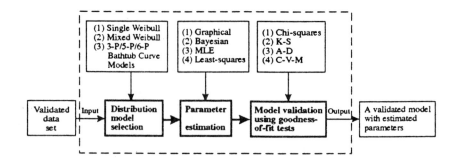

Fig. 14.4 – Parametric burn-in data analysis.

TABLE 14.4– Burn-in model parameter estimation methods.

No.	Method category	Method name	For which model	Chapter /section /subsection
1	Graphical method	Jensen's graphical method (Weibull probability plotting method)	Well-separated mixed-Weibull (including mixed-exponential); Single Weibull ($\beta < 1$)	5.4.1
2	Analytical method	MLE method	Mixed-Weibull	5.4.3
		Least-squares method	Bathtub curve model 1	5.4.4
3	Graphical-analytical method	Bayesian method	Mixed-Weibull (including mixed-exponential)	5.4.2

1. Graphical methods.

2. Analytical methods.

3. Graphical-analytical methods.

Chapters that cover these methods are listed in Table 14.4.

For detailed description of Weibull probability plotting method, the readers are referred to [2, Chapter 6; 3, Chapter 12]. For other parameter estimation methods, such as matching moments, modified matching moments, etc., the readers are referred to [3, Chapter 18].

PARAMETRIC BURN-IN QUANTIFICATION

3. Model Validation Using Goodness-Of-Fit Tests

Once the parameters are estimated for the selected model, the next question is whether or not the selected model fits the data. A goodness-of-fit test is a very effective tool to check the validity of the model. The following four goodness-of-fit tests (including their modified versions) are frequently used:

1. Chi-squared test.
2. Kolmogorov-Smirnov (K-S) test.
3. Anderson Darling test.
4. Cramer-von Mises test.

In Example 5-3 of Chapter 5, the K-S test is used to compare the goodness of fit of the mixed-Weibull distribution models using two different parameter estimation methods. For more details of the K-S test, the readers are referred to [3, Chapter 20]. For the other goodness-of-fit test methods, the readers are referred to [3, Chapters 19 and 21].

14.5.2 OPTIMUM BURN-IN TIME DETERMINATION

Having obtained a validated model and obtained the estimates of its parameters, the next task is to determine the optimum burn-in time using this model and its parameter estimates in conjunction with the burn-in objectives. Determining the optimum burn-in time includes the following two steps:

1. Burn-in optimization objective function development.
2. Optimum burn-in time determination.

This procedure is illustrated in Fig. 14.5.

1. Burn-in Optimization Objective Function Development

Depending on the optimization criteria, the burn-in objectives can be

1. for a specified or for the minimum failure rate at the end of burn-in,
2. for a specified post-burn-in failure rate at the end of a mission,
3. for a specified or for the maximum post-burn-in mission reliability,

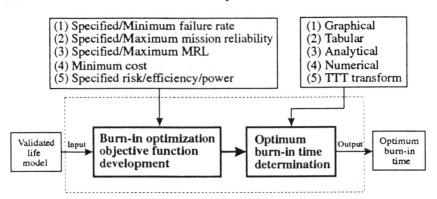

Fig. 14.5 – Procedures for optimum burn-in time determination.

4. for a specified or for the maximum post-burn-in mean residual life (MRL).

5. for the minimum cost,

6. for the maximum profit, and

7. for the specified burn-in risk, burn-in efficiency, or burn-in power function value.

Chapters that cover these objective functions are listed in Table 14.5.

2. Optimum Burn-in Time Determination

Having established an objective function, the next task is to determine the best value for burn-in time such that the burn-in objective will be achieved; i.e., minimized, or maximized, or made equal to the specified value. Various methods are available to accomplish this task. They include the following:

1. **Graphical Method I**: Plot the objective function and read the burn-in time directly from the plot.

2. **Graphical Method II Using The TTT Transforms**: Construct the objective function in terms of the TTT transform and read the burn-in time directly from the TTT transform curve.

3. **Tabulation Method**: Calculate and list in a table the objective function values corresponding to various burn-in times around the neighborhood of the optimum burn-in time and read the optimum burn-in time from the table using linear interpolation.

TABLE 14.5– Various objective functions for burn-in optimization.

No.	Burn-in objective function	Chapter/section /subsection
1	Specified/minimum failure rate at the end of burn-in	7.1; 7.2; 7.3; 7.4; 7.6
2	Specified post-burn-in failure rate at the end of a mission	10.6
3	Specified/maximum post-burn-in mission reliability	7.5; 7.6; 10.4
4	Specified/maximum post-burn-in mean residual life (MRL)	8.7; 10.5; 11.8; 13.6
5	Minimum cost	Chapter 9; 10.11; 10.12.3; 11.7.1; 11.7.2
6	Maximum profit	11.7.3
7	Specified burn-in efficiency	10.7
8	Specified burn-in power	10.8
9	Specified burn-in risk	10.9

4. **Analytical Method I**: Let the objective function be equal to the specified value and solve for the corresponding burn-in time.

5. **Analytical Method II**: Take the derivative of the objective function, let it equal to zero and solve for the optimum burn-in time.

6. **Numerical Optimization Methods**: Minimize or maximize the objective function using such numerical methods as the Quasi-Newton, the Simplex, the Hooke-Jeeves Pattern Moves, the Rosenbrock Pattern Search, and others.

Chapters that cover these methods are listed in Table 14.6.

14.6 NON-PARAMETRIC BURN-IN DATA ANALYSIS AND OPTIMUM BURN-IN TIME DETERMINATION

14.6.1 NON-PARAMETRIC BURN-IN DATA ANALYSIS

Studying burn-in data analytically or graphically without making any distribution assumption is called non-parametric burn-in data analysis. The following three methods are presented in this book:

1. Empirical failure rate function tabulation or plotting method.
2. Empirical MRL function tabulation or plotting method.
3. Scaled TTT function tabulation or plotting method.

Table 14.7 lists the specific chapters and sections that cover these methods. Figure 14.6 shows the typical burn-in data patterns using these three methods; i.e.,

1. a decreasing or a decreasing-and-then-stable pattern for the empirical failure rate plot,
2. an increasing or an increasing-and-then-stable pattern for the empirical MRL plot, and
3. a convex increasing or a convex-and-then-45°-diagonally increasing pattern for the scaled TTT plot.

These graphical patterns correspond to the early-failure period or the early-plus-chance-failure period on the conventional bathtub curve. Generally, the third period, or the so-called wearout failure period, does not appear during a relatively short burn-in.

TABLE 14.6– Optimization methods used for various burn-in objective functions.

No.	Burn-in objective function	Method for optimum burn-in time determination	Chapter/section /subsection
1	Specified failure rate at the end of burn-in	Analytic method I	7.4
		Graphical method I	7.6
2	Minimum failure rate at the end of burn-in	Graphical method I	7.1; 7.6
		Analytic method I	7.2; 7.3
3	Specified post-burn-in failure rate at the end of a mission	Analytic method I	10.6
		Graphical method I	*
4	Specified post-burn-in mission reliability	Analytic method I	7.5; 10.4
		Graphical method I	7.6
5	Specified post-burn-in MRL	Analytic method I	8.7; 10.5
		Graphical method I	*
6	Maximum post-burn-in MRL	Analytic method I	13.6
		Graphical method I	*
7	Minimum Cost	Graphical method II	11.8
		Analytic method II	9.2
		Graphical method I	9.1; 9.3; 9.4; 10.12.3
		Tabular method	9.1; 9.3; 9.4; 10.12.3
		Numerical methods	10.11
		Graphical method II	11.7.1; 11.7.2
8	Maximum profit	Graphical method II	11.7.3
9	Specified burn-in efficiency	Analytic method I	10.7
10	Specified burn-in power	Analytic method I	10.8
11	Specified burn-in risk	Analytic method I	10.9

* For these cases, the Graphical Method I is not used in the book. But it can be done fairly easily by reading the optimum burn-in time directly from the plot. The readers are encouraged to try it themselves.

TABLE 14.7– Non-parametric burn-in data analysis methods and their locations.

No.	Non-parametric methods		Chapter/section
1	Empirical failure rate	Function tabulation	Chapter 6
		Plot	Chapter 6
2	Empirical MRL	Function tabulation	8.4
		Plot	8.8
3	Scaled TTT	Function tabulation	11.5
		Plot	11.9

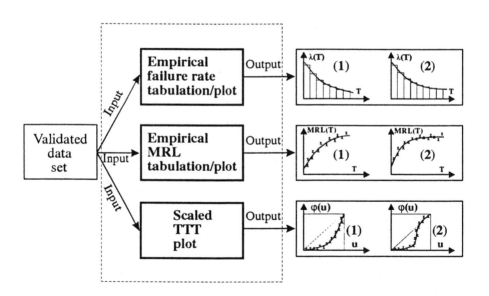

Fig. 14.6– Non-parametric burn-in data analysis methods and the typical patterns of their output results.

TABLE 14.8– Objective functions for the non-parametric burn-in optimization.

No.	Objective function	Chapter/section
1	Specified/minimum failure rate at the end of burn-in	Chapter 6
2	Specified/maximum post-burn-in MRL	8.7; 8.8; 11.8; 11.9
3	Minimum cost	11.7.1; 11.7.2; 11.7.3; 11.9

14.6.2 BURN-IN TIME OPTIMIZATION OBJECTIVE FUNCTION DEVELOPMENT

Corresponding to the optimization criteria and the available non-parametric methods, the burn-in objectives can be

1. for a specified or for the minimum failure rate at the end of burn-in,

2. for a specified or for the maximum post-burn-in MRL, and

3. for the minimum cost.

Chapters that cover these objective functions, which can be optimized using the presented non-parametric methods, are listed in Table 14.8.

14.6.3 OPTIMUM BURN-IN TIME DETERMINATION

Having established an objective function, the next task is to determine, using the non-parametric method, the best value for the burn-in time such that the burn-in objective will be achieved; i.e., minimized, or maximized, or made equal to the specified value. Table 14.9 lists the chapters and sections that cover these non-parametric optimization methods.

14.7 BURN-IN TIME JUSTIFICATION AND ADJUSTMENT

The optimum burn-in times determined using the parametric methods summarized in Section 14.5, or using the non-parametric methods

TABLE 14.9– Non-parametric optimization methods used for various burn-in objective functions.

No.	Burn-in objective function	Method for optimum burn-in time determination	Chapter /section /subsection
1	Specified /minimum failure rate at the end of burn-in	Empirical failure rate plot method	Chapter 6
		Empirical failure rate function tabulation method*	Chapter 6 **
2	Specified post-burn-in MRL	Empirical MRL function tabulation method*	8.8 **
		Empirical MRL plot method	8.8
3	Maximum post-burn-in MRL	Empirical MRL function tabulation method*	8.8
		Empirical MRL plot method	8.8
		Scaled TTT plot method	11.9
4	Minimum cost	Scaled TTT plot method	11.9

* Tabulation method: list the objective function values in a table around the neighborhood of the optimum burn-in time and read the optimum burn-in time from the table using linear interpolation.

** The tabulated objective function values are in the examples of these sections. But the optimum burn-in times are determined graphically, not directly from the tables. The readers are encouraged to try the tabulation method themselves.

BURN-IN TIME JUSTIFICATION AND ADJUSTMENT 637

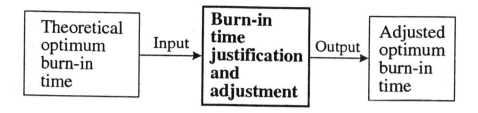

Fig. 14.7– Burn-in time justification and adjustment.

summarized in Section 14.6 are theoretical values. They need to be justified and are subject to adjustment if other factors need to be considered. This is shown in Fig. 14.7.

For example, a burn-in time may be optimum because it minimizes the failure rate function, or maximizes the post-burn-in MRL function, or maximizes the post-burn-in mission reliability, etc. But it may be too expensive. Then, an optimization of the original objective function with an added cost constraint function may be conducted to come up with a new but more justified burn-in time.

For another example, a burn-in time may be optimum because it minimizes the total expected cost per unit time of burn-in. However, it may be too short to satisfy a specified failure rate goal, or a post-burn-in MRL goal, or a post-burn-in mission reliability goal, etc. Then, an optimization of the cost objective function with an added failure rate (or MRL, or mission reliability) constraint may be conducted to come up with a new, but more justified burn-in time.

These are called the constrained burn-in optimizations. For constrained optimization methods, the readers are referred to [4].

Sometimes, the objective function is not very sensitive around the neighborhood of its optimum point, such as the one shown in Fig. 14.8 where the cost objective function is very flat around its minimum point T_b^*. This flatness or shallowness offers us a very good justification for adjusting the burn-in time to meet the production and delivery scheduling constraints.

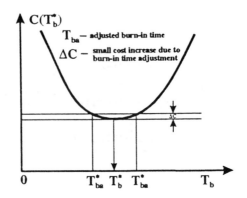

Fig. 14.8– Cost function flatness and burn-in time adjustment.

14.8 ACCELERATED BURN-IN AND BURN-IN TIME CONVERSION FROM ONE STRESS TO ANOTHER STRESS

It should be pointed out that an optimum burn-in time obtained using any one of the methods presented in the preceding sections is only valid for the stress type and level under which the original burn-in or life test was conducted and the times-to-failure data were collected. If a future burn-in is to be conducted at a different stress level, say at a higher stress level, then a new burn-in time needs to be determined using the following equation:

$$T_{b2}^* = \frac{T_{b1}^*}{A_F(S_1, S_2)},$$

where

S_1 = previous stress level,
S_2 = new stress level,
T_{b1}^* = optimum burn-in time at previous stress level, S_1,
T_{b2}^* = optimum burn-in time at new stress level, S_2,

and

$A_F(S_1, S_2)$ = acceleration factor due to stress change from S_1 to S_2.

This procedure is illustrated in Fig. 14.9. The following four (4) burn-in stress types are introduced and their acceleration factors quantified in this book:

ACCELERATED BURN-IN

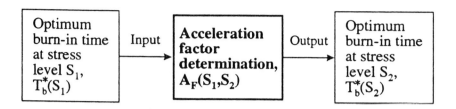

Fig. 14.9– Burn-in time conversion from one stress level to another stress level.

TABLE 14.10– Locations where the acceleration factors for the various stress types can be found.

No.	Stress type	Acceleration model	Chapter/section
1	Constant temperature	Arrhenius (including modified) model	12.2
2	Voltage bias	Inverse power law	12.3
3	Constant temp. plus voltage bias	Combination (including modified) model	12.4
4	Temperature cycling	Combined average equivalence model	13.3; 13.4

1. Constant temperature.
2. Voltage bias.
3. Constant temperature plus voltage bias.
4. Temperature cycling.

Depending on the stress types used for burn-in, the expressions of the acceleration factors vary. Table 14.10 lists the locations where the acceleration factors for these four stress types can be found.

For the detailed definition, quantification and application of each stress type and its acceleration model, the readers are referred to Chapters 12 and 13, respectively.

REFERENCES

1. Leemis, L. M. and Beneke, M., "Burn-in Models and Methods: A Review," *IIE Transactions*, Vol. 22, No. 2, pp. 172-180, 1990.

2. Kececioglu, D., *Reliability Engineering Handbook*, Vol. 1, Prentice Hall, Upper Saddle River, NJ, 720 pp., 1991, Fifth Printing, 1996.

3. Kececioglu, D., *Reliability & Life Testing Handbook*, Vol. 1, Prentice Hall, Upper Saddle River, NJ, 917 pp., 1993.

4. Press, W. H., et al., *Numerical Recipes*, Cambridge University Press, 963 pp., 1992.

Index

A

AC, 56
Accelerated aging, 29
Accelerated burn-in, 394, 527, 638
  using elevated temperature, 528
  using elevated temperature and voltage bias, 565
  using elevated voltage bias, 563
  using temperature cycling, 573
Accelerated test, 29
Acceleration factor, 29, 412, 577
  due to reaction rate stress, 579
    first modified form, 580
    second modified form, 580
  due to temperature change rate stress, 579
  for the Arrhenius model, 528
  for various stress types, 639
  summary, 639
Acceleration spectral density, 29
Acceleration test, 30
Acceptance test, 30
Acoustic vibration, 30
Activation energy, 30, 528, 533, 543
  experimental determination, 534
  of main subpopulation, 536
  of various components, 536
  of various failure mechanisms, 536
  of weak subpopulation, 536
  physical meaning, 533
  provided, 542
  required, 542
  temperature dependency, 529, 541, 542, 565, 566
Adjustment of burn-in time, 635
AFC, 56
Ag, 56
Age replacement policy, 437
Aging, 30
Aging acceleration factor, 542
Aging properties analysis, 437
AGREE, 56
AI, 56
AIF, 56
AIM, 56
Airborne, 30
Airborne inhabited (AI), 30
Airborne uninhabited (AU), 30
Al, 56
ALC, 56
ALSTTL, 56
Am, 57
Ambient environment, 30
Ambient test, 31
Anderson Darling test, 629
AOQ, 57
AOQL, 57
AQL, 7, 57
AQR, 57
ARL, 57
Arrhenius Law, 579

Arrhenius model, 527, 528
  acceleration factor, 528
  caution, 532
  modification, 544
  modification by Pugacz & Muraszkiewicz, 553
  modified, 559
  pitfall, 541
Arrival quality, 31
As, 57
ASICs, 57
Assembly level, 31
ASTTL, 57
ATE, 57
Attribute, 31
Attributes testing, 31
Au, 57
AU (Airborne Uninhabited), 57
AUF (Airborne Uninhabited Fighter), 57
Autopsy, 31
Average outgoing quality level (AOQL), 31
Axis, 31

B

B.S., 57
Bad component, 378, 405, 406
Bad subpopulation, 536
  activation energy, 536
Barlow, 416, 437
Bartholomew, 437
Basquin's exponent, 39
Batch, 31
Bathtub curve, 31, 69, 70
  decomposition, 71
  from stress-strength interference, 90
Bathtub curve models, 117
Bathtub Model 1, 162, 229, 236, 280, 283
Bathtub-shaped failure rate curve, 274
Bayesian approach
  burn-in goodness measures, 425
  cost model, 421
  for burn-in time determination, 416
  joint prior distribution, 417
  optimum burn-in duration, 421
  why, 416
Bayesian method, 125, 148, 158
Bazovsky model, 527
BBBB, 437
Belonging probability, 128, 145
  posterior, 126, 138
  prior, 126, 135
Benign environment, 32
Bergman, 437, 472, 488
Bergman-Klefsjo's cost model, 472
Beta prior family, 417
Bimodal distribution, 32
  for initial strength, 101
Bimodal initial strength distribution, 96
Bimodal life distribution, 221
Bimodal mixed distribution, 536
Bimodal mixed population parameter estimation, 119
Bimodal mixed-exponential distribution, 375, 377, 457, 594
Bimodal mixed-Weibull distribution, 237, 281, 284, 457
  for initial strength, 101
Bimodal strength distribution, 88, 94
Bimodal times-to-failure distribution, 96, 101
Bimodality, 72, 106
Binney, 578
BIT, 57
bit, 57
BITE, 57

REFERENCES 643

BITS, 9
Black-box testing, 32
Board level, 32
BOC, 57
Bogey test, 32
Boltzmann's constant, 528
Branching processes, 256
Bremner, 437
Broad-band vibration, 32
Brunk, 437
Bulk defect, 32
Burn-in, 32
   accelerated, 527, 573, 638
   acronyms, 56
   classifications, 15
   definitions, 15
   documents, 15, 19
   failure pattern, 84
   failure phenomena, 69
   how, 5
   terminologies, 29–56
   test conditions, 15
      by MIL-STD-883C, 19
   test temperature, 25
   versus ESS, 16
   why, 1
Burn-in time conversion, 638
Burn-in chamber, 32
Burn-in cost, 337
Burn-in cost categories, 338
Burn-in data analysis
   initial, 624, 625
   non-parametric, 632
   non-parametric methods, 632
   parametric, 626
Burn-in data collection, 622, 624
Burn-in data format
   grouped, 135, 156, 262, 265, 460, 463, 467
   individual, 261, 263, 458, 461, 465
   summary, 624
   time-terminated, 155
   ungrouped, 126, 155, 261, 263, 458, 461, 465
Burn-in effectiveness, 32
Burn-in efficiency, 393
Burn-in goodness measures
   Bayesian approach, 425
Burn-in interval, 398
Burn-in math models
   bathtub curve models, 117
   mixed-Weibull model, 115
   parameter estimation, 115
   selection criteria, 119
Burn-in methods, 16
   effectiveness, 16, 19
   versus failure mechanisms, 19
Burn-in model selection, 119
   criteria, 119, 626
Burn-in optimization
   general procedure, 621
   guidelines, 621
   non-parametric, 635
   objective functions, 629, 630, 635
Burn-in period, see Early failure period
Burn-in power function, 398
Burn-in quantification
   general procedure, 621
   guidelines, 621
Burn-in residue, 393, 394, 416
Burn-in risk, 398
   definition, 398
Burn-in strength, 426
Burn-in time adjustment, 635
Burn-in time conversion, 570
Burn-in time determination
   based on mean residual life, 255
   Bayesian approach, 416
   directly from MRL function, 286
   directly from MRL plot, 286, 287

for a specified burn-in efficiency, 393
for a specified burn-in power, 396
for a specified burn-in risk, 398
for a specified failure rate, 233, 388
for a specified MRL, 278, 286, 384
for a specified reliability, 235, 382
for the maximum MRL, 282, 287, 492
for the maximum profit, 472
for the minimum cost, 337, 404, 472
for zero-slope failure rate curve, 229
  with error, 229
from higher stress level, 569
graphical approach, 238
  using failure rate function, 238
  using reliability function, 239
non-parametric method, 632, 635
parametric method, 626, 629, 630
quick-calculation approach, 211
subpopulation truncation approach, 221
using bimodal distribution, 221
using TTT transform, 437, 472, 492
Burn-in time justification, 635
Burn-in time reduction, 527, 638
  using elevated temperature, 528, 531, 559
  using elevated temperature and voltage bias, 565
  using elevated voltage bias, 563
  using temperature cycling, 573

C

Capacity deterioration, 91
Capacity variability, 91
Caution for Arrhenius model, 532
CCCs, 57
CCD, 57
CDE, 57
CERDIPs, 57
CFR, 441, 444
Chamber, 33
Chance failure period, see Useful life period
Chance failures, 215
Characteristic life, 565
Chen, 90, 106
Chi-squared test, 629
CIM, 57
Circuit board, 33
Cl, 57
Class B device, 33
Class S device, 33
Classical approach, 416
CML, 57
CMOS, 21, 57, 72, 120, 123, 536, 543
CMOS/SOS, 58
CND, 58
COB, 58
COE, 58
Coefficient of thermal expansion
  of bond pad, 604
  of case, 609
  of die, 607, 608
  of substrate, 602, 604, 607–609
  of wire, 602, 604
Coefficient of variation, 453
Coffin-Manson coefficient, 608
Coffin-Manson relation, 608

Cold cracking, 33
Cold soak, 33
Cold solder connection, 33
Cold solder joint, 33
Cold starts, 33
Collection of data, 622, 624
Combination Model, 527, 565
Combination model
    conventional, 565
    modified, 566
Combined environmental test, 33
Common cause failures, 34
Comparison
    of Burn-in and ESS, 16
    on parameter estimates, 135, 148, 154, 158
    on the goodness of fit, 154
Complete data, 458
Component defect, 34
Component level, 34
Concave, 448, 469
Condition monitoring, 43
Conditional cost, 421
Conditional exponential pdf, 420
Conditional probability, 136, 137
Conditioning, 30, 34
Constrained optimization methods, 637
Contraction, 578
Control chart, 34
Control limit, 34
Conventional Combination Model, 565
Conversion of burn-in times, 570, 638
Convex, 448, 469
Cooling rate stress, 578
COQ, 58
Corrective action effectiveness, 34
Corrective actions, 34
Correlation coefficient, 555
    for modified Arrhenius model, 555

COS, 58
Cost
    composition, 337
    consideration in burn-in, 337
    minimization, 337, 404, 472
    versus burn-in, 337
Cost effectiveness, 34
Cost model
    Bayesian approach, 421
    minimization, 410
    of Bergman-Klefsjo, 472
    of Weiss and Dishon, 481
    two-level burn-in, 409
Cost Model 1, 338
Cost Model 2, 344
Cost Model 3, 352
Cost Model 4, 363
Cost of failures
    during burn-in, 402
Cost of performance, 34
Cr, 58
Crack growth, 106
Crack size, 607
    critical, 607
    initial, 607
Cramer-von Mises test, 629
Crazing, 34
Creep, 34
Criteria
    for burn-in model selection, 626
Cu, 58
CVD, 58

D

Damage, 34
Damaging overstress, 35
Data analysis
    initial, 624, 625
    non-parametric, 632
    non-parametric methods, 632
    parametric, 626
Data collection, 622, 624
Data format

grouped, 135, 156, 262, 265, 460, 463, 467
individual, 261, 263, 458, 461, 465
summary, 624
time-terminated, 155
ungrouped, 126, 155, 261, 263, 458, 461, 465
Davidon-Fletcher-Power, 412
DC, 58
DE, 58
Debugging period, *see* Early failure period
Decision costs, 417
Defect, 35
Defect density, 35
Defect-free, 35
Defective, 35
Defective unit, 35
Definitions
 for burn-in, 15
Degradation, 35
Degradation factor, 35
Degradation failure, 35
Dependent failure, 35
Dependent failure data analysis, 437
Derating, 35
Derating factor, 35
Design adequacy, 354
Design defect, 35
Design deficiency, 35
Design fault, 36
Design margin, 36
Design of experiments (DOE), 36
Design weakness, 36
Designed test, 36
Desired structure
 for objective function, 468
Detectable failure, 36
Detection efficiency, 36
Deterioration Monitoring Model, 527

Device hours, 212
DFR, 58, 441, 444
DFRA, 58
Diagnostic coverage, 39
DID, 58
Die and substrate adhesion
 thermal fatigue life prediction, 607
Die fracture, 606
Die thermal fatigue life prediction, 606
Dietrich, 106
Differential equation, 469
DIFR, 442, 444
DIMRL, 260
DIPs, 58
Dishon, 481, 488
Distributed strength, 85
Distributed stress, 85
Distribution-Free Tolerance Limits Model, 527
DMOS, 58
DMRL, 58, 259
DOA, 58
Documents
 for burn-in, 15, 19
DOE, 58
Double-S-shaped, 75
DR, 58
DRAM, 20, 21, 58
DTC, 58
DTL, 58
DUT, 58
Dwell, 577
Dynamic burn-in, 6, 17, 36
Dynamic programming, 256

E

EAPROM, 58
Early failure period, 70
EBIC, 59
ECL, 59
EDAX, 59
EED, 59

EEPROMs, 59
Effectiveness
  of burn-in methods, 16, 19
Efficiency
  of a burn-in, 393
EIA, 8
Electromigration, 36
Elevated temperature, 528
EMI, 59
EML algorithm, 155, 158
Empirical failure rate, 215
Empirical mean residual life, 260
Empirical MRL function, 260,
  286, 293, 297, 302, 308
  for a censored sample, 263
  for a complete sample, 261
Empirical MRL plot, 286, 287,
  293, 299, 302, 310
Endurance strength, 36
Energy barrier, 534
Engelmaier, 610
Environmental cycle, 37
Environmental sensitivity, 37
Environmental stress screening, 37
Environmental testing, 37
EOS, 59
EPROMs, 59
Equipment, 52
Equipment power on-off cycle, 37
Equivalent acceleration factor, 579
  using the modified Arrhenius model, 580, 584
Error
  on zero slope, 229
ESD, 10, 59
ESS, 59
  versus burn-in, 16
ESS profile, 37
ESSEH, 37, 59
ETM, 59
eV, 59

Expansion, 578
Expected number of failures, 402
Expected number of renewals,
  see Expected number of
  failures
Experimental determination
  of activation energy, 534
Exponential distribution, 268
  scaled TTT transform, 442
  TTT transform, 445
Exponential prior pdf, 417
Exponential process, see Exponential renewal process
Exponential renewal process, 409
Extreme-value distribution, 117
Eyring Model, 527

F

Factorial experiment, 38
Factory failure cost savings, 38
Failure categories
  in microelectronic devices, 574
Failure humps, 79
Failure mechanism
  activation energy, 536
  temperature dependency, 574
Failure mechanisms, 1
  versus burn-in methods, 19
  versus process steps, 1
  versus technologies, 1
Failure pattern
  from stress-strength interference, 88
  physical explanation, 84
Failure physics model, 106
Failure process
  during burn-in, 69
Failure rate
  plot, 215
  specified value, 233, 388
  zero slope, 229
Failure rate curve
  derived from SSI, 106

Failure rate goal, 233, 388
Failure-free period, 38
Failure-free test, 38
Failure-terminated data, 465
Fallout, 38
FAMECA/FMECA, 59
FAMOS, 59
Fatigue ductility coefficient, 38, 610
　of die, 608
　of substrate, 609
Fatigue ductility exponent, 39, 610
Fatigue failures, 39
Fatigue resistance, 39
Fatigue strength, 36, 39
Fatigue strength coefficient, 39
Fatigue strength deviation, 39
Fatigue strength exponent, 39
Fatigue strength limit, 39
Fault detection, 39
Fault detection coverage, 39
Fault detection efficiency, 39
Fault isolation, 39
FBT, 59
FET, 59
FFAT, 59
FFT, 59
Field, 39, 344
Field failure cost savings, 39
Field replaceable unit (FRU), 40
Field warranty rate, 40
Final assembly level, 40
FIT, 59
Five-parameter model, 118
Fixed sinusoidal vibration, 40
Fixture, 40
Fixturing, 40
FL, 59
FLOTOX, 59
FMA, 59
FMEA, 59
FOM, 59
FPHP, 60

FPPM, 60
FRACAS, 60
Fractional defective, 40
Fractional factorial design, 40
Fractional rank plotting method, 125, 139, 148
Fractional replicates, 40
Fracture mechanics, 106
Fracture of die, 606
FRB, 60
Freak distribution, 88
Freak failures, 40
Freak subpopulation, 72
Frequency factor, 528
FRU, 60
FTA, 60
FTCs, 60
FTTL, 60
Functional test, 40

G

Ga, 60
GaAs, 20, 21, 60
Gamma distribution, 267, 356
　scaled TTT transform, 442
　TTT transform, 449, 450
GB, 60
General procedure
　for burn-in optimization, 621
　for burn-in quantification, 621
Generalized Eyring Model, 527
Generalized Gamma distribution, 352, 356
Geometrical properties
　of TTT transforms, 443
GF, 60
Ghost, 40
GIDEP, 60
GM, 60
Go/No-Go, 40
Go/No-Go testing, 40
Good component, 378, 380, 393, 394, 406, 407, 409, 417, 421, 425–427

Good components, 405
Good subpopulation, 536
   activation energy, 536
Goodness-of-fit tests, 629
Graphical methods
   for burn-in time determination, 238
Graphical optimization, 468, 469
GRB, 60
GRF, 60
GRM, 60
Ground benign environment (GRB), 40
Ground fixed environment (GRF), 41
Ground mobile environment (GRM), 41
Grouped data, 156, 262, 265
Grouped data set, 460, 463, 467
Gruber, 578
Guidelines
   for burn-in optimization, 621
   for burn-in quantification, 621

H

Hard failure, 41
HAST, 60
HC, 60
HCMOS, 60
HD, 60
HDSCs, 60
Heat endurance, 41
Heat shock, 41
Heating rate stress, 578
HEMT, 60
High reliability program, 41
High resistance joint, 33
High temperature reverse bias, 18
High temperature reverse bias (HTRB), 41
High-temperature operating test, 18
High-voltage stress burn-in, 6

High-voltage stress tests, 18, 41
History
   of burn-in, 6
HMOS, 60
Homogeneous Poisson process, 345
Hooke-Jeeves Pattern Moves Method, 547
Hot start, 41
HTCMOS, 60
HTOT, 19, 61
HTRB, 7, 18, 61
HTTL, 61
Humps
   on failure rate curve, 79

I

I^3L, 61
I-V, 61
i.i.d., 155
I/O, 61
IC, 61
ICA, 61
ICs, 61
ICT, 61
IDFR, 442, 444
IDMRL, 61, 260
IEEE, 61
IES, 61
IFR, 61, 441, 444
IFRA, 61
IIL, 61
IMPATT, 61
Imperfection, 42
IMRL, 61, 260
Incipient failure, 42
Incomplete gamma function, 359, 360, 448–450
Individual data set, 261, 263, 458, 461, 465
Infant failures, 42
Infant mortalities, 4
Infant mortality period, *see* Early failure period

Infant mortality subpopulation, 75
Infimum, 440
Inherent defect, 42
Initial data analysis, 624, 625
  plotting techniques, 625
Initial failure period, *see* Early failure period
Initial parameter estimates, 141
Initial strength, 91
  bimodal distribution, 94
Interference of stress and strength, 85
Intermittent failure, 42
Intrinsic failure period, *see* Useful life period
Inverse Laplace transform, 403
Inverse Power Law, 527, 563
ISL, 61

J

Jacobian, 98, 107
JEDEC, 7
Jensen, 88, 94, 106, 338, 340
Jensen's graphical method, 120, 123, 141, 148, 158
Joint prior distribution, 417
Justification of burn-in time, 635

K

K-S test, 140, 148, 629
Kececioglu, 81, 125
Klefsjo, 472, 488
Kolmogorov-Smirnov, 629
Kuo, 15, 363

L

Laplace transform, 403
Latency, 42
Latency time, 42
Latent defect, 35, 42
LBS, 61
LCC, 61
LCCCs, 61
LCMs, 61
LCRU, 61
Least-squares estimates, 535, 547, 555
  for modified Arrhenius model, 547, 555
Least-squares method, 162
Least-squares parameter estimates, 348, 349
LED, 62
Level of assembly, 42
Lewis, 90, 106
Life test, 42
Life-aging, 30
Life-cycle cost, 404
Life-cycle cost model, 407, 409
Life-cycle testing, 42
Likelihood function, 155–157
Limiting final value, 403
Limiting quality level (LQL), 43
Line replaceable unit (LRU), 43
Load variability, 91
Log-log Stress-Life Model, 527
Lognormal distribution, 270
  scaled TTT transform, 442
  TTT transform, 453, 455
Loll, 106
Loss, 416
Lot, 43
Lot tolerance percent defective, 43
LQL, 62
LRU, 62
LSC, 62
LSCS, 62
LSI, 21, 62
LSTTL, 62, 542, 543
LTPD, 43, 62
LTTL, 62

M

Main distribution, 88
Main effect, 43

Main subpopulation, 119, 536, 537
  activation energy, 536
Malfunction, 43
Manufacturing defects, 43
Margin of safety, 43
Marginal failure, 43
Math models
  for burn-in failure process, 115
Maximization, 282, 285–289, 468, 469, 473
Maximum MRL, 282, 287, 492
Maximum profit, 472
Maxwell distribution, 352
MCMs, 62
Mean life, 530
Mean residual life, 43, 255
  after burn-in, 274, 276, 278
  empirical function, 260
  for exponential distribution, 268
  for Gamma distribution, 267
  for lognormal distribution, 270
  for mixed distribution, 272
  for Rayleigh distribution, 271
  for truncated normal distribution, 269
  for Weibull distribution, 266
  math defination, 256
  versus failure rate, 256
  versus life distribution, 258
Mechanical degradation, 578
Median life, 530
Memories, 21
Microelectronic device
  failure categories, 574
Microprocessors, 21
MIL-HDBK-217, 19, 380
MIL-HDBK-217D, 543
MIL-STD-781, 19
MIL-STD-785, 19
MIL-STD-883, 5, 8, 19

MIL-STD-883C, 15, 19
Military standards, 420
Minimization, 239, 305, 312, 365, 404, 412, 413, 416, 424, 468, 469, 473, 481
  DFP method, 412
  of cost model, 410
Minimum cost, 337, 404, 472
MIPS, 62
Mixed distribution, 272, 536
  TTT transform, 456
Mixed-exponential distribution, 375, 377, 406, 594
  TTT transform, 457
Mixed-Weibull distribution, 237, 281, 284, 305
  TTT transform, 457
Mixed-Weibull population, 115, 155
MLE, 62, 155, 156
MNOS, 62
Model identification, 437
Model selection, 626
  criteria, 626
Model validation, 629
Modification of Arrhenius model, 544
Modified acceleration factor, 556
Modified Arrhenius model, 544, 559
  model application, 590
  parameter estimation, 555
Modified Combination model, 566
Modulus of elasticity
  of pad material, 604
  of substrate material, 604
  of wire, 605
Monitoring, 43
MOS, 21, 62, 542, 543
MRL, 62, *see* Mean residual life
  after burn-in, 274, 276, 278
  empirical function, 260
  for exponential distribution, 268

for Gamma distribution, 267
for lognormal distribution, 270
for mixed distribution, 272
for Rayleigh distribution, 271
for truncated normal distribution, 269
for Weibull distribution, 266
math defination, 256
maximization, 492
specified value, 384
versus failure rate, 256
versus life distribution, 258
MRL function, 286
MRL plot, 286
MSI, 21, 62
MTBF, 62
specified, 593
MTBR, 62
MTBUR, 62
Multimodal, 75, 88, 111
Multiple failures, 44
Multiple fault condition, 44
Multiple-S-shaped, 75, 83

N

Nachlas, 578
NBU, 62
NBUE, 62, 260
Next higher level effect, 44
NFF, 62
NMOS, 62, 543
Non-environmental stress screening failures, 44
Non-parametric Acceleration Model, 527
Non-parametric data analysis, 632
Non-parametric methods
for burn-in data analysis, 632
summary, 632
Non-postmortem data, 155, 156
Noncensored data, 458
Nonchargeable failure, 44
Nonconformance, 44

Nonconforming unit, 44
Nonconformity, 44
Nonhomogeneous Poisson process, 344, 345
Nonrelevant failures, 44
Normal distribution, 269
scaled TTT transform, 442
TTT transform, 451, 453
Notching, 44
NPL, 62
NTF, 62
Number of failures
during burn-in, 402
NWU, 62
NWUE, 63, 260

O

Objective function
desired structure, 468
development, 629, 635
for burn-in optimization, 629, 630, 635
summary, 630, 635
OC, 63
OEM, 63
On-off cycling, 44
Operational cycling, 45
Operational readiness, 354
Optimization, 437
graphical procedure, 468, 469
objective functions, 630
of burn-in, 375
using TTT transform, 468, 469
Optimization methods
for various objective functions, 632
non-parametric, 635
summary, 632, 635
Optimum acceleration factor, 432
Optimum Acceleration Model, 527
Optimum burn-in time

based on mean residual life, 255
Bayesian approach, 416, 421, 429, 432
determined from higher stress level, 569
directly from MRL function, 286
directly from MRL plot, 286, 287
for a specified burn-in efficiency, 393
for a specified burn-in power, 396
for a specified burn-in risk, 398
for a specified failure rate, 233, 388
for a specified MRL, 278, 286, 384
for a specified reliability, 235, 382
for specified MTBF, 593
for the maximum MRL, 282, 287, 492
for the maximum profit, 472
for the minimum cost, 337, 404, 472
for zero-slope failure rate curve, 229
  with error, 229
graphical approach, 238
  using failure rate function, 238
  using reliability function, 239
non-parametric determination, 632, 635
parametric determination, 626, 629, 630
quick-calculation approach, 211
subpopulation truncation approach, 221
using bimodal distribution, 221
using TTT transform, 437, 472, 492
Optimum thermal cycles
  for specified MTBF, 593
Orthogonal axis vibration, 44
Outliers, 625
Overload-Stress Reliability Model, 527

P

PA, 63
Parameter drift, 44
Parameter estimates
  initial, 141
Parameter estimation, 626
  Bayesian method, 125, 158
  EML algorithm, 155, 158
  for bimodal mixed population, 119
  for burn-in math models, 115
  for modified Arrhenius model, 555
  Jensen's graphical method, 120, 158
  least-squares method, 162, 547
Parameter estimation methods
  for burn-in models, 628
  summary, 628
Parametric data analysis, 626
Pareto distribution
  scaled TTT transform, 442
Paris Law, 106, 111, 606
Part fraction defectives, 45
Patent defect, 45
PCB, 63
PCC, 63
PED, 63
PEP, 63
Percent defective, 45
Percent-Life Model, 527

Performance monitoring (PM), 43
Performance requirements, 337
Periodic conformance test, 45
Petersen, 88, 94, 106
Peterson, 338, 340
PFMEA, 63
PGA, 63
Physical explanation
    for burn-in failure pattern, 84
Physical insight
    for burn-in failures, 69
Physical performance, 337, 352
PICs, 63
PIN, 63
PIND, 63
Pitfall of Arrhenius model, 541
PLCCs, 63
Plesser, 344
Plotting techniques
    for burn-in data analysis, 625
    summary, 625
PM, 63
PMOS/pMOS, 63
Poisson distribution, 427
Poisson process, 91, 340, 402
Poisson's ratio, 604
Posterior probability, 126, 138
Postmortem data, 155
Power cycling, 45
Power function of a burn-in, 396, 398
PPM, 63
PQFPs, 63
Precipitation of defects, 45
Predefect-free, 45
Prestress screening, 45
Prestress test, 45
Preventive replacement, 256
Printed circuit board (PCB), 45
Printed wiring board (PWB), 45
Prior distribution in Bayesian approach, 417

Prior probability, 126, 135
Producibility, 45
Product assurance, 45
Product maturing, 30
Production flaw, 45
Production lot, 46
Production reliability test, 46
Production sampling, 46
Productivity, 46
Profit
    maximization, 472
    Weiss-Dishon's model, 488
PROMs, 63
Proof-of-design test, 46
Proportionality constant, 528
Provided activation energy, 542
Ps, 63
PSD, 63
PSG, 63
PSR, 63
Pt, 63
PTHs, 63
Pugacz-Muraszkiewicz, 553
PWA, 64
PWB, 64

Q

QA, 64
QC, 64
QE, 64
QFD, 64
QFP, 64
QPL, 64
Quadratic fitting, 287
Qualification test, 46
Quality assurance, 46
Quality defect, 46
Quality engineering (QE), 46
Quality function deployment (QFD), 46
Quality management, 46
Quantification
    of burn-in, 375
Quasi-Newton Method, 547

Quasi-random vibration(QRV), 46
Quick calculation approach for burn-in time determination, 211

R

R&R, 64
RAM, 21, 64
Ramps, *see* Temperature ramps
Random defect, 42
Random failure period, *see* Useful life period
Random vibration, 46
RAS, 64
Rayleigh distribution, 271
RCM, 64
Reaction rate, 528
Reaction rate stress, 579
REDR, 64
Relevant failure, 47
Reliability defects, 47
Reliability demonstration, 47
Reliability function
    for burn-in time determination, 239
Reliability goal, 235, 382
Remaining defect density, 426, 427
Renewal process of the burn-in, 402
Renewal rate, 402
Renewal rate function, *see* Renewal rate
Renewal theory, 256, 402, 403, 407
Repair time, 47
Required activation energy, 542
Residual life, 255
Residual strength, 106
Resonant dwell, 47
RFI, 64
RGM, 64
RH, 64

Ring oscillator, 24
RISE, 64
Risk
    of a burn-in, 398
ROI, 64
Roller-Coaster curve, 79, 81, 83
ROM, 64
Rosenbrock Pattern Search Method, 547
RQL, 64
RSER, 64
RTD, 64
RTOK, 64
RTV, 64
Run-in, 47
Running-in period, *see* Early failure period
Ryerson, 70, 72

S

S-N curve, 50
S-shaped, 119, 123, 141, 223, 225, 340, 442, 444, 469, 627
S-shaped *cdf*, 72, 94, 101
Safety allowance, 47
Safety factor, 47
Safety margin, 47
Sample test, 47
SAW, 64
Scaled TTT plot, 438
    for a given sample, 442
    for complete data, 458
    for failure-terminated data, 465
    for time-terminated data, 461
Scaled TTT statistics, 438
Scaled TTT transform, 438, 440
    for exponential distribution, 442
    for gamma distribution, 442
    for lognormal distribution, 442
    for normal distribution, 442

for Pareto distribution, 442
geometrical properties, 443
of exponential distribution, 445
of Gamma distribution, 449, 450
of truncated distribution, 451, 453
of Weibull distribution, 446, 448
SCPs, 65
Screen effectiveness, 47
Screen level, 47
Screen parameters, 47
Screenable latent defect, 48
Screening, 48
Screening attrition costs, 48
Screening complexity, 48
Screening cost effectiveness, 48
Screening regimen, 48
Screening sequence, 49
Screening strength, 49
Screening-induced degradation, 48
Search method
   non-linear, 412, 413
   one dimensional, 412, 413
Seasoning, 49
Secondary damage effects, 49
Secondary failure, 49
Selection and placement, 49
Selection criteria
   for burn-in models, 119
SEM, 65
SEMs, 65
Sensitivity analysis, 437
Sensor, 49
Separation plotting method, 125, 148
SER, 65
Shakedown, 49
Shaker, 49
Shape parameter
   for cooling profile, 578

for heating profile, 578
Shear modulus, 604
Shen, 106
Shock test, 49
Shop replaceable assembly, 49
Shop replaceable unit (SRU), 49
Si, 65
Si_3N_4, 65
Simplex Method, 547
SIMs, 65
Sine fixed frequency vibration, 50
Sine vibration, 50
Sine wave, swept frequency vibration, 50
Single-frequency vibration, 50
SiO_2, 65
SIP, 65
Six-parameter model, 118
Smith, 106
Sn, 65
Sneak circuit analysis, 50
Sneak condition, 50
Sneak indicators, 50
Sneak labels, 50
Sneak paths, 50
Sneak software analysis, 50
Sneak timing, 50
Soak, *see* Dwell
Soft failure, 50
Solder joint
   thermal fatigue life prediction, 609
Solid failure, 50
Sonic fatigue, 50
Sonic vibration, 51
SOP, 65
SOR, 65
Spatial temperature gradient, 573
Specified burn-in efficiency, 393
Specified burn-in power, 396
Specified burn-in risk, 398
Specified failure rate goal, 233, 388

REFERENCES

Specified MRL, 278, 286, 384
Specified MTBF goal, 593
Specified reliability goal, 235, 382
Spectrum analyzer bandwidth, 51
SQL, 65
SRAM, 65
SRD, 65
SRU, 65
SS, 65
SSI, 21, 65
Stable failure period, *see* Useful life period
Standard environmental profile, 51
Standard subpopulation, 119
Static burn-in, 6, 16
Static/Steady-state burn-in, 51
STATISTICA, 548
Steady-state burn-in, 51
Steady-state temperature, 573
Step stress, 51
Step-Stress Model, 527
Strain, 51
Strain gage, 51
Strength, 51
Strength deterioration, 96
Stress, 51
Stress screen degradation, 51
Stress screening, 37, 51
Stress-strength classifications, 51
Stress-strength interference, 52, 84
    behind bathtub curve, 90
    versus failure pattern, 88
Stress-strength interference concept, 84
Stress-strength interference model, 85
Strong subpopulation, 125, 131, 148, 398–400
    activation energy, 536
Structure
    for objective function, 468

STTL, 65
Subassembly level, 52
Subpopulation truncation approach, 221
Substandard, 1
Substandard item, 52
Substandard subpopulation, 119
    activation energy, 536
Summary of thermal fatigue life model, 601
Sun, 125
Survival analysis, 256
Swept sinusoidal vibration, 52
System, 52
System effectiveness, 353, 354

T

TAAF, 65
TAB, 65
TAC, 65
Tailored environmental profile, 52
Tailoring, 52
Tangent line, 469
Target flaw, 52
TCM, 65
TDBI, 6, 65
TDDB, 65
TEC, 65
TEM, 66
Temperature
    elevated, 528, 565
    steady state, 573
Temperature chamber, 33
Temperature change rate, 573
Temperature change rate stress, 579
Temperature cycling, 52
    for burn-in, 573
    why, 573
Temperature dependency
    of activation energy, 529, 541, 542, 565, 566
    of failure mechanisms, 574

Temperature dwell, 577
Temperature gradient, 573
Temperature profile, 577
    a cycle with uniform ramp, 578
    a typical cycle, 577
    general model, 577
Temperature ramps, 577
Temperature range, 573
Temperature shock, 53
Temperature soak, 53
Temperature stabilization period, 53
Temperature-Humidity Model, 527
Tensile modulus
    of adhesive, 607
    of substrate, 607
Test, 53
Test conditions
    for burn-in, 15, 19
Test detection efficiency, 36, 53
Test during burn-in, 6, 17, 53
Test specification, 53
Test strength, 53
Test temperature
    for burn-in, 25
Test-to-failure, 53
THB, 66
Thermal aging, 53
Thermal conductivity, 54
Thermal cycling, 54
    acceleration factor, 577
    equivalent acceleration factor, 580, 584
    temperature profile, 577
Thermal endurance, 54
Thermal equilibrium, 54
Thermal fatigue, 54
Thermal fatigue life prediction
    a summary, 601
    of die, 606
    of die and substrate adhesion, 607
    of solder joint, 609
    of wire, 602
    of wire bond, 603
Thermal gradient stresses, 54
Thermal rating, 54
Thermal resistance, 54
Thermal shock, 54
Thermal survey, 54
Three-parameter model, 117
Ti, 66
Ties, 458, 459, 461, 462, 465, 466
Time-terminated, 155
Time-terminated data, 461
Tolerance, 54
Tolerance limits, 54
Total device hours, 212
Total Time on Test, 437
Trade-offs, 409
Transducer, 55
Transparent failure, 55
Treatment, 55
Treatment combinations, 55
Trimodal life distribution, 75
Trimodally-mixed Weibull, 305
Troubleshooting time, 55
Truncated distribution, 223, 225
Truncated normal distribution, 269
    TTT transform, 451, 453
TS, 66
TTL, 21, 66, 542, 543
TTL-LS, 66
TTT, 66, see Total Time on Test
TTT plot, 437, 438
    for a given sample, 442
    for complete data, 458
    for failure-terminated data, 465
    for time-terminated data, 461
TTT statistics, 438
TTT transform, 437, 438, 440
    for exponential distribution, 442
    for gamma distribution, 442

for lognormal distribution, 442
for normal distribution, 442
for Pareto distribution, 442
geometrical properties, 443
of exponential distribution, 445
of Gamma distribution, 449, 450
of Lognormal distribution, 453, 455
of mixed distribution, 456
of mixed-exponential distribution, 457
of mixed-Weibull distribution, 457
of truncated normal distribution, 451, 453
of Weibull distribution, 446, 448
Two-level burn-in cost model, 409
Two-parameter Weibull model, 117
TWT, 66

U

ULSI, 66
Ultrasonic vibration, 55
Ungrouped data, 155, 261, 263
Ungrouped data set, 458, 461, 465
Unified field (failure) theory, 79
Uniform prior *pdf*, 417
Unilateral tolerance, 55
Unit, 55
Useful life period, 70
Utility, 416
UUT, 66
UV, 66

V

Variable data, 55
Vibration, 55

Vibration survey, 55
VLSI, 66
VMOS, 66
Voltage bias stress
 for burn-in, 563, 565
VSR, 66

W

Warranty, 55, 365
Warranty claim, 55
Warranty maintenance, 55
Warranty period, 56, 414
Washburn, 352
Weak component, 378, 380, 393, 394
Weak subpopulation, 72, 119, 123, 125, 128, 131, 145, 148, 398–400, 536, 537
 activation energy, 536
Wearout, 56
Wearout failure, 56
Wearout failure period, 56, 71
Wearout life, 56
Wearout period, 56
Weibull distribution, 266, 597
 TTT transform, 446, 448
Weibull Stress-Life Model, 527
Weiss, 481, 488
Weiss-Dishon's cost model, 481
Weiss-Dishon's profit model, 488
Well-mixed mixture, 155
Well-separated mixture, 120, 125, 155
Wire bond thermal fatigue life prediction, 603
Wire thermal fatigue life prediction, 602
Wong, 79, 81, 83, 106
Workmanship defects, 56
WPP, 123, 131, 141, 148
WUC, 66

Y

Yang, 363

Yield, 56, 427

Z

Zero-slope failure rate curve, 229
    with error, 229

ABOUT THE AUTHORS

Dr. Dimitri B. Kececioglu, P.E., a Fulbright Scholar and a Fellow of SAE, received his B.S.M.E. from Robert College, Istanbul, Turkey in 1942, and his M.S. in Industrial Engineering in 1948 and his Ph.D. in Engineering Mechanics in 1953, the two latter degrees from Purdue University, Lafayette, Indiana. He is currently Professor in the Department of Aerospace and Mechanical Engineering at The University of Arizona; Professor-in-Charge of a unique over 10-course Reliability Engineering program leading to the Master of Science degree in Reliability Engineering; Director of the Annual Reliability Engineering and Management Institute; Director of the Annual Reliability Testing Institute; Director of the Applied Reliability Engineering and Product Assurance Institute for Engineers and Managers; and a Reliability and Maintainability Engineering consultant. This book is based on the following extensive experience of the author in Reliability and Maintainability Engineering and in Life Testing:

1. He initiated and was the Director of the Corporate Reliability Engineering Program at the Allis-Chalmers Manufacturing Co., Milwaukee, Wisconsin, from 1960 to 1963.

2. He started the Reliability Engineering Instructional Program at The University of Arizona in 1963, which now has more than 10 courses in it. A Master's Degree with a Reliability Engineering Option is currently being offered in the Aerospace and Mechanical Engineering Department at The University of Arizona. He started this option in 1969. A Master's Degree in Reliability Engineering is also being offered in the Systems and Industrial Engineering Department at The University of Arizona. This degree started in January 1987. Since 1991 this became an all encompassing Master's Degree in Reliability and Quality Engineering.

3. He conceived and directed the first two Summer Institutes for College Teachers in Reliability Engineering ever to be supported by the National Science Foundation. The first was in 1965 and the second in 1966, for 30 college and university faculty each summer. These faculty started teaching reliability engineering courses at their respective universities and/or incorporating reliability engineering concepts into their courses.

4. He helped initiate The Professional Certificate Award in Reliability and Quality Engineering program at The University of Arizona in 1991. This is a 15-unit program. The certificate's requirements are met via videotapes of the EXTENDED UNIVERSITY organization of The University of Arizona. No par-

ticipant needs to be present on the campus of The University of Arizona to get this certificate.

5. In 1963 he conceived, initiated, and has directed since then the now internationally famous *Annual Reliability Engineering and Management Institute* at The University of Arizona.

6. In 1975 he conceived, initiated, and has directed since then the now internationally famous *Annual Reliability Testing Institute* at The University of Arizona.

7. In 1992 he conceived, initiated, and has directed since then the popular Applied Reliability Engineering and Product Assurance Institute for Engineers and Managers.

8. He has lectured extensively and conducted over 350 training courses, short courses and seminars worldwide, and has exposed over 10,000 reliability, maintainability, test, design, and product assurance engineers to the concepts in this book, and his other 12 books listed in Item 14.

9. He has been the principal investigator of mechanical reliability research for the NASA-Lewis Research Center, the Office of Naval Research, and the Naval Weapons Engineering Support Activity for ten years.

10. He has been consulted extensively by over 85 industries and government agencies worldwide on reliability engineering, reliability and life testing, maintainability engineering, and mechanical reliability matters. He is considered to be the Deming of Reliability Engineering and Product Assurance.

11. He has been active in the Reliability and Maintainability Symposia and Conferences dealing with reliability engineering since 1963.

12. He founded the Tucson Chapter of the Society of Reliability Engineers in 1974 and was its first President. He also founded the first and very active Student Chapter of the Society of Reliability Engineers at The University of Arizona.

13. He has authored or co-authored over 130 papers and articles, of which over 120 are in all areas of reliability engineering.

14. In addition to this book, he has authored or contributed to the following books:

(1) *Bibliography on Plasticity - Theory and Applications*, Dimitri B. Kececioglu, published by the American Society of Mechanical Engineers, New York, 191 pp., 1950.

(2) *Manufacturing, Planning and Estimating Handbook*, Dimitri B. Kececioglu and Lawrence Karvonen contributed part of Chapter 19, pp. 19-1 to 19-12, published by McGraw-Hill Book Co., Inc., New York, 864 pp., 1963.

(3) *Introduction to Probabilistic Design for Reliability*, Dimitri B. Kececioglu, published by the United States Army Management Engineering Training Agency, Rock Island, Illinois, contributed Chapter 7 of 109 pp., and Chapter 8 of 137 pp., May 1974.

(4) *Manual of Product Assurance Films on Reliability Engineering and Management, Reliability Testing, Maintainability, and Quality Control*, Dimitri B. Kececioglu, printed by Dr. Dimitri B. Kececioglu, 7340 N. La Oesta Avenue, Tucson, AZ 85704-3119, 178 pp., 1976.

(5) *Manual of Product Assurance Films and Videotapes*, Dimitri B. Kececioglu, printed by Dimitri B. Kececioglu, 7340 N. La Oesta Avenue, Tucson, AZ 85704-3119, 327 pp., 1980.

(6) *Reliability Engineering Handbook*, Vol. 1, Dimitri B. Kececioglu, Prentice Hall, Inc., Upper Saddle River, NJ 07458, 720 pp., May 1991, and its Fifth Printing in July 1995.

(7) *Reliability Engineering Handbook*, Vol. 2, Dimitri B. Kececioglu, Prentice Hall, Inc., Upper Saddle River, NJ 07458, 568 pp., June 1991, and its Fifth Printing in December 1996.

(8) The 1992-1994 *Reliability, Maintainability and Availability Software Handbook*, Dimitri B. Kececioglu and Pantelis Vassiliou, 7340 N. La Oesta Avenue, Tucson, AZ 85704-3119, 118 pp., November 1992.

(9) *Reliability & Life Testing Handbook*, Vol. 1, Dimitri B. Kececioglu, Prentice Hall, Inc., Upper Saddle River, NJ 07458, 960 pp., 1993.

(10) *Reliability & Life Testing Handbook*, Vol. 2, Dimitri B. Kececioglu, Prentice Hall, Inc., Upper Saddle River, NJ 07458, 900 pp., 1994.

(11) *Maintainability, Availability and Operational Readiness Engineering Handbook*, Vol. 1, Dimitri B. Kececioglu, Prentice Hall, Inc., Upper Saddle River, NJ 07458, 1995.

(12) *Environmental Stress Screening – Its Quantification, Optimization and Management*, Dimitri B. Kececioglu and Feng-

Bin Sun, Prentice Hall, Inc., Upper Saddle River, NJ 07458, 520 pp., 1995.

15. He has received over 80 prestigious awards and has been recognized for his valuable contributions to the field of reliability and maintainability engineering and testing. Among these are the following: (1) Fulbright Scholar in 1971; (2) Ralph Teetor Award of the Society of Automotive Engineers as "Outstanding Engineering Educator" in 1977; (3) Certificate of Excellence by the Society of Reliability Engineers for his "personal contributions made toward the advancement of the philosophy and principles of reliability engineering" in 1978; (4) ASQC-Reliability Division, Reliability Education Advancement Award for his "outstanding contributions to the development and presentation of meritorious reliability educational programs" in 1980; (5) ASQC Allen Chop Award for his "outstanding contributions to Reliability Science and Technology" in 1981; (6) The University of Arizona College of Engineering Anderson Prize for "engineering the Master's Degree program in the Reliability Engineering Option" in 1983; (7) Designation of "Senior Extension Teacher" by Dr. Leonard Freeman, Dean, Continuing Education and University Extension, University of California, Los Angeles in 1983; (8) Honorary Member, Golden Key National Honor Society in 1984; (9) Honorary Professor, Shanghai University of Technology in 1984; (10) Honorary Professor, Phi Kappa Phi Honor Society in 1988; (11) The American Hellenic Educational Progressive Association "Academy of Achievement Award in Education" in 1992; (12) On the occasion of "The 30th Annual Reliability Engineering and Management Institute," the President of The University of Arizona, Dr. Manuel T. Pacheco, presented him a plaque inscribed: "Your reputation as an outstanding teacher and advocate of reliability and quality engineering is well established in the international engineering community. In your capacity as Director of this Institute, as well as the Reliability Testing Institute, you have provided the forum in which many hundreds of our nation's engineers and students of engineering have received training in reliability and quality engineering.

I particularly acknowledge your efforts in establishing and developing funding for the endowment which bears your name and which will support worthy students in the future. The "Dr. Dimitri Basil Kececioglu Reliability Engineering Research Fellowships Endowment Fund" will help to ensure that The University of Arizona remains in the forefront of engineering education and continues to provide engineering graduates to support our nation's industries. In this highly competitive world the quality and the reliability of American products are essential to retaining our position of world economic leadership. The University of Arizona is proud to be an important part of that effort and can take justifiable pride in your own very significant contribution."

16. He conceived and established *The Dr. Dimitri Basil Kececioglu Reliability Engineering Research Fellowships Endowment Fund* in 1987. The cosponsors of his institutes, mentioned in Items 5 and 6, have contributed generously to this fund, which has crossed the $335,000 mark.

17. He was elected to the prestigious Fellow Member grade of the Society of Automotive Engineers in 1997 for his exceptional professional distinction and important technical achievement.

18. He has been granted five patents.

Dr. Feng-Bin Sun received his B.S.M.E. specializing in Electromechanical Structural Design from the Southeast University, the former Nanjing Institute of Technology, Nanjing, China, in 1984. He earned his M.S.M.E. specializing in Reliability Engineering from the Shanghai University, the former Shanghai University of Technology, Shanghai, China, in 1987. He received his Ph.D. in Reliability Engineering from The University of Arizona in 1997. He is currently a Staff Reliability Engineer in the Corporate Reliability Engineering Department of Quantum Corporation, Milpitas, California.

His introduction to Reliability Engineering was at the Shanghai University of Technology during 1984-1987, where he got his M.S.M.E. specializing in Reliability Engineering and enjoyed his first practical reliability and maintainability experience in industry: a big iron and steel company. The following extensive experience has qualified him to contribute major segments of this book:

1. During 1985-1986, he worked at the Shanghai First Iron and Steel Plant, Shanghai, China, where he conducted extensive research on the reliability and maintainability of a steel rolling production line.

2. During 1986-1990, he offered over 20 lectures and training courses in Reliability and Maintainability Engineering for managers, chief and senior engineers, design and quality engineers from China's Ministry of Electromechanical Industry and its subordinate factories and research institutes. He also lectured at and was consulted by two big metallurgical and chemical companies in Shanghai, China, on reliability, preventive maintenance decision making and spares provisioning.

3. As an assistant Professor at the Shanghai University of Technology during 1987-1990, he taught several courses on Reliability and Maintainability including Reliability Engineering and Management, Reliability Life Testing, Maintainability Engineering and Probabilistic Mechanical Design. These courses were open to both undergraduate and graduate students.

4. During 1987-1990, he supervised 15 undergraduate and 4 graduate students in the Mechanical Engineering Department, Shanghai University of Technology, for their graduate projects and theses in reliability engineering.

5. During 1988-1990, he cooperated with the Beijing Machine Tool Research Institute, Beijing, China, and was the person-in-charge of a research grant on Reliability Assessment of Numerical Control Systems with emphasis on burn-in and ESS data analysis,

which was funded by the Ministry of China's Electromechanical Industry.

6. During 1988-1990, he was in charge of another research grant on replacement policies for large scale industrial systems which was funded by China's National Natural Science Foundation. He proposed an unified approach of system classification for the purpose of maintenance and replacement, and advanced the multimode replacement policy idea for large scale industrial systems.

7. During 1989-1990, he cooperated with the Shanghai Machine Tool Research Institute and was the person-in-charge of a research grant on reliability evaluation of Numerically Controlled Machine Tools, which was funded by the Shanghai Municipal Bureau of Electromechanical Industry.

8. As a guest lecturer under Dr. Dimitri B. Kececioglu in the Department of Aerospace and Mechanical Engineering at The University of Arizona, he has been lecturing on Reliability Engineering, Reliability Engineering and Quality Analysis, Reliability & Life Testing and Maintainability Engineering since 1991.

9. In 1991 he conducted research with Dr. Dimitri B. Kececioglu on *MTBF* demonstration test plans of electronic units for ITT Power Systems Corporation.

10. In 1992 he conducted research with Dr. Dimitri B. Kececioglu on burn-in test planning and statistical analysis of multi-time-terminated burn-in data for Brooktree Corporation.

11. During the summer of 1996, he worked in the Corporate Reliability Engineering Department of Quantum Corporation at Milpitas, California. He offered four lectures on Reliability Statistics and Applications in Quantum Reliability Assessment Program. He worked on hard-disk drive reliability growth analysis, life data evaluation, sampling plan, DOE analysis of environmental stress testing with different head-media combinations, etc.

12. He has co-authored over 12 papers in various areas of reliability, maintainability and quality engineering, such as maintenance and replacement policies, system reliability, maintainability and availability evaluation, burn-in and *ESS* data analysis, mechanical reliability, quality loss functions, etc.

13. In addition to this book, he co-authored or contributed to the following books:

(1) *Maintainability Design*, by Zhi-Zhong Mou, contributed two chapters on advanced replacement policies and a practical application in metallurgical industry, published by Shanghai Science and Technology Press, Shanghai, China, 300 pp., 1990.

(2) *Reliability Design*, co-authored with Zhi-Zhong Mou, et al, contributed Chapter 10 on reliability design for electromechanical equipment, pp. 238-272, published by China Machinery Industry Press, Beijing, China, 366 pp., 1993.

(3) *Reliability & Life Testing Handbook*, by Dimitri B. Kececioglu, contributed four chapters on modified matching moment parameter estimation, modified K-S, Anderson-Darling and Cramer-von Mises goodness-of-fit tests, proposed jointly with Dr. Dimitri B. Kececioglu a new outlier test method called *Rank limit method*, published by Prentice Hall, Inc., Upper Saddle River, NJ 07458, Vol. 1, 960 pp., 1993.

(4) *Advanced Methods in Reliability Testing*, by Dimitri B. Kececioglu, contributed three chapters on burn-in, *ESS*, and reliability growth, respectively, to be published in 1997 by Prentice Hall, Inc., Upper Saddle River, NJ 07458.

(5) *Maintainability, Availability and Operational Readiness Engineering Handbook*, by Dimitri B. Kececioglu, contributed two chapters on renewal theory and additional replacement policies, respectively, Vol. 1 Prentice Hall, Inc., Upper Saddle River, NJ 07458, 1995.

(6) *Environmental Stress Screening – Its Quantification, Optimization and Management*, co-authored with Dr. Dimitri B. Kececioglu, Prentice Hall, Inc., Upper Saddle River, NJ 07458, 520 pp., 1995.